Process Validation in Manufacturing of Biopharmaceuticals

The fourth edition of *Process Validation in Manufacturing of Biopharmaceuticals* is a practical and comprehensive resource illustrating the different approaches for successful validation of biopharmaceutical processes. A pivotal text in its field, this new edition provides guidelines and current practices, contains industrial case studies, and is expanded to include in-depth analysis of the new process validation (PV) guidance from the US FDA.

Key Features:

- Offers readers a thorough understanding of the key concepts that form the basis of a good process validation program for biopharmaceuticals.
- Includes case studies from the various industry leaders that demonstrate application of these concepts.
- Discusses the use of modern tools such as multivariate analysis for facilitating a process validation exercise.
- Covers process characterization techniques for scaling down unit operations in biopharmaceutical manufacturing, including chromatography, chemical modification reactions, ultrafiltration, and microfiltration, and practical methods to test raw materials and in-process samples.

Providing a thorough understanding of the key concepts that form the basis of a good process validation program, this book will help readers ensure that PV is carried out and exceeds expectations. Fully illustrated, this is a much-needed practical guide for biopharmaceutical manufacturers.

Biotechnology and Bioprocessing Series

Series Editor
Anurag S. Rathore

Process Validation in Manufacturing of Biopharmaceuticals, 4th Edition
Edited by Anurag S. Rathore, Harold Baseman, Scott Rudge

Biotransformations and Bioprocesses
Edited by Mukesh Doble, Anil Kumar Kruthiventi, Vilas Ganjanan Gaikar

A Lifecycle Approach to Knowledge Excellence in the Biopharmaceutical Industry
Edited by Nuala Calnan, Martin J Lipa, Paige E. Kane, Jose C. Menezes

Process Validation in Manufacturing of Biopharmaceuticals, 3rd Edition
Edited by Anurag S. Rathore, Gail Sofer

PAT Applied in Biopharmaceutical Process Development And Manufacturing: An Enabling Tool for Quality-by-Design
Edited by Cenk Undey, Duncan Low, Jose C. Menezes, Mel Koch

Process Synthesis for Fuel Ethanol Production
Edited by C.A. Cardona, O.J. Sanchez, L.F. Gutierrez

Process Scale Bioseparations for the Biopharmaceutical Industry
Edited by Abhinav A. Shukla, Mark R. Etzel, Shishir Gadam

Cell Culture Technology for Pharmaceutical and Cell-Based Therapies
Edited by Sadettin Ozturk, Wei-Shou Hu

Metabolic Engineering
Edited by Sang Yup Lee, E. Terry Papoutsakis

Bioreactor System Design
Edited by Juan A. Asenjo

Cell Adhesion in Bioprocessing and Biotechnology
Edited by Martin Hjortso

Recombinant Microbes for Industrial and Agricultural Applications
Edited by Yoshikatsu Murooka

Protein Purification Process Engineering
Edited by Roger Harrison

Insect Cell Culture Engineering
Edited by Mattheus F. A. Goosen

Biosensor Principles and Applications
Edited by Loic J. Blum and Pierre R. Coulet

Extractive Bioconversions
Edited by Bo Mattiasson and Olle Holst

Large-Scale Mammalian Cell Culture Technology
Edited by Anthony S. Lubiniecki

Process Validation in Manufacturing of Biopharmaceuticals

Fourth Edition

Edited by Anurag Singh Rathore,
Hal Baseman, and Scott Rudge

CRC Press
Taylor & Francis Group
Boca Raton London New York

CRC Press is an imprint of the
Taylor & Francis Group, an **informa** business

Fourth edition published 2024
by CRC Press
6000 Broken Sound Parkway NW, Suite 300, Boca Raton, FL 33487–2742

and by CRC Press
4 Park Square, Milton Park, Abingdon, Oxon, OX14 4RN

CRC Press is an imprint of Taylor & Francis Group, LLC

© 2024 selection and editorial matter, Anurag S. Rathore, Hal Baseman, Scott Rudge; individual chapters, the contributors

Third Edition published by CRC Press, 2013
Second Edition published by CRC Press, 2005
First Edition published by CRC Press, 2000

Reasonable efforts have been made to publish reliable data and information, but the author and publisher cannot assume responsibility for the validity of all materials or the consequences of their use. The authors and publishers have attempted to trace the copyright holders of all material reproduced in this publication and apologize to copyright holders if permission to publish in this form has not been obtained. If any copyright material has not been acknowledged please write and let us know so we may rectify in any future reprint.

Except as permitted under U.S. Copyright Law, no part of this book may be reprinted, reproduced, transmitted, or utilized in any form by any electronic, mechanical, or other means, now known or hereafter invented, including photocopying, microfilming, and recording, or in any information storage or retrieval system, without written permission from the publishers.

For permission to photocopy or use material electronically from this work, access www.copyright.com or contact the Copyright Clearance Center, Inc. (CCC), 222 Rosewood Drive, Danvers, MA 01923, 978–750–8400. For works that are not available on CCC please contact mpkbookspermissions@tandf.co.uk

Trademark notice: Product or corporate names may be trademarks or registered trademarks and are used only for identification and explanation without intent to infringe.

ISBN: 978-0-367-69762-4 (hbk)
ISBN: 978-0-367-69763-1 (pbk)
ISBN: 978-1-003-14313-0 (ebk)

DOI: 10.1201/9781003143130

Typeset in Palatino
by Apex CoVantage, LLC

Contents

Author Biographies

Anurag S. Rathore is Institute Chair Professor at the Department of Chemical Engineering, Indian Institute of Technology, Delhi, India. He is also the coordinator for the DBT COE for Biopharmaceutical Technology. His areas of interest include process development, scale-up, technology transfer, process validation, biosimilars, continuous processing, medical image analysis, process analytical technology, and quality by design. He is presently serving as the Editor-in-Chief of *Preparative Biochemistry and Biotechnology* and Associate Editor for *Journal of Chemical Technology and Biotechnology*. He also serves on the editorial advisory boards for *Biotechnology Progress, Biotechnology Journal, Electrophoresis, BioPharm International, Journal of Chromatography B, Journal of Chromatography Open, Pharmaceutical Technology Europe*, and *Separation and Purification Reviews*. Dr. Rathore has edited books titled *Preparative Chromatography for Separation of Proteins and Peptides* (2017), *Quality by Design for Biopharmaceuticals: Perspectives and Case Studies* (2009), *Elements of Biopharmaceutical Production* (2007), *Process Validation* (2005), *Electrokinetic Phenomena* (2004), and *Scale-up and Optimization in Preparative Chromatography* (2003). He has a PhD in chemical engineering from Yale University.

Hal Baseman is the Chief Operating Officer at ValSource Inc., a consulting services firm. Hal has held various positions in the biopharmaceutical manufacturing and consulting industry and has been involved in sterile product manufacturing for over 45 years. He is the past chair of the PDA Board of Directors, the past co-chair of the PDA Science Advisory Board, past chair or lead on several PDA groups and efforts, including the PDA Process Validation Interest Group, the task forces for PDA Technical Reports on Process Validation, Aseptic Process Simulation, and Quality Risk Management, and the PDA Aseptic Processing Points to Consider. He is currently a member of the PDA TRI (Training and Research Institute faculty, co-lead for the Kilmer Conference Regulatory Innovation Collaboration team, chair of the SfSAP (Society for Sterility Assurance Professionals) Aseptic Processing Team, and co-chair of the PDA Annex 1 revision response team. Hal has presented and published numerous articles and book chapters on topics related to sterile product manufacturing, risk management, and validation. He has edited books on process validation and quality risk management. Hal holds a bachelor of science in biology from Ursinus College and a master of business administration in management from La Salle University.

Scott Rudge is Principal Consultant and a member of the Syner-G team since January 2023. He is an experienced pharmaceutical professional with 33 years of progressive experience directing process engineering, process development, product development, and overall pharmaceutical operations. Scott was a co-founder of RMC Pharmaceutical Solutions in 2004, which was acquired by Syner-G Biopharma Group in January 2023. Scott has built and led several organizations in the biopharmaceutical industry, including process engineering, process development, and product development teams at Synergen, Amylin Pharmaceuticals, and FeRx, where he reported to the CEO. Ultimately, Scott formed his own consulting company, RMC Pharmaceutical Solutions, with the vision to provide product development services to biotech and pharmaceutical companies of all sizes and all stages of development. Scott has remained active as a thought leader and innovator in biotechnology and pharmaceutical development. He is a co-author on the award winning textbook Bioseparation Science and Engineering, co-author on the new Bioprocess chapter in Perry's Chemical Engineering Handbook on biotechnology, and co-editor on the CRC Process Validation in Manufacturing of Biopharmaceuticals, and he has been a reviewer for the National Academy of Science's ten-year strategic goals for separation sciences in 2020, in addition to many publications and patents. Scott earned his BS in chemical engineering from Worcester Polytechnic Institute and his MS and PhD in chemical engineering from Purdue University.

Contributors

Frank Agbogbo

Tiffany Baker

Hal Baseman

Marc Bisschops

Jennifer Campbell

Audrey Chang

Brenda Carrillo Conde

Arch Creasy

Carole Davis

Phil DeSantis

Nargisse El Hajjami

Allison Farrand

Renija George

Ranga Godavarti

Jin Guo

John Haury

Vishwanath Hebbi

Stephen Kolodziej

Frank Kotch

Ganesh Krishnamoorthy

Guillaume Lesage

Renato Lorenzi

Patrick Mains

John McEntire

Raymond W. Nims

Carolyn Phillips

Esther Presente

Jennifer Pulkowski

Ravali Raju

Anurag S. Rathore

Pascale Richert

Nadine Ritter

Scott Rudge

Mark Schofield

James E. Seely

Robert J. Seely

Tatenda Shopera

Russ Shpritzer

Gail Sofer

Anissa Souidi

Maria Wik

Nichole Wood

Xiaolu Zheng

1

The Evolution of Modern Process Validation: Commentary on the U.S. Food and Drug Administration's 2011 Guidance for Industry, Process Validation Principles and Practices

Hal Baseman

The chapter provides insight into the background, intent, and content of the current 2011 guidance, based in part on responses to the 2008 draft guidance, Parenteral Drug Association (PDA) sponsored workshops, and related discussions, as well as discussions and updates since the issuance of the guidance. The FDA process validation guidance was published in and in use since 2011. Understanding of the principles of the guidance are needed to design and implement effective biopharmaceutical process validation approaches.

The publication of the US FDA 2011 *Guidance for Industry: Process Validation: General Principles and Practices*, commonly referred to as the Process Validation Guidance is a significant step in the evolution of process validation of drug and biopharmaceutical products and forms the basis for much of the approaches presented in the chapters of this book. The guidance was issued as a draft in December of 2008 and the final version was published on January 24, 2011. The guidance presents a relatively non-prescriptive, science-/risk-based, lifecycle approach to process validation. It places emphasis on process understanding and design as the primary means to gain confidence in the reliability and outcome of the process, rather than on process monitoring or product testing (1).

The approach presented in the FDA guidance is largely consistent with global regulatory guidance and expectations that followed. These include key process validation guidance published by the European Commission in 2015 with Annex 15, Qualification and Validation, and in 2016 with their Guideline on Process Validation for the Manufacture of Biotechnology-Derived Active Substances and Data to Be Provided in the Regulatory Submission in 2016. This role of process design and life cycle during process validation is also noted in ICH Q8, Q9, Q10, and Q11 (2–5).

1.1 The Lifecycle Approach

The FDA and other process validation guidance divides process validation into three lifecycle stages as shown in Figure 1.1, Process Design, Process Qualification, and Continued or On-going Process Verification.

DOI: 10.1201/9781003143130-1

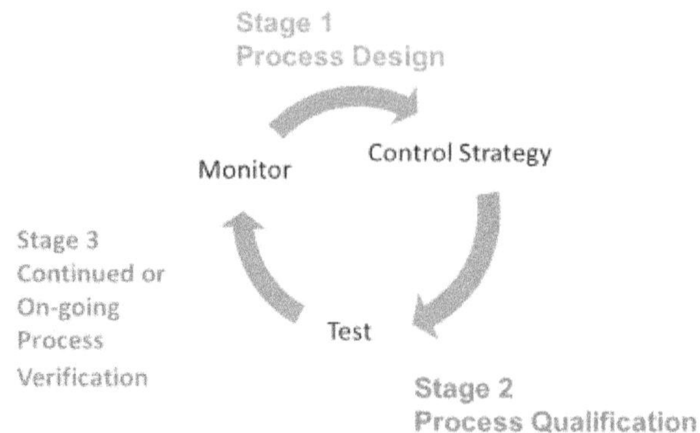

FIGURE 1.1
Process validation lifecycle approach.

A lifecycle approach is based on the need to fully understand the relationship of product attributes, process requirements, process variables, and process controls. A key principle is process failure as a result of unforeseen and unaddressed or under-addressed process variables. These variables can adversely affect the outcome of the process, resulting in unreliable performance and the potential for process failure. Understanding of the link between process steps, controls, parameters, and conditions on the reliability of the process and product quality attributes is required to ensure proper process performance and prevent process failure.

The basis of the lifecycle approach is that there should be a linkage or line of sight between product quality and process validation studies. As illustrated in Figure 1.2, the information needed to establish that link is gleaned from all stages of the process life cycle.

- The product is defined through specifications of its critical product attributes.
- The process to designed to achieve or maintain those product attributes.
- The variables that pose a risk to process performance and product quality are addressed with process controls.
- The qualification and validation studies and data from those studies provide confidence that those controls are effective.
- The monitoring of process parameters and outcomes are evaluated throughout commercial manufacturing to ensure that the process remains in a state of control.

The decisions related to the level of information and data needed, the controls, testing, monitoring, and evaluation criteria employed should be based on scientifically valid principles and with consideration of risk to process performance, product quality, and patient safety (2).

The lifecycle approach starts in Stage 1 with identifying product critical quality attributes or CQAs. These are the attributes of the product that define its quality and usefulness or benefit to the patient. The process is defined by the steps or conditions required to reliably achieve or maintain these CQAs. The parameters of process step performance are defined as critical process parameters or CPAs. The process variables are those actions, events, or conditions that may adversely affect the performance of the process steps or the

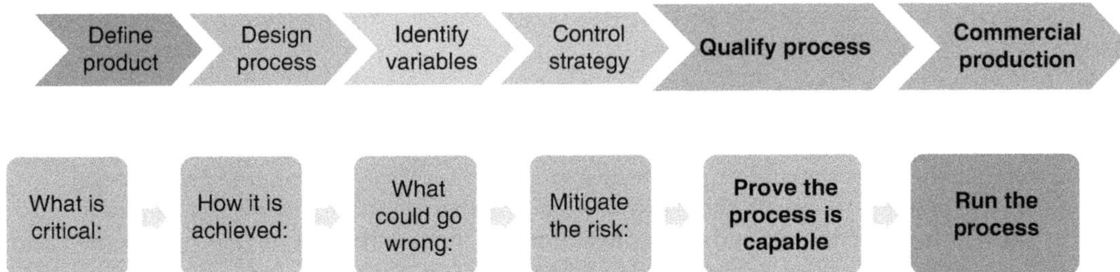

FIGURE 1.2
Line of sight process validation control and validation.

outcome of the process in regard to the achievement or maintenance of the CQAs. Process control measures are used to address these variables and prevent the failure of the process. The objective of process validation is to ensure that the process works in a reliable manner and consistently results in product meeting the defined CQAs.

During Stage 2, process qualification includes operational review, studies, tests, monitoring, measurements, and data evaluation that demonstrates or proves adequate levels of process control design and effectiveness. The first part of Stage 2 addresses the qualification of the facility, equipment, and instrumentation required to run the process. The second part, process performance qualification, addresses the qualification of the process itself.

In Stage 3, the effectiveness of the process is monitored to ensure that proper performance is maintained and that process variables remain adequately controlled. Where performance becomes unreliable or significant variables are noted, the process may be changed to address these issues.

1.2 Process Validation and Guidance Background

Knowing the background of validation is helpful to understand the evolution of modern process validation. Concepts of modern drug manufacturing process validation may be the result of an incident that occurred in the 1970s, the 1970–1971 outbreaks of *E. cloacae* and *Erwinia* contamination in large-volume parenteral bottles at several hospitals (6). The outbreaks may have been caused by moisture seeping into a space between the cap and stopper of the bottles during cooling after sterilization with contamination released when the containers were pierced during opening and use. Routine sterility testing of the solution would not necessarily have uncovered the contamination, as the contaminated areas may have been discarded during the testing procedure. Therefore, testing of product quality would not have been enough to assure product quality or patient safety. A systematic approach to confirm the capability and effectiveness of the process steps and conditions that affect product quality to assure product quality and safety was needed.

Companies or individuals wishing to distribute drug products that affect the health of patients have an obligation to assure that those products are safe and effective. In the US that requirement is presented in section 501(a)(2)(B) of the *Federal Food, Drug, and Cosmetic Act* (21 U.S.C. 351(a)(2)(B)), which states,

> A drug . . . shall be deemed to be adulterated . . . if . . . the methods used in, or the facilities or controls used for, its manufacture, processing, packing, or holding do not conform to or are not operated or administered in conformity with current good manufacturing practice to assure that such drug meets the requirements of this Act as to safety and has the identity and strength, and meets the quality and purity characteristics, which it purports or is represented to possess.

$$(7)$$

To meet this requirement, the US FDA published regulations describing current good manufacturing practice (cGMP) for finished pharmaceuticals. These include 21 CFR part 211.100 of the Current Good Manufacturing Practice regulations, which states, "There shall be written procedures for product and process control designed to *assure* that drug products have identity, strength, quality, and purity they purport or are represented to possess." Other cGMP regulations define the various aspects of validation. For example, § 211.110(a), Sampling and testing of in-process materials and drug products, requires that control procedures "be established to monitor the output and to validate the performance of those manufacturing processes that may be responsible for causing variability in the characteristics of in-process material and the drug product" (8; emphasis added).

1.3 Process Validation and the Assurance of Quality

What 21 CFR part 211.100 states is that testing of product is not enough to assure product quality. One must also demonstrate process capability and control. The requirement for assurance of process control can be met by one of two means.

Observation or Monitoring. Where the outcome of a process, a condition, or a product quality attribute can be observed, inspection or testing can provide assurance. For this to be effective, the attribute or outcome must be apparent for all product units, for example dimensions on the medical device portion of combination product or printing on a biopharmaceutical product label. For observation to be effective all units or product must be sampled and inspected, and the inspection must be fully reliable.

Prediction. Where the outcome of a process, a condition, or a product quality attribute cannot be fully observed, inspected, or tested, assurance must be provided by predicting the outcome. The prediction of a condition, outcome, or attribute is based on other information that can be observed or calculated, for example environmental monitoring, statistical sampling, personnel training, confirmation of equipment design, or temperature mapping. For predication to be effective there must be good correlation between what is observed and the predicted outcome.

Process validation is assurance of the capability of a process to produce quality product. It is achieved through the combination of observation of what can be seen and measured and the prediction of what cannot be fully observed, based on conditions and aspects that can be observed. The effectiveness of the validation depends on the strength of the correlation between what can be observed and the desired outcome. Where a strong correlation exists, testing and observation are effective. Where the correlation is weaker, more

FIGURE 1.3
Verification of process control.

dependence on surrogate means must be used. The correlation is based on process understanding that is gained from sound process design. Thus, there is an undeniable importance to the information that is gained from process design and prior knowledge.

As Figure 1.3 illustrates, the best means to ensure process control is with proper and carefully planned process design control rather than monitoring and testing. It is important for companies to understand this key principle presented in the guidance.

The verification of process control can be achieved through each of the four means noted in the figure. However, not all are equal in value. Product testing is limited by the ability to reflect the capability of the process. Product testing from a given batch is designed to show the attributes of the product at the time of the testing. Variables in the process that may not be apparent or occurring at that time may not be reflected in this testing. In addition, product testing is often destructive and therefore limited in the number of samples that might be feasible to test. Process monitoring allows for process output verification. However, even if the monitoring is continuous, it is only as effective as the correlation between the test results and the desired outcome of the process. That correlation largely is determined by the ability to catch and address the effect of process variables. Reliance on testing and monitoring alone will have limited effectiveness. Reliance on well-designed lifecycle validation studies will be more effective. Of course, validation studies alone cannot assure that the process is effective. The confidence of process effectiveness starts and is largely dependent on sound and proper process and process control design.

1.4 Evolution of Modern Process Validation

Validation efforts in the 1970s focused on contamination control, specifically sterility assurance and sterilization. The validation studies included moist and dry heat temperature mapping and biological challenges of autoclaves, dry heat ovens and tunnels, and media fills. This emphasis on contamination control and sterilization made sense on a relative risk to patient safety basis. In the 1980s, validation programs included equipment and facility qualification, cleaning validation, mixing studies, sanitization/disinfection effectiveness validation, analytical test method validation, computer system valuation, environmental monitoring qualification, hold studies, clean in place and sterilize in place studies, and container formation, ethylene oxide, gamma radiation, and electron-beam sterilization validation.

In the 1990s, more detailed design qualifications, installation and operational qualification studies, validation planning, packaging qualification, and container integrity studies were performed to a greater degree. Over the ensuing years, the focus included the validation of manufacturing processes for all dosage forms and manufacturing technologies. Throughout this period, it appears that much of process validation was based on evaluating samples of product taken from a series of consecutive commercial-scale or near-commercial-scale "demonstration" batches of product. If these batches did not meet all of the prescribed quality specifications, then an investigation and evaluation was performed. The evaluation of results became a combination of deviation investigation and justification, rather than a presentation of process control supporting evidence. In other words, more time was spent explaining or defending what was wrong than presenting why it provided assurance of process control and product quality. It was apparent that for processes to work properly, the equipment and systems that supported those processes needed to be suitable for use and function in a reliable manner. Equipment, facility, and system qualification provided that assurance.

In 1987 the FDA issued its Process Validation Guidance. The 1987 guidance is a presentation of industry best practices, as well as regulatory expectations. This guidance codified practices and regulatory expectations, including equipment qualification and process validation. Europeans would soon follow with different emphasis but basically the same format for validation. Qualification and validation became an important topic and business endeavor. If a company wanted to market a drug product, it needed to provide validation documentation, and the facility could not manufacture commercial product until validation was complete. Doing anything and everything to assure regulatory compliance and product approval made business sense. Validation seemed to shift from a method of proving the capability and reliability of the process to a means of meeting regulators' expectations.

In some cases, validation was becoming a checklist item, a demonstration of process control through the passage of multiple (usually three) production-scale batches. Reliance solely on multiple-batch testing led to insufficient consideration of the sources of that variability and the impact that variability might have on process performance and product quality. There was insufficient interaction with the process development people to determine process variability or weakness. There was insufficient feedback after the completion of the process validation studies on process performance once the process was released for commercial manufacturing use. Further, since the equipment and systems were qualified early, often before critical process parameters were identified, the parameters tested were not always the ones needed to verify that the equipment was capable of reliable process performance.

In the 1990s companies were found to have deficiencies with various quality-related systems, resulting in warning letters and consent decrees. Over 25 consent decrees were entered into from 1990 to 2004 (9). In many cases, the response to the consent decree was to create very robust validation and process control programs to meet regulatory expectations that proved to be difficult to sustain in operations. Adding to the problem, in the 2000s many companies faced financial pressures. There was increased scrutiny on the cost of drug manufacturing placed on the industry. The result was a closer look at validation-related costs. As the decade closed, the FDA appeared to increase its enforcement policies. The increased emphasis on cost and additional regulatory compliance pressure meant that there needed to be a change in approach to process validation in the industry.

At first, the response by regulators appeared to be more prescriptive. One way to ensure more effective validation approaches was to tell companies what was expected. This was often accomplished by proclamation or regulating from the podium of conferences, through the content of regulatory citations and warning letters, through formal regulatory agency guidance, through answers to industry questions posted on the FDA websites, and through individual meetings and discussions. However, at times the message appeared unclear and answers conflicting. Understanding process validation expectations became more of a challenge as global regulatory and process validation requirements were realized. This resulted in criticism from the industry and regulators that process validation approaches were not always effective or based on sound scientific principles. In addition, traditional or "one size fits all" validation approaches did not always address newer technologies and often hindered innovation. Some in the industry suggested that regulators should set the ground rules that companies provide assurance for safe and effective products but leave the method for providing that assurance to the company manufacturing process experts.

1.5 The 2008 Draft Process Validation Guidance and Industry Reaction (10, 11)

In 2006, the industry began to hear that the FDA was going to revise its 1987 Guidance on the Principles of Process Validation. Throughout 2007 and 2008 the industry waited for the release of a Process Validation Guidance draft. In anticipation of the release of the draft guidance, the PDA organized a task force of industry experts to be ready when the draft guidance was released. Trade associations and companies requested information from the FDA on the content of the revised guidance. The answer given from one source was that it is all about identifying and controlling process variability.

On November 17, 2008, the FDA released the draft revision to Process Validation: General Principles and Practices Guidance for Industry. This was meant to be a shift from the documentation focus of the 1987 guidance to a less prescriptive, more risk- and science-based approach. The draft revision was released the week before the winter holiday season, initially allowing for 60 days to submit comments (an extension was later granted). Excessive comments were not anticipated, because the revision was a reflection of what was believed to be current industry practice and the document was non-prescriptive and general in nature. Nonetheless, by the end of December of 2008, the PDA task force had collected over 400 industry comments. It should be noted that the PDA committee did not agree with all of the comments but felt that the fact that companies raised these issues meant that there was some confusion in the industry as to the intent of the guidance and therefore there was a need for further clarification. The comments received by PDA were reviewed and organized into six major categories (12, 13).

1.5.1 Wording and Terminology

The use and definition of specific terms and language had the most questions and comments. Collectively, these comments promoted the value of including a document glossary, as well as the desire to use terminology consistent with ICH and other FDA regulatory guidance definitions in order to reduce potential misinterpretation. This was important, because the guidance would be used internationally and the interpretation of terminology should be consistent.

1.5.2 Approach and Assurance for Commercial Distribution

These comments noted the need for clarification of expectations for determining the level of assurance required to initiate commercial product manufacture and release batches for commercial distribution. Related to this issue was the concern that a limited number of developmental batches would not be sufficient to develop a statistically sound rationale for commercial product distribution. For example, comments noted that the draft guidance indicated that extensive testing on early commercial batches might be required, without providing the criteria for reducing this level of testing. Further, the guidance was interpreted as not adequately promoting risk assessment as a means to reduce the number of samples and level of monitoring on relatively low-risk processes and steps.

1.5.3 Viral and Impurity Clearance

Objections were raised relative to the expectation that viral and impurity clearance studies performed at small scale should be performed under full CGMP conditions. The objections noted the additional burden this expectation would place on companies.

1.5.4 Concurrent Release

The comments indicated that there was some confusion over when and how concurrent release could be used. Several concerns were raised on the recommendation for stability testing of all concurrently released batches being overly prescriptive and that the release of batches was not in the scope of the guidance.

1.5.5 Scope and Legacy Systems (Processes)

Scope—Questions were raised on whether the guidance covered clinical product supplies, investigational medical products, blood products, *in vitro* diagnostic products, and vaccine products. These also included questions related to whether processes such as cleaning, sterilization, sanitization, holding, and distribution of commercial products were included in the scope of the guidance.

Legacy processes—Clarification was sought regarding how the guidance should be applied to the validation of existing processes or processes previously validated using the 1987 Process Validation Guidance.

1.5.6 Qualification, Documentation, Organization, and Regulatory Impact

Qualification—Concerns were expressed regarding the expectation to demonstrate the capability of equipment to maintain operating ranges over anticipated production times, especially where extended processing times are encountered.

Documentation and organization—Clarification was sought on the difference between qualification plans and protocols. There were recommendations to remove language that prescribes organizational dynamics and personnel activities such as having a variety of disciplines and project plans, as well as trending production line operator's errors.

Regulatory impact—There were comments requesting clarification of expectations related to information and format of information to be included in regulatory submission. There were concerns raised over apparent or perceived conflicts between the guidance and some regulatory submission requirements.

The PDA comments were submitted to the FDA in a letter and report on March 3, 2009. The FDA received these comments along with others from companies, other trade associations, individuals, and other interested parties. Throughout 2009, the PDA conducted a series of workshops designed to present the draft guidance and obtain additional comments from industry. The workshops included participation by one or more of the draft guidance authors and were held in San Francisco, Chicago, Las Vegas, Bethesda, Munich, and Puerto Rico (13). The workshop discussions indicated that questions remained on the intent of the authors and raised concerns over how to comply with its recommendations. As one consultant put it,

> the problem is that we just spent the past 30 years being told that process validation was a matter of good documentation practice. Now you are saying it is based on science. That is quite an adjustment in philosophy to overcome 30 years of thinking and practice in a matter of months.

In workshops the FDA representatives noted that the principles presented in the 2008 draft and 2011 guidance were not new. In fact, the FDA representatives at the workshops emphasized that the requirements specified in the 2008 draft guidance are consistent with the 1978 cGMPs (Parts 210, 211). The FDA asserted that the principles and recommendations in the 2008 draft guidance is a reflection of what the industry was already doing. More than one FDA representative expressed surprise when industry attendees responded that this might not be the case with all companies.

Leading up to the release of the draft, there was a perception in the industry that regulators considered three conformance batches as the primary criteria for demonstrating process control. However, the revised 2008 draft guidance was silent on the number of batches required to validate the process and did not necessarily note number of batches as validation acceptance criteria. Instead, the guidance spoke to the need for data, which would come from batches. Despite much discussion on the subject, this perceived change in the three-batch expectation raised questions, and likely still raises questions, as to the correct number of batches required. For example, the March 2015 version of EU Annex 15: Qualification and Validation states in part:

> 5.19. The number of batches manufactured and the number of samples taken should be based on quality risk management principles, allow the normal range of variation and trends to be established and provide sufficient data for evaluation. Each manufacturer must determine and justify the number of batches necessary to demonstrate a high level of assurance that the process is capable of consistently delivering quality product.
>
> 5.20. Without prejudice to 5.19, it is generally considered acceptable that a minimum of three consecutive batches manufactured under routine conditions could constitute a validation of the process. An alternative number of batches may be justified taking into account whether standard methods of manufacture are used and whether similar products or processes are already used at the site. An initial validation exercise with three batches may need to be supplemented with further data obtained from subsequent batches as part of an on-going process verification exercise.
>
> (14)

The PDA review committee expressed concern that despite intentions and language to the contrary, the mention of three even as a minimum number would add to the confusion and be interpreted by companies as the only acceptable number.

PDA is concerned that the reference to a specific minimum number of batches (in this case three) needed to validate the process will lead some companies to incorrectly forego the efforts to develop this process understanding. Some firms will opt instead for running three batches without the applying principles of quality risk management, including planned data evaluation and justification, thus negating the best intentions of the Annex. For that reason, PDA recommends removing the reference to three batches, or any number, and instead reinforcing the evaluation of data and determination of number of batches needed to provide that data.

(15)

1.6 The Process Validation Guidance Scope and Purpose

In the previous 1987 Process Validation Guidance, the FDA noted that the document outlined what are acceptable elements of process validation for human and veterinary drug products and medical devices. In the 2008 draft guidance and 2011 final version guidance, the FDA added biological and active pharmaceutical ingredients but omitted medical devices. The 2008 and 2011 guidance versions changed *acceptable elements* to *appropriate elements*, but kept much of the remaining scope the same. The 2011 final guidance notes that guidance on medical device process validation is provided in a separate document, Quality Management Systems—Process Validation, edition 2.

The 2011 guidance scope is applicable to active drug substances and active pharmaceutical ingredients (APIs), not just finished pharmaceutical dosage forms. During the early 1990s, some firms in the industry debated the applicability of process validation principles and approaches to the manufacture of bulk drug products or bulk drug chemicals. Some argued that these were well-established processes with long-standing history of achieving reliable and predictable results and should therefore not be covered by process validation requirements. Today few would use such arguments, and the requirement for validation for API and drug substance processes is generally accepted.

The 1987 guidance has an "Introduction" section. The section presents the regulatory basis for process validation, as a requirement in the CGMP regulations for finished pharmaceuticals. The section further notes that industry firms have requested guidance on what the FDA expects them to do to assure compliance with this requirement for process validation. This is an important statement, because it (1) establishes that process validation is a requirement, not just a recommendation, and (2) it makes it a request of the industry to provide an approach, rather than the regulators unilaterally dictating an approach. The 1987 guidance goes on to state that the recommendations are not intended to be all-inclusive, and that a variety of methods and techniques (are and) may be used to assure compliance with the requirement of process validation. There is an acknowledgment that many methods are and may be employed. This leads one to imagine that the agency reviewed many of these methods and are presenting what they feel is a compilation of current industry best practices, rather than a carefully devised treatise on process control.

The 2008 draft guidance mentions the 1987 version and goes on to point out that since this version the FDA has obtained additional experience through regulatory oversight that allows for an update of the recommendations to industry on the topic of process validation. This is an interesting passage. The authors do not say that the industry has gained experience or that new and more effective methods have been developed for process validation or even that new technologies and challenges have emerged that warrant changes in the guidance. The authors focus on the agency's regulatory enforcement experience as the impetus for the new guidance.

This should not be surprising. Over the years the preferred method of process validation for many companies has been to depend on the monitoring of process parameters and testing of product from multiple batches, usually three consecutive full-scale commercial runs. At times these runs did result in some out-of-specification results or deviations. In these cases companies often spent time explaining why the deviation did not indicate an invalid or failed process. In some cases, processes continued to exhibit failures during commercial production. These resulted in modification to the process or incorporation of additional control measures, i.e., enhanced product inspection. However, the question that remained was, if process validation provides a high degree of assurance that the process works reliably and predictably, then how does one explain process failure?

These failures were often explained as anomalies, isolated failures of procedure or systems, or just unforeseen events. Actually, the failures are the result of unforeseen or inadequately controlled process variables. This suggests a lack of full process understanding and process variables. There were several reasons why this may not be the case, including:

- Incomplete process understanding when process validation protocol was developed.
- Insufficient knowledge of critical process parameters (CPPs) when user requirement specifications (URS) and process control strategies were developed, and process equipment designed and qualified.
- Undocumented or inadequately addressed changes to the process because of additional information gained after URS and control strategy development.
- Focus on speed to market resulting in back loaded process development and validation approaches.
- Poor communication between process design, validation, and operation functions.
- Inefficient transfer of knowledge from process design to validation to operations and back to process design functions.

1.7　Final Version of FDA 2011 Process Validation: General Principles and Practices Guidance for Industry

In this part of the chapter, a section-by-section analysis of the 2011 final version guidance will be presented. It does so by first presenting the text of the guidance in *italics*, with commentary indented after that section. Most of the commentary is based on comments received during the 2008–2009 review period, along with discussions that occurred at PDA Process Validation workshops, and related discussions and meetings (4, 6, 8, 9, 10, 11).

I Introduction

This guidance outlines the general principles and approaches that FDA considers appropriate elements of process validation for the manufacture of human and animal drug and biological products, including active pharmaceutical ingredients (APIs or drug substances), collectively referred to in this guidance as drugs or products. This guidance incorporates principles and approaches that all manufacturers can use to validate manufacturing processes.

The word *appropriate* replaces the word *acceptable* in the 1987 guidance. Appropriate is a broader term that indicates the approach is sound and effective, rather than just compliant. Active pharmaceutical ingredients and biologicals have been added to clarify that API manufacture does require validation. Medical devices have been removed, but a reference does point the reader to an acceptable guidance source.

It should be noted that advanced therapy medicinal products (ATMPs), such as cell and gene therapy products, were not in widespread development or use in 2011 and are not mentioned in the guidance. However, many of these manufacturers use the principles presented in the guidance as a basis for their process validation programs.

This guidance aligns process validation activities with a product lifecycle concept and with existing FDA guidance, including the FDA/International Conference on Harmonization (ICH) guidances for industry, Q8(R2) Pharmaceutical Development, Q9 Quality Risk Management, and Q10 Pharmaceutical Quality System. Although this guidance does not repeat the concepts and principles explained in those guidance, FDA encourages the use of modern pharmaceutical development concepts, quality risk management, and quality systems at all stages of the manufacturing process lifecycle.

Harmonization to international guidance is mentioned in this section. This is an addition to the 2008 guidance. One of the industry comments relayed to FDA by the PDA in 2009 brought up concerns over the lack of alignment with ICH and other international guidance. The alignment with ICH Q9, Quality Risk Management, is especially important in that it sets the groundwork for the linkage of process parameters and conditions with the achievement of and risk to product attributes. Quality risk management is mentioned in the guidance. Risk management and risk assessment appear twice as frequently in the 2011 final guidance as in the 2008 draft guidance. The limited mention of risk management and assessment in the 2008 guidance was brought up by attendees at PDA workshops. One of the answers given by FDA at those workshops was that the authors felt risk principles were so engrained in our current practice that it did not have to be mentioned repeatedly. However, the 2011 guidance does reinforce the use of risk approaches to a higher degree than did the 2008 guidance.

Note: It would be difficult to effectively design a process validation approach without employing some level of risk assessment and quality risk management. The key principle of the guidance is the identification and control of process variables that could adversely affect product quality. To be able to do this, one must incorporate an assessment of the risk posed by process steps and the potential for variables that increase such risk. One must assess process- and product-related risk, taking into consideration the impact to product quality of process failure, the likelihood of failure, and the ability to detect the failure in time to prevent patient harm.

The lifecycle concept links product and process development, qualification of the commercial manufacturing process, and maintenance of the process in a state of control during routine commercial production. This guidance supports process improvement and innovation through sound science.

The 2011 guidance provides a method for defining and developing a lifecycle approach to process validation. A lifecycle approach is not new. However, truly effective lifecycle approaches were not always well implemented. As noted earlier in the chapter, the 2011 guidance recommends that companies develop an approach that links, relies on, and transfers information from process development to process qualification to commercial manufacturing.

The 2011 guidance mentions process improvement and innovation as acceptable outcomes, even key objectives of process validation. In the PDA workshops, participants raised questions and concerns over the reaction of regulators to changes or improvements made after process qualification. They expressed concern that regulators would see this as an admission of allowing an imperfect process to be released into commercial production and therefore be penalized. FDA participants responded that would not be the case, especially if the improvement was the result of additional information becoming available due to commercial manufacture and if the initial process was sound and effective, but perhaps not yet optimal. However, companies should consider taking steps to better identify areas of improvement prior to completion of process design stages.

This guidance does not specify what information should be included as part of a regulatory submission. Interested persons can refer to the appropriate guidance or contact the appropriate Center in determining the type of information to include in a submission.

Industry comments made through the PDA in 2009 and at PDA-sponsored workshops raised concerns that the 2008 guidance did not address regulatory submissions and would potentially be in conflict with regulatory guidance on the inclusion of process validation related information in submission. The FDA participants in the workshops recommended that companies consult individual Center rules for guidance on submissions but noted that conflicts should be minimal. A prudent approach would be to consider a plan that presents the most stringent regulatory expectation.

This guidance also does not specifically discuss the validation of automated process control systems (i.e., computer hardware and software interfaces), which are commonly integrated into modern drug manufacturing equipment. This guidance is relevant, however, to the validation of processes that include automated equipment in processing.

The 2011 guidance does not address computer software validation, but because most items of equipment and instruments used to manufacture and test covered products rely on some level of automation, assuring the functionality and reliability of those systems would be part of this guidance. It is important to note that the principles of the process validation life cycle are logical and can, and in many instances should, be considered as the basis for other types of validation programs not mentioned specifically in the guidance.

II Background

In the Federal Register of May 11, 1987 (52 FR 17638), FDA issued a notice announcing the availability of a guidance entitled Guideline on General Principles of Process Validation (the 1987 guidance). Since then, we have obtained additional experience through our regulatory oversight that allows us to update our recommendations to industry on this topic. This revised guidance conveys FDA's current thinking on process validation and is consistent with basic principles first introduced in the 1987 guidance. The revised guidance also provides recommendations that reflect some of the goals of FDA's initiative entitled "Pharmaceutical CGMPs for the 21st Century—A Risk-Based Approach," particularly with regard to the use of technological advances in pharmaceutical manufacturing, as well as implementation of modern risk management and quality system tools and concepts. This revised guidance replaces the 1987 guidance.

There are three important points to highlight in this paragraph:

First, the FDA is revising the 1987 guidance in part because of knowledge they have gained from their experience with industry approaches over the last 20 years. This includes good experience, as in identifying effective approaches, and bad experience, as in failures of process validation and ineffective process validation programs. In other words, what works and what doesn't work. This is a logical outcome of a guidance issued 20 years ago to determine best practices and remedy bad habits. Second, the FDA promotes and reinforces the use of risk-based approaches. Third, the FDA recognizes that new technologies have changed the way process validation should be approached. Old or traditional approaches may not be adequate or appropriate for new technologies and manufacturing methods. A relatively non-prescriptive approach can help encourage companies to develop improved methods for assuring new technology process control.

FDA participants in workshops told industry participants that they would be expected to answer the question: "How does what you do to validate a process demonstrate or assure that process is under control?" It will be up to the industry to show or prove to the FDA that their approach to process validation—monitoring, sampling, and testing—meets that objective. Just saying this is what we did in the past or this is what we did at the other company will not be adequate.

FDA has the authority and responsibility to inspect and evaluate process validation performed by manufacturers. The CGMP regulations for validating pharmaceutical (drug) manufacturing require that drug products be produced with a high degree of assurance of meeting all the attributes they are intended to possess (21 CFR 211.100(a) and 211.110(a)).

A related point noted at PDA workshops and in discussions with FDA representatives was that the FDA would conduct post-approval inspections of manufacturing facilities six to 12 months after the approval to see how companies were complying with process validation expectations and requirement.

A Process Validation and Drug Quality

Effective process validation contributes significantly to assuring drug quality. The basic principle of quality assurance is that a drug should be produced that is fit for its intended use. This principle incorporates the understanding that the following conditions exist:

Quality, safety, and efficacy are designed or built into the product.

Quality cannot be adequately assured merely by in-process and finished-product inspection or testing.

The 2011 guidance reinforces that end or in-process testing cannot solely assure product quality. Therefore, process validation is not only a valuable exercise but an essential one in the manufacture of regulated products. Later in the guidance process analytical technology (PAT) is mentioned. PAT has the potential to allow for complete product and intermediate testing that should assure product quality. However, most drug manufacturing processes at the time of the 2011 guidance issuance did not include this type of real-time testing methodology (16).

B Approach to Process Validation

For purposes of this guidance, process validation is defined as the collection and evaluation of data, from the process design stage through commercial production, which establishes scientific evidence that a process is capable of consistently delivering quality product. Process validation involves a series of activities taking place over the lifecycle of the product and process. This guidance describes process validation activities in three stages.

The 2011 guidance presents a new definition of process validation that incorporates some advanced concepts. In 1987, the US FDA defined process validation as *establishing documented evidence which provides a high degree of assurance that a specific process will consistently produce a product meeting its pre-determined specifications and quality characteristics.* The key principles were documentation, providing a high degree of assurance, and meeting predetermined quality specifications. In the 2011 guidance, the FDA modified the definition of process validation . . . *as the collection and evaluation of data, from the process design stage through commercial production, which establishes scientific evidence that a process is capable of consistently delivering quality products.*

The definitions are similar, but the emphasis has shifted to evaluation of data, from process design through production, and the establishment of scientific evidence of process capability. In the previous 1987 guidance definition, emphasis was placed on the need for documentation and consistency as a means to provide that high degree of assurance that the process was fit for use. In the 2011 guidance, the emphasis shifts to the use of scientific data obtained from several sources beginning with process design and not ending with the completion of "validation" studies to prove that a process is capable of a desired outcome. The distinction in the wording of the definitions is important in that it signals a change in emphasis and expectation by FDA. In the 1987 definition, *documented evidence* is synonymous with collection of data. In the 2011 definition, *evaluation of data* denotes an expectation that there will be a careful analysis of the data to reach a conclusion. It will be this evaluation and the rationale for reaching the conclusion, rather than just the confirmation that data was obtained that regulators will be most interested in seeing. It can also be interpreted as an expectation for sound statistical analysis and rationale.

The addition of *design stage through commercial production* further defines the mechanism for developing a true and effective lifecycle approach. The addition of *capability* to the 2011 definition adds the concept of being able to predict outcome through the analysis of the manufacturing process rather than merely testing product and parameters. Finally, there is a change from defining desired outcome in 1987 as producing *a product meeting its pre-determined specifications and quality characteristics* to *delivering quality product* in the 2011 version. The inclusion of delivery indicates that the process should include more than just manufacture. It includes the handling, storage, and distribution of regulated products.

Stage 1—Process Design: *The commercial manufacturing process is defined during this stage based on knowledge gained through development and scale-up activities.*

Stage 2—Process Qualification: *During this stage, the process design is evaluated to determine if the process is capable of reproducible commercial manufacturing.*

Stage 3—Continued Process Verification: *Ongoing assurance is gained during routine production that the process remains in a state of control.*

This guidance describes activities typical of each stage, but in practice, some activities might occur in multiple stages.

This section summarizes and defines the lifecycle approach. The recommended approach is divided into stages beginning with the design or development of the process and not ending with conclusion of qualification studies. All information of value should be considered. The lifecycle approach is based on the principle that process validation is a continuum of process evaluation—a journey and not a destination.

Before any batch from the process is commercially distributed for use by consumers, a manufacturer should have gained a high degree of assurance in the performance of the manufacturing process such that it will consistently produce APIs and drug products meeting those attributes relating to identity, strength, quality, purity, and potency. The assurance should be obtained from objective information and data from laboratory-, pilot-, and/or commercial-scale studies. Information and data should demonstrate that the commercial manufacturing process is capable of consistently producing acceptable quality products within commercial manufacturing conditions.

This section reinforces the expectation that companies will review data to make decisions related to the appropriateness and effectiveness of processes, based on sound scientific rationale. It links the language of 21 CFR part 211.100 to the principles of process validation.

It is notable that the guidance does not include the term *worst case*, a commonly used term and concept in process validation. Instead, it recommends that information should demonstrate that the process is capable of producing acceptable product *within commercial manufacturing conditions*. This is a change from the 2008 draft which recommended that process also demonstrate that it can produce acceptable product under conditions which *pose a high risk of process failure*. Process conditions that pose a high risk of process failure should be mitigated, so that the risk is reduced. Where acceptable operating conditions are variable, data should demonstrate that the process functions adequately at all anticipated conditions.

A successful validation program depends upon information and knowledge from product and process development. This knowledge and understanding are the basis for establishing an approach to controlling the manufacturing process that results in products with the desired quality attributes. Manufacturers should:

Understand the sources of variation

Detect the presence and degree of variation

Understand the impact of variation on the process and ultimately on product attributes

Control the variation in a manner commensurate with the risk it represents to the process and product

Each manufacturer should judge whether it has gained sufficient understanding to provide a high degree of assurance in its manufacturing process to justify commercial distribution of the product. Focusing exclusively on qualification efforts without also understanding the manufacturing process and associated variations may not lead to adequate assurance of quality. After establishing and confirming the process, manufacturers must maintain the process in a state of control over the life of the process, even as materials, equipment, production environment, personnel, and manufacturing procedures change.

This section reinforces the expectation that companies will use process development information to develop the process validation protocol and plan. The company should focus on the variation in process and process outcome. Failures are the result of unforeseen or under-addressed process variation. It is this variation that needs to be understood and controlled to assure consistent process performance and product quality. This is an important statement. Efforts which focused on qualification, without fully understanding the relationship of process variables to performance and product quality, often result in failures.

Manufacturers should use ongoing programs to collect and analyze product and process data to evaluate the state of control of the process. These programs may identify process or product problems or opportunities for process improvements that can be evaluated and implemented through some of the activities described in Stages 1 and 2.

This reinforces the concept of a lifecycle approach, where the validation does not end with the process qualification testing. Valuable information can be obtained as the process is run commercially, and manufacturers should use all information available to make informed decisions about the validity and reliability of the process. Variance in performance during ongoing testing may be a result of under-addressed process variation or unforeseen changes to the process. There may also be an opportunity for process improvement. Changes that result from these improvements should be incorporated in a qualification and validation plan.

Manufacturers of legacy products can take advantage of the knowledge gained from the original process development and qualification work as well as manufacturing experience to continually improve their processes. Implementation of the recommendations in this guidance for legacy products and processes would likely begin with the activities described in Stage 3.

This section discusses legacy products and processes. In comments sent to the PDA and raised in workshops, some in industry expressed concerns related to how the guidance should be used to validate legacy processes. A legacy process is one which has previously been validated prior to the publication of the 2011 guidance. Historical performance can be used in place of prospective testing and monitoring. If the previously validated legacy process is robust, less complex, better understood, showing little or no variation, showing little or no out-of-specification (OOS) results, and showing little or no process failures, deviations, or investigations, then the need for performing another set of process validation studies may not be necessary.

It is important to note that all processes must be under control, as demonstrated by continued acceptable performance. Therefore, Stage 3 ongoing process monitoring, and verification is expected. If the process was not as well understood, more complex, and resulted in continued failures, deviations, and out-of-specification results, then this might indicate that the process validation was not effective and that additional studies and information were necessary. These might include revisiting the source of variation understanding and process control strategy (process design) as well as additional process qualification studies.

III Statutory and Regulatory Requirements for Process Validation

Section III presents the regulatory rationale and FDA justification for requiring process validation. The section references the Food, Drug, and Cosmetic Act (21 U.S.C. 351(a)(2)(B) section 501(a) (2)(B), as well as 21 CFR parts 211.100(a), 211.110(a)(b), 211.160(b)(3), 211.165(a)(c)(d)211.180(e), 211.42, 211.63, and 211.68.

. . . the CGMP regulations regarding sampling set forth a number of requirements for validation: samples must represent the batch under analysis (§ 211.160(b)(3)); the sampling plan must result in statistical confidence (§ 211.165(c) and (d)); and the batch must meet its predetermined specifications (§ 211.165(a)). . . .

Most notable in this section are references to the requirement for the use of statistical analysis in designing study and sampling plans. When the guidance mentions scientific evidence in its definition of process validation, it largely refers to evidence from the statistical analysis of data. Comments from industry received by the PDA and expressed at workshops indicated concern related to the strong expectation for the use of statistical analysis and statistically trained individuals to prepare process validation plans and analyze the results. A point of concern for the industry was that the use of statistically sound sampling plans would result in extensive sampling and testing at Stage 2 Process Qualification and that that would carry into Stage 3 Continued Process Verification for an extended, if not perpetual, period. This would in turn result in burdensome and costly process requirements.

In workshop and industry presentations, including the PDA FDA joint conference held in Washington, DC, in September of 2010, the FDA speakers spent a good deal of time presenting and defending their expectation for more statistical analysis. The FDA response to these concerns was that extended sampling is a matter of risk management, but that decisions related to assurance of process control must be based on reliable scientific information. Therefore, statistically sound study and sampling plans are a logical requirement. However, it is also important to note that as pointed out in one of the PDA workshops, statistics is but one tool for evaluation of the process reliability. Prior knowledge, historical performance, scientific principles, published studies, and other relevant information may be used as well.

This section also reinforces and justifies the recommendations for continued process verification, noting that *ongoing feedback about product quality and process performance is an essential feature of process maintenance.* The section justifies the requirement for equipment and facility qualification, citing parts 211.42 and 211.63 to note a requirement that facility and equipment be suitable for use. Finally, the section makes a summary statement: *the CGMP regulations require that manufacturing processes be designed and controlled to assure that in-process materials and the finished product meet predetermined quality requirements and do so consistently and reliably.* This is the regulatory basis and requirement of process validation.

IV Recommendations

In the following sections, we describe general considerations for process validation, the recommended stages of process validation, and specific activities for each stage in the product lifecycle.

A General Considerations for Process Validation

In all stages of the product lifecycle, good project management and good archiving that capture scientific knowledge will make the process validation program more effective and efficient. The following practices should ensure uniform collection and assessment of information about the process and enhance the accessibility of such information later in the product lifecycle.

The guidance mentions that scientific knowledge will make the process validation program more effective and efficient. Making it more effective means that it will accomplish the goal of providing confidence that the process is well controlled and will perform in a predictable manner. However, *efficiency* is a different term. It means that the program will utilize resources and costs to an optimal level. This may be a response to concerns raised that the recommendations into the revised versions of the guidance would add cost and burden to the process, without a corresponding benefit. The recommendations that follow are designed to help companies identify, analyze and transfer information accurately and reliably.

We recommend an integrated team approach to process validation that includes expertise from a variety of disciplines (e.g., process engineering, industrial pharmacy, analytical chemistry, microbiology, statistics, manufacturing, and quality assurance). Project plans, along with the full support of senior management, are essential elements for success.

> This appears to reflect industry best practices. The isolated validation department function coming in at the end to be the police, confirming that all is right and in order, has been replaced by some companies with a more flexible and temporary validation team approach, incorporating people from several involved functions.

Throughout the product lifecycle, various studies can be initiated to discover, observe, correlate, or confirm information about the product and process. All studies should be planned and conducted according to sound scientific principles, appropriately documented, and approved in accordance with the established procedure appropriate for the stage of the lifecycle.

The terms attribute(s) (e.g., quality, product, component) and parameter(s) (e.g., process, operating, and equipment) are not categorized with respect to criticality in this guidance. With a lifecycle approach to process validation that employs risk based decision making throughout that lifecycle, the perception of criticality as a continuum rather than a binary state is more useful. All attributes and parameters should be evaluated in terms of their roles in the process and impact on the product or in-process material, and reevaluated as new information becomes available. The degree of control over those attributes or parameters should be commensurate with their risk to the process and process output. In other words, a higher degree of control is appropriate for attributes or parameters that pose a higher risk. The Agency recognizes that terminology usage can vary and expects that each manufacturer will communicate the meaning and intent of its terminology and categorization to the Agency.

> This section discusses the implementation of a multivariate, risk assessment approach. The approach takes into consideration the interaction of process elements and the shifting of criteria and risk. This increase in process understanding is gained over time through evaluation of collected information. This section hints to the need for continued process risk review, assessment, and communication, rather than one-time assessment. As such it is an important point that counteracts the checklist approach to risk assessment, where assessments are performed, information gleaned, and then put on the shelf, so to speak.

Many products are single-source or involve complicated manufacturing processes. Homogeneity within a batch and consistency between batches are goals of process validation activities. Validation offers assurance that a process is reasonably protected against sources of variability that could affect production output, cause supply problems, and negatively affect public health.

> This is an interesting paragraph and may signal a shift in philosophy. It has been moved from the Statutory and Regulatory Requirements section in the 2008 draft to the General Consideration section of the 2011 final version. The 1987 version focused our validation efforts on processes and process elements that affected product quality. That was reinforced in the late 1990s and into the 2000s as companies tried to streamline qualification and validation efforts (i.e., ASTM E-2500–07 and ISPE ISPE's Commissioning and Qualification Baseline® Guide).
>
> In this section, the FDA recommends that validation be used to also qualify those factors that could affect product output and supply. They make the link between product supply and public health. This is logical, but again in the past if product did not get to a patient and therefore did not harm them, then that was enough. Risk to product availability is noted as a potential harm. Therefore, we are now to be concerned with public health as well as product quality. How this will affect process qualification is yet to be seen. However, the sustainability and reliability of process output may someday be a process qualification concern.

B Stage 1—Process Design

Process design is the activity of defining the commercial manufacturing process that will be reflected in planned master production and control records. The goal of this stage is to design a process suitable for routine commercial manufacturing that can consistently deliver a product that meets its quality attributes.

1 Building and Capturing Process Knowledge and Understanding

Generally, early process design experiments do not need to be performed under the CGMP conditions required for drugs intended for commercial distribution that are manufactured during Stage 2 (process qualification) and Stage 3 (continued process verification). They should, however, be conducted in accordance with sound scientific methods and principles, including good documentation practices. This recommendation is consistent with ICH Q10 Pharmaceutical Quality System. Decisions and justification of the controls should be sufficiently documented and internally reviewed to verify and preserve their value for use or adaptation later in the lifecycle of the process and product.

Although often performed at small-scale laboratories, most viral inactivation and impurity clearance studies cannot be considered early process design experiments. Viral and impurity clearance studies intended to evaluate and estimate product quality at commercial scale should have a level of quality unit oversight that will ensure that the studies follow sound scientific methods and principles, and the conclusions are supported by the data.

> The 2008 draft stated that viral and impurity clearance studies should be performed under CGMP conditions even when performed at small scale, since they do have a direct impact on drug safety. PDA received industry comments raising concerns about that expectation. The 2011 final version modified the language to focus on quality oversight and scientific reliability of data rather than strict adherence to CGMPs. This appears to be a modification in response to industry-raised concerns.

Product development activities provide key inputs to the process design stage, such as the intended dosage form, the quality attributes, and a general manufacturing pathway. Process information available from product development activities can be leveraged in the process design stage. The functionality and limitations of commercial manufacturing equipment should be considered in the process design, as well as predicted contributions to variability posed by different component lots, production operators, environmental conditions, and measurement systems in the production setting. However, the full spectrum of input variability typical of commercial production is not generally known at this stage. Laboratory or pilot-scale models designed to be representative of the commercial process can be used to estimate variability.

> This section mentions leveraging of process information at early process design stages. During PDA-sponsored workshops, the idea that the more effort is put in and information gathered during the process design stages, the less effort and information is needed during the latter stages. This is reasonable. If the result is process understanding, then extensive work performed during early development phases to identify and capture that knowledge, or to find and correct weakness in process, should result in a better-understood process going to process qualification and continued process verification stages. Therefore, less information might be needed to better understand the process at those mid and later stages.

Designing an efficient process with an effective process control approach is dependent on the process knowledge and understanding obtained. Design of Experiment (DOE) studies can help develop process knowledge by revealing relationships, including multivariate interactions, between the variable inputs (e.g., component characteristics or process parameters) and the resulting outputs (e.g., in-process material, intermediates, or the final product). Risk analysis tools can be used to screen potential variables for DOE studies to minimize the total number of experiments conducted while maximizing knowledge gained. The results of DOE studies can provide justification for establishing ranges of incoming component quality, equipment parameters, and in-process material quality attributes. FDA does not generally expect manufacturers to develop and test the process until it fails.

There are three important ideas presented in this paragraph:

1. The FDA expects experimentation to obtain useful information, and not just data collection.
2. The FDA will allow for risk assessments to help determine the information needed and scope of experiments. This should help limit the range of experiments needed.
3. The FDA will allow testing of systems and processes within the expected performance and parameter ranges, rather than expect full or edge-of-failure testing.

Other activities, such as experiments or demonstrations at laboratory or pilot scale, also assist in evaluation of certain conditions and prediction of performance of the commercial process. These activities also provide information that can be used to model or simulate the commercial process. Computer-based or virtual simulations of certain unit operations or dynamics can provide process understanding and help avoid problems at commercial scale. It is important to understand the degree to which models represent the commercial process, including any differences that might exist, as this may have an impact on the relevance of information derived from the models.

> This section notes that computer modeling and simulation of unit operations may be acceptable in lieu of commercial-scale empirical data. However, it is also implied that the company should recognize and articulate the correlation between the simulation data and commercial-scale performance.

It is essential that activities and studies resulting in process understanding be documented. Documentation should reflect the basis for decisions made about the process. For example, manufacturers should document the variables studied for a unit operation and the rationale for those variables identified as significant. This information is useful during the process qualification and continued process verification stages, including when the design is revised or the strategy for control is refined or changed.

> This paragraph reinforces the expectation for good documentation of early stage information gathering and experimentation. Information needed to support control strategies and process qualification testing should be reliable. Procedures and training should be put in place to ensure an accurate and efficient supply of information between the departments developing processes and those qualifying processes.

2 Establishing a Strategy for Process Control

Process knowledge and understanding is the basis for establishing an approach to process control for each unit operation and the process overall. Strategies for process control can be designed to reduce input variation, adjust for input variation during manufacturing (and so reduce its impact on the output), or combine both approaches.

Process controls address variability to assure quality of the product. Controls can consist of material analysis and equipment monitoring at significant processing points (§ 211.110(c)).

Decisions regarding the type and extent of process controls can be aided by earlier risk assessments, then enhanced and improved as process experience is gained.

> Parts of this section reinforce the expectation of risk assessments at early phases of process development and the periodic review of those assessments. There is an expectation that risk assessments be somewhat living documents subject to review, repeat, and change.

FDA expects controls to include both examination of material quality and equipment monitoring. Special attention to control the process through operational limits and in-process monitoring is essential in two possible scenarios:

1. *When the product attribute is not readily measurable due to limitations of sampling or delectability (e.g., viral clearance or microbial contamination) or*
2. *When intermediates and products cannot be highly characterized and well-defined quality attributes cannot be identified.*

> This section notes the need for ways to predict process outcome, when product attribute observation is limited. As noted earlier, validation becomes more important as the ability to directly observe process outcome (product characteristics and attributes) decreases. This is especially important for aseptic processes and sterile product manufacturing processes, where observation of sterility is difficult if not practically impossible. However, for in-process monitoring to be effective, a strong correlation and causational relationship must be established between the in-process observation and the desired product attribute outcome.

These controls are established in the master production and control records (see § 211.186(a) and (b) (9)). More advanced strategies, which may involve the use of process analytical technology (PAT), can include timely analysis and control loops to adjust the processing conditions so that the output remains constant. Manufacturing systems of this type can provide a higher degree of process control than non-PAT systems. In the case of a strategy using PAT, the approach to process qualification will differ from that used in other process designs. Further information on PAT processes can be found in FDA's guidance for industry on PAT—A Framework for Innovative Pharmaceutical Development, Manufacturing, and Quality Assurance.

> This section notes that PAT represents a high degree of verification, in that the product attribute or outcome is inspected on a constant basis. Therefore, processes controlled by PAT systems may not require the same level of process qualification. The distinction between traditional processes and more advanced continuous verification and control processes is also addressed in the EU Annex 15 stating, *For products developed by a quality by design approach, where it has been scientifically established during development that the established control strategy provides a high degree of assurance of product quality, then continuous process verification can be used as an alternative to traditional process validation. . . . There should be a science based control strategy for the required attributes for incoming materials, critical quality attributes and critical process parameters to confirm product realisation. This should also include regular evaluation of the control strategy. Process Analytical Technology and multivariate statistical process control may be used as tools.*

[reference annex 14]

The planned commercial production and control records, which contain the operational limits and overall strategy for process control, should be carried forward to the next stage for confirmation.

Again, this sentence reinforces the expectation and importance of gleaning information from early process design stages to form basis of test parameters and approaches in process qualification stages.

C Stage 2—Process Qualification

During the process qualification (PQ) stage of process validation, the process design is evaluated to determine if it is capable of reproducible commercial manufacture. This stage has two elements: (1) design of the facility and qualification of the equipment and utilities and (2) process performance qualification (PPQ). During Stage 2, CGMP-compliant procedures must be followed. Successful completion of Stage 2 is necessary before commercial distribution. Products manufactured during this stage, if acceptable, can be released for distribution.

This section introduces the term *process performance qualification* or PPQ. This is a clarification added as a response to confusion over the use of process qualification PQ to denote the entire Stage 2 and performance qualification PQ, which was the second part of Stage 2. The last sentence of the paragraph clarifies that product manufactured during qualification can be used for commercial distribution. This should not be confused with recommendations related to concurrent release of product, which is explained later in the guidance.

1 Design of a Facility and Qualification of Utilities and Equipment

Proper design of a manufacturing facility is required under part 211, subpart C, of the CGMP regulations on Buildings and Facilities. It is essential that activities performed to assure proper facility design and commissioning precede PPQ. Here, the term qualification refers to activities undertaken to demonstrate that utilities and equipment are suitable for their intended use and perform properly. These activities necessarily precede manufacturing products at the commercial scale.

This section clarifies the expectation that equipment and facilities are proven to be fit for use or qualified. Equipment qualification precedes process qualification. The section largely refutes questions raised by some that the FDA did not require formal equipment qualification and therefore it did not have to be performed.

Qualification of utilities and equipment generally includes the following activities:

Selecting utilities and equipment construction materials, operating principles, and performance characteristics based on whether they are appropriate for their specific uses.

Verifying that utility systems and equipment are built and installed in compliance with the design specifications (e.g., built as designed with proper materials, capacity, and functions, and properly connected and calibrated).

Verifying that utility systems and equipment operate in accordance with the process requirements in all anticipated operating ranges. This should include challenging the equipment or system functions while under load comparable to that expected during routine production. It should also include the performance of interventions, stoppage, and start-up as is expected during routine production. Operating ranges should be shown capable of being held as long as would be necessary during routine production.

Equipment should be qualified to perform through normal operating ranges and conditions, not necessarily at the extremes of equipment capability. This is an important clarification. In the past, some have proposed full operating capability testing to show that the equipment functions as designed. While this approach may have merit for establishing that the equipment is in top working condition, unless the operation beyond anticipated process ranges is needed to provide confidence of consistent and reliable performance at process ranges, it should not be necessary during formal qualification.

Qualification of utilities and equipment can be covered under individual plans or as part of an overall project plan. The plan should consider the requirements of use and can incorporate risk management to prioritize certain activities and to identify a level of effort in both the performance and documentation of qualification activities. The plan should identify the following items:

1. *the studies or tests to use,*

2. *the criteria appropriate to assess outcomes,*

3. *the timing of qualification activities,*

4. *the responsibilities of relevant departments and the quality unit, and*

5. *the procedures for documenting and approving the qualification.*

The project plan should also include the firm's requirements for the evaluation of changes. Qualification activities should be documented and summarized in a report with conclusions that address criteria in the plan. The quality control unit must review and approve the qualification plan and report (§ 211.22).

This section presents an outline of expectations for the content of the process qualification plan or protocol. Three interesting points are made in ascending order of use. The guidance allows for the use of risk management principles to prioritize activities, suggests a final report to summarize and state conclusion, and requires the approval of the quality unit for plan and report.

Note: During the PDA workshops, questions were asked whether the guidance endorsed, promoted the use of, or was consistent with the principles presented in ASTM E-2500–07, *Standard Guide for Specification, Design, and Verification of Pharmaceutical and Biopharmaceutical Manufacturing Systems and Equipment*. The response was that the guidance does not promote any one particular approach to qualification. It does recommend qualification of equipment as a precursor to process qualification and uses the term *"qualification"* rather than the ASTM-E2500–07 term *"verification."* However, the FDA speakers indicated that there are many ways to achieve this objective, including those endorsed by ASTM, ISPE, PDA, and other organizations. It is left to the companies to decide which approach works best for them and their facility. The ASTM standard guide is mentioned in the reference section of the 2011 guidance but otherwise is not discussed in the body of the text.

V Process Performance Qualification

The process performance qualification (PPQ) is the second element of Stage 2, process qualification. The PPQ combines the actual facility, utilities, equipment (each now qualified), and the trained personnel with the commercial manufacturing process, control procedures, and components to produce commercial batches. A successful PPQ will confirm the process design and demonstrate that the commercial manufacturing process performs as expected.

The PPQ should involve commercial-scale batches manufactured by the trained production personnel qualified to operate the equipment and perform the process, using the actual facility and equipment utilizing manufacturing control procedures. In other words, all should be ready at this stage, including completion of training and final procedures written. Therefore, this is not the time to shake down the process, establish parameters, or learn the manufacturing lines.

Success at this stage signals an important milestone in the product lifecycle. A manufacturer must successfully complete PPQ before commencing commercial distribution of the drug product. The decision to begin commercial distribution should be supported by data from commercial-scale batches. Data from laboratory and pilot studies can provide additional assurance that the commercial manufacturing process performs as expected.

Note the inclusion of the word *milestone*. This reinforces that the guidance presents a lifecycle approach where process validation is a journey to a destination or event.

The approach to PPQ should be based on sound science and the manufacturer's overall level of product and process understanding and demonstrable control. The cumulative data from all relevant studies (e.g., designed experiments; laboratory, pilot, and commercial batches) should be used to establish the manufacturing conditions in the PPQ. To understand the commercial process sufficiently, the manufacturer will need to consider the effects of scale. However, it is not typically necessary to explore the entire operating range at commercial scale if assurance can be provided by process design data. Previous credible experience with sufficiently similar products and processes can also be helpful. In addition, we strongly recommend firms employ objective measures (e.g., statistical metrics) wherever feasible and meaningful to achieve adequate assurance.

This section reinforces that decisions related to validity for the process cannot just rely on results of the PQ or PPQ studies; rather, they need to be based on process design and control-related information. The section allows for prior process knowledge, historical data from similar processes, and information from smaller scale batches to support the design of PPQ studies and decisions to release the process for commercial manufacture. The section suggests information gathered from similar processes may be used to support the design and decisions. The section implies that the more information developed in the early stages of process development, the better the process will be understood at the PPQ stage and therefore the less comprehensive these studies will need to be. It is also important to note that this section strongly reinforces the expectation for statistical analysis in designing and evaluating these studies.

These principles were generally found to be reasonable and logical. However, some concerns were raised at workshops. Companies may be reluctant to use data from similar batches, because it would open these products to further regulatory agency scrutiny. Companies may be reluctant to do more during early process development stages because that requires an upfront investment, often before the product is approved. Companies may be concerned with the use of statistical analysis, because it would increase the number of samples and tests required.

In most cases, PPQ will have a higher level of sampling, additional testing, and greater scrutiny of process performance than would be typical of routine commercial production. The level of monitoring and testing should be sufficient to confirm uniform product quality throughout the batch. The increased level of scrutiny, testing, and sampling should continue through the process verification stage as appropriate, to establish levels and frequency of routine sampling and monitoring for the particular product and process. Considerations for the duration of the heightened sampling and monitoring period could include, but are not limited to, volume of production, process complexity, level of process understanding, and experience with similar products and processes.

This section continues the discussion on statistically sound sampling methods. It also notes the expectation that this level of testing will carry for an indeterminate period of time into commercial manufacture. The level of sampling and, in this case, the additional commercial manufacturing testing may depend on a level of process understanding from process development work done and experience with similar processes. The section notes that process complexity can be a factor. This is a risk assessment principle. The more complex the process, the less likely it will be that it is fully understood, therefore the need to demonstrate understanding with additional testing and sampling.

Industry comments on the 2008 guidance and at PDA workshops indicated a level of concern over the expectation for more testing during PPQ and especially for more testing during commercial manufacturing. Companies were concerned over the burden of additional testing, not knowing the amount of testing, and not knowing for how long the testing would be expected. The answer the FDA seemed to give to these concerns was that the amount and length of time is up to the individual companies but should be based on risk to product quality and confidence in the process, as demonstrated through statistical analysis. Companies need to better understand the process and its variables. They should be the ones to decide how to design the process validation approach. Again, companies must be able to demonstrate that what they do satisfies the objective of proving the process is under control.

The extent to which some materials, such as column resins or molecular filtration media, can be re-used without adversely affecting product quality can be assessed in relevant laboratory studies. The usable lifetimes of such materials should be confirmed by an ongoing PPQ protocol during commercial manufacture.

A manufacturing process that uses PAT may warrant a different PPQ approach. PAT processes are designed to measure in real time the attributes of an in-process material and then adjust the process in a timely control loop so the process maintains the desired quality of the output material. The process design stage and the process qualification stage should focus on the measurement system and control loop for the measured attribute. Regardless, the goal of validating any manufacturing process is the same: to establish scientific evidence that the process is reproducible and will consistently deliver quality products.

This section addresses approaches to the validation of processes employing PAT. It notes that proper use of PAT will accomplish the objective of confirming process control though increased and perhaps complete process observation. Therefore, there is a reduced need for prediction of outcome or validation. However, in this case the burden shifts from looking at process variables and control to confirming that the PAT monitoring and control systems work reliably. This shifts the emphasis to the equipment/ instrument PQ stage.

2 PPQ Protocol

A written protocol that specifies the manufacturing conditions, controls, testing, and expected outcomes is essential for this stage of process validation. We recommend that the protocol discuss the following elements:

The manufacturing conditions, including operating parameters, processing limits, and component (raw material) inputs.

The data to be collected and when and how it will be evaluated.

Tests to be performed (in-process, release, characterization) and acceptance criteria for each significant processing step.

The sampling plan, including sampling points, number of samples, and the frequency of sampling for each unit operation and attribute. The number of samples should be adequate to provide sufficient statistical confidence of quality both within a batch and between batches. The confidence level selected can be based on risk analysis as it relates to the particular attribute under examination. Sampling during this stage should be more extensive than is typical during routine production.

Criteria and process performance indicators that allow for a science- and risk-based decision about the ability of the process to consistently produce quality products. The criteria should include:

A description of the statistical methods to be used in analyzing all collected data (e.g., statistical metrics defining both intra-batch and inter-batch variability).

Provision for addressing deviations from expected conditions and handling of nonconforming data. Data should not be excluded from further consideration in terms of PPQ without a documented, science-based justification.

Design of facilities and the qualification of utilities and equipment, personnel training and qualification, and verification of material sources (components and container/closures), if not previously accomplished.

Status of the validation of analytical methods used in measuring the process, in-process materials, and the product.

Review and approval of the protocol by appropriate departments and the quality unit.

This section presents expectations for information contained in the written protocol. Much of it reflects what most companies would consider as standard industry practice, but some of the risk and statistical analysis expectations may indicate a higher level of planning and evaluation than many companies currently provide in protocols. The section sets an expectation for a scientifically sound basis for the sampling plan based on a predetermined level of confidence. The company should be able to present how this plan provides confidence that the process will consistently produce a product or outcome that meets the desired level of quality.

Many of the comments and questions received by PDA during the formal comment gathering period and workshops had to do with these statistical expectations, including how to determine the correct level of confidence. Companies should understand and be able to articulate how the analysis of samples demonstrates that the process is producing the desired outcome in a consistent and reliable manner. Further, the section indicates that this analysis would form the basis for more extensive sampling that would carry over to Stage 3, continued process verification.

The section also allows for the use of risk-based criteria to be used to determine the sampling plan and confidence levels. The risk factors would include risk to product quality and patient safety. Therefore, process steps, or attributes that pose the highest risk, would require more analysis to reach a level of process control confidence. The confidence level should not change, but the burden of proof might. This could also be based on the level of understanding and knowledge of process performance.

3 PPQ Protocol Execution and Report

Execution of the PPQ protocol should not begin until the protocol has been reviewed and approved by all appropriate departments, including the quality unit. Any departures from the protocol must be made according to established procedure or provisions in the protocol. Such departures must be justified and approved by all appropriate departments and the quality unit before implementation (§ 211.100).

The commercial manufacturing process and routine procedures must be followed during PPQ protocol execution (§§ 211.100(b) and 211.110(a)). The PPQ lots should be manufactured under normal conditions by the personnel routinely expected to perform each step of each unit operation in the process. Normal operating conditions should include the utility systems (e.g., air handling and water purification), material, personnel, environment, and manufacturing procedures.

A report documenting and assessing adherence to the written PPQ protocol should be prepared in a timely manner after the completion of the protocol. This report should:

Discuss and cross-reference all aspects of the protocol.

Summarize data collected and analyze the data, as specified by the protocol.

Evaluate any unexpected observations and additional data not specified in the protocol.

Summarize and discuss all manufacturing nonconformance such as deviations, aberrant test results, or other information that has bearing on the validity of the process.

Describe in sufficient detail any corrective actions or changes that should be made to existing procedures and controls.

State a clear conclusion as to whether the data indicates the process met the conditions established in the protocol and whether the process is considered to be in a state of control. If not, the report should state what should be accomplished before such a conclusion can be reached. This conclusion should be based on a documented justification for the approval of the process, and release of lots produced by it to the market in consideration of the entire compilation of knowledge and information gained from the design stage through the process qualification stage.

Include all appropriate department and quality unit review and approvals.

This section presents standard execution and report writing expectations. The final point notes the need for a conclusion and decision to move from Stage 2 to Stage 3 based not only on the results of the PPQ studies, but also on information obtained from Stage 1 Process Design, consistent with the lifecycle approach. The final report should present the justification for moving into commercial manufacture and release of product to the market.

D Stage 3—Continued Process Verification

The goal of the third validation stage is continual assurance that the process remains in a state of control (the validated state) during commercial manufacture. A system or systems for detecting unplanned departures from the process as designed is essential to accomplish this goal. Adherence to the CGMP requirements, specifically, the collection and evaluation of information and data about the performance of the process, will allow detection of undesired process variability. Evaluating the performance of the process identifies problems and determines whether action must be taken to correct, anticipate, and prevent problems so that the process remains in control (§ 211.180(e)).

This section describes the expectations for Stage 3—Continued Process Verification. This stage is designated as ongoing process verification in EMA Annex 15. Stage 3 was one of the most discussed elements of the guidance and the section that may most exemplify the concept of lifecycle approach. Stage 3 is a key element that suggests that information being generated throughout the process life cycle will be helpful in confirming and improving process control. The section notes the need for a system to detect unplanned excursions in process performance or undesired process variability, meaning deviations and out-of-specification outcomes. In other words, the process may not be performing as intended, and the decision to proceed with commercial distribution may need to be revisited.

This section refers to actions taken to assure the process remains in control. The guidance suggests that as additional knowledge is gained during commercial manufacture, it is acceptable that the company use that knowledge to further improve the process. Where actions can be taken to prevent problems, the process is improved. Any changes to the process would require qualification.

An ongoing program to collect and analyze product and process data that relate to product quality must be established (§ 211.180(e)). The data collected should include relevant process trends and quality of incoming materials or components, in-process material, and finished products. The data should be statistically trended and reviewed by trained personnel. The information collected should verify that the quality attributes are being appropriately controlled throughout the process.

We recommend that a statistician or person with adequate training in statistical process control techniques develop the data collection plan and statistical methods and procedures used in measuring and evaluating process stability and process capability. Procedures should describe how trending and calculations are to be performed and should guard against overreaction to individual events as well as against failure to detect unintended process variability. Production data should be collected to evaluate process stability and capability. The quality unit should review this information. If properly carried out, these efforts can identify variability in the process and/or signal potential process improvements.

> This section presents the expectation that a scientifically sound approach and plan to the collection and analysis of data should determine if the process is performing as expected in commercial manufacture. Questions were raised at the PDA workshops related to the extent and presentation of this analysis. The use of the annual product review (APR) was raised as a means to assess and present post-PPQ process data. However, the speakers indicated that the FDA was looking for more than just the information presented in the APR. While the APRs may be a valuable source and good starting point, the APR is designed to demonstrate product quality and not necessarily evidence of process control. The section also recommends that the company should consider the trends that the data may indicate, rather than reacting to individual events.

Good process design and development should anticipate significant sources of variability and establish appropriate detection, control, and/or mitigation strategies, as well as appropriate alert and action limits. However, a process is likely to encounter sources of variation that were not previously detected or to which the process was not previously exposed. Many tools and techniques, some statistical and others more qualitative, can be used to detect variation, characterize it, and determine the root cause. We recommend that the manufacturer use quantitative, statistical methods whenever appropriate and feasible. Scrutiny of intra-batch as well as inter-batch variation is part of a comprehensive continued process verification program under § 211.180(e).

> This section notes that unanticipated process variability may be uncovered during commercial manufacturing. These unanticipated events and conditions can cause out-of-specification results and process failures. Continued or ongoing process verification should be used to identify and address those areas of previously undiscovered or under-controlled variability.

We recommend continued monitoring and sampling of process parameters and quality attributes at the level established during the process qualification stage until sufficient data are available to generate significant variability estimates. These estimates can provide the basis for establishing levels and frequency of routine sampling and monitoring for the particular product and process. Monitoring can then be adjusted to a statistically appropriate and representative level. Process variability should be periodically assessed, and monitoring adjusted accordingly.

This section addresses one of the areas of greatest concern by industry comments and workshop discussion—continued, enhanced sampling and testing. In the 2008 draft guidance, the qualifying sentence read: *Once the variability is known, sampling and/or monitoring should be adjusted to a statistically appropriate and representative level.* The language was changed in the 2011 version to infer that the data obtained from this more extensive sampling should be used to establish the ongoing and routine sampling program. In other words, the additional sampling is needed to establish areas and rates of variability. Once knowing these areas and rates, the company can better design a sampling program which will assure that the process remains under suitable control.

Variation can also be detected by the timely assessment of defect complaints, out-of-specification findings, process deviation reports, process yield variations, batch records, incoming raw material records, and adverse event reports. Production line operators and quality unit staff should be encouraged to provide feedback on process performance. We recommend that the quality unit meet periodically with production staff to evaluate data, discuss possible trends or undesirable process variation, and coordinate any correction or follow-up actions by production.

This section presents sources of information and feedback on process performance. Industry comments questioned the prescriptiveness of the mention of direct feedback from production line and quality unit personnel. However, production staff can be a valuable source of process performance information—which can go beyond what may be captured in batch records, logbooks, and more formal means of observation. The message appears to be using all of the means at your disposal to gather information that can be of scientific value to help make the decisions related to process performance and effectiveness of control measures.

Data gathered during this stage might suggest ways to improve and/or optimize the process by altering some aspect of the process or product, such as the operating conditions (ranges and set-points), process controls, component, or in-process material characteristics. A description of the planned change, a well-justified rationale for the change, an implementation plan, and quality unit approval before implementation must be documented (§ 211.100). Depending on how the proposed change might affect product quality, additional process design and process qualification activities could be warranted.

This section discusses Stage 3 process evaluation as a path for post-PPQ process improvement. It presents a basic change control approach, where the need for change or improvement is identified, evaluated, and if warranted approved and implemented. Once this change is implemented, additional testing or qualification may be needed. This provides a continuum or cycle of process qualification, improvement, and qualification.

Maintenance of the facility, utilities, and equipment is another important aspect of ensuring that a process remains in control. Once established, qualification status must be maintained through routine monitoring, maintenance, and calibration procedures and schedules (21 CFR part 211, subparts C and D). The equipment and facility qualification data should be assessed periodically to determine whether re-qualification should be performed and the extent of that re-qualification. Maintenance and calibration frequency should be adjusted based on feedback from these activities.

This section discusses the need to maintain the process and systems that support the process in their qualified state. An assessment of the maintenance should be performed to confirm that the systems remain in operation or in a state largely as they were when the Stage 2 qualifications were performed. If there is evidence or doubt that the systems have been maintained properly, then requalification may be warranted. This would also signal the need to adjust maintenance schedules and evaluation to ensure compliance with requirements for keeping systems fit for use and in good operating condition.

VI Concurrent Release of PPQ Batches

In most cases, the PPQ study needs to be completed successfully and a high degree of assurance in the process achieved before commercial distribution of a product. In special situations, the PPQ protocol can be designed to release a PPQ batch for distribution before complete execution of the protocol steps and activities, i.e., concurrent release. FDA expects that concurrent release will be used rarely.

> This section resulted in considerable comments and discussion at and after workshops. The section discusses situations where product from PPQ batches may need to be released for distribution of product prior to the completion of all PPQ batches. It does not address the release of Stage 3 batches or the release of batches produced during the completed and approved Stage 2 PPQ. Those by definition are released for commercial distribution as part of the ongoing evaluation and adjustment of the process validation program. Instead, it discusses the release of batches produced under controlled conditions prior to the completion of the PPQ.

Concurrent release might be appropriate for processes used infrequently for various reasons, such as to manufacture drugs for which there is limited demand (e.g., orphan drugs, minor use and minor species veterinary drugs) or which have short half lives (e.g., radiopharmaceuticals, including positron emission tomography drugs). Concurrent release might also be appropriate for drugs that are medically necessary and are being manufactured in coordination with the Agency to alleviate a short supply.

> This section cautions that this procedure should be used only on rare occasions to allow for distribution of certain products where the risk to patient safety is outweighed by the risk of not having the products available when needed or in a suitable condition to be properly used.

Conclusions about a commercial manufacturing process can only be made after the PPQ protocol is fully executed and the data are fully evaluated. If Stage 2 qualification is not successful (i.e., does not demonstrate that the process as designed is capable of reproducible performance at commercial scale), then additional design studies and qualification may be necessary. The new product and process understanding obtained from the unsuccessful qualification study(ies) can have negative implications if any lot was already distributed. Full execution of Stages 1 and 2 of process validation is intended to preclude or minimize that outcome.

Circumstances and rationale for concurrent release should be fully described in the PPQ protocol. Even when process performance assessment based on the PPQ protocol is still outstanding, any lot released concurrently must comply with all CGMPs, regulatory approval requirements, and PPQ protocol lot release criteria. Lot release under a PPQ protocol is based upon meeting confidence levels appropriate for each quality attribute of the drug.

> The section maintains that the company must provide a level of assurance that the process used for concurrent release product is well controlled and that the products are safe and effective.

When warranted and used, concurrent release should be accompanied by a system for careful oversight of the distributed batch to facilitate rapid customer feedback. For example, customer complaints and defect reports should be rapidly assessed to determine root cause and whether the process should be improved or changed. Concurrently released lots must also be assessed in light of any negative PPQ study finding or conclusions and appropriate corrective action must be taken (§§ 211.100(a), 211.180(e), and 211.192). We recommend that each batch in a concurrent release program be evaluated for inclusion in the stability program. It is important that stability test data be promptly evaluated to ensure rapid detection and correction of any problems.

> The section notes the need for caution when attempting to release product before the completion of the Stage 2 PQ and PPQ program. It suggests additional testing and scrutiny of these batches to provide assurance of product quality. The section suggests placing these batches in the stability program to assure a higher level of scrutiny. An objection was raised in industry comments to the suggestion of including the concurrent released batches in the company's stability program, since the stability program was not designed to evaluate process performance. However, the statement remained in the final guidance, indicating that the authors felt the stability program at least provided some additional data that could provide confidence and monitoring of product quality in the absence of a completed PPQ protocol.

VII Documentation

Documentation at each stage of the process validation lifecycle is essential for effective communication in complex, lengthy, and multidisciplinary projects. Documentation is important so that knowledge gained about a product and process is accessible and comprehensible to others involved in each stage of the lifecycle. Information transparency and accessibility are fundamental tenets of the scientific method. They are also essential to enabling organizational units responsible and accountable for the process to make informed, science-based decisions that ultimately support the release of a product to commerce.

> This section presents the basis for good documentation, including the scientific rationale for making decisions related to the suitability and reliability of the process. The company must understand that basis and be able to articulate it to regulators. There is also the need to memorialize the rationale, as well as the testing criteria. Good documentation will accomplish this need and will limit the potential for individual interpretation of process validation study intent.

The degree and type of documentation required by CGMP vary during the validation lifecycle. Documentation requirements are greatest during Stage 2, process qualification, and Stage 3, continued process verification. Studies during these stages must conform to CGMPs and must be approved by the quality unit in accordance with the regulations (see §§ 211.22 and 211.100). Viral and impurity clearance studies, even when performed at small scale, also require quality unit oversight.

CGMP documents for commercial manufacturing (i.e., the initial commercial master batch production and control record (§ 211.186) and supporting procedures) are key outputs of Stage 1, process design. We recommend that firms diagram the process flow for the full-scale process. Process flow diagrams should describe each unit operation, its placement in the overall process, monitoring and control points, and the component, as well as other processing material inputs (e.g., processing aids) and expected outputs (i.e., in-process materials and finished product). It is also useful to generate and preserve process flow diagrams of the various scales as the process design progresses to facilitate comparison and decision making about their comparability.

> This section presents recommendations for the content of process validation related documentation. Some industry comments on the 2008 draft guidance noted concern

over the prescriptiveness of the recommendations, while questions raised at workshops indicated a need for further clarification of FDA expectations on documentation. In other words, some felt it was too prescriptive and some felt it was not prescriptive enough.

VIII Analytical Methodology

Process knowledge depends on accurate and precise measuring techniques used to test and examine the quality of drug components, in-process materials, and finished products. Validated analytical methods are not necessarily required during product- and process-development activities or when used in characterization studies. Nevertheless, analytical methods should be scientifically sound (e.g., specific, sensitive, and accurate) and provide results that are reliable. There should be assurance of proper equipment function for laboratory experiments. Procedures for analytical method and equipment maintenance, documentation practices, and calibration practices supporting process-development efforts should be documented or described. New analytical technology and modifications to existing technology are continually being developed and can be used to characterize the process or the product. Use of these methods is particularly appropriate when they reduce risk by providing greater understanding or control of product quality. However, analytical methods supporting commercial batch release must follow CGMPs in parts 210 and 211. Clinical supply production should follow the CGMPs appropriate for the particular phase of clinical studies.

This section discusses expectations related to the level of control and qualification needed for analytical methods used to test and support the process validation program. The section does not discuss expectations for analytical method validation beyond its use during the process validation. It does mention that any analytical methods used for the release of commercial or patient use product must of course be validated. Analytical methods used to support any stages of the process validation, including Stage 1 process design and Stage 2 process qualification, must be reliable and scientifically sound, and the information obtained must be accurate, because this information will provide the basis for decisions related to the assurance of control and process performance capability.

1.8 Final Thoughts

The 2011 final version of the FDA Process Validation Guidance signaled an important change in expectations in the regulated drug product industry. The lifecycle approach to process validation represents a shift from a dogmatic, regimented process control verification to a sequentially organized, scientific, and risk-based method, as illustrated in Figure 1.4. Assuring process control and process validation relies on information from the earliest sources, process design and development, and should not conclude until the process has been tested in actual manufacturing for a period of time. Although processes such as sterilization, aseptic processing, cleaning, and automated controls are not specifically mentioned in the guidance, the principles presented in the lifecycle approach present a framework for a sound and logical means to validate any process.

A key element of the lifecycle approach is that assurance that the process will perform to the level required to produce product meeting prescribed quality attributes is attained from multiple sources prior to the PPQ. Process design presents the requirements for the process and controls required to ensure proper process performance. The PPQ tests then

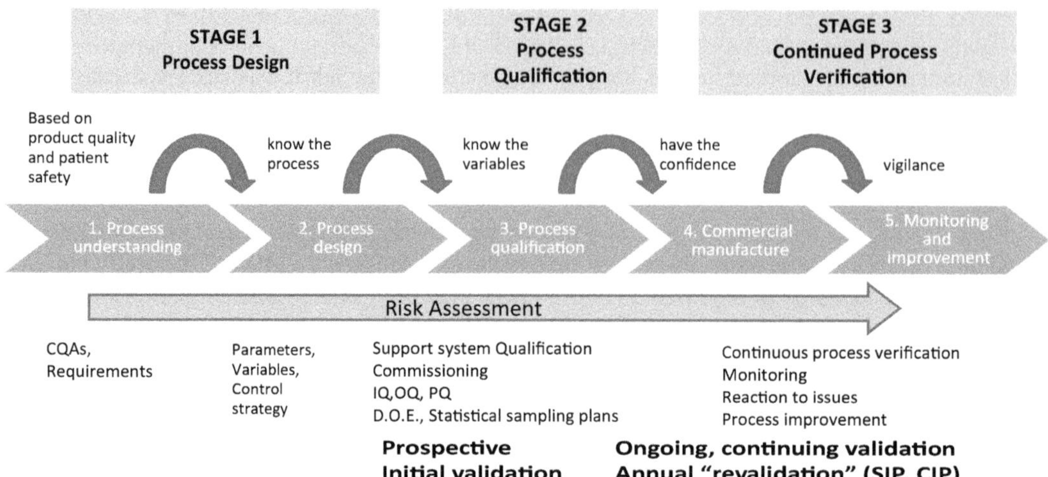

FIGURE 1.4
The process validation process.

provide confidence that all process variables have been addressed and the process is well designed, operational, and reliable. However, processes are complex and have process variables, and the assurance of process validity is not complete until the process has been performed over time. The process must be monitored to ensure that it remains in a qualified and adequately controlled state. Where this is found not to be the case, controls may be modified or additional controls put in place. These changes to process controls represent enhanced controls and an improved process. As noted in Figure 1.4, information obtained throughout the process life cycle provides valuable insight into the performance of the process and the control elements that must be included in the process validation approach.

References

(1) Guidance for Industry, *Process Validation: General Principles and Practices*, Rockville, MD: U.S. Department of Health and Human Services, Food and Drug Administration, January 2011, Revision 1.

(2) International Conference on Harmonization, *Q8 Pharmaceutical Development*, Rockville, MD: U.S. Department of Health and Human Services Food and Drug Administration, 2009.

(3) International Conference on Harmonization, *Q9 Quality Risk Management*, Rockville, MD: U.S. Department of Health and Human Services Food and Drug Administration, 2006.

(4) International Conference on Harmonization, *Q10 Pharmaceutical Quality*, Rockville, MD: U.S. Department of Health and Human Services Food and Drug Administration, 2009.

(5) International Conference on Harmonization, *Q11 Development and Manufacture of Drug Substances*, Rockville, MD: U.S. Department of Health and Human Services Food and Drug Administration, 2012.

(6) Heller, William M., and A. K. Harvey. Whitney Award Lecture, Originally published in *Am J Hosp Pharm* 29: 550–555, 1972, American Society of Health-System Pharmacists, Inc. Posted with permission 2011, ASHP Research and Education Foundation.

(7) U.S. Food, Drug, and Cosmetic Act, Section 501(a)(2)(B) of the Act (21 U.S.C. 351(a)(2)(B))

(8) 21 CFR PART 211 Current Good Manufacturing Practice for Finished Pharmaceuticals
(9) *After the Consent Decree—An Uphill Battle for Affected Companies*, an independent research paper at Fairleigh Dickenson University, Teaneck, NJ MBA Program, Michele Burd and Suggy S. Chrai, June 1, 2004, reprinted on BioPharm International website, 2011
(10) *Guideline on General Principles of Process Validation*, May 1987, U.S. Food and Drug Administration.
(11) *DRAFT Guidance for Industry, Process Validation: General Principles and Practices*, U.S. Department of Health and Human Services Food and Drug Administration, November 2008.
(12) *PDA Submits over 400 Comments on FDA Draft Revise Validation Guidance*, PDA Letter, vol. XLV, issue #3, March 2009.
(13) Proceedings of PDA—*FDAs New Guidance on Process Validation Workshops* held in San Francisco, CA on March 4, 2009, Munich, Germany on March 9, 2009, Las Vegas, NV on April 23, 2009, Chicago, Ill on June 8–9, 2009, October 26–27, Bethesda, MD, San Juan, PR on November 20, 2009, and April 13–14, 2011, San Antonio, TX. Proceedings of *PDA-FDA Joint Regulatory Conference* held in Washington, DC, on September 16, 2009. Proceedings of *PDA-FDA Joint Regulatory Conference* held in Washington, DC, on September 15, 2010.
(14) EudraLex Volume 4, EU Guidelines for Good Manufacturing Practice for Medicinal Products for Human and Veterinary Use, Annex 15: Qualification and Validation. March 30, 2015
(15) PDA comments to the European Commission Health and Consumers Directorate –General, Brussels sanco-pharmaceuticals-d6@ec.europa.eu Ref: EudraLex Volume 4 EU Guidelines for GMP Annex 15: Qualification and Validation, dated May 29, 2014. www.pda.org/docs/default-source/website-document-library/scientific-and-regulatory-affairs/regulatory-comments-resources/2014/eu-gmp-annex-15-qualification-and-validation-may-29–2014.pdf?sfvrsn=77ccae8e_8.
(16) US FDA, Docket No. FDA-2003-D-0032, Center for Veterinary Medicine, Office of Regulatory Affairs, Center for Drug Evaluation and Research, *Guidane for Industry: A Framework for Innovative Pharmaceutical Development, Manufacturing, and Quality Assurance*, October 2004.

2

Applications of Failure Modes and Effects Analysis to Biotechnology Manufacturing Processes

Robert J. Seely and John Haury

2.1 Introduction

Failure modes and effects analysis (FMEA) is a very powerful risk assessment tool widely used in a variety of manufacturing industries and business practices. Like many risk analysis procedures, FMEA provides a rigorous methodology for identifying, evaluating, and documenting potential modes of product or process failure (1–3). In contrast to the other methods, an FMEA results in a numerical ranking of each potential failure, aiding the prioritization of follow-up investigations and implementation of corrections or controls to mitigate the failure (4). FMEA is a useful tool in guiding and documenting the thinking process when operating parameters are evaluated for criticality or when a process is transferred to a different manufacturing site. It is a systematic, rigorous method for ranking parameters into (potentially) high-risk categories and for defining which variables need further process characterization (3).

The risk assessment is based on assigning a ranking of 1 to 10 (low to high) to three critical criteria: (1) the severity of a failure, (2) the expected frequency of occurrence, and (3) the likelihood of detecting the failure. The product of the three scores results in a risk priority number (RPN), which can vary between 1 and 1000. It is important to evaluate the potential failure with all three criteria because the effects may either multiply or offset one another. That is, a failure may be very severe, but if the occurrence is low and the detectability is high the resulting RPN is low. The primary benefits of this tool are that it provides a rational approach to evaluating a process, and it generates a ranked order of parameters requiring characterization, hence a shortening of the total list of operating variables to be studied. In addition, it provides a sound documenting mechanism to record the group decision-making process.

2.2 Risk Analysis Methods

The concept of risk implies a degree of uncertainty regarding the outcome of an event, process, project, behavior, or decision. To evaluate and measure this uncertainty requires a systematic framework involving the elements of probability, consequences,

DOI: 10.1201/9781003143130-2

detectability, and recoverability or correctability. There are a wide variety of assessment methodologies available, ranging from risk avoidance to risk acceptance (5). While they all share the common elements given earlier, they have distinguishing characteristics that accommodate specific applications. For example, FMEA and preliminary hazard analysis are inductive (or inferential). Inductive logic starts with particular instances and infers that the general cause exists (with a given probability). This logic is based on the question, "Given a particular situation, what is the likely general system causing it?" FMEA, for example, allows the prioritization of those causes for preventive action. Other methods such as fault tree analysis and success tree analysis are deductive. They proceed from a general premise to derive or predict consequential results. Deductive logic asks the question, "What general system components or scenarios must go right or wrong in order to cause a particular consequence?" Within these categories, some methods are qualitative, some are quantitative, and some rely heavily on probabilistic theory (4, 5).

2.3 Two Applications of FMEA

This chapter will describe a streamlined application of FMEA to two main aspects of bioprocessing: process characterization and process transfer. Process characterization is the portion of process development that examines the ranges to be specified in the manufacturing procedures, robustness of the process, and for a limited number of critical parameters, the edge of failure. In a recombinant protein process there may be several hundred operational parameters and it is not practical, or necessary, to test the high and low value of every range. The FMEA method can be an effective tool to evaluate every variable, first as a paper exercise, then by follow-up study of the variables ranked as high risk if failure were to occur.

The transfer of a process from one site to another has been found to be another area where FMEA can provide a structured thinking process to help ensure success. Process transfers invariably involve some changes in equipment, processing, raw material sources, water quality, personnel, and environmental conditions. Here the FMEA target is to identify any changes in the two processes, however slight. Many of the operational parameters will remain exactly the same as in the established process, and there may be a good deal of historical data to support their associated ranges. The variables that are identified by the group as being different or potentially different are the ones that should be subjected to the FMEA analysis, and the resulting high RPN parameters further evaluated. Additional lab studies performed by process development are often suitable for the evaluation.

These two applications of FMEA demonstrate a usable, value-added method to identify potential problems before they occur. The method is readily adaptable to a variety of other applications in the biotechnology industry and is simplified such that the readers can readily apply the techniques to their particular processes. In addition, FMEA is an effective mechanism for promoting teamwork and facilitating discussions throughout the development cycle and between departments. The benefits of such applications very much offset the person-hours required to execute the analysis (4, 5).

2.4 FMEA Worksheet

The most efficient way to capture an FMEA exercise is to use a simple spreadsheet, as shown in Table 2.1. The first column is to prospectively identify and list each and every parameter that is to be evaluated. For a recombinant protein production process, this list might consist of all the input variables for performing a manufacturing process. Here we list every control parameter specified in a manufacturing procedure (batch record), such as the setting of flow rate, temperature, mixing speed and time, and pH. These are the operating set points that are staged by an operator or by a computer controller to perform a specific unit operation, such as fermentation, centrifugation, and chromatography.

Once this list is completed, and there may be several dozen variables for a given operation, the evaluation team begins to discuss and identify potential modes of failure and their respective causes and effects. Based on the causes and effects, the team can then decide

TABLE 2.1

Example FMEA Worksheet

Failure Modes and Effects Analysis				Page ______ of ______					
Process:									
Unit Op:									
Leader:									
Date:									
Operational Parameter	**Failure Mode(s)**	**Cause(s)**	**Effect(s)**	**Follow-up By**	**S**	**O**	**D**	**RPN**	**Recommended Follow-up**

on a numerical scoring of the severity of the (potential) failure, the possible frequency of occurrence, and the current ability to detect the failure (S, O, and D, respectively). These numerical assignments are somewhat subjective but are also based on historical experience with the process or related processes, scientific judgment, and an understanding of equipment capability (6, 7). Working definitions of the SOD criteria and examples are presented in the next sections.

The values for S, O, and D are arrived at by interactive discussions of an interdisciplinary team. It is critical to have the system experts present, as well as plant manufacturing personnel, development scientists, and scale-up engineers (3, 7). In addition, representatives from Quality Control and Quality Assurance may be called in for portions of the assessment that pertain to their roles. From the scores assigned, the RPN is calculated and the results can be graphically displayed as a Pareto chart (8). Typically, the RPN values fall into clusters of very high, moderate, and very low. The RPN scale is 1 to 1000. At what point the "high" risk variables require further examination and additional characterization data need to be generated is often difficult to predetermine. This is due to factors such as the subjectivity involved in assigning S, O, and D values and team-to-team differences in consistently utilizing the definitions for SOD. Thus, rather than setting a prospective cutoff between "high" and "low" RPN, we rely on clustering of the values. The clustering can be readily seen in a Pareto chart, where an obvious set of high-ranking numbers can be visually distinguished from the obviously low values. Alternatively, one can choose to further evaluate the top ranking 30% or 50% initially and evaluate some or all of the remaining variables as time and resources permit.

The purpose of the FMEA is to collectively evaluate potential failures, prioritize them on a consensus basis, and document the evaluation process. From there, the Process Team must decide which of the variables require dedication of future efforts. As the top-ranking risk variables are investigated and corrected or controlled to reduce their risk of failure, individual follow-up reports will be written to document the actions taken.

2.5 Evaluation Criteria: Severity, Occurrence, and Detection

Table 2.2 gives some example definitions for the levels, 1 to 10, of the three criteria (courtesy of Ref. 9). The definitions usually need to be modified to fit a particular FMEA application. Those for a medical device, where design needs and tolerances are fairly well established, are different from a biological process, where the effects of excursion of a manufacturing operating range may not be known. The definitions should be discussed as a team before the FMEA is begun. Even when a rating system is clearly defined, there may be disagreements as to the numerical values for SOD of a particular parameter. Further discussions, moderated by a trained facilitator, can bring consensus to the group (2, 3, 7). Some examples of SOD assignments to manufacturing processes are given after the general discussions presented next.

Severity. The severity rating is a measure of the seriousness of a particular failure. The severity may be clear from previous experiences. Often it must be estimated based on what the outcome might be; for example, yield loss, total batch loss, validation failure, or the need to perform an extensive investigation before further process or product release can occur. Out-of-compliance issues and patient safety are also major concerns. The examples given in Table 2.2 are presented as generic, and careful thought should be given to

TABLE 2.2

Example Ratings for Severity, Occurrence, and Detectability. Occurrence here is based on 50 runs per year. CpK is the process capability index (10).

Scale	Severity	Occurrence	Detectability
10	Hazardous, without warning. May endanger machine or assembly operator. Noncompliance with government regulation. Fails final product specs >90% of the time. Product lost or completely unrecoverable.	>25 lots/yr >50% CpK <0.33	Almost impossible to detect. No known control(s) available to detect failure mode.
9	Hazardous, with warning. May endanger machine or assembly operator. Fails in-process performance parameters 100% of the time and final product specs 50% of the time. Over 50% impact on step and overall yield.	10–20 lots/yr ~ 25%–40% CpK ≥0.33	Very remote likelihood current control(s) will detect failure mode. Occasionally we check for defects.
8	Very high. Major disruption to product line. 100% of product may have to be scrapped. Fails in-process performance parameters ~75% of the time and final product specs >25% of the time. Approx. 50% impact on step yield and over 25% impact on overall yield.	6–9 lots/yr ~ 15% CpK ≥0.51	Remote likelihood current control(s) will detect failure mode. Systematic sampling and inspection.
7	High. Major disruption to production line. Product may have to be sorted and a portion scrapped. Fails in-process performance parameters ~50% of the time. Final product purity specs fail 10% of the time. 30%–40% step yield and >20% overall yield impact.	5 lots/yr ~ 10% CpK ≥0.67	Very low likelihood current control(s) will detect failure mode. All units are manually inspected.
6	Moderate. Minor disruption to production line. May fail in-process performance parameters in ~25% instances. May fail product specs 5% of the time. Approx. 25% step yield and >10% overall yield impact.	2–3 lots/yr ~ 5% CpK ≥0.83	Low likelihood current control(s) will detect failure mode. Manual inspection with mistake-proofing.
5	Low. Minor disruption to production line. 100% of product may have to be reworked. Runs on edge of in-process performance parameters and may fail these in ~10% instances. ~10% impact on step yield and ~5% impact on overall yield.	1 lot per year ~ 2% CpK ≥1.00	Moderate likelihood current control(s) will detect failure mode. SPC monitoring and manual inspection.
4	Very low. Minor disruption to production line. Measurable effect on in-process performance parameters, but will not exceed in-process control. Measurable effect on step yield (5%).	1 lot every other year ~ 1% CpK ≥1.17	Moderately high likelihood current control(s) will detect failure mode. SPC with immediate reaction to special causes.
3	Minor disruption to production line. A portion of the product may have to be reworked online. Slightly measurable impact on in-process performance parameters. Slight but measurable impact on step yield (<3%).	1 lot every 3–5 years ~ 0.5% CpK ≥1.33	High likelihood current control(s) will detect failure mode. SPC with 100% inspection for special causes.
2	Very minor disruption to production line. In-process impact may go unnoticed.	1 lot every >5 years ~ 0.2% CpK ≥1.50	Very high likelihood current control(s) will detect failure mode. All units are automatically inspected.
1	No effect on performance. Not noticed.	Never or > every 10 years CpK ≥1.67	Almost certain current control(s) will detect failure mode. Defect is obvious and cannot affect anyone.

individual FMEA targets, especially for severity scoring. During the assessment of the final RPN ranks, items with very high severity rating should be considered for further study regardless of their overall RPN (7, p. 39).

The severity of a failure can be assessed in several ways. Patient safety should always be a primary concern, but the severity may be a major issue before product is ever released for distribution. In some instances, plant personnel safety might be the driving concern. Business issues such as cost or productivity, regulatory compliance, and consistent process control are other effects of operation failure. The FMEA may be geared toward one specific concern or a mixture as long as the target is agreed upon by the FMEA team.

Occurrence. This is a measure of how frequently the failure might occur. If the excursion of a variable (operating temperature, for example) has occurred often in the past or may occur often at a new facility, additional controls may be needed. The occurrence is also assessed with respect to severity and detectability. If the severity is rated low and there are detection measures in place, the overall RPN may be low and this particular variable might not be studied until a later time in the development cycle. The examples for occurrence in Table 2.2 demonstrate that for a biological process, occurrence is somewhat easier to define than severity. The table offers three occurrence measurements: the failure rate based on number of lots per year, the percentage of batches, and the capability of the process (9).

Detectability. Detection is a significant criterion to include in the evaluation of risk. Even if a given failure has serious consequences and it might occur often, if there are adequate detection modes in place that provide time for corrective action, the overall RPN might be low. However, there are several classes of detectability. The degree of detection just described is ideal; however, the failure may be detected but not in time for immediate correction. Often the batch of material is being processed at a later operational step before the failure, or excursion, is noticed. Detection may be noted from continuously logged data but no alarms are in place, or the results from analytical data require an extended period of time. Detection in these cases is still considered "good", and an intermediate rating of 4 to 6 might be appropriate. If the failure cannot be detected before the product is shipped, or even worse before it is used, the rating should be very high.

2.6 Example of FMEA Applied to Process Transfer

Application of FMEA to process characterization and an example has been presented previously (11). An example of application to a validated, commercial process being transferred to a new manufacturing site is given in Table 2.3. The entries are not self-explanatory, and the example is shown to illustrate a few noteworthy features. First, the spreadsheet is a streamlined version of those provided in the references for the more "classical" FMEA applications. For the biotechnology process there are so very many modes of potential failure that are largely unknown that it is the best use of the team's time to strive to quickly identify and rank the ones that are known. For a process transfer, as mentioned previously, the focus is on what aspects (processing variables, equipment, materials) are different from the site of origin to the site of transfer.

In the spreadsheet it can be seen that several variables were identified as being possibly different but not known for certain at the time of the FMEA. A simple "follow-up" was noted with an associated responsible person (see Shake platform throw). If it is found that

TABLE 2.3

Snapshot of the FMEA Process for a Particular Transfer Being Made

Unit Operation: Seed Train

Date:	Page____of____	Responsibility	Failure/Problem						
Parameter	Follow-up		Potential Failure Mode	Severity	Potential Cause of Failure	Occurrence	Current Controls	Detection	RPN
Shake platform throw	If throws are the same in both machines, no issue.	I. L.							0
Shake temp controls	Temperature mapping studies will be performed.	I. L.	Difference in growth profile	5	Different heat distribution profile could modify growth curve. Potential to fail PV criterion.	2	Temperature mapping studies	2	20
Tubing materials	Same material required by MP.	J. M.							
Pumps	Will be same as B-7.	J. M.							
Glassware same	Same as B-7.	J. M.							0
Innoc. vial handling	Handling of vials from freezer to plant will be equivalent to B-7.	B.D.							0
Fermenter R-0180 seal/pressure		B. W.	Leakage	4	Design	2		5	40
				4	Assembly & maintenance				
				4	Utility failure				
		B. W.	Contamination	10	Design	2		5	100
				10	Assembly & maintenance				
				10	Utility failure				
				10	Inadeq. ster. cycle				

the item is indeed going to be different, then that person is responsible for investigating the degree of difference, the SOD it might have (offline), and defining what investigations or corrective action is necessary. The time may not be available to reassemble the team and review such follow-up activities.

A second feature of the table is shown by the fermenter seal/pressure. Here it was noted that the shaft seal will be of a different material and it may fail by not holding pressure in either direction. If the seal fails to prevent incoming air, as compared to exhaust gas, the failure effect of contamination will be much more severe, leading to a severity rating of 10. Further, detectability was given a 5 because, even though it would be detected very quickly during operation, it would necessitate an unacceptably long shutdown and replacement time. The corrective action to this item was to expedite delivery and field testing to ensure the seal is adequate.

The example shown in Table 2.3 is a very small snapshot of the FMEA process for the particular transfer being made, but it does show that the FMEA concept is useful to quickly identify, catalog, and assign risk priorities to the variables that will be or are suspected of being different between two sites. It also affords a structured methodology for a cross sectional team to (1) reevaluate the possible changes and help ensure nothing was overlooked in the transfer and (2) document that that was done.

2.7 Next Steps

Once the initial FMEA exercise has been completed, there remains the critical part of follow-up. The FMEA has resulted in (1) documentation that every element of the process (for characterization or for transfer) has been evaluated by a team and (2) a prioritization of parameters or issues that now need to be addressed. Based on the RPNs, as visualized in Pareto fashion, the team should agree on a cutoff value for studying the "high" RPNs first, and perhaps some or all of the others as time and resources permit. As previously mentioned, this cutoff can be made prospectively, although for biological process this may be difficult and is not necessary, or it can be made retrospectively based on the results. The cutoff can be based on criteria such as obvious clustering, the top 25%, 33%, 50%, and so on, or the follow-up studies can be performed one at a time, working from high to low, as time permits. Whatever the team decision is, it should be recorded in the FMEA report.

The report can now be written and the initial FMEA can be closed out. The follow-up items identified in the FMEA are to be addressed by the responsible person or team and should be documented in subsequent, separate technical reports. The closure of all these items may take several months and typically is done by individuals from different departments. Because of these factors, it is important to finalize the initial FMEA and get it into the hands of the team members who need to act on the identified issues and the team leader who will be responsible for ensuring timely completion.

We find the best person to write the report is the facilitator. Even though he or she may not be fully aware of the physical/chemical aspects the operating parameters discussed, he or she will have been present during the entire meeting(s). Other team members may come and go as the topics of expertise change. Also, it is crucial to the success of an FMEA to make it as easy as possible on the team and team leader. By assisting in organizing and moderating the FMEA, including the defining SOD and writing the initial report, the facilitator can assume many tasks from the team and leader. The report is essentially the "minutes" of the FMEA meeting and should contain the following elements:

FMEA scope and date

List of team members, by name and organization

Definitions of SOD determined by the team

Cutoff RPN, if known

Completed worksheet(s)

Pareto chart(s)

Future work: The readers should be reminded that the identified individuals are to further investigate the items assigned to them and write subsequent technical reports.

It may be useful to write one final report, once all items are closed, to summarize all the follow-up reports, listing them by title, author, and report number, and perhaps provide a brief outline of the issue and corrective action taken. For a process characterization FMEA, the final report could list the final key parameters and a discussion of why some were determined to be non-key and thus need not be validated. Such summaries will aid in retrieval for nonconformance investigations, proposed process changes, or questions that might arise during an inspection or other regulatory review. The resulting compilation of documents, and the resolution of potential problems before they occur, should represent a body of work that was value-added and can be utilized for the life of the product.

References

1. C. DeSain and C.V. Sutton. Risk Management Basics. Cleveland: Advanstar, 2000.
2. R. Kieffer, S. Bureau, and A. Borgmann. Applications of failure mode effect analysis in the pharmaceutical industry. Pharm. Technol. Europe. Sept.: 36–49, 1997.
3. R.E. McDermott, R.J. Mikulak, and M.R. Beauregard. The Basics of FMEA. Portland, OR: Productivity, 1996.
4. A. Shani. Using Failure Mode and Effect Analysis to Improve Manufacturing Processes. Medical Device & Diagnostic Ind. July: 47–51, 1993.
5. B.M. Ayyub, Risk Analysis in Engineering and Economics. Boca Raton: Chapman & Hall/ CRC, 2003.
6. R.T. Clemen. Making Hard Decisions. 2nd ed. Pacific Grove: Duxbury Press, 1995, pp. 5–8.
7. D.H. Stamatis. Failure Mode and Effect Analysis; FMEA from Theory to Execution. 2nd ed. Milwaukee: ASQ Quality Press, 2003.
8. J.T. Burr. SPC Tools for Everyone. Milwaukee: ASQ Quality Press, 1993, pp. 8–12.
9. R.G. Kieffer. Validation, Risk-Benefit Analysis. PDA J. Pharmaceutical Sci. &. Technol. 49: 249–252, 1995.
10. D.P. Stockdale Associates. D. Stockdale, President. 10 Reata. Orange, CA: Rancho Santa Margarita, 92688.
11. J.E. Seely and R.J. Seely, A Rational, Step-Wise Approach to Process Characterization, BioPharm Int. 16: 24–34, 2003.

3

Process Characterization

James E. Seely

3.1 Introduction

Although considered to be a significant time and resource commitment from Process Development, process characterization has been shown to be valuable in ensuring validation and manufacturing success. Given the expense of producing biopharmaceuticals at large scale, process characterization gives an excellent return on investment over the lifetime of a product or process. Inadequate process characterization can result in costly lot failures and incidents, failed validation runs, and difficult inspections (1).

The overall goal of adequate process characterization for commercial manufacturing processes is to ensure efficient and successful process validation and assure consistent process performance (2). More specifically, process characterization provides:

i. An understanding of the role of each process step, such as an understanding of where impurities are cleared during a particular purification step.
ii. An understanding of the impact of process inputs (operating parameters) on process outputs (performance parameters) and identification of key operating and performance parameters.
iii. Assurance that process delivers consistent product yields and purity within all operating ranges.
iv. Acceptance criteria for in-process performance parameters.

In addition, although not a primary reason for doing process characterization, these studies will frequently uncover areas for subtle process improvements in terms of process consistency, product yields, and product purity.

In this chapter we present an outline and some examples for how to carry out thorough and consistent process characterization. The proposed methods could provide a framework for carrying out this work. A good portion of this chapter will describe "precharacterization" studies. These studies are used to help define the scope of the actual experimental characterization work. They also lay the foundation for the experimental studies by demonstrating the adequacy of scaled-down process models and analytical methods. A framework and examples for doing experimental process characterization work will also be presented.

Finally, we will discuss future directions and challenges as our approach to process characterization evolves.

3.2 Resources and Timing for Process Characterization Studies

The driver for the timing of process characterization is the start of conformance/validation lots. Process characterization should be completed in time such that the information gained from these studies can be used to support operating ranges and acceptance criteria for validation protocols. Thorough process characterization may add as much as a year to the overall process development time, so the completion of commercial process development work and initiation of process characterization studies should be timed with this factor in mind (2). Thorough process characterization requires a fully integrated process characterization team (~ 8–12 people) including upstream and downstream processing, analytical departments, and representatives from pilot and full-scale manufacturing. Resource planning from the analytical departments is especially important, since a single characterization run may generate several samples for analysis.

3.3 Pre-Characterization Work

There are three key aspects to pre-characterization work: (1) historical data review and risk assessment, (2) scale-down model qualification, and (3) analytical method qualification.

3.3.1 Historical Data Review and Risk Assessment

Retrospective review of historical data and risk assessment analysis can be used to determine operating parameters that need to be examined experimentally as part of process characterization. Lab notebooks, technical reports, process histories, run summaries, and manufacturing records, as well as a list of the operating parameters and the provisional operating ranges for each unit operation, can be used by the process characterization team to determine knowledge gaps in the process. Information from the operating ranges tested during process development can help identify those parameters that are most likely to impact the process (2–4). In particular, experimental design studies (DOE) from the commercial process develop work can be useful for identifying key parameters or even in determining operating ranges in certain instances, since these experiments are carried out over a range of operating parameters and may yield information about operating parameter interactions.

Once data mining is completed, a risk assessment analysis can be carried out on each unit operation where the effect and likelihood of an excursion from each operating parameter range is addressed. Hazard analysis and critical control points (HACCP) (5), failure modes and effects analysis (FMEA) (6–9), cause-and-effect diagrams (10), and other risk assessment tools can be used for these purposes. The FMEA tool assigns a numerical rating to the *severity* of an excursion of an operating parameter, the *frequency* of an excursion, and the *ability to detect* the excursion before it has an impact on the product (6–9). The combined risk factor (risk priority number or RPN) is a multiple of these three variables, giving a rating scale from 1 to 1000 if a 1–10 numerical rating is used (6–9). These data are usually presented in the form of a Pareto chart (6–10) and those operating parameters below a predetermined threshold are considered non-key and will not be examined in the characterization experiments. It is a good idea to involve not only scientists who developed the process in the FMEA exercise, but also quality and plant engineers since they can bring insight into the likelihood of certain process excursions and the ability to detect them. There may be significant differences between the commercial and first-in-human processes and there may be relatively little historical data on the commercial process. Therefore, it may be important to draw on any development and historical data from both the first-in-human and commercial processes.

Probably the biggest challenges in doing FMEA is coming up with a risk category definition system that is not totally subjective, is consistent, and everyone can agree on. There are a number of generic risk category definitions available (6–9, 11). A custom-made risk category definition system that we have used is shown in Table 3.1. One way to better

TABLE 3.1

Custom-Made Risk Category Rating Definitions for FMEA

1-10 Scale	Severity	Occurrence	Detection
10 "BAD"	Fails final product specs > 90% of the time or product lost or completely unrecoverable.	>50% > 25 times per year	No way to detect defect.
9	Fails in-process performance parameters 100% of the time and final product specs > 50% of the time and/or over 50% impact on step and overall yield.	~ 30 - 40% 15 – 20 times per year	Unit sampling and inspection. Defect not detected until after impact on process.
8	Fails in-process performance parameters ~ 75% of the time and final product specs > 25% of the time and/or approx 50% impact on step yield and over 25% impact on overall yield.	~ 20% 10 times per year	Unit sampling and inspection. Defect can be detected prior to impacting process.
7	Fails in-process performance parameters ~ 50% of the time. Final product purity specs failed 10% of the time. (And/or) 30 – 40% step yield and > 20% overall yield impact.	~10% 5 times per year	All units are manually inspected. Defect not detected until after impact on process.
6	May fail in-process performance parameters in ~ 25% instances. May fail final product specs 5% of the time. (And/or) approx 25% step yield and > 10% overall yield impact.	~ 5% 2-3 times per year	All units automatically controlled. Defect not detected until after impact on process.
5	Runs on edge of in-process performance parameters and may fail these in ~ 10% instances and / or ~ 10% impact on step yield and measurable impact on overall yield (~5%).	~ 2% Once a year	All units automatically controlled with secondary manual inspection. Defect not detected until after impact on process.
4	Measurable effect on in-process performance parameters, but will not exceed in-process control limits and / or more measurable effect on step yield (~5%)	~ 1 % Once every 2-3 years	All units manually inspected. Defect detected prior to impact on process.
3	Slightly measurable impact on in-process quality attribute parameters and/or slight but measurable impact on step yld (< 3%).	~ 0.5% Once every 5 years	All units automatically inspected. Defect detected prior to impact on the process.
2	Measurable effect on non-key, non-quality attribute in-process performance parameter (i.e. pool volume, peak postion).	~ 0.2% Once every 10 years	All units automatically controlled with secondary manual control. Defect detected prior to impact on the process.
1 "GOOD"	Not noticed; no affect on performance	Never	Defect is obvious and would always be detected prior to starting process.

TABLE 3.2

Examples of Tightest and Preferred Operating Ranges (+/−) for Operating Parameters

(FMEA considers the severity of an excursion that is 2-3 times outside the preferred operating range. The test range for initial screening experiments is ~1.5 – 2 times outside the preferred operating range.)

Parameter	Tightest operating range	Preferred operating range	Test range for screening
pH	0.1	0.2	0.3 or 0.4
Time	5%	10%	15–20%
Temperature	1°C	2°C	3 or 4 °C
Flow Rate	5%	10%	15%
Volume	2%	5%	10%
OD	5%	10%	15%

define the rating system for FMEA is to consider the preferred operating range for each operating parameter in manufacturing. For example, although it may be possible to run a process at +/−0.1 pH units, operationally the process may be more robust if it can be run at +/−0.2 units. Examples of some preferred operating ranges for different operating parameters are shown in Table 3.2. For the FMEA exercise, we can improve the signal-to-noise ratio of our analysis if we assume that we are considering the severity running the process at about three times outside the normal operating range for a given operating parameter. For considering the frequency of operating parameter excursions and the ability to detect them, we can increase our sensitivity by considering excursions that are just outside the tightest controllable operating range (Table 3.2).

Case Study 1

FMEA analysis for removal of a detergent from a protein preparation using an ion-exchange chromatography method is shown in Table 3.3. Scientists who developed the process determined the severity of an excursion that is about two to three times outside the preferred operating range. Manufacturing and plant engineers determined the frequency of excursions outside of the tightest operating range. Quality control and manufacturing provided information about the ability to detect these excursions, as well as, perhaps more importantly, the ability to react to this excursion before it has product impact. For column loading, the frequency and detection scores were quite low and were the same whether loading was too high or too low. However, with regard to severity, too high of a loading was deemed to have a much greater severity since it had the possibility to result in inadequate detergent removal. This resulted in a much lower RPN score for underloading than overloading. Likewise, high and low flow rates have the same frequency and detection level, but high flow rates can lead to inadequate removal of detergent resulting in protein aggregation, giving it a higher overall RPN score. For those parameters that do not have a specified range (such as stop collect in this instance), some judgment has to be made as to how much of an excursion would have a serious impact. In the case of this chromatography step, missing the stop collect by a significant amount could result in detergent breakthrough, again resulting in potential product aggregation and loss of activity.

A Pareto plot of the different operating parameters vs. their respective RPN scores is shown in Figure 3.1. In most cases with this chromatography step, those with the highest severity

TABLE 3.3

FMEA Analysis for Removal of Detergent from a Protein Using Ion-Exchange Chromatography (Case Study 4.1)

Operational Parameter	Function or Requirement	Potential Failure Mode	Potential Effect of Failure	Severity Score (1-10)	Potential Causes of Failure	Frequency Score (1-10)	Current Controls	Detection Score (1-10)	RPN
Ion Exchange Lot	Detergent Retention	Bad Resin	Insufficient Removal of Detergent	7	Poor supplier, QC failure	2	Monitored in QC. Defect detected prior to use	2	**28**
Load Volume	Remove Detergent	< 20 CV's	Loss of protein on resin	3	Miscalculation, operator error	3	MP's	3	**27**
"	"	> 60 CV's	Detergent in product	7	"	3	"	3	**63**
Start Collect A280	Capture product	Too early start collect	Dilute pool	2	Poor PM, Incorrect standardization	3	Calibration, SOP	5	**30**
"	"	Too late start collect	Lower yield	5	"	3	"	5	**75**
Stop Collect A280	Capture product w/o Detergent	Too early stop collect	Lower yield	5	"	3	"	5	**75**
"	"	Too late stop collect	Detergent in product	7	"	3	"	5	**105**
Flow rate	Remove Detergent	Low flow rate	Longer process time, possible loss of product on resin	3	Pump failure, power outage, operator error, incorrect calibration	4	SOP, MP, Calibration	3	**36**
"	"	High flow rate	Inadequate detergent removal	6	"	4	"	3	**72**

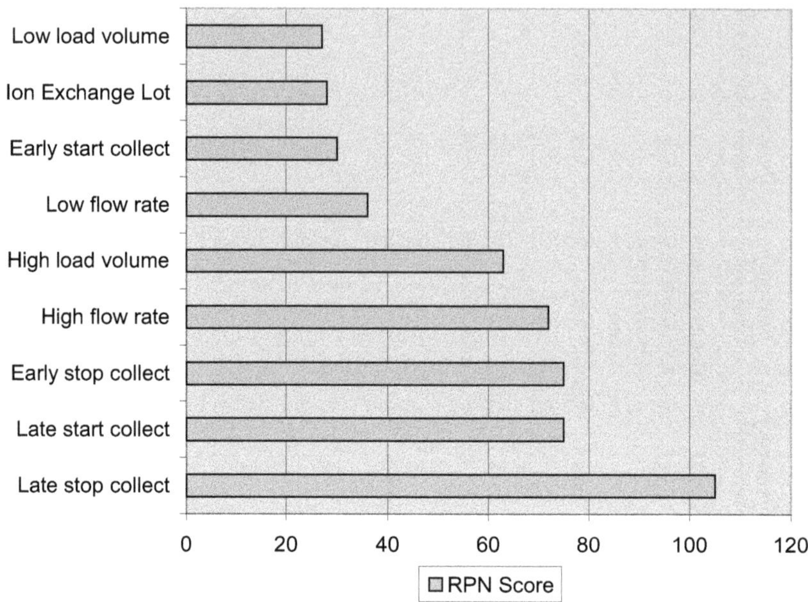

FIGURE 3.1
Pareto chart of RPN scores from detergent removal step (Case study 1).

impact had the highest RPN scores and are the parameters that warrant further study. However, another outcome of this exercise may be the identification of more redundant controls in manufacturing that can serve to either increase the ability to detect an excursion or decrease its frequency. For example, adding a pH check of the column eluate post-equilibration can give us extra assurance that the column is adequately equilibrated and decrease the frequency of equilibration errors by increasing our ability to detect them. In this case it might bring the frequency and detection scores both down to "2," and perhaps make it unnecessary to examine equilibration pH in our process characterization studies. Likewise, having an additional way of checking the elution flow rate could increase our ability to detect an excursion here and, hence, decrease its frequency. However, the potential severity of running at too high of a flow rate makes this excursion something we would want to investigate as part of our characterization studies, in any case.

3.3.2 Scale-Down Model Qualification

The development of a representative scale-down model of each unit operation is the Achilles' heel of good process characterization work. If we can't mimic, to a reasonable degree, the large-scale manufacturing process with our small-scale studies, any bench-scale process characterization work becomes meaningless (2). Some unit operations, such as homogenization and centrifugation steps, are much more difficult to scale down and typically have to be run in a pilot plant setting. Other operations, such as chromatography, ultrafiltration, and microbial fermentation operations, can usually be run at a bench scale or at smaller scales.

In general, the approach to scale-down model qualification is to run all operating parameters at the center of the operating range used for clinical/large-scale manufacturing. If the process has yet to be run in clinical manufacturing, the operating parameters should mirror those of the largest pilot-scale runs. If no appropriate large-scale data are available, an earlier manufacturing process may be used; however, the scale-down model has to give data consistent with the large-scale runs. It is a good idea to write a protocol that describes the scale-down model and how it is controlled and provides the acceptance criteria for each unit operation in terms of performance parameters. Generally, a numerical indication that the small-scale process is acceptable, such as a t-test or some other statistical means, should be used, or there may be a simple requirement that the mean of the data from the scale-down run is within the historical range of the large-scale data. Doing replicate runs can improve the level of statistical rigor for the scale-down model, since it is good to know something about the variability that occurs with the unit operation at small scale (3). This can help in determining the number of replicates that are needed for the actual characterization studies.

The following chapter in this book provides points to consider and strategies for scaling down a number of different process steps, including chromatography, chemical modification reactions, ultrafiltration, microfiltration, and several other unit operations. Some points to consider for scaling down microbial fermentation and cell culture processes are described next.

3.3.2.1 Scaling Down Fermentation and Cell Culture

Comprehensive characterization of microbial fermentation work requires many experiments and is usually carried out at the 10 L scale with some supporting experiments run at larger scales. In our experience, outputs from 10 L microbial fermentation experiments have been fairly representative of we have seen at manufacturing scale. However, scaling down mammalian cell culture has proven to be more difficult. Although we have been able to determine what impact a change in operating parameters might have on a cell culture process, we frequently have no good way of predicting the magnitude of that response from bench-scale data. Even so, process characterization experiments can still give insights into which operating parameters are the most important to control for cell culture processes.

Some of the points to consider when scaling down fermentation and cell culture processes include the following:

- Use the most current manufacturing procedures for scaled-down process.
- Use released GMP materials whenever possible. This includes master or working cell bank vials.
- Sterilization times of media and feed should match manufacturing scale. In some instances it may be important to extend the heating time to make sure there is no effect of the sterilization on the media. Make sure media mass change from pre- to post-sterilization is the same at both scales.
- Use same size and shape of shaker flask (baffled or unbaffled), as well as incubator conditions (shaker speed, throw, etc.).
- Maintain constant inoculum ratios for seed and production fermentation between scales.
- Operational parameters (pH, temperature, backpressure, dissolved oxygen (DO) set point, etc.) should be identical to those specified in the manufacturing process. Ensure that the DO calibrations are compatible.

- Total airflow should be scaled down on the scale factor to ensure similar sweeping of CO_2 from the fermentation. The overall oxygen control strategy should be equivalent to that used at scale (i.e., order of cascade, backpressures employed, etc.). A maximum agitation for the small scale should be selected which mimics as close as possible power input limitations at scale. This might be accomplished with theoretical calculation or through equipment design.
- Addition order for all the ingredients should be in the same order at both scales. Also, make sure the hold times for all ingredients are within the same historical ranges.
- Antifoam strategy or total amounts of antifoam should match manufacturing scale. Antifoam usage typically increases with scale due to higher superficial gas velocities.
- Both scales should have the same feed rates and step times.
- Make sure you have good calibration of all DO probes, pH meters, spectrophotometers, etc., and that they match between scales.

Some of the key performance parameters to monitor include:

- Growth curves and final OD
- Growth rates
- Titer
- % solids
- Feed/acid base usage
- Nutrient profiles
- Times (total seed fermentation time, main fermentation time, etc.)
- Genetic stability
- Cell viability
- Product quality

All of the aforementioned are good to monitor as a part of the scale-down work, but the ones that are deemed critical may be on a case-by-case basis. Certainly, titer and product quality are two important performance parameters that should be considered. In addition, cell viability and percent solids could impact subsequent cell processing and purification steps. Therefore, it may be necessary to process one or two more steps downstream to determine the impact of these parameters on subsequent processing steps or product quality.

3.3.2.2 Analytical Methods Qualification

It is important to ensure that the analytical methods for process characterization studies are robust and representative of what will later be used for the commercial process. Ideally, the methods would be validated; however, for a product in early Phase III clinical stage this may not always be possible. In these instances, the method should be qualified and be developed to the point where there is a high degree of confidence that it can be validated at some point in the future. It is important, therefore, to allow enough time for method qualification prior to process characterization work. This can require anywhere from 3 to 12 months depending on the assay, the complexity of the protein and protein matrices, and so on.

Points to consider for method qualification depend on what the assay is used for. For all assays, critical variables and nominal target values should be defined. For product quantification assays the sample handling and preparation, particularly for in-process samples, should be established. There should be minimum interference from matrix components and adequate resolution. The linearity and range of the analysis should be determined, as well as the limit of quantification, if applicable. Reproducibility should be established based on a statistical equivalence between labs, and the relative standard deviation should be less than or equal to 5% for chromatographic methods, if possible. Any critical assay variables should be identified and the nominal target value defined. For product immunoassays, there should be minimal matrix interference and the antibody specificity should be confirmed. Linearity, precision, and ranges should be established on quantitative immunoassays for both product and process-related impurities (i.e., host cell proteins). For identification immunoassays linearity and precision are not required, but the limit of detection and limit of quantitation should be confirmed.

As mentioned earlier, the analytical group supporting process characterization, whether in a process development department or in quality control, will be one of the hardest hit from a resource standpoint for characterization studies. A single purification run may produce as many as four or five samples for analysis. It is important that the analytical group supporting the characterization activity is staffed appropriately so that sample turnaround does not become too much of a rate-limiting step for completing process characterization.

3.4 Process Characterization Studies

3.4.1 Impurity Clearance

Much of this data may be available prior to the process characterization studies. An understanding of what each process step delivers in terms of yield, impurity clearance, or in-process pool quality should be an outcome of this work. In the case of cell culture, this may be a titer or some qualitative assessment of the product (such as the degree of glycosylation or sialyation). For a chromatography step it would not only be what impurities are cleared during the step, but also at what point are they cleared, i.e., in the wash step, before the product elution, after the product elution, during the regeneration step, and so forth. For a diafiltration step one might examine conductivity or pH after different turnover volumes. The outcome of this work would be the identification of key performance parameters for each process step, which can help in the design of further characterization experiments. How these performance parameters are affected by excursions from the operating ranges will be the objective of the next set of characterization studies.

Case Study 2

This study was designed to track impurity clearance across a cation exchange capture/purification step for a recombinant protein made in *E. coli*. The impurities tracked included DNA, endotoxin, *E. coli* proteins (ECPs), product charge variants (measured by cation-exchange HPLC), and oxidized methionine (as measured by reversed-phase HPLC). The clearance of these impurities was tracked by collecting fractions starting with the load and going through the stop collect. (In some instances we would also collect eluate from the regeneration step; however, all impurities had cleared prior to this step in this case.) Clearance of these impurities relative to the product peak is shown in Figures 3.2A–2D. Endotoxin clearance was virtually identical to the DNA clearance and is not shown. From this data we can provide a rationale for pool criteria. In addition, these studies can be used for addressing certain nonconformances, such as inadequate wash volumes, early pool collections, and the like. This information is also used to identify the quality-indicating performance parameters that should be monitored for this process step, which would include all of the impurities tested in this example.

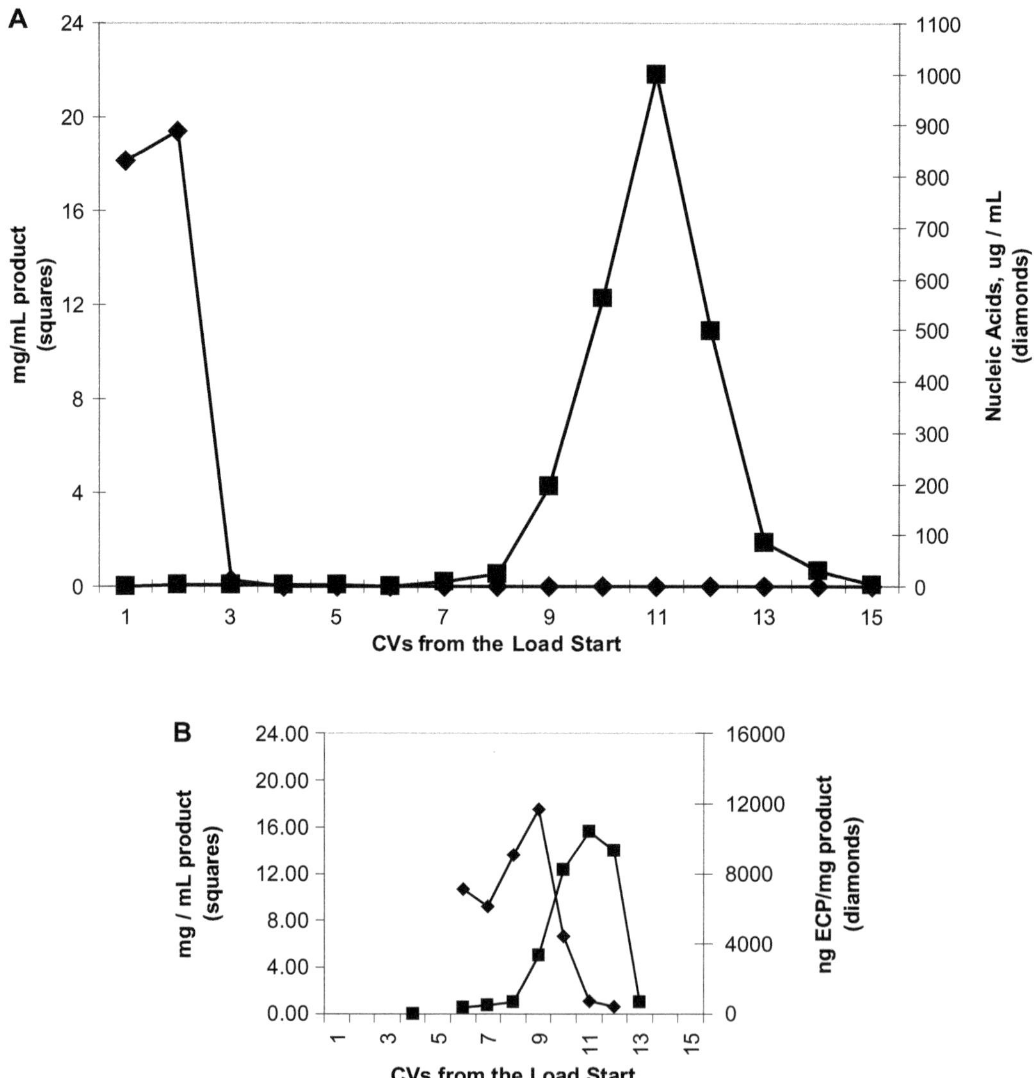

FIGURE 3.2

Impurity clearance study (Case Study 2) examining the removal of DNA (A), *E. coli* proteins (B), reversed-phase impurities (C), and cation-exchange impurities (D).

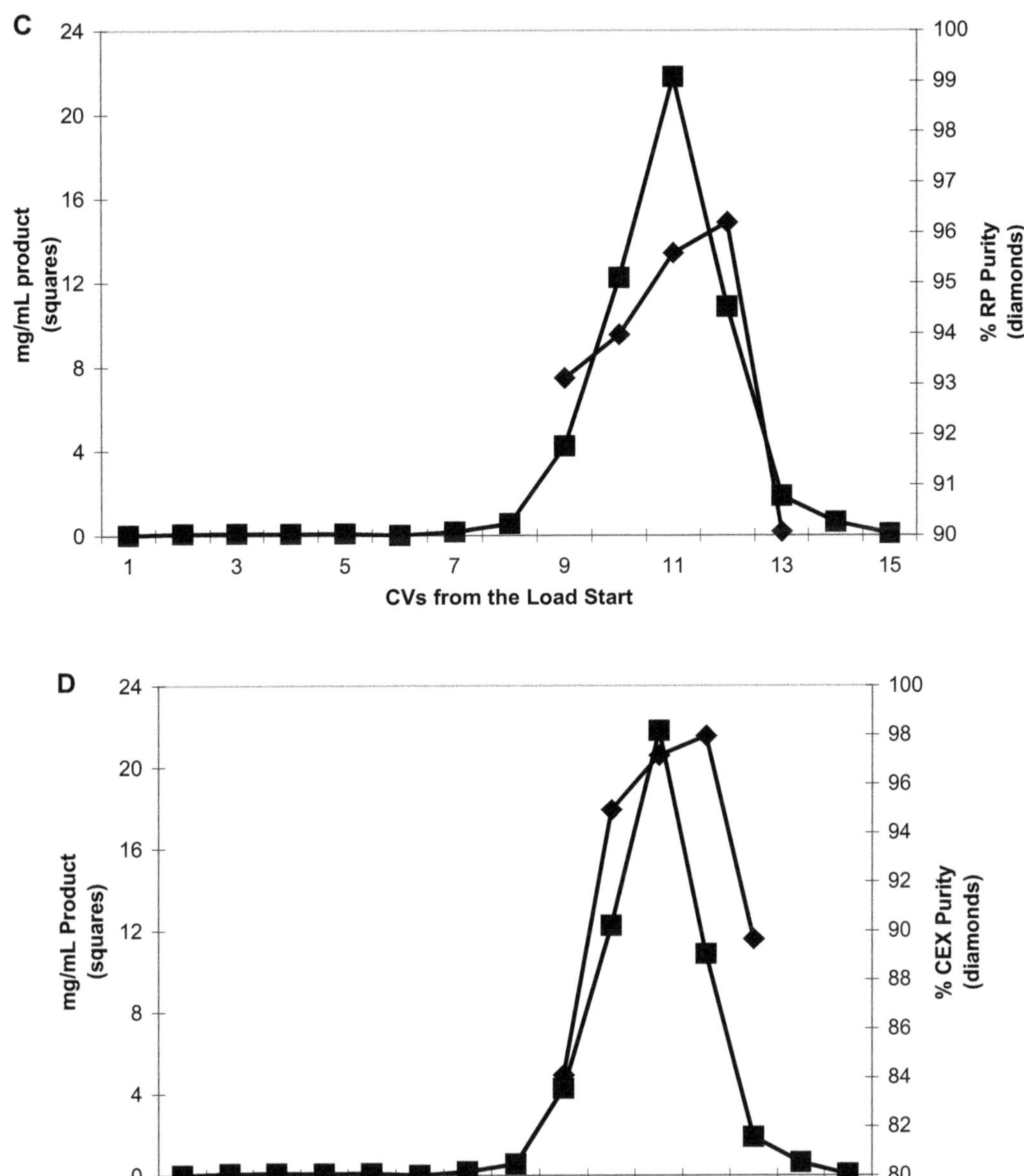

FIGURE 3.2 *(Continued)*

3.4.2 Screening Experiments

Screening experiments are designed to eliminate the less critical parameters from further, more rigorous, process characterization work. Experimental design (DOE) approaches can be used for these studies to easily screen a number of operating parameters (12, 13), and there are a number of DOE software packages that can aid in the design and interpretation of these experiments (14, 15). Fractional factorial, Plackett-Burman, and D-Optimal DOE designs can be used for screening (2, 12, 13). An example of a simple Resolution III fractional factorial design for a fermentation process is shown in Table 3.4. In this design only nine experiments are required to screen six operating parameters. Results can be analyzed using Pareto charts or regression analysis (12, 13). While these studies do not allow us to see interactions between parameters, they enable us to determine the main effects and identify what parameters have the greatest effect on key performance parameters. In some cases, it may be simpler to do "one-off" studies where only a single operating parameter is tested and all other parameters are held at their center point. While this approach can require more experiments, it can give results that are more easily interpreted.

Typically, we assume that the performance response over the operating parameter range tested will be linear, therefore only two-level (a high and a low value) studies are used for these experiments. In addition, the two levels we prefer to test are at ~1.5–2 times the *preferred* operating range in manufacturing (2). Of course the "preferred" operating range may be different for different unit operations and with different processes. (Examples of these ranges for several commonly used operating parameters are shown in Table 3.2). This approach can enable us to get a clear indication of the effect of an operating parameter on unit operation performance. The range is wide enough to see an effect should one exist, yet not so wide as to make a performance failure inevitable. Because of this, it can give information on process robustness, that is, the ability of the process to run outside the prescribed operating range. This can be useful for closing out manufacturing incidents and excursions from set operating ranges. In addition, information from these studies can be used to tell manufacturing which parameters need tighter control or narrower ranges and which require less attention. In some instances they may be able to run at a more easily preferred operating range and in others they may have to tighten the range (Table 3.3). Finally, these studies will clearly separate key from non-key parameters. If an operating parameter tested over this range has no significant effect on process performance, we may designate it a non-key parameter. However, even if excursions from these operating parameters have no product impact, we still may monitor them to ensure consistent process control. Those parameters that have a significant, measurable effect are identified as key parameters that should be tested in the next set of characterization experiments.

For some operations, particularly purification steps where several contaminants are removed, there will be more than one performance parameter, and some will be more important than others. For example, an operating parameter that has a large effect on

TABLE 3.4

Resolution III Fractional Factorial Design for a Fermentation Process

Run	pH	Temp	Start Feed OD	Inducer volume	Feed rate (kg/hr)	Induction time (hr)
1	6.7	27	17	9.6	97	17
2	6.7	27	23	10.4	72	13
3	6.7	33	17	10.4	72	17
4	6.7	33	23	9.6	97	13
5	7.3	27	17	10.4	97	13
6	7.3	27	23	9.6	72	17
7	7.3	33	17	9.6	72	13
8	7.3	33	23	10.4	97	17
9	7	30	20	10	84	15

TABLE 3.5

Fractional-Factorial Screening Study to Examine Operating Parameters from a Reversed-Phase Column (Case Study 3)

Run No.	pH	Temperature	Protein load	Resin type*	Bed height	Flow rate
1	6	4	4	NE	3	50
2	6	4	4	E	15	100
3	6	4	20	NE	15	100
4	6	4	20	E	3	50
5	7.5	4	4	NE	3	100
6	6.4	7	8	E	8.5	75.5
7	7.5	4	4	E	15	50
8	7.5	4	20	NE	15	50
9	7.5	4	20	E	3	100
10	6	22	4	NE	15	50
11	6	22	4	E	3	100
12	6.4	7	8	E	8.5	75.5
13	6	22	20	NE	3	100
14	6	22	20	E	15	50
15	7.5	22	4	NE	15	100
16	7.5	22	4	E	3	50
17	7.5	22	20	NE	3	50
18	7.5	22	20	E	15	100
19	6.4	7	8	E	8.5	75.5

*E = elutriated, NE = non-elutriated

pool volume and a minimal effect on product purity would be considered to be of less importance than a parameter where the reverse was true. In these cases the impact of the operating parameters will have to be weighed relative to their effect on the different performance parameters.

Case Study 3

This study involves a reversed-phase column that is used to remove several product-related variants as well as host cell proteins from a recombinant glycoprotein. Risk assessment analysis determined that there were six potential key parameters that could affect the performance of the chromatography step: pH, bed height, column load factor, temperature, resin type (elutriated vs. non-elutriated), and flow rate. A near-resolution IV fractional factorial design was set up to screen these operating parameters with regard to eight different process performance parameters study, which included % yield, product variants 1, 2, and 3, host cell proteins, pool volume, retention time, and peak asymmetry (Table 3.5). (This experiment was a bit unusual in that the ranges tested were somewhat wider than what we usually test.) The relative impact of each operating parameter on each performance parameter was determined using Pareto plots, and the data are summarized in Table 3.6. The importance of the performance parameter was multiplied times the impact of the operating parameter to give the rating for a given operating parameter/performance parameter pair (i.e., temperature/yield had a relative effect of 6 × 2 = 12). The total score for each operating parameter is the sum of the effect on all of the performance parameters

TABLE 3.6

Impact of Operating Parameters on Different Performance Parameters from a Reversed-Phase Column (Case Study 3)

Output	Relative size of effect (Pareto plot)						Relative total effect (size of effect x impact)					
	pH	temperature	Load	Bed height	Flow rate	Resin type	pH	temperature	Load	Bed height	Flow rate	Resin type
Product purity (column pool)	1	1		1.5			6.0	6.0		9.0		
Yield		2			1			12.0			6.0	
Product variant 2	1	0.5	2	1.2			5.0	2.5	10.0	6.0	0.0	0.0
Product variant 1	1	1	2	1			3.0	3.0	6.0	3.0	0.0	0.0
Pool volume	1		2	1		1	3.0		6.0	3.0		3.0
Host cell protein	1		2.4				2.0	0.0	4.8	0.0	0.0	0.0
Peak position	1	2		1			2.0	4.0		2.0		
Peak asymmetry		1	1	1.5				1.0	1.0	1.5		
TOTAL SCORE							21.0	28.5	27.8	24.5	6.0	3.0
Ease of control (1 = easiest)							1.5	1.5	3.0	2.0	1.0	1.0
ADJUSTED SCORE							**32**	43	83	49	6	3
(score x control factor) RANK							**4**	3	1	2	5	6
include in next DOE?							**yes**	**yes**	**yes**	**yes**	**no**	**no**

(i.e., total score for pH was 21.0, temperature was 28.5, etc.). Another factor was added to account for the "ease of control" for each operating parameters to give the final adjusted score ("ADJUSTED SCORE"). From this analysis, it is evident that load factor has the largest overall impact on the process; pH, temperature, and bed height have some effect; and flow rate and resin type have a minimal role. Therefore, for our next set of experiments, we would only want to focus on load rate, pH, temperature, and bed height.

3.4.3 Interactions between Key Parameters: i.e., the Next Round of Process Characterization Experiments

From the screening experiments we know that the operating parameters being tested at this juncture have some effect on process performance. Therefore, we typically test these parameters only to the edge of their normal or preferred operating ranges. As in the screening experiments, a DOE approach may be used. Depending on the number of variables to be tested, a full factorial, fractional factorial, or other designs could be used (12–15). In general, however, we will want to use a design where the effect of any suspected interactions can be determined, which will necessarily mean a higher resolution (IV or V) experimental design (12–15). Although we typically assume the performance response to be linear over the operating ranges tested, some judgment has to be made as to whether this is a valid assumption. In those cases where nonlinearity is suspected, multilevel experimental designs should be used. In addition, we generally look for no more than two-factor interactions, since interactions with more than two variables are quite rare and would require extensive studies to detect. One of the outcomes of these experiments may be the identification of other process weak spots, due to interactions or additive effects between parameters that were not observed in the initial screening experiments. In some cases certain operating ranges may have to be readjusted. The ultimate deliverable from this set of experiments is to provide assurance that the process provides consistent yields and product quality attributes within the confines of all combinations of the operating limits.

Case Study 4

In order to more accurately characterize the behavior of the chromatography experiment from Case Study 3 and to determine if there were interactions between key parameters that could result in process failure, we undertook a second, more rigorous DOE with the four key inputs identified from the DOE screening studies. An experimental design with the fewest number of experiments that would provide a model capable of discerning nonlinear behavior and two-factor interactions was a 20-run D-optimal design with each of the four inputs set at three levels: low, intermediate, and high (Table 3.7). The input settings (high and low) were narrowed compared to those in the first study since we only wanted to test to the edge of the proposed operating ranges in this set of experiments. Column performance was monitored by seven output parameters: step yield, product purity, product variants 1 and 2, an additional assay for product variant 3, host cell proteins, and pool volume.

The results for the study are shown in Table 3.8. The most important take-home message from this study is that, with the exception of host cell proteins, all of the process outputs fell within the historical ranges of what had been observed for this process. Based on this we can conclude that within the confines and various combinations of these operating ranges, the process delivers a product of adequate purity and yield. There were several instances where host cell proteins exceeded the historical upper limit (~4000 ppm). Based on the Pareto plot for host cell protein (Figure 3.3), it is apparent that the main factor

TABLE 3.7

A 20-Run, 3-Level D-Optimal Design Study (Case Study 4)

Run no.	pH	Temperature	Protein load rate	Bed height
1	5.8	3	5	4.5
2	7.0	3	5	10
3	5.8	3	5	15
4	6.4	3	10	15
5	7.0	3	15	4.5
6	6.4	3	15	4.5
7	5.8	3	15	10
8	7.0	3	15	15
9	7.0	7	5	4.5
10	5.8	7	5	15
11	5.8	7	10	4.5
12	6.4	7	15	10
13	6.4	11	5	4.5
14	5.8	11	5	10
15	7.0	11	5	15
16	7.0	11	10	10
17	5.8	11	15	4.5
18	7.0	11	15	4.5
19	7.0	11	15	15
20	5.8	11	15	15

TABLE 3.8

Results from D-Optimal Experimental Design Study for Reversed-Phase Column (Case Study 4)

Run no.	Yield	Pool volume CVs	Product purity	Product Variant 1	Product Variant 2	Product Variant 3	Host cell protein
1	38.6	2.2	90.5	0.57	0.81	0.5	2077
2	38.7	1.8	93.2	0.67	0.68	0.7	817
3	39.5	1.6	91.1	0.34	N/D*	0.3	1524
4	39.8	2.1	91.1	0.57	0.66	0.5	2596
5	43.2	5.3	90.1	1.29	0.93	1.3	10835
6	36.7	3.9	90.5	1.03	0.85	1.1	4599
7	39.0	2.7	90.1	0.72	0.80	0.8	3658
8	41.3	3.1	91.1	1.04	0.82	1.0	3357
9	41.9	2.4	91.3	0.86	0.78	0.9	265
10	37.0	1.5	91.8	0.27	0.57	0.3	938
11	38.2	2.6	91.0	0.72	0.78	0.8	256
12	36.2	2.7	91.8	0.92	0.80	0.9	5163
13	38.6	2.0	91.7	0.31	0.80	0.5	39
14	33.4	1.5	92.9	0.35	0.67	0.3	137
15	39.4	1.7	92.9	0.42	0.81	0.4	68
16	39.4	2.4	91.7	0.73	0.75	0.9	443
17	35.3	3.0	91.0	0.80	0.77	0.9	4138
18	38.9	4.5	91.8	1.14	0.80	1.4	4861

TABLE 3.8 *(Continued)*

Results from D-Optimal Experimental Design Study for Reversed-Phase Column (Case Study 4)

Run no.	Yield	Pool volume CVs	Product purity	Product Variant 1	Product Variant 2	Product Variant 3	Host cell protein
19	40.4	3.0	91.0	0.83	0.93	1.0	**2506**
20	33.1	2.2	91.6	0.31	0.86	0.4	**4662**
ctrl 1^	39.1	2.0	91.8	0.36	0.84	0.5	**2653**
ctrl 2^	37.4	2.2	91.7	0.53	0.80	0.5	**3659**
ctrl 3^	36.8	2.2	92.9	0.49	0.85	0.6	N/D*

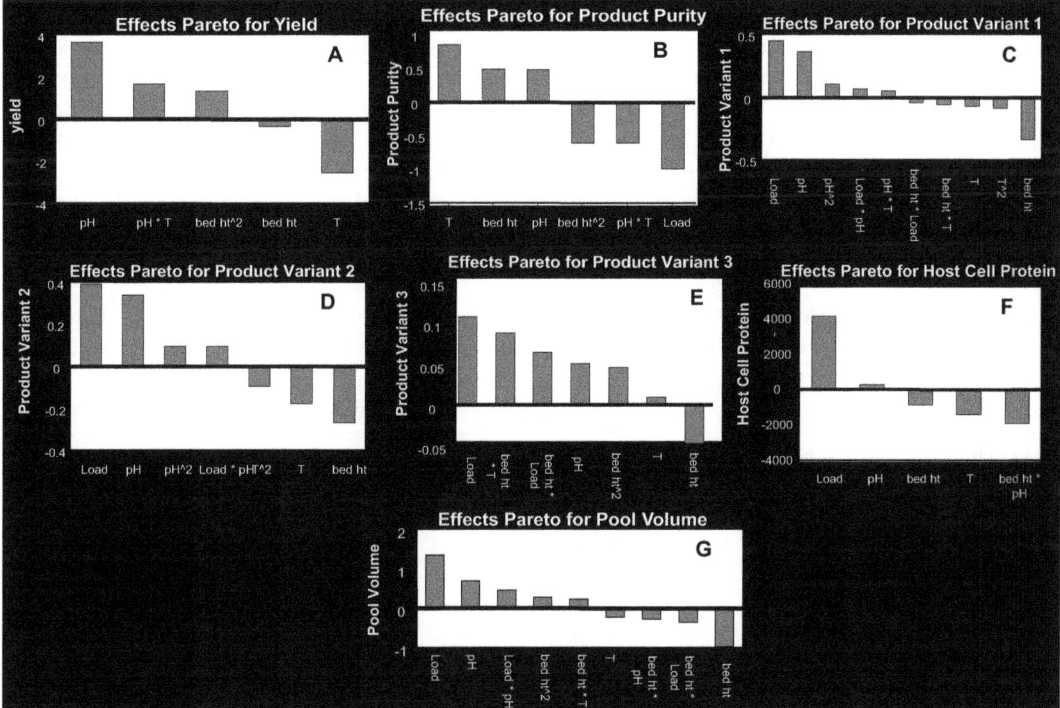

FIGURE 3.3
Pareto plots from D-optimal study of reversed-phase column (Case Study 4).

affecting host cell protein content in the product pool is the column load factor. In order to ensure that we are consistently below the historical upper limit for host cell proteins, we set the load factor at no greater than 10. The Pareto plots also gave some hints as to how we might improve step yields (increasing pH and decreasing temperature) or increase product purity (also by decreasing load).

3.4.3.1 Combining Variables

In some instances it may be possible to combine variables in such a way as to more quickly explore the "operating space" of a given unit operation (2, 12). This will greatly cut down on the amount of experimentation required. Sometimes, however, it may give you results that are difficult to interpret if there is a process failure and it only assumes that there are additive effects between operating parameters. We typically use this approach when we have some degree of confidence in our operating ranges and only need confirmatory information that the process will perform properly over these ranges. One example would be to combine pH and conductivity as inputs to an ion-exchange chromatography step. An even more extreme example is shown in the following case study:

Case Study 5

This study was from a cation-exchange chromatography step using a linear salt gradient. From an initial screening study, seven operating parameters were determined to have a significant effect on the performance of a cation-exchange step. The low salt buffer conductivity and pH, the high salt conductivity and pH, the feed conductivity and pH, and the temperature. These operating parameters were combined to the edges of their operating range in such a way as to obtain the very earliest or latest retention times, as well as the largest or smallest pool volumes (Table 3.9). One way of looking at this is as a two-factor full-factorial design, with retention time and pool volume as inputs or combined variables, and step yields, quality attributes, and so on as outputs. Table 3.10 shows the effect of pool volume as an input variable. There is a slight effect on yield (but nothing outside of what

TABLE 3.9

Combination of Operating Parameters for a Cation-Exchange Step to Obtain Earliest and Latest Retention Times and Minimum and Maximum Pool Volumes (Case Study 5)

Operating Parameter	Earliest Retention Time	Latest Retention Time	Minimum Pool Volume	Maximum Pool Volume	Control
Equil pH	Low	High	High	Low	Center
Equil conductivity	High	Low	Low	High	Center
Elution pH	Low	High	Low	High	Center
Elution conductivity	High	Low	High	Low	Center
Small ion capacity	Low	High	Center	Center	Center
Temperature	High	Low	Center	Center	Center
Feed pH	Low	High	Center	Center	Center
Feed conductivity	High	Low	Center	Center	Center

TABLE 3.10

Effect of Cation-Exchange Pool Volume on Product Purity and Yields (Case Study 5)

Pool volumes	Runs	Product concentration	Pool Volume (mL)	% Yield	% RP Purity	% CEX Purity	*E.Coli* Proteins
Low	1	20.8	195.6 (1.47 CV)	102.2	95.3	95.1	3776
	2	20.8	194.6 (1.47 CV)	105.7	95.2	95.0	3878
Center	1	14.8	257.3 (1.94 CV)	95.7	94.4	95.9	3526
	2	14.6	259.3 (1.95 CV)	97.0	94.4	95.2	3833
High	1	11.2	313.1 (2.36 CV)	90.4	96.2	95.2	3144
	2	11.7	300.1 (2.26 CV)	89.8	94.7	96.3	3037

had been observed historically) and no effect on quality attributes over the range of possible pool volumes from this process. Table 3.11 shows the results of an experiment where retention time is set up as an input variable. At the earliest retention times, the yield drops dramatically. Also, for both the late and early retention times there is a slight increase in *E. coli* proteins. When the column feed pH and conductivity are kept at their center points, the yield effect is greatly reduced, as is the effect on *E. coli* proteins (Table 3.12). We concluded from these experiments that we needed better control of our feed pH and conductivity and lowered the center point of the column feed pH by 0.1 units.

The advantage of this study design is that with no more than six experiments we were able to explore the extreme operating corners of the operating parameters for this unit operation. The disadvantage is that any nonlinear responses or significant interactions between variables could be lost. As with any design, one has to determine the risks and benefits of combining variables or using a lower-powered study.

3.4.4 Key and Critical Parameters

Operating parameters that have a significant effect on the performance of a unit operation are considered key parameters. More specifically, this would include those parameters from the initial screening experiments that have a significant, measurable effect on process performance parameters (particularly product quality attributes, impurities, and step yields). These parameters should be included as the key operating parameters for full-scale validation runs, although in some instances there may be additional parameters included in the full-scale validation runs as key parameters as well.

TABLE 3.11

Effect of Cation-Exchange Retention Time on Product Purity and Yield (Case Study 5)

Retention Times	Runs	Product concentration (mg/mL)	% Yield	% RP Purity	% CEX Purity	*E.Coli* Proteins
Early	1	7.7	37.9	92.9	94.8	4682
	2	7.0	36.2	94.5	95.2	5195
Center	1	12.7	87.0	94.8	95.7	4068
	2	12.7	88.2	95.5	95.6	3387
Late	1	14.1	102.7	93.6	94.4	4289
	2	13.5	102.0	93.6	94.1	5085

TABLE 3.12

Effect of Cation-Exchange Retention Time on Product Yields and Purity When Column Feed pH and Conductivity Are Run at Their Center Points

Retention Times	Runs	% Yield	% RP Purity	% CEX Purity	*E.Coli.* Proteins
Early	1	84.1	95.1	95.8	4590
	2	83.5	94.8	96.5	3788
Center	1	94.2	94.2	96.3	3229
	2	91.0	94.2	95.8	3520
Late	1	98.6	95.0	95.6	4050
	2	98.4	95.5	95.4	4519

In addition to key parameters, there may be a subset of "critical" operating parameters that have an even greater effect on product quality. The FDA clarifies the definition of critical parameters as "process parameters that must be controlled within established operating ranges to ensure that the API or intermediate will meet specifications for quality and purity" (16). Most parameters, even most key parameters, can be run slightly outside their prescribed operating range without resulting in a failed product or product intermediate specification. However, there may be a handful of operating parameters for a given process for which an excursion outside the prescribed range would result in such a failure. Generally, for a robust process there should be very few of these. Those that are identified as critical should be highly characterized in terms of defining their edge of failure and identification of any interactions with other operating parameters that could result in process or product failures.

3.4.5 Setting Acceptance Criteria for In-Process Performance Parameters: Using Feed Quality as a Process Input

A number of factors need to be considered for setting acceptance criteria for in-process performance parameters. Setting the criteria too stringently can lead to unnecessary validation failures. Setting the criteria to loosely may result in product that would ultimately fail the final product release specifications.

Statistical analysis of pilot, clinical, and commercial-scale manufacturing data is one of the most common ways of setting acceptance criteria on key performance parameters. Frequently, however, the datasets may not be representative (due to process differences between pilot and large scale) or be limited due to an inadequate number of manufacturing runs. In these cases, basing acceptance criteria on a historical range or a statistical evaluation of the data, such as three standard deviations or tolerance intervals (16), may not give the appropriate acceptable ranges for process performance parameters. Therefore, it is important to use process characterization data to determine the process capability and redundancy so that acceptance criteria is based on what the process can actually deliver.

In most process characterization studies, representative feed material should be used, whether this is material from a seed fermenter feeding a production fermenter, a column feed, or a feed material going into an ultrafiltration step, and so on. However, to really test the "top-to-bottom" robustness of the process the effect of feed quality on each unit operation should be tested. This will give us an understanding of the downstream sensitivity to upstream process excursions. For fermentation this might include different seed fermenter cell densities. For cell harvesting it might include different fermentation media OD or viscosity. In the case of chromatography steps, the purity of product from the previous step and/or load factor should be considered (although we usually run all process characterization experiments at the upper end of the loading range). All other operating parameters (pH, temperature, etc.) are run at the center of their respective ranges, because the likelihood of having both an operating parameter excursion as well as a feed quality excursion is remote.

These experiments can be used to set performance parameter acceptance criteria for each unit operation. One way to address this is to run a unit operation under conditions where it fails to perform adequately. The pool from the failed unit operation is processed further downstream to see if subsequent process steps can make up for the poor performance from the failed unit operation and allow for the product to stay within specifications. These experiments give information on process redundancy with regard to different key performance parameters (2).

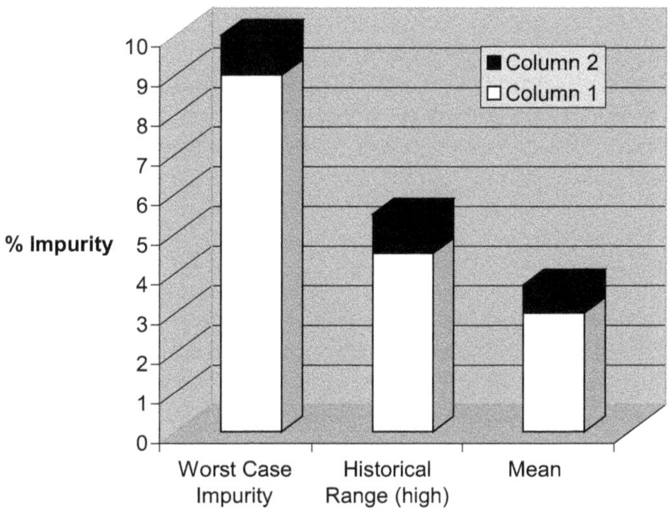

FIGURE 3.4
Impurity clearance showing process redundancy between columns 1 and 2.

An example of this is shown in Figure 3.4. In this instance column 1 is run in such a way that a product-related variant was present at values higher than the historical range. When this column 1 pool is processed through column 2, however, the amount of this variant falls within the historical range for the column 2 pool. Therefore, rather than setting the acceptance criteria for the column 1 pool based on historical ranges for the product-related variant, a wider acceptance criteria can be set due to the process redundancy between columns 1 and 2. There may be other instances where a pool has to be processed though more than one additional step downstream to determine the process redundancy, but the principle is the same. This method of setting acceptance criteria is scientifically based and can give a more realistic indication of what the process can actually deliver than can statistical analysis of historical data (standard deviation, tolerance intervals, etc. (16)), particularly since at the time the process characterization is carried out there may be very little if any historical data to set a statistically valid acceptance criteria. Setting the acceptance criteria for process validation studies based what the process can actually deliver will result in less validation failures due to acceptance criteria being set incorrectly. Of course, as the process matures and more lots are run, the acceptance criteria can more easily be set based on statistical analysis.

3.4.6 Finishing Up: Reports, Follow-Up, etc.

Process characterization reports should be written for each unit operation. These reports include the results from the clearance studies, the screening and interactions studies, as well as the feed quality studies. Key operating parameters and their respective ranges are identified, as are acceptance criteria for all key in-process performance parameters. A rationale for why certain parameters were identified as non-key is included as well. Data from these reports will be used to support the validation studies, and the reports should be completed prior to writing the validation protocols.

After the characterization work is completed, it may be valuable to go back and repeat the FMEA exercise (17). The severity factor for each operating parameter will be known at this point and this could, in some instances, dramatically change the outcome of the risk priority number. In this way, plant engineers can best devote their time to those unit operations and operating parameters that require the greatest control and detection.

3.4.7 Future Challenges

Process characterization requires a significant commitment of time and resources, but the payoff in terms of better process understanding, improved success rate in manufacturing, and avoidance of costly regulatory delays makes it a very worthwhile investment. The approaches described in this chapter can provide information used for setting operating ranges and performance parameter acceptance criteria and give an indication of overall process robustness.

There are a number of challenges we continue to face in process development and other departments involved in carrying out process characterization studies. Some of these include:

- Developing appropriate scale-down models for cell culture, centrifugation, and other difficult-to-scale upstream steps: Finding a representative scale that will still be amenable to the high run number generally required for thorough process characterization work can be a real challenge. For cell culture, one approach may be to use small-scale bioreactors to identify key operating parameters and then to use larger, more representative bioreactors for confirming ranges or studying interactions between key parameters. Even so, this entails a great deal of work and may still give information that is of marginal value.

- Use of appropriate analytical methods and analytical method turnaround time: Having "mature" analytical methods available in time for starting process characterization studies can be a real problem. The availability of more generic analytical method platforms so that methods can be more easily qualified would help ensure that process characterization studies are started at the appropriate time. In the absence of this, analytical resources will have to be spent "at risk" earlier in the product development cycle in order to ensure timely qualification of analytical methods. Also, since sample analysis can be a major bottleneck for completing process characterization work, development of more rapid and automated methods can help in sample turnaround time and in planning of subsequent process characterization experiments.

- Appropriate resources for process characterization: Many companies are still dialing in the appropriate resource requirements for process characterization. It is important to tailor the process characterization requirements for each product to key business drivers in order to have the most efficient use of resources. All products will have certain requirements with regard to regulatory commitments, such as providing data that the process will provide consistent yields and product quality attributes within the normal operating ranges. For products with less intensive run rates or less cost of goods issues, broader yield ranges or higher process excursion rates may be acceptable. In these instances it may be possible to reduce the scope of some of the characterization work by reducing the number of variables to examine or only testing to the edge of the normal operating range for certain parameters. Another problem encountered is how to deal with products that have accelerated development timelines that may not allow time for thorough characterization studies. In these cases a more "bare-bones" approach to characterization might be used prior to conformance runs, with more thorough studies being carried out later (but still in time for the biologics license application (BLA) filing). Definitions around what these "bare-bones" studies might be are still somewhat ambiguous.

It is hoped we will be able to better address these and other challenges as the strategies for doing process characterization continue to evolve in the biopharmaceutical industry.

References

1. G. Bobrowicz, The Compliance Costs of Hasty Process Development. *BioPharm* 12 (8): 35–38, 1999.
2. J. Seely and R. Seely, A Rational, Step-Wise Approach to Process Characterization. *BioPharm International* 16 (8): 24–34, 2003.
3. A. Gardner and T. Smith, Identification and Establishment of Operating Ranges of Critical Process Variables. In: *Pharmaceutical Process Validation*, G. Sofer and D. Zabriskie, eds. Marcel Dekker Press, New York, 1999, pp. 61–76.
4. A.S. Rathore, G.V. Johnson, J.J. Buckley, D.M. Boyle, and M.E. Gustafson, Process Characterization of the Chromatography Steps in the Purification Process of a Recombinant Escherichia Coli-Expressed Protein. *Biotechnology Applied Biochemical* 37: 51–61, 2003.
5. A. Armbruster and T. Feldsien, Applying HACCP to Pharmaceutical Process Validation. *BioPharm* 13 (10): 170–178, 2000.
6. R. Kieffer, et al., Applications of Failure Mode and Effects Analysis in the Pharmaceutical Industry. *Pharm Tech Europe*, 36–49, September 1997.
7. P. Nobel, Reduction of Risk and the Evaluation of Quality Assurance. *PDA Journal of Pharmaceutical Science and Technology*, 55 (4): 235–239, 2001.
8. A. Sahni, Using Failure Mode and Effects Analysis to Improve Manufacturing Processes. *Medical Device and Diagnostics Industry*, 47–51, July1993.
9. R. McDermott, et al., *The Basics of FMEA*. Productivity, Inc. P.O. Box 13390, Portland, OR 97213.
10. J.T. Burr, *SPC Tools for Everyone*. ASQC Quality Press, Milwaukee, WI, 1993.
11. Rath and Strong Consultants, *Six Sigma Pocket Guide*. Division of Aon Worldwide, Lexington MA, 2001, pp. 26–31.
12. B. Kelley, Establishing Process Robustness Using Designed Experiments. In *Pharmaceutical Process Validation*, G. Sofer and D. Zabriskie, eds. Marcel Dekker Press, New York, 1999, pp. 29–60.
13. P. Haaland, *Experimental Design in Biotechnology*. Marcel Dekker, Inc., New York, 1989.
14. J.M. Juran and A.B. Godfry, *Juran's Quality Handbook*, 5th edition. McGraw-Hill, New York, 1999, pp. 47.1–47.77.
15. D.C. Montgomery, *Design and Analysis of Experiments*, 5th edition. John Wiley and Sons, New York, 2001.
16. FDA Guidance for Industry, *Manufacturing, Processing or Holding of Active Pharmaceutical Ingredients*. FDA Guidance for Industry, Rockville, MD, August 1996.
17. R. Seely, L. Munyakazi and J. Haury, Statistical Tools for Setting in Process Acceptance Criteria. *BioPharm*, 14 (10): 28–34, 2001.

4

Scale-Down Models for Microbial and Mammalian Cell Culture Processes: Approaches and Applications

Ravali Raju, Jin Guo, Tatenda Shopera, Xiaolu Zheng, Jennifer Pulkowski, and Nichole Wood

4.1 Introduction

A typical biopharmaceutical manufacturing process is complex, comprising multiple upstream and downstream unit operations with the main goal of delivering products with consistent quality for patient efficacy and safety. A key component of the regulatory filing to enable manufacture for any biopharmaceutical product is a complete process validation package. Process validation ensures that the drug is fit for intended use and provides assurance of drug quality, safety, and efficacy. It requires that each step of the manufacturing process be controlled to ensure that the finished product meets all the desired product quality attributes and specifications and is not limited to the in-process or finished product inspection and testing (1–3).

Process validation for drugs (finished pharmaceuticals and components) is a legally enforceable requirement under section 501(a)(2)(B) of the Act (21 U.S.C. 351(a)(2)(B)). FDA regulations describing current good manufacturing practice (CGMP) for finished pharmaceuticals are provided in 21 CFR parts 210 and 211 (1, 4). The CGMP regulations require that manufacturing processes be designed and controlled to assure that in-process materials and the finished product meet predetermined quality requirements and do so consistently and reliably. This offers assurance that the process is robust and protected from sources of variability that could impact drug manufacture, supply chain, or public health (1).

Regulatory agencies have published general guidelines to aid in developing validation strategies. In 1987, the FDA released a guidance on process validation that put emphasis on data from commercial-scale validation batches (5). In 2011, the FDA published updated guidance that defined process validation as "the collection and evaluation of data, from the process design stage throughout production, which establishes scientific evidence that a process is capable of consistently delivering quality products." (1, 4). This revised guidance is in line with the FDA's initiative "Pharmaceutical CGMPs for the 21st Century—A Risk-Based Approach," which promotes the concepts of technological advancements, risk mitigation and quality system tools for pharmaceutical manufacturing. This initiative emphasizes that quality should be built into a process by design (i.e., quality by design) by leveraging development studies and manufacturing experience to establish a design space, specifications, and manufacturing controls (6).

DOI: 10.1201/9781003143130-4

The process validation guidance aligns process validation activities with a product lifecycle concept and with existing FDA guidance, including the FDA/International Conference on Harmonization (ICH) guidance, Q8(R2) Pharmaceutical Development (7), Q9 Quality Risk Management (8), Q10 Pharmaceutical Quality System (9), and Q11 Development and Manufacture of Drug Substances (10). The process validation activities, described in three stages in the new guidance, are (1) process design, (2) process qualification, and (3) continued process verification (1).

The goal of the process design stage is to design a commercial manufacturing process that can consistently deliver a product with the desired quality attributes by leveraging a quality by design (QbD) approach. In combination with quality risk management as described in ICH Q9 (8), this allows for a systematic approach towards quality risk management which is based on scientific knowledge and ultimately linked to quality and patient safety. Risk assessments are conducted through tools like failure modes and effects analysis (FMEA) to identify the critical process parameters, critical performance attributes, and control ranges (11, 12). The criticality of all the parameters and attributes impacting the process and the product are evaluated, and the control strategy is implemented. This may include defining design space for the process or for individual unit operations. A design space is defined by ICH Q8 as "the multidimensional combination and interaction of input variables (e.g., material attributes) and process parameters that have been demonstrated to provide assurance of quality." Per ICH Q8, working within the design space is not considered to be a process change, while movement out of the design space is a change and would normally initiate a regulatory post-approval change process. The control strategy for parameters or attributes is commensurate with their risk to the process, with tighter or additional controls for those that pose a higher risk. In 2003, FDA introduced a process analytical technology (PAT) initiative to implement process monitoring and control to improve process and product consistency in pharmaceutical manufacturing. While small-scale studies need not be performed under CGMP conditions, they should be in accordance with sound scientific principles and methods with good documentation practices according to ICH Q10 Pharmaceutical Quality System.

Scale-down models (SDMs) are laboratory- or pilot-scale models of unit operations that can be designed to be representative of the commercial manufacturing process (4). Small-scale models can be developed and used to support process development studies. The development of a model should account for scale effects and be representative of the proposed commercial process. "A scientifically justified model can enable a prediction of quality and can be used to support the extrapolation of operating conditions across multiple scales and equipment" (ICH Q11 Step 4) (10). Scale-down models can be used to estimate process and product variability over broad ranges of conditions that may not be possible at the commercial scale. They can be useful in gaining process knowledge and understanding as well as designing effective process control strategies. Following the ICH Q9 guidelines, input parameters and ranges can be prioritized based on risk assessments (8). They can be used to perform design of experiment (DOE) studies to develop process knowledge through interactions between the variable inputs and process outputs (7). The results can provide justification for establishing ranges of raw material attributes, equipment and process parameters, and in-process quality attributes. Computer-based or virtual simulations of certain unit operations or dynamics can provide process understanding and help avoid problems at the commercial scale (8). Process knowledge and understanding is the basis for establishing an approach to process control for each unit operation and the process overall. Strategies for process control can be designed to reduce variability and/or enable adjustment during manufacturing (i.e., adaptive manufacturing control) to ensure a robust process with consistent product quality (1, 5, 7, 8).

During the process qualification stage, the process is evaluated to determine if it is suitable for commercial manufacture under CGMP compliance. The two key elements of this stage are (1) facility design and equipment qualification and (2) process performance qualification (PPQ) (1, 4). Laboratory-scale and pilot-scale data from the SDMs can provide assurance of process comparability at different scales and confirm that the commercial manufacturing process performs as expected. Data from the process design stage and process characterization can further justify operating ranges to ensure successful commercial production (1, 4).

The goal of the continued process verification (CPV) stage is to assure that the process remains in a state of control throughout commercial production (1, 4). Process performance data are continuously collected and evaluated to detect variability and determine if any action or risk mitigation is necessary to maintain process control. Manufacturers typically use a combination of qualitative and quantitative statistical methods to monitor intra-batch and inter-batch variability under a comprehensive CPV program (1, 4, 13). SDMs are valuable in assessing impact from process deviations, manufacturing deviations, equipment or vendor changes, process changes, and raw material changes/challenges as they may evolve through the product life cycle to ensure process comparability (1, 3, 4, 13). A more detailed discussion of the various applications of SDMs is provided later in this chapter.

SDMs for each unit operation across the stages of the product life cycle have unique principles involved during their design. The Process Validation Guidance promotes an increase in use of data from representative and predictive SDMs of commercial process throughout the life cycle of a product (1, 3, 4, 13). This chapter is intended to provide a perspective on the general scale-down principles, scale-down parameters, and primary endpoints used in the design of SDMs for a variety of upstream unit operations for both mammalian and microbial biopharmaceutical manufacturing processes.

4.2 Design of Scale-Down Models

SDMs that represent full-scale performance are crucial to successful process design and characterization. They allow for extensive experimental studies to be conducted at bench scale due to cost and feasibility limitations of performing such work at large scale (11, 12). Process understanding from large-scale clinical campaigns or platform knowledge serves as a prerequisite to initiate the development of the SDM, which in turn should reproduce the important outputs or attributes of the manufacturing dataset. SDMs should not only represent large-scale performance at set-point conditions, but also be representative in operational ranges and even beyond to ensure the successful application of process characterization at manufacturing scale and eventually achieve an accurate and reliable design space (14).

Approaches to develop and assess SDM involve (1) determining the scale-down strategy and assessment criteria, (2) performing bench-scale studies, and (3) assessing the SDM to demonstrate comparable performance with respect to process performance and product quality. During the development of SDM, it is important to define the scale-down strategy by identifying the appropriate operating parameters, performance parameters, and quality attributes (if applicable) for the given unit operation. The formality of the scale-down model assessment or qualification can vary widely. Assessments may be retrospective in nature, with SDM data already generated as part of process design and process characterization being used to compare against the relevant large-scale data, or a prospective approach may be taken. For the prospective case, it is crucial to have available representative large-scale runs to which the SDM can be compared. These prospective assessments or qualifications may be done with or without predefined acceptance criteria in pre-approved protocols, with or without formal approvals from the quality assurance (QA) functional area/department within a company. While the details of defining and assessing SDMs vary across the industry, the underlying principles remain the same.

To initiate the development of SDM, commercial manufacturing datasets may be available as a baseline of scale-up and scale-down. If manufacturing data are limited or unavailable, pilot-scale data can provide a preliminary understanding of the process. However, this will increase the risk that process characterization studies may be found to be non-representative and may require reevaluation of the SDM after large-scale data are available, along with re-execution of process characterization studies with the improved SDM.

Bench-scale bioreactors are usually the primary choice for the seed bioreactor and production bioreactor unit operations. Other culture systems such as shake flasks and spinner flasks can be used to evaluate process parameters or raw materials used in the inoculum stages of the process. Novel high-throughput systems such as Ambr® 15 and Ambr® 250 systems (Sartorius) have also been successfully applied to develop processes for therapeutic protein products in Pfizer Inc. and will be discussed later in this chapter.

Once the SDM is established, the model needs to be assessed (i.e., qualified) to demonstrate comparable performance to large scale. Qualification of SDMs involves assessing the comparability of outputs across scales and proving that the outputs are within acceptable ranges. Gaps and inconsistency across scales should be identified and investigated to optimize the SDM. An ideal SDM should predict performance of commercial-scale processes at the operational set-point conditions, but also adequately reflect process variability in response to fluctuations in the operating parameters (15–17). If offsets are still observed after SDM optimization, the cause of the offset should ideally be identified and understood/characterized. These offsets in the SDM may then be considered to enable prediction of large-scale performance. In these cases, while the SDM may not be ideal, the SDM would still be suitable for its primary purpose of predicting large-scale performance. In addition, SDMs should be continuously monitored and refined during the product life cycle to address changes during process improvement or troubleshooting activities.

4.2.1 Principles of Scaling-Up/Scaling-Down Processes

4.2.1.1 Scale-Dependent and Scale-Independent Parameters

Once the operational process parameters are selected and prioritized by FMEA, they are divided into scale-dependent and scale-independent variables (**Table** 1). Scale-independent variables (e.g., pH, temperature, dissolved oxygen, medium composition, seeding density/split ratios, feeding schedule, culture duration, cell age) should be operated at the same target values in the scale-down bioreactors as the commercial counterparts. On the other hand, scale-dependent parameters (e.g., agitation parameters, aeration parameters, mixing time, working volume, feed volume/rate, pressure) should be scaled down linearly or proportionally. Operating conditions at the small-scale can be adjusted based on small-scale studies which accommodate scale differences and match the process performance at full scale. SDMs should maintain similar design and control strategy as compared to the full-scale vessels. Features of the stirred tank bioreactor design (e.g., mixing, aeration, mass

TABLE 4.1

Summary of scale-dependent and scale-independent parameters

Scale-independent parameters	Scale-dependent parameters
pH	Agitation parameters
Temperature	Aeration parameters
Dissolved oxygen	Mixing time
Medium composition	Working volume
Seeding density/Split ratios	Feed volume/rate
Feeding schedule	Pressure
Culture duration	
Cell age	

transfer) and process control strategies (e.g., pH, dissolved oxygen, temperature, nutrient addition) should be comparable to ensure similar performance across scales (16, 18).

Success in scaling-up contributes to the development and implementation of SDMs. A successful scale-up strategy requires consideration of both engineering design features and empirical correlations. Performance differences across scales in stirred tank bioreactors may be attributed to the following aspects including bioreactor design, agitation differences (mixing/blend time, localized inhomogeneities, impeller shear, etc.) and aeration differences (superficial gas velocity, mass transfer (k_La), gas hold-up volume, CO_2 stripping, bubble shear, etc.). Focusing on addressing these aspects can enable successful scale-up/scale-down of processes.

4.2.1.2 Bioreactor Design

A variety of bioreactor specifications need to be considered, including aspect ratio (height to diameter ratio), design and location of impellers, gas spargers and probes, ancillary equipment, and gas mixture composition. While aspect ratio may be kept similar, it is common that aspect ratios are different due to limited options/availability of equipment. General requirements for impellers and their operation are to provide adequate liquid blending and gas dispersion. More impellers may be needed as the height relative to the width of the bioreactor increases. Additionally, impeller placement relative to sparger location may affect mass transfer. For small-scale vessels, spargers such as open or drilled pipes or microporous (e.g., sintered steel) spargers may support effective dissolved oxygen and dissolved carbon dioxide (CO_2) control, while large-scale bioreactors usually demand employment of open or drilled pipes alone or in combination with microporous spargers in order to ensure adequate removal of carbon dioxide. The design and use of fluid addition ports needs to ensure adequate mixing during the addition of base/acid titrants or feeds to limit pH and concentration gradients (19, 20). The addition rate of these solutions may need to be considered to overcome localized inhomogeneity that may affect cell culture performance (e.g., localized zone of high pH near the point of base titrant addition).

In general, a proportional duplication across scales may be impractical as limited by equipment availability. To enhance process understanding and provide additional insights when developing scale-down systems or scaling-up/scaling-down processes, computer-based and virtual simulations like computational fluid dynamics (CFD) can serve as additional tools for equipment characterization (21).

4.2.1.3 Mixing Regime

Mixing time describes a measure of time to reach a certain degree of homogeneity. Highly efficient mixing is usually observed in bench-scale bioreactors, leading to homogenous cell culture environments. Mixing time can differ while scaling up, sometimes resulting in inhomogeneous cell microenvironments with gradients in pH, DO, dissolved CO_2 and nutrients. Numerous factors have been used to characterize mixing and agitation in stirred tank bioreactors. Two parameters are usually used to evaluate the bioreactor performance: power per volume (P/V) and mixing time (T_m). P/V is a function of agitation speed, working volume and impeller design. It directly impacts mixing and oxygen mass transfer. Maximum local P/V may be calculated by considering the impeller sweep volume, and this maximum local P/V correlates with the highest specific energy dissipation rates in the vessel. Higher agitation speed facilitates better mixing and gas dispersion as P/V and mixing time are proportional to the agitation speed (22).

$$\frac{P}{V} = \frac{P_0 \rho N^3 D^5}{V} \tag{4.1}$$

Where: P = power (W), V = liquid volume in the bioreactor (m³), P_0 = power number of the impeller, ρ = density of the culture (kg/m³), N = agitation speed (s⁻¹), D = impeller diameter (m).

4.2.1.4 Oxygen and Carbon Dioxide Mass Transfer

Oxygen supply is a critical part of bioreactor design and operation. Sufficient mixing and oxygen transfer are important to avoid DO gradients and local depletion in the vessel. Operating parameters such as agitation and sparging rates need to be regulated to ensure comparable aeration and DO control when scaling up. High CO_2 partial pressure along with the elevated medium osmolarity may inhibit cell growth and protein production as well as alter the IgG glycosylation (23). As high dCO_2 may impact product yields and quality, CO_2 stripping needs to be considered during scale-up. Gas flow rates and compositions can be tuned to maintain DO within acceptable ranges and ensure sufficient CO_2 stripping (24). Bubble bursting at the gas-liquid interface can cause significant damage to mammalian cells but can be minimized with the presence of various surfactants. The oxygen transfer rate (OTR) determines the capability of a bioreactor to deliver oxygen, as expressed below (25, 26).

$$OTR = \frac{d[O_2]}{dt} = k_L a \left([O_2]^*_{gas} - [O_2] \right) \tag{4.2}$$

Where: $k_L a$ is the volumetric mass transfer coefficient, $[O_2]$ is the dissolved oxygen concentration in the bulk liquid, and $[O_2]^*_{gas}$ is the equilibrium dissolved oxygen concentration in the liquid. $k_L a$ can be dynamically measured by supplying oxygen or air to the bioreactor which is previously depleted or partially depleted of oxygen by air or nitrogen sparging, respectively (27).

Maximum oxygen uptake rate (OUR_{max}) is described below. To ensure adequate oxygen supply, the OTR must be higher than maximum oxygen consumption. Desired OTR can be achieved by adjusting the $k_L a$ and oxygen concentration.

$$OUR_{max} = \left(Q_{O2max}\right) \cdot VCC_{max} \qquad (4.3)$$

Where, $VCCmax$ = maximum viable cell concentration, $QO2max$ = maximum specific oxygen uptake rate.

$$k_L a = k \left(\frac{P}{V}\right)^{\alpha} \left(v_s\right)^{\beta} \qquad (4.4)$$

Where, P/V = energy dissipation rate, vs = superficial gas velocity, k, α, and β = constants depending on bioreactor geometry specifications and medium composition.

4.2.2 Criteria to Scale-Down SDM Parameters

Scale-dependent parameters need to be investigated via small-scale studies to decide the proper scale-down criteria, while the scale-independent parameters remain constant. Among the scale-dependent variables, scale changes associated with agitation and aeration are high risk factors due to differences in vessel geometry, gas sparging and control, liquid surface area to volume ratio, etc. Therefore, they must be evaluated to ensure adequate mixing and aeration.

Power per unit volume (P/V) as a traditional scaling criterion, can be used as a scale-independent factor as compared to the scale-dependent agitation rate (e.g., impeller revolutions per minute). Other criteria such as mixing time, impeller speed (or shear rate), Reynolds number and volumetric mass transfer coefficient ($k_L a$) are less commonly employed since they are not always the limiting factors, or such scaling strategies result in impractical scaled values (20). Sometimes these criteria may be substituted with other parameters based on the same physical principle (e.g., oxygen transfer rate instead of $k_L a$) (28). During the scaling down process, the most important scaling criterion should remain constant or within an acceptable range.

Power per volume (P/V) is a common scaling criterion, usually used for stirred tank processes. P/V defines the amount of power inputted into the liquid volume of bioreactor. P/V strategy aims to provide similar overall shear environment. If P/V is maintained constant, typically maximum tip speed of the impeller increases during scale-up. The decreased pumping per volume during scaling up results in longer mixing times (22). Energy dissipation rate (EDR) can be used to evaluate the scale-down process. It is hypothesized that mammalian cell damage associated with shear sensitivity depends on the local hydrodynamic energy dissipation, cell proximity to the solid or gas-liquid interfaces and cell type (29).

Impeller tip speed is a less common scaling criterion since maintaining consistent impeller tip speed often leads to impractical scale values. Historically, a constant impeller tip speed usually leads to "similar" shear stress or shear rate. However, to maintain a constant tip speed when scale increases, P/V is often reduced which could impact mass transfer efficiency. This can be overcome by applying more impellers in the larger bioreactors to keep both tip speed and P/V constant; but above certain scales, this may not be practical (22). Furthermore, high local energy dissipation rate rather than tip speed has been shown to be more reliable to surrogate for shear stress on cells.

Mixing time (tm) describes the time needed for the composition to reach a certain homogeneity in a mixed bioreactor after the addition of a tracer. Mixing time indicates the characteristics of mixing and flow in a bioreactor and can be used as a criterion during scale-up/scale-down. The mixing time of small-scale bioreactors (typically in terms of seconds) are much shorter than manufacturing scale (typically in terms of minutes) at similar power input (i.e., P/V) (30). Scaling down based on mixing time may result in significant decrease in P/V and thus poor gas transfer.

$$t_m = \frac{V}{N_f NT^3} \tag{4.5}$$

V is liquid volume, N_f is the pumping number, N is impeller speed, T is impeller diameter

Reynolds number (Re) is only occasionally used, but such a scaling strategy fails to consider the impact of aeration. Scaling down based on a constant Re is not applicable to operation in the turbulent flow regime of baffled bioreactors, as power number of the impeller which affects P/V is independent of Re number (30). Other dimensionless numbers are only occasionally used as scaling criteria due to technically impractical equipment design or condition predictions.

Volumetric mass transfer coefficient (kLa) is also used as a scaling criterion to reveal the gas-transfer characteristics of an SDM and provide similar oxygen transfer across scales (24, 31). Choosing k_La as the criterion, the SDM can be tuned to match the k_La of full-scale vessels by changing the impeller speed. However, this will impact mixing conditions and may yield differences in dCO_2 profiles across scales. K_La can be measured experimentally or predicted through empirical correlations. In some case (e.g., cell lines that are sensitive to gas sparge rate and bubble size), a SDM based on the k_La criterion can be substituted with the use of oxygen transfer rate to represent endpoint sparge rate at large scale and address the performance difference across scales. Additionally, it is important to maintain CO2 partial pressure within the range established at large scale, as high CO2 level may impact cell growth, protein production and quality (14, 32). Optimal sparger design and control strategy is crucial to yield sufficient oxygen supply, CO_2 removal, and ultimately ensure comparable performance to large scale. However, it is impossible to maintain a constant criterion as scale increases due to limitations of energy and mass transfer.

Computational fluid dynamics (CFD) can serve as an additional tool to understand the mixing characteristics at different agitation rates. Furthermore, using a CFD approach can provide an estimation or confirmation of an impeller power number (33, 34).

4.2.3 Scale-Down Model Assessment

Once the SDM is established, it must be assessed (i.e., qualified) to determine the comparability of its performance to that of the large-scale process. Procedures can vary across the industry, but the basic principle is to establish that the scale-down model is representative (i.e., predictive) of its full-scale counterparts. The assessment package typically consists of considerations of bioreactor design and evaluations of culture performance (e.g., cell growth, culture viability, product titer) and relevant product quality attributes (4). It is ideal to maintain similar bioreactor design (e.g., geometry and equipment specifications) to the production-scale counterpart (35). In case that proportional duplication is impossible due to equipment limitations, computer based virtual simulations may be considered appropriate and have been approved by FDA to predict input values and facilitate process understanding (36, 37). Statistical tools such as process capability index (Cpk) can be used to demonstrate that bioreactors used for the SDM show comparable variability with their full-scale counterparts.

Two main deliverables are defined for SDMs: process performance and relevant product quality attributes. To determine the appropriate metrics for assessment, important performance parameters, relevant quality attributes and their acceptable ranges need to be defined first. A SDM model should show comparable process performance with commercial-scale processes. In-process performance parameters, such as viable cell density (VCD), culture viability, product titer (and/or other suitable measurements of productivity such as volumetric productivity or cell specific productivity), metabolism indicators (e.g., osmolality, glucose, lactate) and residuals (e.g., host cell proteins, nucleic acids) should be monitored. More importantly, SDM should deliver materials with comparable quality as the corresponding large-scale products. Assays to analyze the quality attributes should be performed in an identical manner across scales. The materials should be tested based on the same in-process and product-release assays and specifications intended for use in commercial manufacturing. Determination of acceptance criteria should rely on scientific knowledge of factors that could potentially impact product quality and specifications as well as statistical distribution at the set-point conditions (16, 38). Bench scale runs should be performed at the center points of each parameter except for those scale-dependent parameters. It is ideal to perform them in parallel with the same operational conditions to avoid differences due to raw materials, culture media, or inoculum sources.

Statistical tools like equivalence tests can be used to compare the outputs, or simpler methods as described below may be applied to check if the SDM results are within the acceptance ranges of the large-scale process. When non-equivalency is observed for a particular data point, justification is needed to identify the potential root causes before concluding that the dataset is inequivalent. To qualify a scale-down model, several statistical tools are available such as the Student's t test, equivalence test, multivariate data analysis (MVA), quality range approach, and risk analysis (14, 15, 39).

The conventional Student's t test is a practical and standard statistical tool to assess the comparability of the SDM dataset against the full-scale profiles. But one should note that the acceptance of the null hypothesis may indicate insufficient data to reject and fail to prove no difference (40). The two one-sided t tests (TOST), also known as the equivalence test, has also been applied to qualify SDMs. The null hypothesis means the difference between two groups is equal or greater than a tolerance interval bounds predefined by manufacturing data. If sufficient statistical evidence rejects the null hypothesis of inequivalence, the two datasets are within a meaningful margin and thus statistically equivalent (41). Outputs such as growth, metabolism, productivity, and product quality may be evaluated by TOST to determine the equivalence of performance statistically on an individual output basis.

Multivariate data analysis (MVA) enables process comparison across scales by using techniques to reduce the dimensionality of large multidimensional datasets which are comprised of different process parameters and product quality attributes. Methodologies such as principal component analysis (PCA) and partial least squares (PLS) analysis are powerful tools to capture interactions between critical process parameters and critical quality attributes within the process design space (42–44). Multivariate modeling is thus more sensitive than univariate methods because it can capture observations that don't fit the predicted response and result in fewer false-positive signals (18). The development of MVA models can be based on the combination of manufacturing and pilot profiles to assess the comparability of the small-scale data. Alternatively, bench-scale data can be utilized to develop MVA models first and then compared with manufacturing dataset. The qualification of SDMs provides confidence in performing process characterization studies to support QbD, as the bench-scale models are shown to adequately represent full scale performance.

4.2.4 Scale-Down Model Limitations

Although SDMs offer the capability to study a variety of process parameters at set-point conditions, and even beyond the normal operating ranges it is sometimes impractical for SDMs to fully represent more expensive and sophisticated large-scale systems. It is important to acknowledge the limitations of a specific SDM during the development process. As the FDA has noted, any differences between small-scale and full-scale process "may have an impact on the relevance of information derived from the models" (4).

It is usually impossible to maintain all scale-related factors simultaneously at exactly the same levels at different scales due to differences in bioreactor design, size, and operation. For example, vessel geometry and shear stress patterns may be difficult to match across scales. Despite the efforts to choose the appropriate scale-down strategy, the resulting agitation rate may not represent the same exact mixing and hydrodynamic features of the large-scale production bioreactor and thus may lead to varying shear/turbulence patterns and mixing times.

In addition, aeration is also difficult to scale down when designing the SDM due to different bioreactor geometry (e.g., liquid surface area-to-volume ratios) and control strategy. For example, the SDM may show lower dCO_2 as compared to large scale if the scaling strategy is not optimized. The variation in CO_2 stripping efficiency is usually attributed to different bioreactor geometry design, for example high surface area to volume ratio, low gas bubble residence time, and high maximum volumetric gas flow rate in small-scale bioreactors. To solve this issue, aeration strategies (e.g., gas sparging and/or overlay gas approaches) typically need to be adjusted to ensure comparable dCO_2 profiles across scales. For example, the sparge gas composition may be modulated to change the driving force associated with the oxygen transfer rate and consequently alter total sparge gas flow and the associated CO_2 stripping effects. Furthermore, an alternative way is to add CO_2 to the sparge gas or to the overlay gas of the small-scale bioreactors to increase dCO_2/reduce CO_2 stripping and mimic large-scale dCO_2 profiles (16). Differences in sparging and agitation strategies may lead to unmatched shear stresses across scales and thus impact the performance especially for shear-sensitive cell lines. CFD models can be valuable in understanding the impact of aeration and shear stress on the mixing characteristics across different scales. They are particularly useful in understanding mechanical shear (e.g., impeller shear) as opposed to bubble shear that would be associated with aeration.

The guidelines and principles governing scale-down of various upstream processes have some common features. The following sections describe examples of processes currently used in industry, focusing on general guidelines and relevant parameters affecting scale-down and primary endpoints used to assess the performance of the SDM. As discussed in ICH Q8(R2), a critical parameter is defined as a process parameter whose variability has an impact on a critical quality attribute and therefore should be monitored or controlled to ensure the process produces the desired quality. All process parameters that fall outside of the definition for critical process parameters are noncritical. Noncritical process parameters may be further subdivided into key and non-key process parameters. A key parameter is an input process parameter that should be controlled within a narrow range and is essential for process performance. A key parameter does not affect critical product quality attributes but may affect performance (such as yield or process time) within the control limits. The characterization of different parameters using risk assessments and process and product knowledge is discussed in detail in ICH Q9 (2b) and PDA TR 42 (13).

4.3 Microbial and Mammalian Upstream Processes

4.3.1 Process Overview

Bioreactors are the key workhorses in upstream cell culture processes. The goal of a successful bioreactor process is to provide an optimal environment for cells to grow exponentially and generate the product of interest, typically in the stationary growth phase. A typical mammalian upstream process comprises vial thaw, seed train passaging and expansion, seed (N-X) bioreactors, and the production bioreactor. Recombinant protein productivity can be influenced by several factors such as the cell line, expression system, media composition, pH, temperature, and dissolved oxygen (DO) levels (45, 46).

Microbes such as yeast and *E. coli* are widely used for production of recombinant proteins due to their robustness and ease of culture. In contrast to the mammalian bioprocess, microbial fermentations typically have short culture durations because their doubling time may be as short as 20–30 minutes. Therefore, it is important to monitor critical process parameters such as pH, DO, cell density, and feed rate as frequently as possible or in real time (47, 48).

4.3.1.1 Vial Thaw

Vial thaw is the first step of the upstream process. Typically, cell vials generated through designated cell banking procedures are thawed via standardized protocols. As the stability of the cell line is the main concern for a robust and consistent process (49, 50), it is important to compare the generation number of the cell banks used during large-scale manufacturing and those used during small-scale process development. As media composition is often optimized for a better process, sometimes the cell banks will be frozen

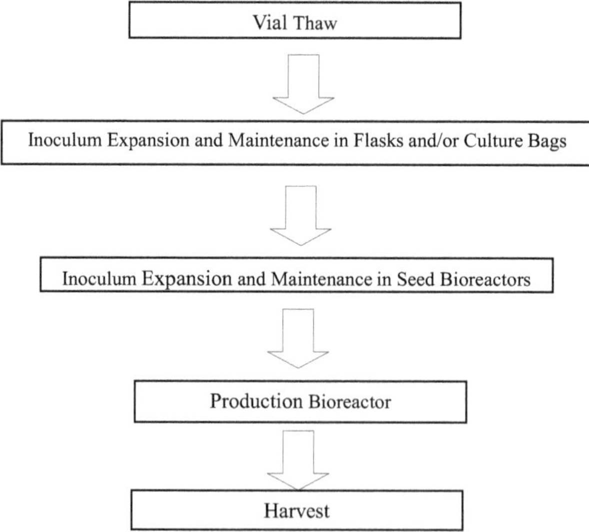

FIGURE 4.1
Overview of mammalian or microbial upstream process.

in a different media. For this reason, it may be important to investigate the effect of media composition on efficiency of vial thaw and any changes to the characteristics of the resulting culture that is used for the remainder of the upstream process.

Compared to mammalian cell bioprocesses, microbial bioprocesses generally require fewer expansion steps since the split ratio is on the order of 1:100 for microbial versus 1:10 for mammalian cultures. Depending on the scale, stirred tank seed bioreactors may not be required for a microbial process. Optical density is the main criteria for evaluating microbial culture performance. In both mammalian and microbial process, the shaker agitation speed and the agitation motion target range needs to be specified. If a different shaker is used, the agitation speed should be calculated accordingly (51).

4.3.1.2 Inoculum Expansion and Maintenance in Flasks and/or Culture Bags

4.3.1.2.1 Flasks and Culture Bags

After cells are successfully thawed and transferred to a small shake flask, the culture is then expanded stepwise to ever larger volume flasks or culture bags. Single-use culture bags can be mixed by one-directional motion rockers. The rocking motion induces waves in the culture to provide efficient mixing and gas transfer. For this reason, parameters such as rocker angle and rock rate are key factors to ensure proper mixing. The rate and composition of the gases in the culture bag overlay can be manipulated to maintain an optimal oxygen transfer rate and appropriate dissolved carbon dioxide level and pH (51). Cell growth and viability are the main criteria for evaluating the culture performance at this stage.

4.3.1.2.2 Seed Bioreactors

Cell cultures are further expanded into larger volume vessels such as stirred tank bioreactors. Depending upon the volume of the final production bioreactor, the expansion phase in N-X bioreactors may involve multiple steps. Most N-X bioreactors are operated in a batch or fed-batch mode, with the addition of some nutrients depending on the specific cell line, initial medium composition, and degree of expansion. Process intensification of the production bioreactor often incorporates increasing the inoculation density, which in turn requires intensification of the N-1 bioreactor (52). In such cases the N-1 culture may require multiple concentrated nutrient feeds, or in case of very high cell densities, continuous perfusion can be used to provide a healthy inoculum source of sufficient density (53).

4.3.1.3 Production Bioreactor

Mammalian cell production bioreactors are typically operated in fed-batch mode with typical culture duration between 10 and 16 days. The product accumulates in the supernatant, and cell and particulate removal is usually performed before further downstream processing. Automated high-throughput small-scale disposable bioreactors have been used to study the effects of a wide range of environmental variable culture performance (54–57). Continuous perfusion bioreactors have also been used for the production of biologics, especially in the case of unstable proteins. In contrast to fed-batch culture, perfusion cell culture can be extended for many weeks or even months by continuously feeding the cells with fresh media while keeping cells in the bioreactor and removing spent media through a cell retention device (58).

For microbial processes, there are three main types: batch, fed batch, and continuous. The process is usually determined based on the product properties. Batch mode is primarily used for native production of the cell host such as polysaccharides from different pathogen hosts. Protein and plasmid production usually require a prolonged fed-batch process for higher yield. A common scale-down model at Pfizer is a 10–30 L fermentation tank instead of 2–5 L bioreactors used for mammalian cultures. In some cases, automated high-throughput small-scale disposable bioreactors are also used to study the effects of a wide range of culture parameters (59). Continuous fermentation is widely used for production of metabolic products such as ethanol (60). Depending on the desired product and choice of hosts, microbial fermentations are performed under either aerobic or anaerobic conditions.

4.3.2 Practical Considerations in the Development of Scale-Down Models

Many vessel parameters such as tubing, geometry, and position of gas pipes play a crucial role in culture performance. These elements, though difficult to maintain with identical geometry and materials of construction between large-scale and small-scale, should be kept as similar as possible. One common example of equipment differences is the use of stainless-steel pipes and vessels in large-scale manufacturing compared to plastic tubing and glass vessels at small-scale. It has been demonstrated that trace metals, such as manganese, leaching from stainless steel surfaces can alter the glycosylation profile of antibodies or other therapeutic proteins produced by mammalian cells in such systems (61). Since such leaching may be different due to materials of construction and surface area to volume ratios, careful selection of the materials used in SDMs may be critical for some molecules.

Procedures for preparation of basal media and nutrient feed for scale-down studies should be according to established protocols used in large-scale manufacturing, since subtle changes in nutrient composition, pH, and osmolarity could lead to altered cell growth profiles. For example, the usage of filters with the proper capacity plays an especially important role in the first passage of cells post vial thaw. Oversized (excess capacity) filters can result in higher levels of leached toxic filter residues that can inhibit cell growth, particularly at low inoculation densities (62). Variability of raw materials can also introduce increased process variability during biologic manufacturing (63). This is especially true for microbial fermentation processes which often use complex media components like yeast extract but is also a reason that many mammalian cultures processes have moved away from ill-defined media additives such as various hydrolysates and other complex peptide mixtures. Even in the case of defined media components, minor variabilities in trace metal levels can significantly impact process performance and product quality (64). In some cases, direct addition of nontoxic trace element solutions can saturate such effects to reduce process variability due to changes in raw material lots.

Care must be taken to ensure that nutrient feed rates match across scales. The working volume doesn't change significantly at the small-scale due to relatively large sampling volumes, whereas sampling volume is essentially negligible at large scale. The working volume of any culture of course increases with additions of base titrant, glucose, and nutrient feeds. The feeding strategy therefore must match the feed rate to the instantaneous working volume of the bioreactor and be consistent upon scale-up/down. While the volume additions of antifoam are typically small at all scales, higher levels of antifoam may be required at large scale to avoid risks of foaming out.

Viable cell density and viability are the two most important indicators of a successful mammalian cell culture process. They can be measured using automated imaging cell counters such as ViCell™ (Beckman Coulter), NovaFlex® (Nova Biomedical), and Cedex Bio® Analyzer (Roche). Accurate and consistent cell counting is crucial for a robust process, especially when key process decisions are based on VCD and/or viability. For example, temperature shifts, feed initiation, and harvest criteria are often based on the VCD or viability (65). Similarly, optical density (OD600), which is a spectrophotometric measurement, is typically used to monitor growth and enable process decisions for microbial processes, including temperature shifts or induction time in the case of protein production in *E. coli*. Therefore, equipment for measuring indicators of mammalian or microbial cell growth should be aligned between small-scale and large-scale facilities.

The gassing strategy can play an important role in cell growth, productivity, and product quality across different scales. It is therefore important to evaluate the dissolved gas profiles as a tool for comparison at different scales, with pCO_2 and pO_2 being the key factors. Obtaining accurate measurements of the dissolved gas levels can be difficult, particularly at large scale or at high cell densities. Factors such as test sample volumes, sample-handling practices, testing devices, and equipment differences can all influence the measured values. The profiles of dissolved gases can provide important insights in cell culture process performance, and harmonizing the sampling strategies at different scales will minimize discrepancies or offsets between the SDM and large-scale manufacturing (66). Similar to mammalian processes, the gassing strategy in microbial processes is crucial to cell growth and productivity and scale differences in these parameters should be minimized. One of the challenges for developing a representative SDM for a microbial process is the different operating pressures between small-scale and large-scale fermentors. The operating pressure will impact the concentration of dissolved O_2, CO_2, and pH, therefore the scale-down model should best mimic the configurations and capacities of the gassing delivery systems at different manufacturing facilities (67).

Mixing is a key parameter in the production process to ensure nutrients and cells are evenly distributed throughout the bioreactor and to control other critical process parameters such as DO, pCO_2, pH, and temperature. The most common bioreactor configuration is the stirred tank, where mixing is created via an impeller mounted on a turning shaft. The configuration of the impeller such as shape, size, location, and the number of impellers can all play a crucial role in generating the desired process profiles (60, 68).

Other parameters that are critical for upstream cell culture processes include basal media composition, temperature, pH (using either a set point or a deadband control strategy), DO set point, agitation, pressure, culture duration, nutrient feed, feed rate and timing, feeding strategy for glucose, supplemental feeds, antifoam, and any additional inducers. Compared to mammalian processes, the process of producing recombinant proteins in *E. coli* requires additional considerations. When the recombinant protein is expressed using a plasmid under the control of an inducible promoter, production is initiated by the addition of an inducer such as isopropyl ß-D-1-thiogalactopyranoside (IPTG). Cell growth and temperature at induction, as well as the inducer concentration, are common parameters evaluated in SDMs for microbial processes with inducible protein production (69, 70).

The successful performance of any cell culture process irrespective of whether it is mammalian or microbial across different scales is typically measured through primary endpoints of growth, product yield, and product quality.

4.4 Advancements in High-Throughput Technologies

Scale-down models are necessary in order to experimentally test process optimization and robustness, as large-scale experiments would be time intensive and costly, as discussed in earlier sections. However, standard scale-down models are still limited by experimental capacity and the resources required to perform experiments. High-throughput technologies allow for larger, more in-depth experimentation at a fraction of the resources and time.

Technologies such as microtiter plates, shake flasks, roller bottles, and tube bioreactors allow for high-throughput experimentation, but key disadvantages limit their ability to accurately match performance at large scale (71–73). These traditional high-throughput technologies do not allow online culture performance monitoring, such as pH and dissolved gases. Additionally, there is limited ability to control parameters such as pH without frequent human manipulation.

Recent advancements and technologies allow for better and more appropriate process parameter control and performance monitoring. Bioreactors such as the DASbox® minibioreactor system (Eppendorf) (74) and the Pharyx microbioreactors (Erbi Biosystems) (75) provide end users with the ability to perform more sophisticated high-throughput experiments. The Sartorius Ambr® systems are available for higher-throughput cell culture studies and include Ambr® 15, Ambr® 250 modular, and Ambr® 250 high-throughput, each with their own benefits and limitations.

The Ambr® 15 is equipped with either 24 or 48 vessels, split into culture stations of 12 vessels each, with a working volume of 10 to 15 milliliters per vessel. The Ambr® 15 provides online pH and dissolved oxygen monitoring and control for each individual vessel through liquid additions into each vessel or gas manipulation with each vessel's sparger. Due to the design of the Ambr® 15, agitation and temperature are controlled per culture station, i.e., a block of 12 vessels. A robotic liquid handler can be programmed with the associated computer and software to provide automated control over the cultures. The liquid handler can operate on a feedback control loop, such as with pH control, or on a set time basis, such as with media supplement additions (76). Many operations can occur without the end user being present, allowing for greater process control to mimic large-scale manufacturing compared with older high-throughput technologies. Two variations of the Ambr® 15, Ambr® 15 cell culture and Ambr® 15 fermentation, tailor the system to cell culture or microbial applications. Studies show process performance in the Ambr® 15 match standard small-scale bioreactors (55, 77–80), supporting the use of this system for cell line screening and early process development.

End users have expanded the use of the Ambr® 15 into perfusion applications. Standard perfusion setups with tangential flow filtration (TFF) or alternating tangential flow filtration (ATF) cannot be adapted into the Ambr® 15 due to equipment restrictions. Many have mimicked a cell retention device using gravity settling (81) or intermittent centrifugation (56). Gravity settling within the Ambr® 15 system allows for partially cell-free media removal, but the operation is limited because of the variable cell density in the liquid removal aliquot and detrimental effects from settling the cells for too long (low DO and pH perturbations). The intermittent vessel centrifugation method to allow for the removal of cell-free media is burdensome and restricted since manual operations are involved.

An advantage of the Ambr® 15 system is its low working volume, decreasing the cost of goods per experimental condition; however, this can become a limiting factor as additional culture is required for analytical assays and downstream support. This is addressed with the introduction of the Ambr® 250 systems.

The Ambr® 250 systems limit the experimental output compared to the Ambr® 15, but the increased working volumes allow for more analytical testing per vessel. The Ambr® 250 modular comes in sets of two vessels, up to eight vessels per system. The Ambr® 250 modular differs in that it is installed on a benchtop, decreasing the amount of laboratory space required, but a disadvantage is the removal of a liquid handler. Without the liquid handler, manual sampling is required, and liquid supplements are loaded into disposable liquid reservoirs and added by syringe pumps. The Ambr® 250 high-throughput is a 12- or 24-vessel system, residing in a biosafety cabinet like the Ambr® 15. Liquid additions are performed by the liquid handler. Both Ambr® 250 models have a working volume range of 100 to 250 milliliters (54, 82). Studies have shown comparable performance between the Ambr® 250 and large-scale bioreactors for both mammalian and microbial cultures (83) and have further shown the ability to use the Ambr® 250 for late-stage process characterization (80, 84, 85).

Sartorius continues to develop their Ambr® systems with recent introductions of perfusion capability within their Ambr® 250 high-throughput model and Raman spectroscopy in their Ambr® 15 and Ambr® 250 high-throughput systems. The Ambr® 250 high-throughput perfusion system allows for individual vessel manipulation of perfusion rates and cell bleeds. Both TFF and ATF applications are available in the perfusion system. The Raman spectroscopy capability, BioPAT® Spectro, coupled with these high-throughput systems allows for fast modeling time, and the models can be transferred for use with Sartorius's Flexsafe STR bags in large-scale manufacturing (86).

4.5 Applications of Scale-Down Models

4.5.1 Overview of SDM Applications

During the early clinical process development phase, it is critical to identify high-producing cell lines or clones, which also satisfy other selection conditions (e.g., product quality, high densities, and viabilities). In general, hundreds to thousands of clones with many cell phenotypes are generated during cell line development, which necessitate the use of SDMs to accelerate evaluation and selection of a manufacturing cell line. For a bioprocess, cell culture medium and feed supplements consist of a wide range of interacting components, which are carefully formulated to meet cellular metabolic demands. Because of a large design space, multidimensional DOE studies (Abu-Absi et al., 2010) are typically performed to allow for an extensive screening and optimization of nutrient concentrations (87). These complex DOE experiments are conducted using various small-scale cultivation systems ranging from 96 deep well plates to miniature bioreactors to reduce time and costs.

SDMs that represent large-scale process performance play an important role in achieving reliable process characterization. They facilitate process understanding by revealing the relationship between control parameters and process performance. For example, Li et al. developed a 2 L bioreactor SDM to represent a 2000-L commercial-scale process (11). The 2 L SDM was qualitatively evaluated by comparing performance across the two scales in terms of growth, productivity, culture environment (e.g., pH, CO_2, DO), metabolite profiles, and product quality. Additionally, the 2 L SDM was used to determine appropriate parameter operating ranges and acceptance criteria. A qualified SDM can be used to evaluate raw materials and new cell banks. To this end, DiCesare et al. successfully developed a 12 L SDM to match the process performance at 2000 L (38). The SDM was statistically qualified by comparing growth, metabolite profiles, productivity, and product quality attributes. Furthermore, the SDM was used to evaluate new GMP working cell banks before implementation on a large scale to ensure the expected performance was achieved with the newly generated cell banks.

Recently, various SDMs (e.g., 24-well plates, shake flasks, and a benchtop bioreactor) were used to develop a robust valinomycin fermentation process (88). Despite the presence of glucose and oxygen oscillatory behavior, the process was robust against these fluctuations and the process performance was not impacted. Furthermore, the microbial fermentation process was successfully scaled up to a 15 L scale, which demonstrated the potential of the valinomycin production process to be transferred to a larger industrial scale. In another landmark microbial fermentation study where SDMs were used to increase process understanding and control, variations in environmental conditions significantly impacted process performance at large scale (89). From this study, it was clear that large-scale performance was mainly impacted by the presence of micro-environmental gradients at large scale. To explore and understand the impact of gradients and their characteristics (e.g., glucose, pH, DO, CO_2) on cellular behavior at large scale, SDMs are typically used. For example, Lara et al. (90) used a two-compartment SDM to simulate gradients at large scale by subjecting *E. coli* expressing a recombinant protein to fluctuating dissolved oxygen conditions. Authors found that exposing cells to changing environments forces cells to adapt to different conditions by changing their metabolism at both molecular (transcription) and macroscopic (e.g., kinetic and stoichiometric parameters, by-product formation, etc.) levels. Additionally, several studies have been performed in SDMs to understand the impact of gradients and stresses on process performance in various organisms ranging from filamentous fungi to mammalian cell culture (91–95).

Despite the development of a robust manufacturing process that consistently produces a drug substance with desired quality attributes, process deviations are common at large scale. In general, process deviations that occur in a manufacturing process are also reproduced in satellite runs, which indicates the degree to which SDMs predict and represent a commercial-scale process (96). SDMs are used to perform root-cause analysis when process deviations occur. For example, lot-to-lot or vendor-to-vendor variability in raw materials could cause detrimental nutrient imbalances for cell culture leading to process excursions and unexpected quality attributes. SDMs enable understanding of such material attributes and their impact on product quality and process. In addition, leaching of trace metals and chemicals from stainless steel vessels and single-use vessels, respectively, can cause process deviations by significantly impacting cell culture behavior and product quality attributes. To troubleshoot how metal or chemical leaching can be controlled to ensure desired process performance, multidimensional experiments are performed using SDMs, which further increase understanding of the process control strategies.

4.5.2 Selected Case Studies

4.5.2.1 *Evaluation of Scale-Down Models for Mammalian Cell Culture Processes*

The goal of this study was to evaluate the performance of various SDMs and assess how they could enable rapid development of a robust manufacturing process in our workflow. In this work, miniature bioreactor systems (Ambr® 250 and Ambr® 15) were used to assess the extent to which they could mimic manufacturing processes of a therapeutic protein (therapeutic proteins one; TP1). Ambr® 250 does not only allow for high-throughput upstream process development, but it also generates more material that is required for downstream process development than Ambr® 15. These Ambr® 250 and Ambr® 15 SDMs were compared to their previously qualified 2 L benchtop counterpart as well as large-scale bioreactor (i.e., 1000 L XDR single-use bioreactor). While either qualitative or statistical assessments (or both) are typically conducted to evaluate the extent to which SDMs represent and predict cell culture performance at full scale (96–98), here we used a qualitative method to assess the performance of SDMs for simplicity. Specifically, we compared process parameters such as growth, production, metabolite profiles, culture conditions (e.g., pH, DO, and CO_2), and product quality attributes. As shown in Figure 4.2, Ambr® 250 and Ambr® 15 miniature bioreactor systems perform comparably to their 2 L benchtop counterpart. In addition, performances of SDMs were also validated at a 1000 L large-scale bioreactor. Other parameters such as metabolite profiles, culture conditions (e.g., pH, DO, and CO_2), and product quality attributes were also comparable across the four scales evaluated (data not shown). These qualified SDMs have the potential to streamline and accelerate our process development workflow, which will enable us to rapidly bring therapeutic breakthroughs that change patients' lives.

4.5.2.2 Qualification of a Lab-Scale Model for Microbial Process Characterization

As described earlier, the use of SDMs to support process characterization requires scientific justification that the SDM is representative of the large-scale manufacturing process. In a microbial process example, the fermentation step was scaled down ~73 fold from the commercial scale using the 30 L Biostat® Dfermentor (Sartorius). In order to qualify the SDM, attributes that impacted the fermentation step were classified as primary attributes while those that did not were deemed as secondary attributes. In this example, titer, optical density, and percent solids at the time of harvest were identified as primary attributes. These primary attributes were compared between the SDM and the relevant manufacturing-scale (2200 L) process history. Secondary attributes were tracked to ensure consistent performance but were not fundamental to the qualification of the SDM.

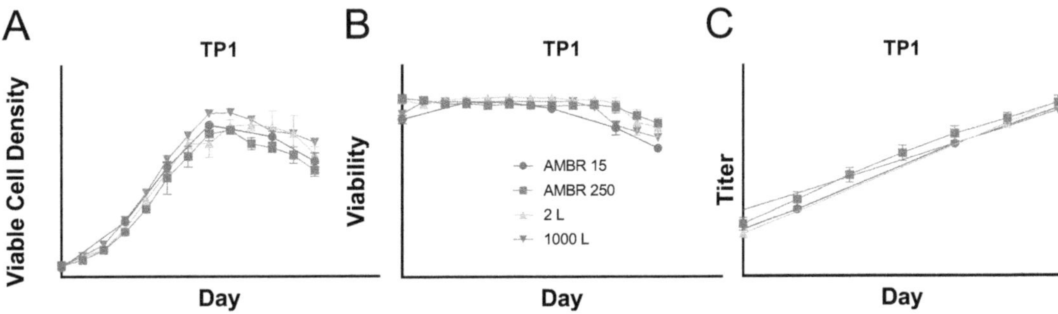

FIGURE 4.2
SDM development for a therapeutic protein (TP1) manufacturing process. Qualitative analysis of process performance across four various scales: (1) Ambr® 15 (filled circles; $n = 2$); (2) Ambr® 250 (filled squares; $n = 3$); (3) 2 L benchtop bioreactor (filled diamonds; $n = 2$); and (4) 1000 L XDR single-use bioreactor (filled triangles; $n = 1$). Three process parameters are described above: (A) viable cell density (growth); (B) cell viability; and (C) production. Other process parameters (e.g., metabolite profiles, culture conditions (pH, DO, CO2), and product quality attributes) are also comparable across scales (data not shown). Data (2 L bioreactor, $n = 2$, Ambr15®, $n = 2$, and Ambr250®, $n = 3$) are averages of indicated replicates and error bars denote a standard deviation of one.

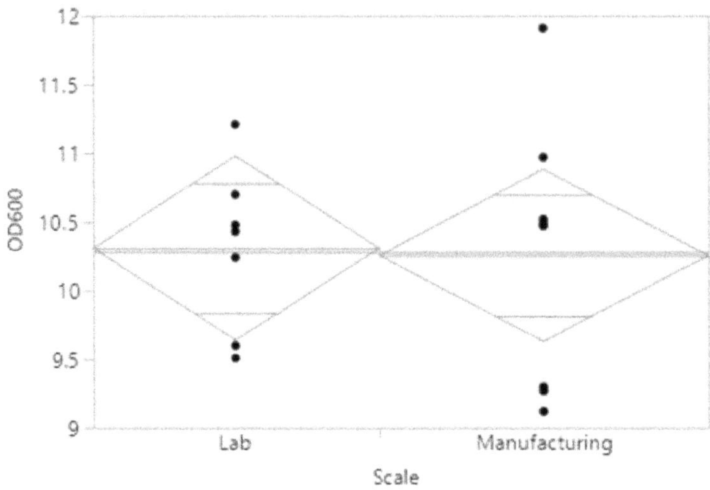

FIGURE 4.3
SDM qualification for a microbial fermentation process. Optical density (OD600 nm) at harvest was evaluated between the lab scale and manufacturing scale.

A comparison of the optical densities (OD600 nm) obtained at harvest from the scale-down model and the relevant manufacturing process history are shown in Figure 4.3 (results for titer and % solids not shown). The qualification criterion for optical density was evaluated using an ANOVA with a 95% confidence interval to determine statistical significance between scales. The optical density of the lab-scale center points proved to have no significant difference compared to the manufacturing scale, as seen in Figure 4.3 (p-value probability = 0.9012). Because there was no significant difference between scales, the lab scale was qualified on this basis.

4.5.2.3 Use of Scale-Down Models to Investigate Deviations at Manufacturing Scale

SDMs can also support troubleshooting of large-scale processes and investigation of process or scale-dependent deviations. For example, Williamson et al. (61) developed an SDM based on Mobius 3 L single-use vessels (Millipore Sigma) to meet the target glycan profile at both pilot and commercial scales. Significant scale-dependent discrepancies in the terminal galactosylation profile of a therapeutic protein were found using 2 L glass bioreactors and 100 L and 6000 L stainless steel bioreactors (Figure 4.4A). Through an extensive fishbone cause-and-effect analysis, the root cause was determined to be varying degrees of manganese leaching from the stainless-steel surfaces in the bioreactors at different scales. The 2 L glass SDM showed a higher degree of exposure to stainless steel surface and greater levels of terminal glycosylation (61). Single-use vessels were employed instead to minimize stainless steel surface exposure, which resulted in comparable levels of terminal galactosylation to their large-scale counterparts (Figure 4.4B). These findings reveal the impact of metal leaching on cell culture processes and highlight the importance of SDMs on improving process understanding and control.

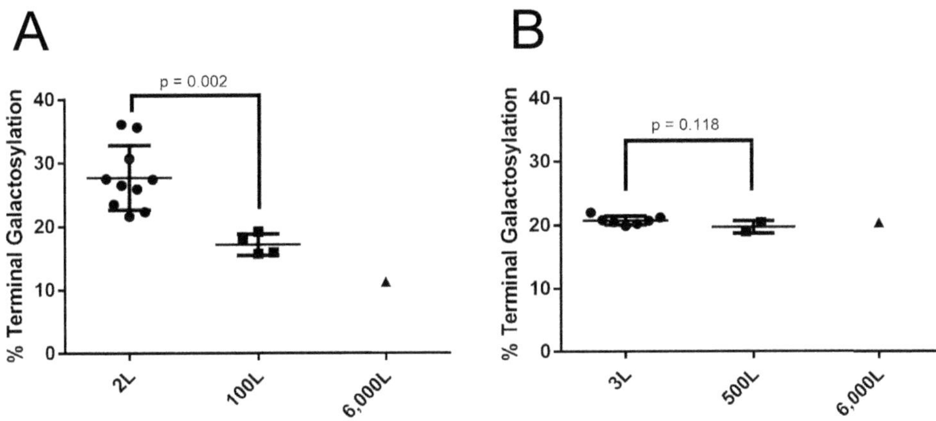

FIGURE 4.4

Using SDMs to improve process understanding and control. (A) Variability of N-glycan terminal galactosylation levels in therapeutic protein at different scales of production (2 L, n = 10; 100 L, n = 4; 6000 L, n = 1), mean with one standard deviation. Significant difference observed between 2 L and 100 L scale, P = 0.002, statistical analysis not run on 6000-L due to n = 1. (B) Variability of N-glycan terminal galactosylation levels in therapeutic protein produced at different scales (3 L, n = 7; 500 L, n = 2; 6000 L, n = 1), mean with one standard deviation. No significant difference observed between an SDM and 500 L scale, P = 0.118, statistical analysis not run on 6000 L due to n = 1. Note: Figures modified from (61).

4.6 Summary

In the biopharmaceutical industry, SDMs are indispensable and powerful tools that are employed to rapidly develop and support robust manufacturing processes. These SDMs are particularly used to enhance process understanding and control through systematic characterization of the design space, linking critical quality attributes to process parameters and identifying both key process parameters and critical process parameters. Once SDMs are scientifically justified, they are typically designed to either fully or partially mimic a drug manufacturing process at laboratory scale. During qualification, process scientists typically compare SDM performance to its large-scale counterpart using qualitative and quantitative methods. Data from a small-scale model that fully represents large-scale does not only provide assurance that the design space selected in the development stage will result in a predictable and reliable manufacturing process, but it also demonstrates the potential to use SDMs to investigate large-scale process excursions. Additionally, this provides confidence to use data from such representative SDMs throughout product life cycle as promoted by regulatory agencies.

SDMs have various applications in upstream process development ranging from selection of high-producing cell lines to troubleshooting manufacturing processes. Specifically, they are used to screen cell lines, develop media and feeds, establish ranges of raw materials, equipment, parameters, and in-process material quality attributes, and investigate deviations at large scale. In this chapter, we have further provided examples of SDM use cases for both microbial and mammalian cell culture processes across various scales (Figures 4.2–4.4). These case studies demonstrate the power and effectiveness of SDMs to predict process performance at manufacturing scale.

In summary, we have provided a general perspective on the scale-down or scale-up principles, scale-down parameters, and primary endpoints used in the design of SDMs for a variety of upstream unit operations for both mammalian and microbial manufacturing processes.

Acknowledgments

We thank Greg Hiller and Ranga Godavarti for helpful comments and Steve Sobacke, Derrick Ruble, and Adine Hickey for their contributions to the book chapter.

References

1. FDA Guidance for Industry, *Process Validation: General Principles and Practices (Draft Guidance)*. U.S. Food and Drug Administration, 2008.
2. US Food and Drug Administration, *Points to Consider in the Manufacture and Testing of Monoclonal Antibody Products for Human Use*. Center for Biologics Evaluation and Research, FDA, 1997.
3. Center for Biologics Evaluation and Research, FDA, *Guidance for Industry for the Submission of Chemistry, Manufacturing and Controls for a Therapeutic Recombinant DNA-Derived Product or Monoclonal Antibody Product for In Vivo Use*. US FDA, 1996.
4. Guidance for Industry, *Process Validation: General Principles and Practices*, U.S.D.o.H.a.H. Services, et al., Editors. US FDA, 2011.
5. Center for Biologics Evaluation and Research, FDA, *Guideline on General Principles of Process Validation*. US FDA, 1987.
6. Swann, S. K. A. P., Considerations for biotechnology product quality by design. In *Quality by Design for Biopharmaceuticals: Principles and Case Studies*, A. S. Rathore and R. Mhatre, Editors. John Wiley & Sons, Inc., 2009.
7. Center for Drug Evaluation and Research, Center for Biologics Evaluation and Research, *FDA/ICH, Q8(R2) Pharmaceutical Development*. ICH Guidance for Industry, August 2009.
8. Center for Drug Evaluation and Research, Center for Biologics Evaluation and Research, *FDA/ICH, Q9 Quality Risk Management*. ICH Guidance for Industry, November 2005.
9. Center for Drug Evaluation and Research, Center for Biologics Evaluation and Research, *FDA/ICH, Q10 Pharmaceutical Quality Systems*. ICH Guidance for Industry, May 2007.
10. Center for Drug Evaluation and Research, Center for Biologics Evaluation and Research, *FDA/ICH, Q11 Development and Manufacture of Drug Substances*. ICH Guidance for Industry, November 2012.
11. Li, F., et al. *A systematic approach for scale-down model development and characterization of commercial cell culture processes*. Biotechnol Prog, 2006. **22**(3): p. 696–703.
12. Seely, J. E. S. R., *A rational, stepwise approach to process characterization*. BioPharm Int, 2023. **16**(8), p. 24–34.
13. *Guidance for Industry: Statistical Approaches to Establishing Bioequivalence*, U.S.D.o.H.a.H. Services, F.a.D. Administration, and C.f.D.E.a.R. (CDER), Editors. 2001.
14. Tescione, L., et al., *Application of bioreactor design principles and multivariate analysis for development of cell culture scale down models*. Biotechnol Bioeng, 2015. **112**(1): p. 84–97.
15. Li, F., H. Yasunori, Robert Pendleton, Jean Harms, Erin Collins, and Brian Lee, *A systematic approach for scale-down model development and characterization of commercial cell culture processes*. Biotechnol. Prog., 2006. **22**: p. 696–703.

16. Harmsen, M., J. Stofferis, and L. Malphettes, *Development and fine-tuning of a scale down model for process characterization studies of a monoclonal antibody upstream production process.* BMC Proceedings, 2011. **5**(Suppl 8): p. P70.

17. Farrell, P., et al., *Development of a scaled-down aerobic fermentation model for scale-up in recombinant protein vaccine manufacturing.* Vaccine, 2012. **30**(38): p. 5695–8.

18. Group, C. B. W., *A-Mab: A case study in bioprocess development.* International Society for Pharmaceutical Engineering, 2009.

19. Marks, D. M., *Equipment design considerations for large scale cell culture.* Cytotechnology, 2003. **42**: p. 21–33.

20. Nienow, A. W., *Reactor engineering in large scale animal cell culture.* Cytotechnology, 2006. **50**(1–3): p. 9–33.

21. Shimoni, Y., G. Chetan, Marc Jenne, and Venkatesh Srinivasan, *Qualification of scale-down bioreactors: Validation of process changes in commercial production of animal-cell-derived products, part 1—concept.* BioProcess Int, 2014. **12**: p. 38–45.

22. Ju, L. K., and G. G. Chase, *Improved scale-up strategies of bioreactors.* Bioprocess Eng, 1992. **8**(1): p. 49–53.

23. Schmelzer, A. E., and W. M. Miller, *Hyperosmotic stress and elevated $pCO2$ alter monoclonal antibody charge distribution and monosaccharide content.* Biotechnol Prog, 2002. **18**(2): p. 346–53.

24. Garcia-Ochoa, F., and E. Gomez, *Bioreactor scale-up and oxygen transfer rate in microbial processes: An overview.* Biotechnol Adv, 2009. **27**(2): p. 153–76.

25. Burke, F., Scale up and scale down of fermentation processes. In *Practical Fermentation Technology*, B. H. McNeil and M. Linda, Editors. John Wiley & Sons, 2008. p. 231–69.

26. Doran, P. M., *Bioprocess Engineering Principles.* Academic Press, 2013.

27. Bandyopadhyay, B., A. E. Humphrey, H. Taguchi, and Rao Govind, *Dynamic measurement of the volumetric oxygen transfer coefficient in fermentation systems.* Biotechnol Bioeng, 2009. **104**(5): p. 841–53.

28. Venkat, R. V., and J. J. Chalmers, *Characterization of agitation environments in 250 ml spinner vessel, 3 L, and 20 L reactor vessels used for animal cell microcarrier culture.* Cytotechnology, 1996. **22**(1): p. 95–102.

29. Chalmers, J. J., *Cells and bubbles in sparged bioreactors.* Cytotechnology, 1994. **15**(1): p. 311–20.

30. Marques, M. P. C. C., M. S. Joaquim, and Pedro Fernandes, *Bioprocess scale-up: Quest for the parameters to be used as criterion to move from microreactors to lab-scale.* Chem Technol Biotechnol, 2010. **85**(9): p. 1184–98.

31. Alam, H. Z. A. R., *Firdausi, scale-up of stirred and aerated bioengineeringTM bioreactor based on constant mass transfer coefficient.* Jurnal Teknologi. **43**(F): p. 95–110.

32. Zhu, M. M., et al., *Effects of elevated $pCO2$ and osmolality on growth of CHO cells and production of antibody-fusion protein B1: A case study.* Biotechnol Prog, 2005. **21**(1): p. 70–77.

33. Kelly, W. J., *Using computational fluid dynamics to characterize and improve bioreactor performance.* Biotechnol Appl Biochem, 2008. **49**(Pt 4): p. 225–38.

34. Fang, Z., *Applying computational fluid dynamics technology in bioprocesses-part 2.* BioPharm Int, 2010. **23**(5).

35. Gardner, A. S. T., *Identification and establishment of operating ranges of critical process variables.* In *Biopharmaceutical Process Validation*, G. Z. Sofer, Editor. New York, NY: Marcel-Dekker, 2000.

36. Heath, C., and K. Robert, *Cell culture process development: Advances in process engineering.* Biotechnol Prog, 2007. **23**(1): p. 46–51.

37. McKnight, N., Scale-down model qualification and down model qualification and use in process characterization. CMC Strategy Forum, 2013.

38. DiCesare, C., Y. Marcella, Jin Yin, Weichang Zhou, Chris Hwang, Jennifer Tengtrakool, Konstantin Konstantinov, *Development, qualification, and application of a bioreactor scale-down process: Modeling large-scale microcarrier perfusion cell culture.* BioProcess Int, 2016. **14**: p. 18.

39. Tsang, V. L., et al., *Development of a scale down cell culture model using multivariate analysis as a qualification tool.* Biotechnol Prog, 2014. **30**(1): p. 152–60.

40. Moran, E. B., et al., *A systematic approach to the validation of process control parameters for monoclonal antibody production in fed-batch culture of a murine myeloma.* Biotechnol Bioeng, 2000. **69**(3): p. 242–55.

41. Barker, L. E. L., T. Elizabeth, Mary M. McCauley, and Susan Y. Chu, *Assessing equivalence: An alternative to the use of difference tests for measuring disparities in vaccination coverage.* Am J Epidemiol, 2002. **156**(11): p. 1056–61.

42. Pieracci, J. Y.-M., Leveraging multivariate analysis tools to qualify scaled-down models, in *Process Validation in Manufacturing of Biopharmaceuticals*, A. S. S. Rathore, Editor. Boca Raton, FL: CRC Press, 2012. p. 411–40.

43. Gabrielsson, J., L. Nils-Olof, and Torbjörn Lundstedt, *Multivariate methods in pharmaceutical applications.* J Chemometr, 2002. **16**(3): p. 141–60.

44. Xing, Z., L. Zhengjian, Vincent Chow, and Steven S. Lee, *Identifying inhibitory threshold values of repressing metabolites in CHO cell culture using multivariate analysis methods.* Biotechnol Prog, 2008. **24**(3): p. 675–83.

45. Rezaei, M., S. H. Zarkesh-Esfahani, and M. Gharagozloo, *The effect of different media composition and temperatures on the production of recombinant human growth hormone by CHO cells.* Res Pharmaceut Sci, 2013. **8**(3): p. 211–17.

46. Li, F., et al., *Cell culture processes for monoclonal antibody production.* mAbs, 2010. **2**(5): p. 466–77.

47. Crater, J. S., and J. C. Lievense, *Scale-up of industrial microbial processes.* FEMS Microbiol Lett, 2018. **365**(13).

48. Delvigne, F., and H. Noorman, *Scale-up/Scale-down of microbial bioprocesses: A modern light on an old issue.* Microbial biotechnology, 2017. **10**(4): p. 685–7.

49. Xu, J., et al., *Improving titer while maintaining quality of final formulated drug substance via optimization of CHO cell culture conditions in low-iron chemically defined media.* mAbs, 2018. **10**(3): p. 488–99.

50. Porter, A.J., et al., *Strategies for selecting recombinant CHO cell lines for cGMP manufacturing: Improving the efficiency of cell line generation.* Biotechnol Prog, 2010. **26**(5): p. 1455–64.

51. Klöckner, W., and J. Büchs, *Advances in shaking technologies.* Trends in Biotechnology, 2012. **30**(6): p. 307–14.

52. Yongky, A., et al., *Process intensification in fed-batch production bioreactors using non-perfusion seed cultures.* MAbs, 2019. **11**(8): p. 1502–14.

53. Xu, J., et al., *Development of an intensified fed-batch production platform with doubled titers using N-1 perfusion seed for cell culture manufacturing.* Bioresources and Bioprocessing, 2020. **7**(1): p. 17.

54. *Ambr 250 High Throughput.* 2020 [cited 2020 August 19, 2020]; Available from: www.sartorius.com/en/products/fermentation-bioreactors/ambr-multi-parallel-bioreactors/ambr-250-high-throughput.

55. Janakiraman, V., et al., *Application of high-throughput mini-bioreactor system for systematic scaledown modeling, process characterization, and control strategy development.* Biotechnol Prog, 2015. **31**(6): p. 1623–32.

56. Gagliardi, T. M., et al., *Development of a novel, high-throughput screening tool for efficient perfusionbased cell culture process development.* Biotechnol Prog, 2019. **35**(4): p. e2811.

57. Łącki, K. M., *High throughput process development in biomanufacturing.* Current Opinion in Chemical Engineering, 2014. **6**: p. 25–32.

58. Tapia, F., et al., *Bioreactors for high cell density and continuous multi-stage cultivations: Options for process intensification in cell culture-based viral vaccine production.* Applied Microbiology and Biotechnology, 2016. **100**(5): p. 2121–32.

59. Velez-Suberbie, M. L., et al., *High throughput automated microbial bioreactor system used for clone selection and rapid scale-down process optimization.* Biotechnology Progress, 2018. **34**(1): p. 58–68.

60. 14—Fermentation Technologies, *Brewing.* In D. E. Briggs, et al., Editors. Woodhead Publishing, 2004. p. 509–42.

61. Williamson, J., et al., *Scale-dependent manganese leaching from stainless steel impacts terminal galactosylation in monoclonal antibodies.* Biotechnol Prog, 2018. **34**(5): p. 1290–7.

62. Zhu, M. M., et al., *Industrial production of therapeutic proteins: Cell lines, cell culture, and purification.* Handbook of Industrial Chemistry and Biotechnology, 2017: p. 1639–69.

63. Vulto, A. G., and O. A. Jaquez, *The process defines the product: What really matters in biosimilar design and production?* Rheumatology (Oxford, England), 2017. **56**(suppl_4): p. iv14–iv29.

64. Graham, R. J., H. Bhatia, and S. Yoon, *Consequences of trace metal variability and supplementation on Chinese hamster ovary (CHO) cell culture performance: A review of key mechanisms and considerations.* Biotechnol Bioeng, 2019. **116**(12): p. 3446–56.

65. Kim, Y. J., et al., *Quality by design characterization of the perfusion culture process for recombinant FVIII.* Biologicals, 2019. **59**: p. 37–46.

66. Betts, J. P. J., et al., *Impact of aeration strategies on fed-batch cell culture kinetics in a single-use 24-well miniature bioreactor.* Biochemical Engineering Journal, 2014. **82**: p. 105–16.

67. Hewitt, C. J., and A. W. Nienow, The scale-up of microbial batch and fed-batch fermentation processes. In *Advances in Applied Microbiology*. Academic Press, 2007. p. 105–35.

68. Clapp, K. P., A. Castan, and E. K. Lindskog, Chapter 24—upstream processing equipment. In *Biopharmaceutical Processing*, G. Jagschies, Editor. Elsevier, 2018. p. 457–76.

69. Larentis, A. L., et al., *Evaluation of pre-induction temperature, cell growth at induction and IPTG concentration on the expression of a leptospiral protein in E. coli using shaking flasks and microbioreactor.* BMC Research Notes, 2014. **7**(1): p. 671.

70. Briand, L., et al., *A self-inducible heterologous protein expression system in Escherichia coli.* Scientific Reports, 2016. **6**: p. 33037.

71. Bareither, R., and D. Pollard, *A review of advanced small-scale parallel bioreactor technology for accelerated process development: Current state and future need.* Biotechnol Prog, 2011. **27**(1): p. 2–14.

72. Huang, Y. M., and C. Kwiatkowski, *The role of high-throughput mini-bioreactors in process development and process optimization for mammalian cell culture.* Pharmaceutical Bioprocessing, 2015. **3**(6): p. 397–410.

73. Kim, B.J., J. Diao, and M. L. Shuler, *Mini-scale bioprocessing systems for highly parallel animal cell cultures.* Biotechnol Prog, 2012. **28**(3): p. 595–607.

74. Schomberg, M., and C. M. Huether-Franken, *Downscaling in human cell culture.* G.I.T. Laboratory Journal, 2012. **16**: p. 22–23.

75. Mozdzierz, N. J., et al., *A perfusion-capable microfluidic bioreactor for assessing microbial heterologous protein production.* Lab Chip, 2015. **15**(14): p. 2918–22.

76. *Introducing Ambr15 Cell Culture.* 2020 [cited 2020 August 19, 2020]; Available from: www.sartorius.com/en/products/fermentation-bioreactors/ambr-multi-parallel-bioreactors/ambr-15-cell-culture.

77. Hsu, W. T., et al., *Advanced microscale bioreactor system: A representative scale-down model for bench-top bioreactors.* Cytotechnology, 2012. **64**(6): p. 667–78.

78. Moses, S., et al., *Assessment of AMBR^{TM} as a model for high-throughput cell culture process development strategy.* Advances in Bioscience and Biotechnology, 2012. **03**(07): p. 918–27.

79. Nienow, A. W., et al., *The physical characterisation of a microscale parallel bioreactor platform with an industrial CHO cell line expressing an IgG4.* Biochemical Engineering Journal, 2013. **76**: p. 25–36.

80. Sandner, V., et al., *Scale-down model development in ambr systems: An industrial perspective.* Biotechnol J, 2019. **14**(4): p. e1700766.

81. Sewell, D. J., et al., *Enhancing the functionality of a microscale bioreactor system as an industrial process development tool for mammalian perfusion culture.* Biotechnol Bioeng, 2019. **116**(6): p. 1315–25.

82. *Ambr 250 Modular.* 2020 [cited 2020 August 19, 2020]; Available from: www.sartorius.com/en/products/fermentation-bioreactors/ambr-multi-parallel-bioreactors/ambr-250-modular.

83. Bareither, R., et al., *Automated disposable small scale reactor for high throughput bioprocess development: A proof of concept study.* Biotechnol Bioeng, 2013. **110**(12): p. 3126–38.

84. Manahan, M., et al., *Scale-down model qualification of ambr(R) 250 high-throughput mini-bioreactor system for two commercial-scale mAb processes.* Biotechnol Prog, 2019. **35**(6): p. e2870.

85. Xu, P., et al., *Characterization of TAP Ambr 250 disposable bioreactors, as a reliable scale-down model for biologics process development.* Biotechnol Prog, 2017. **33**(2): p. 478–89.

86. *Ambr 250 High Throughput Perfusion.* 2020 [cited 2020 August 19, 2020]; Available from: www.sartorius.com/en/products/fermentation-bioreactors/ambr-multi-parallel-bioreactors/ambr-250-high-throughput-perfusion#id-33520.

87. Abu-Absi, S. F., et al., *Defining process design space for monoclonal antibody cell culture.* Biotechnol Bioeng, 2010. **106**(6): p. 894–905.

88. Li, J., et al., *Scale-up bioprocess development for production of the antibiotic valinomycin in Escherichia coli based on consistent fed-batch cultivations.* Microb Cell Fact, 2015. **14**: p. 83.

89. Bylund, F. E. C., S.-O. Enfors, and G. Larsson, *Substrate gradient formation in the large-scale bioreactor lowers cell yield and increases by-product formation.* Bioprocess Eng 1998(18): p. 171–80.

90. Lara, A. R., et al., *Transcriptional and metabolic response of recombinant Escherichia coli to spatial dissolved oxygen tension gradients simulated in a scale-down system.* Biotechnol Bioeng, 2006. **93**(2): p. 372–85.

91. Maria Papagianni, M. M., and Bjorn Kristiansen, *Design of a tubular loop bioreactor for scaleup and scale-down of fermentation processes.* Biotechnol Prog 2003. **19**: p. 1498–504.

92. Nienow, Alvin W. H. S., J. Hewitt Christopher, Colin R. Thomas, Gareth Lewis, Ashraf Amanullah, Robert Kiss, and Steven J. Meier, *Scale-down studies for assessing the impact of different stress parameters on growth and product quality during animal cell culture.* Chemical Engineering Research and Design, 2013. **91**: p. 2265–74.

93. Nasution, U., W. M. van Gulik, A. Proell, W. A. van Winden, and J. J. Heijnen, *Generating short-term kinetic responses of primary metabolism of Penicillium chrysogenum through glucose perturbation in the bioscope mini reactor.* Metab Eng 8, 2006. **5**: p. 395–405.

94. Nasution, U., et al., *Generating short-term kinetic responses of primary metabolism of Penicillium chrysogenum through glucose perturbation in the bioscope mini reactor.* Metab Eng, 2006. **8**(5): p. 395–405.

95. Lorantfy, B., M. Jazini, and C. Herwig, *Investigation of the physiological response to oxygen limited process conditions of Pichia pastoris Mut(+) strain using a two-compartment scale-down system.* J Biosci Bioeng, 2013. **116**(3): p. 371–9.

96. Hakemeyer, C., et al., *Near-infrared and two-dimensional fluorescence spectroscopy monitoring of monoclonal antibody fermentation media quality: Aged media decreases cell growth.* Biotechnol J, 2013. **8**(7): p. 835–46.

97. Li, F., et al., *A systematic approach for scale-down model development and characterization of commercial cell culture processes.* Biotechnol Prog, 2006. **22**(3): p. 696–703.

98. Cao, Y., et al., *Evaluating manufacturing process profile comparability with multivariate equivalence testing: Case study of cell-culture small scale model transfer.* Biotechnol Prog, 2018. **34**(1): p. 187–95.

5

Scale-Down Models for Purification Processes: Approaches and Applications

Arch Creasy, Brenda Carrillo Conde, Frank Kotch, Russ Shpritzer, Stephen Kolodziej, and Ranga Godavarti

5.1 Introduction

A complete process validation package is a major component of any regulatory filing. Regulatory agencies have published general guidelines to aid in developing validation strategies (Center for Biologics Evaluation and Research, 1996, 1997). The process validation package consists of systematic documentation of protocols, reports, and results from well-planned studies. Key to successful process validation studies is ensuring strong scientific rationale while maintaining current good manufacturing practice (CGMP) compliance.

The US Food and Drug Administration (FDA) has defined process validation as "establishing documented evidence which provides a high degree of assurance that a specific process will consistently produce a product meeting its predetermined specifications and quality characteristics" (US FDA 1987). A revised guidance was published by the FDA in 2011 that defined process validation as "the collection and evaluation of data, from the process design stage through commercial production, which establishes scientific evidence that a process is capable of consistently delivering quality product" (US FDA 2011). Although the general definition is similar, the 2011 guidance differs significantly from the guidance from 1987. The earlier guidance emphasized the importance of data obtained from commercial-scale validation batches. The recent guidance describes validation occurring in three stages over the life cycle of a product. The first stage is process design, which is based on knowledge gained through development and scale-up activities. The second stage is process qualification, where the process design is confirmed as being capable of reproducible commercial manufacturing. The third stage is continued process verification (Cecchini 2008; Kozlowski and Swann 2008; Godavarti et al. 2005; Rathore, Kateja, and Kumar 2018; Subramanian 2017; Jungbauer 2013).

The 2011 Process Validation Guidance issued by FDA indicates that the assurance of the performance of the manufacturing process can be obtained from information and data from commercial- and laboratory-scale studies. For example, scale-down models can be used to estimate process and product variability using a wider range of input conditions than would be feasible to evaluate at commercial scale (Center for Biologics Evaluation and Research 1997; Kozlowski and Swann 2008). Typically, a risk assessment is performed following the guidelines of ICH-Q9 to prioritize the input parameters and ranges that will be tested (Kozlowski and Swann 2008; FDA/ICH 2023). Design of experiment (DOE) studies can help maximize the knowledge gained from a limited number of experiments and reveal relationships between the various inputs and outputs (Cecchini 2008; Banerjee 2010; Kozlowski and Swann 2008). The underlying principles of quality by design (QbD) and risk management are contained in ICH Q8(R2), Q9, and Q10 (FDA/ICH 2009a, 2009b, 2023; CMC Biotech Working Group 2009). The guidance document also indicates that data from computer-based or virtual simulations and experience with sufficiently similar products and processes can be used to predict performance of commercial-scale processes (Kozlowski and Swann 2008; Smiatek, Jung, and Bluhmki 2020; Hahn et al. 2015; Velayudhan 2014).

The application of scale-down models is essential to deliver a complete process validation package in a regulatory filing. A comprehensive summary of the purpose, development, and content of a process validation package is given in the introduction of Chapter 4. This chapter is a continuation of Chapter 4 with a focus on downstream purification processes and their appropriate scale-down models. There are several reasons why scale-down models are necessary for purification processes. In the case of validation of virus inactivation/removal by a purification process, use of appropriate scale-down models prevents the introduction of virus in manufacturing facilities. In addition to being used in viral clearance studies, scale-down models have been used to evaluate the removal of host cell-derived impurities such as nucleic acids and host cell proteins, removal of media additives, and the useful life of chromatographic resins. For licensed processes, scale-down models play an important role in supporting process changes, establishing process comparability, and supporting manufacturing investigations. Many of these examples were discussed in the previous review (Godavarti et al. 2005). In this chapter, a few more approaches to applications of scale-down models are provided.

Figure 5.1 shows a flow diagram for a typical manufacturing process. The fermentation or bioreactor process involves the addition of multiple media additives, which will need to be removed by the purification process. The cells are separated from the conditioned medium through centrifugation, microfiltration, depth filtration, or other cell removal techniques. For a process where the protein of interest is expressed intracellularly, a cell disruption step is typically used followed by a refolding step, if the expressed protein is insoluble. If the expressed protein is extracellular, the cell-free fluid may be concentrated and diafiltered prior to loading the capture purification step. The purification process typically consists of a capture chromatographic column followed by multiple purification/polishing steps. The process could also include viral inactivation steps such as low pH incubation and/or solvent-detergent addition. A nanofiltration step may be included to provide additional viral clearance. The product is finally concentrated and diafiltered into the formulation buffer, followed by sterile filtration to generate active bulk drug substance.

A comprehensive review of approaches to scale-down models of different unit operations was described previously (Godavarti et al. 2005). These unit operations included nontraditional ones such as precipitation and hydroxylamine cleavage reactions to more traditional operations such as chromatography and filtration. During the past decade, significant advances have been made in scaling down some unit operations such as chromatography and filtration to microscale versions using automated robotic-based high-throughput screening methodologies. There has also been an increase in the use of novel modalities in the clinic such as antibody drug conjugates (ADCs), polysaccharide conjugate vaccines, oligonucleotides, mRNA-based vaccines, and gene and cell therapy,

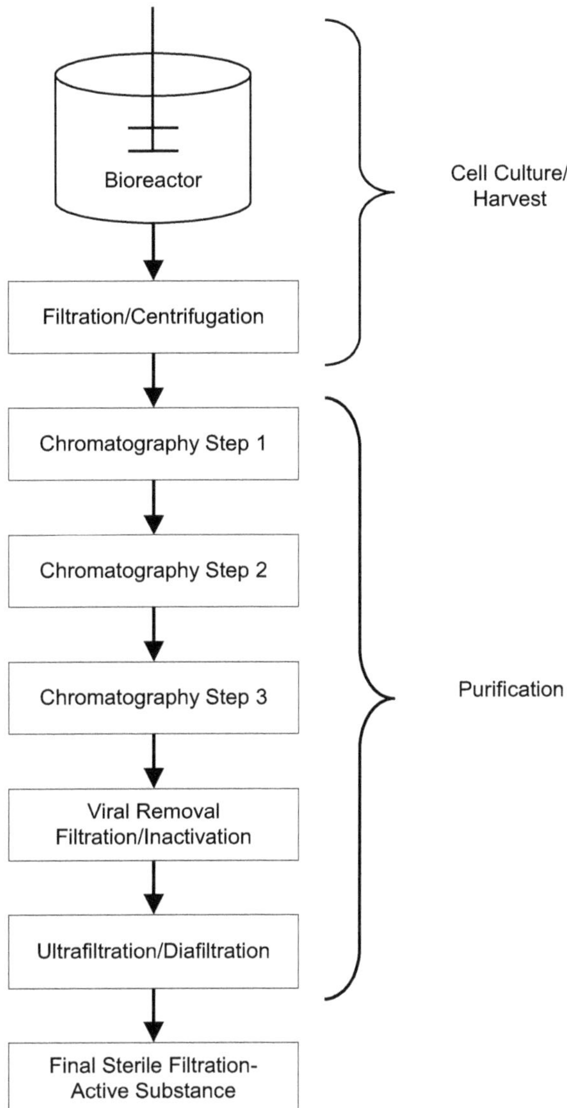

FIGURE 5.1
Typical biopharmaceutical manufacturing process flow diagram illustrating cell culture, cell separation, and purification steps. Multiple bioreactors can be processed per batch. The actual number or sequence of steps depends on the individual processes. Some processes may or may not include additional ultrafiltration/diafiltration steps. Viral inactivation steps may be incorporated where appropriate or necessary.

among others. Further, there has been considerable interest in continuous manufacturing and linking downstream unit operations to continuous cell culture processes such as perfusion processes (Rathore, Kateja, and Kumar 2018; Subramanian 2017; Jungbauer 2013). This chapter will serve as a refresher on scale-down model qualification approaches by updating some of the previously discussed unit operations while also addressing some of these advancements in technologies.

5.2 Design of Scale-Down Systems

5.2.1 Principles

Prior to conducting process validation studies, it is important to ensure that the scale-down model system appropriately reflects the performance of the unit operation at full scale. Several parameters may need to be considered in designing a scale-down model. For example, parameters for chromatographic-based separations would include column dimensions (i.e., bed height), linear flow rates, and buffer conditions. For membrane-based separations these parameters would include flux, transmembrane pressure, and membrane area. It is important to identify the key parameters, scaling principles, and appropriate endpoints for each unit operation prior to designing the scale-down model that would be relevant for comparing the two scales of operation.

5.3 Evaluation of Scale-Up before Scale-Down

Data from laboratory-scale development studies are used to scale up the process to pilot- or full-scale production, and data from these scale-up runs are in turn used to demonstrate the equivalence of scale-down models. Therefore, the first indications of the appropriateness of scaling principles are available upon scale-up. Exact agreement between scales is not necessarily required. If differences in performance are observed between the scales, the scale-down models should be refined, or the scaling differences should be scientifically justified. One could encounter numerous problems upon scale-up, some of which could be related to equipment differences. Impact of hold times on product quality should be evaluated, as there could be longer hold times upon scale-up when large volumes are processed. Longer hold times could lead to differences in product quality such as deamidation, aggregation, and so on, which could consequently lead to changes in product quality such as lower activity. A spectrum of analytical tools may be needed to establish the equivalence of the two scales.

5.4 Scale-Down Limitations: Equipment

A scale-down model not only represents the unit operation itself but also includes the ancillary equipment, which includes pumps, temperature controllers, and agitators, among others. Any equipment limitations that could potentially lead to differences in performance at small scale and large scale should be noted. As an example, for a chromatography unit operation, differences in UV monitors at small scale and large scale could lead to slightly different chromatograms or peak volumes if the peak volume is based on a percent of UV absorbance at peak apex. Another potential difference between the two scales could be length and volume of tubing and other fittings, which could lead to a shift in elution times or altered resolution. Column frits or pre-column filters could absorb product or impurities and should therefore be of comparable materials of construction at both scales to avoid differences in performance (Sofer and Little 1996). Different design of pumps at large scale could introduce shear effects and potentially lead to product denaturation or aggregation. Equipment for temperature control and mixing may also be different between scales and could lead to differences in performance.

5.5 Scale-Down Model of an Entire Process Train

Scale-down models are typically associated with a single unit operation. However, it is often necessary to have a scale-down model for linking unit operations or for an entire process train (Hakemeyer et al. 2016). These models are especially useful while evaluating cell culture process changes or troubleshooting deviations in commercial manufacturing processes. Another application could be in designing process characterization and robustness studies where the robustness of a step is evaluated for its impact on the performance of downstream steps. Most often a scale-down model for an entire process train is not necessarily the individual scale-down models operated sequentially. This is due to possible difference in scales such that the amount of product eluting from the first scale-down model may not provide enough load to the subsequent scale-down model. In such instances a scale-down model with a smaller column diameter may need to be qualified. Establishing a scale-down model for an entire process train could prove to be a valuable tool in mimicking large-scale manufacturing processes or in supporting laboratory-scale bioreactor experiments used in evaluating process options or robustness studies.

5.6 Examples of Scale-Down Models

In the previous version of this chapter, scale-down approaches for several unit operations, including centrifugation, precipitation, microfiltration, ultrafiltration (UF), viral inactivation, and virus removal filtration, were thoroughly reviewed (Godavarti et al. 2005). These unit operations have remained relatively unchanged over the last 10 years and thus are not included in this updated version. We encourage readers to refer to the previous version of this chapter for a comprehensive review. In this section, we highlight several unit operations that have seen development or advancement since the last publishing of this chapter and deserve a further review.

The guidelines and principles governing scale down of various unit operations have both unique and some common features. The following sections describe examples of chromatography, chemical conjugation, and depth filtration, focusing on general guidelines and relevant parameters affecting scale down and primary endpoints used to assess the performance of the scale-down model. As discussed in ICH Q8(R2), a critical process parameter is defined as a process parameter whose variability has an impact on a critical quality attribute and therefore should be monitored or controlled to ensure the process produces the desired quality. All process parameters that fall outside of the definition for critical process parameters are noncritical. Noncritical process parameters may still be controlled within a narrow range to ensure process performance. Within the controlled range, noncritical process parameters do not affect critical product quality attributes but may affect performance such as yield or process time. The characterization of different parameters using risk assessments and process and product knowledge is discussed in detail in ICH Q9 and PDA TR 42 (FDA/ICH 2023; Parenteral Drug Association 2005).

5.7 Chromatography

5.7.1 Description of Chromatographic Techniques

Chromatography is one of the most widely used unit operations in downstream purification processes. The principle of chromatographic separation is based on the differential interactions between the product of interest and impurities for a chromatographic medium. Chromatographic columns are typically operated such that the product of interest binds the column while impurities are recovered in the unbound fraction. Alternatively, the column could be operated in a flow-through mode whereby the impurities bind to the column while the product of interest flows through. In size-exclusion chromatography (SEC), there are no binding interactions involved, as products are separated from impurities according to their size.

A variety of chromatographic techniques are employed in purification processes to achieve the purity required for biopharmaceuticals, enzymes, diagnostics, vaccines, and plasma products. These techniques exploit differences in properties of proteins such as size, surface charge, surface hydrophobicity, and binding specificity. Examples include affinity chromatography, ion-exchange chromatography, hydrophobic interaction chromatography, metal affinity chromatography, and size-exclusion chromatography, among others. Additionally, mixed-mode chromatography resins offer separations that combine more than one mechanism of interaction, often ion exchange and hydrophobic interaction. A typical purification process would comprise a combination of the aforementioned chromatographic steps with orthogonal separation mechanisms. Chromatographic steps are often optimized for each molecule to achieve high yield without compromising purity.

The separation of proteins by chromatography involves multiple interactions between the solute, solvent, and solid chromatographic support matrix. Separation can be influenced by factors such as the nature of the ligand and matrix, solution pH and conductivity, temperature, and size of the beads and pores.

5.7.2 General Scale-Down Principles and Parameters

Chromatographic steps are typically operated in batch mode. In such operations, entire product pools from one step are loaded to subsequent steps either directly or after some sample manipulation such as concentration, diafiltration, pH/salt adjustments, and dilution. In some instances, product pools from multiple cycles on a single chromatographic step may be loaded onto the next step. Automated high-throughput microscale column and well plate methods have been used to study the effects of a wide range of conditions on chromatographic separations and are discussed as small-scale models in section 5.12 (Coffman, Kramarczyk, and Kelley 2008; Chhatre and Titchener-Hooker 2009; Bergander et al. 2008).

5.7.2.1 Scale-Down Validation of Column Chromatography Systems

The equipment of a chromatography system, such as tubing, frits, monitors, and column hardware, plays a crucial role in performance. These elements, though difficult to maintain with identical geometry and materials of construction between large scale and small scale, should be kept as similar as possible. One example of known equipment differences would be the use of stainless steel pipes and column frits in large-scale manufacturing compared to plastic tubing and frits at small scale (Sofer and Little 1996).

Table 5.1 summarizes common scale-down parameters for chromatography systems. Typical scale-down column diameters range from 0.5 cm to 1.6 cm, while the maximum diameter for manufacturing-scale columns are 2.0 m or higher. Therefore, the scale-down factors for chromatography steps may range from 1:100 to over 1:100,000. The residence time of the product must be maintained while scaling down the process step. This can be done by maintaining the bed height and linear velocity and decreasing the column diameter (Smith et al. 1998). All process volumes normalized to column volumes must be the same between the two scales. In some cases, residence time has been maintained by changing the bed height (Yamamoto, Nakanishi, and Matsuno 1988). However, the European Union's Committee for Proprietary Medicinal Products (CPMP) has listed column bed heights as one of the parameters to be compared to show validity of a scale-down model (European Medicines Evaluation Agency 1995). One consideration to keep in mind when scaling down column diameter is wall effects, which could impact chromatographic performance, potentially for column diameters less than 1.0 cm. Yields and product purity between the two scales must be compared (Sofer and Little 1996).

Procedures for preparation of buffers and solutions for scale-down studies should be according to established protocols used in large-scale manufacturing, since subtle changes in ionic strength or pH could lead to altered elution and purity profiles. Further, the quality of buffers and salts used for preparing solutions for scale-down studies should also be consistent with those used in manufacturing. Scale-down studies should also use columns that are packed and qualified with a similar process to large-scale manufacturing. Achieving similar pressure-flow curves between small and large scale can be challenging as bed compression varies with the column diameter and initial bed height. Several

TABLE 5.1

Scale-Down Parameters and Assessment Methods for Chromatographic Steps

Scale-Down Parameters and Scaling Rules[a]	Assessment Methods and Techniques	
HETP and asymmetry factor	Product yield	Product concentration (UV, HPLC, activity)
Bed height	Total protein yield	Total protein concentration (Bradford, UV, HPLC)
Buffer volumes (bed volumes)	Chromatographic profile	Chromatographic profiles for UV, conductivity/pH should be within manufacturing experience
Linear flow rate	Product purity	SDS-PAGE, capillary gel electrophoresis (CGE), HPLC, specific activity
Column loading (g product/L of resin)	Impurity levels	SDS-PAGE, CGE, SEC-HPLC, HCP-ELISA, DNA by qPCR
Elution pool collection criteria	Product isoform distribution	HPLC, imaged capillary electrophoresis (iCE)
Solution pH, conductivity, protein concentration, composition		
Temperature		

[a] Within manufacturing range unless otherwise specified.

groups have developed models to predict pressure-flow curves as a function of column size and aspect ratio that can be useful in scaling (Stickel and Fotopoulos 2001; Prentice et al. 2020; Perez-Almodovar and Carta 2009).

Temperature is another variable that could affect the retention time of proteins on chromatographic resins. Scale-down models should be run at the same temperatures as large-scale manufacturing. Fluctuations in temperature could lead to changes in pH and conductivity of certain buffers, which could affect retention of proteins. Among the various chromatographic techniques, hydrophobic interaction chromatography has been reported to be especially prone to changes in performance due to temperature variations, which could give rise to large changes in product retention or selectivity (Gagnon and Grund 1996).

5.7.3 Primary Endpoints

Several important control variables such as solution pH, conductivity, temperature, and protein concentration should be measured prior to initiating scale-down runs to verify they are comparable to large-scale. Very often column packing can play an important role in the chromatographic separation. Differences in column packing at the two scales may have an impact on the separation and be visualized as differences in the chromatograms. It is therefore important to evaluate the quality of column packing as a tool for comparison at different scales. Height equivalent to a theoretical plate (HETP) and asymmetry factor (As) are typically used to evaluate the quality of column packing (Barry and Chojnacki 1994). A small volume of a concentrated salt solution such as sodium chloride or a UV-absorbing molecule is injected into a column. The resulting conductivity or UV peak near the end of the included volume is used to calculate the HETP and As values (Sofer and Hagel 1997). A range of acceptable values is determined during development and can be compared to values from large-scale columns. When comparing packing quality at the different scales it is important to pay careful attention to factors such as test sample volume, linear flow rate, and equipment differences (tubing length and diameter, monitors, pumps), among others.

Table 5.1 lists several output variables that are used as endpoints to assess the performance of the scale-down model relative to large-scale manufacturing. One of the primary endpoints used in assessing the performance of a scale-down qualification model is an evaluation of chromatograms, to include a qualitative comparison of UV, pH, and/or conductivity profiles. Other variables include product yields (often a quantitative comparison using appropriate statistical methods), product purity (measured by HPLC or other methods), and impurity levels (SDS-PAGE or other methods). Elution pool volumes are determined by the pooling criteria for elution pools. The method of pooling will have an impact on pool volumes and possibly on product purity and should be aligned with the pooling methods used at manufacturing scale.

Many methods exist to compare the results of the scale-down runs to those of the manufacturing-scale runs. A qualitative assessment may be appropriate in some situations and is often used to support early phase viral clearance studies when limited data may be available from small-scale and large-scale runs. During later stages of development when sufficient data are available, a more rigorous statistical assessment may be appropriate. A t-test is commonly used, as it is a simple and fast method for determining whether there is a statistically significant difference between scales. However, there are limitations with this approach. When the sample size is large and the variability is small, even a small difference in the data between scales may lead to wrongly rejecting comparability. Conversely, when the sample size is small and the variability is large, a large difference in data between scales may wrongly be accepted as comparable.

An equivalence test using a two one-sided test (TOST) avoids some of the limitations of a t-test; however, applying TOST requires that acceptance criteria are preset, which can present challenges. Setting tighter acceptance criteria generally requires a larger sample size, while the acceptance criteria will lose practical meaning if they are set too loosely. There are two main options for setting the acceptance criteria. One option is to set it based on the judgement of the process scientist, which is subject to some level of bias. A second option is to set the acceptance criteria based on the variability of the manufacturing data. The use of two or three standard deviations (SD) of the manufacturing data has been reported (Hakemeyer et al. 2016; Nie et al. 2019). The challenge with this method is that the true standard deviation of the manufacturing-scale process is not known due to the relatively small sample size. Additionally, a tight manufacturing-scale dataset can result in an acceptance criterion that is inappropriately tight.

An alternative to an equivalence test is performing a comparison of means with 90% confidence intervals. This method uses the same statistical calculations as TOST but does not use a preset acceptance criterion. A benefit of this method is that it shifts the process scientist judgement from setting of acceptance criteria to the assessment of the differences in means. This also avoids a common situation where the acceptance criterion is failed but arguments are made as to why the results are still valid. A risk of using this method is the potential perception that the qualification is less robust without preset criteria.

Further details on evaluating the suitability of scale-down models are discussed in section 5.11. An example of how one might conduct an assessment for a chromatography system using the comparison of means with confidence intervals method is provided next.

5.7.4 Example of Testing Scale-Down Model Suitability

The following section looks at several example small-scale and manufacturing-scale step yield datasets from five different processes. Table 5.2 shows the yield average, standard deviation (SD), and sample size for the small-scale and manufacturing-scale processes. Table 5.2 also shows the difference in means of the yields and the 90% confidence intervals for each difference.

Process 1 represents a scenario where the small-scale model would clearly be considered representative of manufacturing scale with respect to step yield. The difference in means

TABLE 5.2

Comparison of Average Yields for Small Scale and Manufacturing Processes

| Process # | Small Scale | | | Manufacturing | | | Difference in Means with 90% Confidence Interval | |
	Average Yield	SD	Sample Size	Average Yield	SD	Sample Size	Difference in Means	90% Confidence Interval Range
1	85	2	5	87	3	5	−2.0	(−5, 1)
2	85	5	3	87	5	3	−2.0	(−10.7, 6.7)
3	84	0.5	5	87	2.5	5	−3.0	(−5.4, −0.6)
4	85	5	10	90	5	10	−5.0	(−8.9, −1.1)
5	75	5	10	87	5	10	−12.0	(−15.9, −8.1)

is relatively small (2%), as is the confidence interval (CI). The CI includes zero, indicating that there is no statistically significant offset present.

Process 5 represents a scenario where there is a clear difference between scales. The difference in means is relatively large (12%), and the confidence interval indicates a clear offset, with the small-scale yield 8%–16% lower than manufacturing scale. The reasonable number of runs (10) at small scale and manufacturing scale indicates the result could be statistically significant. This type of result requires an examination of the scale-down model. Several aspects of the scale-down model that could be investigated are listed as follows:

- Was the load or pool material impacted by freezing and thawing or extended holds prior to use and testing?
- Did tubing holdup volume, system configuration, or collection method impact loading volume or product pool collection?
- Were analytics run using the same methods in the same labs as the manufacturing-scale samples?

If the reasons for the offset can be determined, one option would be to modify the model to eliminate any differences. If the offset is an inherent part of the scale-down model, this can be acceptable. According to the EMA Guideline on Process Validation: "Depending on the differences observed and their understanding, approaches to managing these differences could be acceptable if well documented and justified" (European Medicines Agency 2014).

Processes 2, 3, and 4 require a more nuanced evaluation. Process 2 has the same average yields at both small scale and manufacturing scale as Process 1 and the same mean difference; however, the smaller number of runs and larger SD for Process 2 results in a much large CI range. The very small difference in means would likely lead to the conclusion that the scales are comparable, but the generation of data from additional batches at both small scale and manufacturing scale could be considered to gain a better understanding of the true variability of the process yield.

Process 3 also has a relatively small difference in means (3%) but has a very low SD for the small-scale dataset. This results in a statistically significant offset in means, but with a small confidence interval. The phenomenon of a very small SD for scale-down qualification runs might be expected if the runs are performed back-to-back using the same column, load, buffers, and so on. Performing the runs in this manner underestimates the variability that will be seen at manufacturing scale where the load source, buffers, and other variables will change from batch to batch. For Process 3, additional small-scale runs with different load sources and raw materials could be considered to more appropriately mimic the large-scale variability and generate more process understanding. However, as with Process 2, the small difference in means with the existing dataset indicates that, overall, the scale-down model produces results comparable to manufacturing scale.

For Process 4, a slightly larger difference in means is observed (5%) and the CI does not include zero, indicating a statistically significant offset. Like Process 5, the potential root causes of the offset should be explored. Based on that evaluation, a determination can be made to modify the scale-down system, to accept the scale-down model with an offset, or to accept the scale-down model without an offset based on the relatively small size of the offset.

5.8 Conjugation or Chemical Modification Reactions

Many current biotherapeutics involve the chemical modification of macromolecules such as functionalization of proteins or peptides therapeutics (Gunnoo and Madder 2016) or attachment of therapeutic molecules to carrier molecules via covalent crosslinking to generate bioconjugate therapeutics as final products (Li and Mahato 2017; Hermanson 2013). The modification or conjugation of therapeutic agents can be strategically designed to offer several advantages, including (1) increased efficacy through enhanced disease-specific targeting, (2) optimal physical-chemical properties, (3) better safety profile with reduced toxicity, and/or (4) improved pharmacokinetics and/or pharmacodynamic properties (Hu, Berti, and Adamo 2016a; Li and Mahato 2017). Some examples of bioconjugate products are PEGylation for half-life extension of therapeutic proteins, glycoconjugate vaccines, antibody-drug conjugates (ADCs), bispecific antibodies, and lipid-drug conjugates (Stephanopoulos and Francis 2011; Hu, Berti, and Adamo 2016a; Fu and Sakamoto 2007). Previously, scale-down models for some examples of chemical modification reactions (i.e., hydroxylamine cleavage, enzymatic cleavage, and PEGylation) were described (Godavarti et al. 2005). Advancements in scale-down model development and novel conjugation modalities are discussed in this section.

Modifications and conjugation techniques are dependent on the reactive functionalities of the crosslinking or derivatizing reagents and the functional groups present on the target macromolecules to be modified. To effectively design a modification or conjugation strategy, a basic understanding of the mechanisms by which these two interrelated chemistries (e.g., reactive groups and target functionalities) couple is required (Hermanson 2013). The final product quality profile is driven by the chosen conjugation chemistry. For a conjugation process to consistently deliver the target product quality attributes, process parameters influencing the conjugation reaction kinetics (raw materials/reactants quality and concentrations/ratios, temperature, pH, time, etc.) should be properly optimized.

Chemical modification and conjugation manufacturing processes are complex, multistep processes that can involve multiple reaction steps, the isolation of intermediates (e.g., activated polysaccharides), combination of aqueous and organic solvent processing, and multiple purification steps (e.g., chromatography and ultrafiltration/diafiltration (UF/DF)) to remove unreacted materials, reaction side products, process residuals, and/or degradants. Figure 5.2 shows examples of typical manufacturing process flow diagrams (PFDs) for an ADC process in panel A and a glycoconjugate vaccine process in panel B. The following sections are focused on general scale-down principles and parameters for chemical reactors employed in key conjugation reaction steps, including discussion of application of these principles through specific case studies. Scale-down principles for standard purification steps that are commonly used in conjugation processes are discussed throughout other sections of this chapter and include specific considerations for the chemical modifications or conjugation processes.

5.8.1 General Scale-Down Principles and Parameters for Chemical Reactors

Successful process development, scale-up, technology transfer, and robust long-term manufacturing of conjugation reaction steps are all facilitated by an understanding of the effects of reaction kinetics, reactor design, mixing characteristics, reagent addition rate and location, and heat and mass transfer rates on process outputs.

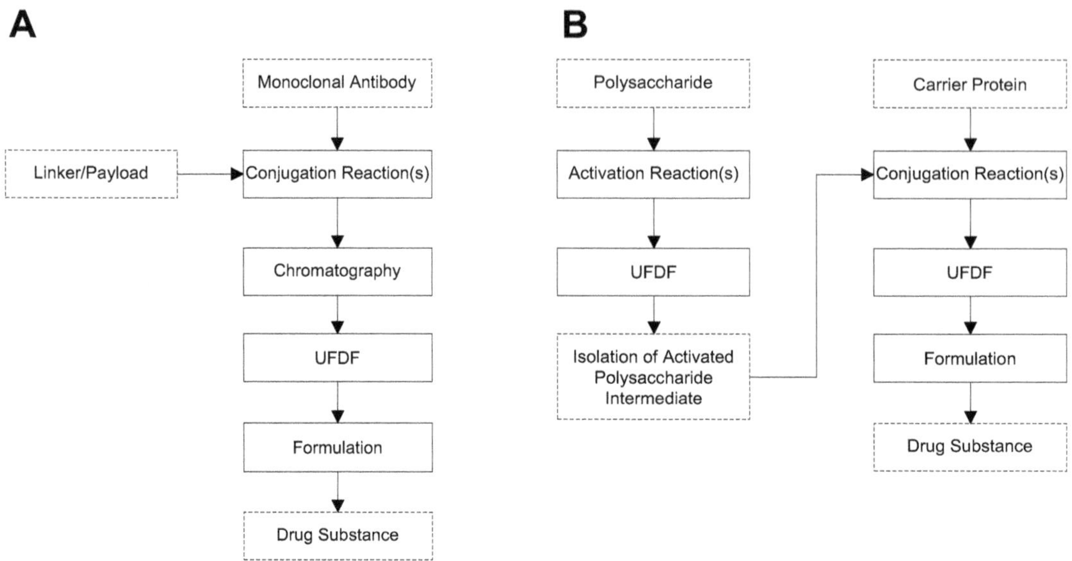

FIGURE 5.2
Process flow diagrams for typical conjugation processes: (A) antibody-drug conjugate process and (B) glycoconjugate vaccine process.

Chemical reactor design and process scaling considerations are dependent on the process objectives, the mechanism(s) controlling the conjugation process outputs, as well as applicable process constraints. These factors need to be identified in order to develop representative scale-down models and to ensure robust performance and control at large scale. For reactor design, initial considerations on the material of construction (MoC) should be taken when selecting a small-scale reactor vessel. A material risk assessment should be performed to identify any potential risks associated with MoC of both large- and small-scale reactors, including compatibility of contact material with reaction streams and potential impact to heat transfer. Rheological characterization of process streams (e.g., viscosity) and kinetics studies, that is, evaluation of the impact of reaction time on product quality attributes and extent of byproduct formation and side reactions, are essential to define target set points for scale-independent parameters (pH, temperature, concentration) and to aid in scale-down/-up considerations for scale-dependent parameters (mixing).

Mixing in reactor vessels is a key factor that can influence control of reaction parameters such as pH and temperature and impact the efficiency of the conjugation reaction. Therefore, achieving the same mixing quality in the scale-down and large-scale systems is an important objective of the scale-up/down process. While maintaining geometric similarities (impeller design and configuration, liquid height/tank diameter ratio, and baffling) between small-scale and larger reactors may facilitate scale-up/down, this is not always possible due to limitations on small-scale availability or to the use of pre-existing equipment at large scale.

There are multiple engineering scaling factors that could be maintained constant for proper scaling of mixing performance; the most typical methods are blend time, power per unit volume (P/V), mixing time, impeller tip speed, and gas-liquid mass transfer coefficient (kLa). Some of these mixing parameters can be measured empirically in small and large reactors to ensure appropriate control—this is the case of blend time (measured via salt or temperature spike studies) (Grenville and Nienow 2004; Brown et al. 2004) or kLa (measured via dynamic gassing out methods) (Brown et al. 2004; Xu et al. 2017). For other cases where detailed information is difficult to obtain experimentally (e.g., characterization of shear rate distribution and velocity profiles), computational fluid dynamics (CFD) models of the full-scale and scale-down reactors can be valuable tools for scaling mixing processes (Marshall and Bakker 2004).

It is not possible to maintain all mixing parameters constant on scale-up/down and therefore, as mentioned earlier, scaling considerations will depend on the process objective. For fast reactions in which the characteristic mixing time must be minimized, scaling is typically based on maintaining blend time. The means of reagent addition (location and rates) may also be important for fast reactions; reactants feeding at the point in the reactor having the highest level of turbulence to ensure quick dispersion is recommended. On the other hand, for slower reactions P/V is commonly used for scaling. For conjugation reactions involving shear-sensitive molecules such as proteins, special attention should be given to the distribution of shear rates related to flow velocities. For reactions requiring special temperature control (e.g., exothermic/endothermic reactions, control of temperature ramp rates), heat transfer considerations in small-scale reactors can be achieved through the use of jacketed reactor vessels, mimicking reactor material of construction and geometry to that of the large-scale reactor, and incorporation appropriate temperature monitoring and control systems.

During early development of chemical modification or conjugation processes, the use of less sophisticated microscale or small-scale systems are common to allow for rapid screening of process parameters. These simple systems may include the use of a stir bar for mixing and the use of a water bath to control temperature. These reactor systems for process development are only as good as their predictability and scalability to larger scales. Identification of representative scaling models for conjugation chemistry and standardization on the use of those models throughout the product development cycle is key for a successful scale-up and technology transfer. Recently, the EasyMax reactor system from Mettler-Toledo has been introduced as a small-scale reactor model for bioconjugation processes. This system allows control of a broad range of reaction parameters such as temperature (jacket control with temperature probes to monitor and control solution temperature), pH, mixing (with overhead magnetic drive agitators), and reagent addition (through dosing pumps). In addition, process analytical technologies can be integrated to enhance understanding of reaction kinetics, potential risks, and formation of bioproducts. Detailed characterization and performance evaluation of this small-scale reactor system have been completed to support scale-up of conjugation processes. Characterization work included level map, transfer rates, mixing time characterization, agitation power measurement, control system response, and mass transfer coefficient measurements. Figure 5.3 shows examples of characterization results for the EasyMax reactor, including determination of blend time at different mixing rates in panel A and evaluation of temperature ramping capabilities in panel B. Performance of this small-scale model has been demonstrated for several conjugation modalities such as ADCs and glycoconjugate vaccines.

5.8.2 Primary Endpoints

Demonstration of small-scale model comparability with manufacturing scale is critical to support process development and characterization and ensures that any data generated at small scale can be related to manufacturing scale. Ideally, the comparison of small-scale model performance to the manufacturing-scale process should be completed in advance

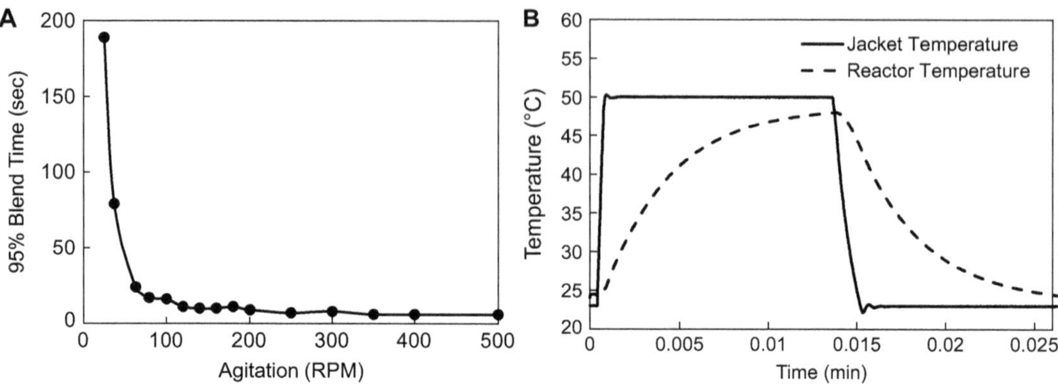

FIGURE 5.3
Examples of small-scale characterization data: (A) blend time as a function of impeller speed as determined by salt spiking studies and (B) evaluation of temperature ramping capabilities. Temperature as a function of time is measured at the reactor jacket and internal temperature probe.

of the implementation of the process into manufacturing scale. In the common scenario where this approach is not feasible, small-scale data are compared with data generated at the largest scale of operation as part of experimental execution to complete qualification of the small-scale model (Sunasara et al. 2018).

For qualification of small-scale reactor models for chemical modification or conjugation processes, it is recommended that comparison of small-scale and manufacturing-scale systems is completed at baseline conditions for scale-independent parameters. Scale-independent parameters such as concentration, pH, and temperature are controlled at the same set points or at least within a range where impact to product output is known to be minimal. For scale-dependent parameters such as agitation rate, the identified scaling mixing parameter (as described in the previous section) should be used to define the mixing rate required to match mixing performance of large-scale reactors. In cases where a mixing scaling variable is not identified, an extra step should be taken to evaluate the effect of mixing on final product quality attributes at small scale to determine an appropriate mixing rate. Table 5.3 summarizes common scale-down parameters and endpoints for conjugation reactors.

An additional consideration for qualification of small-scale model for conjugation reactor processes is to utilize the same or comparable lots of raw materials and intermediates. This limitation could be avoided if well-characterized raw materials or intermediates are being used within the process and, therefore, confidence in lot-to-lot comparability is achieved by evaluating key properties (e.g., purity of derivatization reagents, product profile of drug substance intermediate streams). Execution of small-scale comparability runs in parallel to large-scale runs is another strategy that could mitigate risks of material/intermediates variability. Intermediate process streams from the large-scale process could be taken and processed through the small-scale model, and model qualification can be performed by side-to-side evaluation of product outputs/endpoints.

The outputs or endpoints chosen for scale-down model qualification should reflect the primary purpose of the unit operation. In the case of chemical modification or conjugation reaction operations, key product quality attributes indicative of the conjugation reaction

TABLE 5.3

Scale-Down Parameters and Endpoints for Conjugation and Chemical Modification Reactions

Scale-Down Parameters and Scaling Considerations[a]	Assessment Endpoints and Techniques	
A. Scale-independent parameters	Reaction efficiency (activation, conjugation, capping reactions)[e]	Apparent molecular size, percentage of unconjugated species, concentration of residual active sites
Temperature (includes set points and ramp-up and cool-down rates)[b]	Distribution of conjugation sites	Mass spectrometry, capillary isoelectric focusing (cIEF)
pH (hold points and adjustment endpoints)[b]	Reaction residuals	
Reaction time	Performance attributes (yields, filterability)	Mass balance, Vmax, Pmax
Reactant concentration(s) (e.g., mAb, polysaccharide, carrier protein)		
Input ratios (input ratios of both reactants and reagents)		
Reactant/reagents addition time		
B. Scale-dependent parameters		
Reactor vessel and impeller configuration (reactor geometry, impeller position, presence or absence of baffles, reactor jacket, headspace)[c]		
Mixing (mixing engineering scaling factors such as P/V, blend time, impeller tip speed, and kLa)[d]		
Reactants/reagents additions (rates, location, order of addition)		

[a] Within manufacturing range unless otherwise specified.

[b] Control of these parameters can be impacted by other scale-dependent parameters such as mixing.

[c] If full geometric similarity is not feasible, specific dimensionless numbers could be selected to maintain across scales.

[d] Select engineering scaling factor based on understanding of reaction mechanisms. If reaction mechanism is unknown, evaluate mixing effects on product quality and process performance attributes.

[e] Can be assessed at final drug substance stage and/or intermediate stages.

efficiency, product quality (e.g., attributes that are known to be important to ensure potency), and consistency should be considered. For conjugation processes with process intermediates (e.g., activated polysaccharides), quality attributes of these in-process streams (i.e., in-process tests) can be assessed as part of the small-scale model qualification. In addition to product quality attributes, process performance indicators such as yield and filterability can also be assessed as part of small-scale model demonstration.

The complexity of developing characterization methods to assess the quality of bioconjugates will highly depend on the input molecules to conjugate (e.g., small molecules, proteins, peptides, polysaccharides) and the selected modification/conjugation chemistry. For example, in cases when site-specific technologies can be used, more homogenous products could facilitate their characterization. On the other hand, the presence of multiple sites of attachment in the input molecules (e.g., in polysaccharides and carrier proteins) will result in highly heterogeneous and complex products to characterize. Conjugation efficiency and product quality and consistency are measured by a combination of quality attributes. An increase in the apparent size of the input molecules could be used as an indicator of the efficiency of the conjugation reaction (Chennasamudram and Vann 2014). The amount of unconjugated species (e.g., free saccharide for glycoconjugates, unconjugated linker payload for ADCs) and concentration of residual active sites are also attributes indicative of product quality that should be assessed during small-scale model qualification. Detailed characterization of the distribution of conjugation sites are key to assess product quality and consistency; however, depending on the bioconjugate modality, development of methods to quantify and determine distribution of conjugation sites may be challenging.

In many cases, measurement of outputs to assess scale-down model suitability cannot occur at the process stream immediately after the conjugation reaction step (e.g., crude conjugate) due to potential interference of conjugation reaction matrix. Therefore, characterization may occur after conjugate purification, which is commonly designed for the removal of unreacted species. For scale-down model qualification, impact of downstream purification steps should be also considered for this comparability assessment.

Comparability across scales is determined by assessing level of similarity of product quality attributes and process performance indicators. This level of similarity is assessed by statistical means and, therefore, multiple small-scale runs (a minimum of three runs is recommended) should be completed to enable this analysis. For small-scale model qualification, the acceptance criteria established to determine model equivalency should account for the inherent process and analytical method variability. The minimum requirement for comparability is that the difference in product quality attributes from the scale-down model and the large-scale process are not statistically significant. Common statistical tools to determine comparability include demonstrating small-scale results to be within three standard deviations of the historical process mean or comparing product quality mean values using a one-way ANOVA t-test at a significance level $\alpha = 0.05$.

The following sections discuss two specific case studies for scale-down model development and assessment for common bioconjugates.

5.8.3 Case Study I: Scaling Considerations for ADCs

Antibody-drug conjugates (ADCs) are a rapidly growing class of therapeutics for oncology indications (Khongorzul et al. 2020; Beck et al. 2017) and are being evaluated in other areas such as inflammation (Brandish et al. 2018). An ADC comprises a targeting antibody chemically conjugated to a drug, most often a cytotoxic agent, through a linker. ADCs are prepared using a range of conjugation chemistries, including conventional lysine and cysteine conjugation, site-specific conjugation (e.g., cysteine mutants, enzyme-mediated to glutamine), and conjugation to non-natural amino acids (Beck et al. 2017; Prashad et al. 2017).

ADCs are complex conjugates and present multiple challenges for process development and manufacturing (Stump and Steinmann 2013). The following case study provides one example of applying scale-down strategies to ADCs. However, the strategies discussed are applicable for the development of a range of ADCs (Prashad et al. 2017).

This case study from Hu and coworkers focuses on scale-down principles applied for the development of inotuzumab ozogamicin (InO) (Hu et al. 2017). InO is an ADC comprising a CD22-targeting monoclonal antibody (inotuzumab) covalently linked to the cytotoxic antibiotic calicheamicin (ozogamicin) (Ricart 2011). Manufacture is accomplished in a single-step conjugation by reaction of an activated calicheamicin derivative (succinimidyl ester) with the lysine amino groups of inotuzumab. The final conjugate bears approximately six calicheamicin molecules per antibody on average, therefore having a drug-to-antibody ratio, or DAR, of six.

For the conjugation, a solution of activated calicheamicin derivative in ethanol is added to an agitated buffered solution of inotuzumab. Kinetics experiments monitoring the consumption of unconjugated antibody after addition of the activated calicheamicin derivative determined that the conjugation reaction was quite fast, reaching completion in 2 minutes (Figure 5.4A). Such rapid kinetics requires appropriate mixing to ensure that locally high concentrations of reactants do not result in the formation of separate populations of over-conjugated and under-conjugated/unconjugated species.

To ensure adequate mixing at both laboratory and large scales, blend time was determined by salt-spiking experiments for reactors spanning a range of volumes. As shown in Figure 5.4B, blend times at agitations rates of 50–175 RPM nearly overlay for the 0.1 L scale-down reactor (full circles) and the 10 L large-scale reactor (open circles), confirming equivalent mixing over this broad scale range.

All reactors were constructed of glass to ensure the same contact material across scales. The process is run by heating the antibody solution to the desired temperature and then adding the activated calicheamicin derivative solution in a small volume relative to the volume of antibody solution. For this reason, heat transfer did not need to be considered for the scale-down model.

Based on equivalent mixing, the 0.1 L reactor was qualified as a suitable scale-down model for this conjugation process and was applied to evaluate the impact of agitation rate and location of addition (above and below surface). The combination of low agitation rate and above-surface addition resulted in incomplete conjugation due to poor mixing. When the activated calicheamicin derivative solution was added below surface, conjugation proceeded efficiently across the full range of agitation rates.

The scale-down mixing studies were used to determine an acceptable range for agitation and the need for below-surface addition. A comparison of ADC quality attributes drug-to-antibody ratio (DAR) and aggregation (% Agg) and process performance attribute of yield across scales (Figure 5.5) confirm a scalable process and the suitability of the scale-down model.

The qualified scale-down model was used to run design of experiments (DOE) studies and additional one-factor-at-a-time (OFAT) experiments to establish other parameter and input ranges, such as reaction time, pH, temperature, and activated calicheamicin derivative input, en route to a robust manufacturing process with a well understood design space.

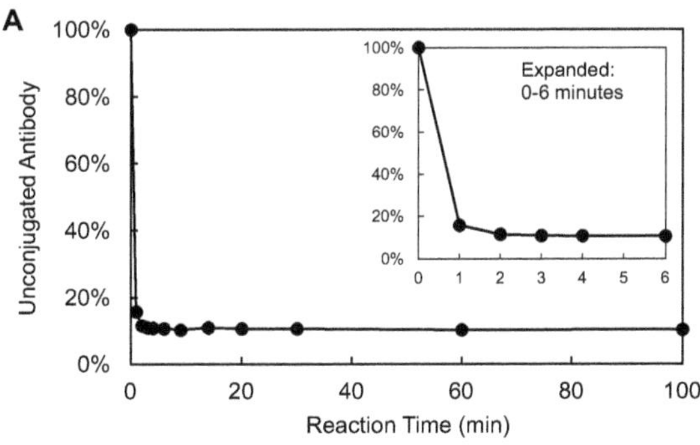
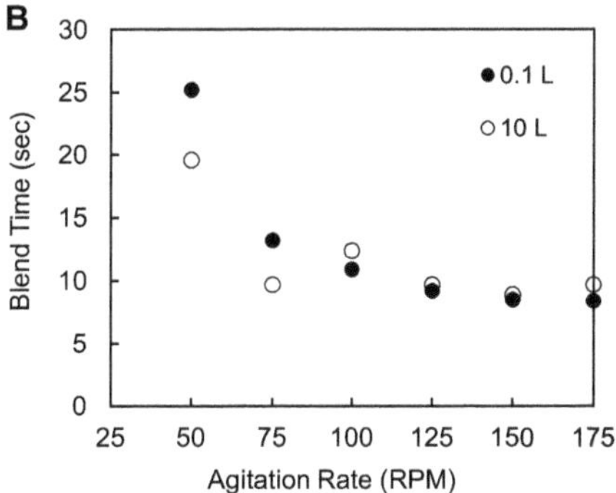

FIGURE 5.4

Conjugation kinetics vs. reaction time (A) and blend time vs. agitation rate (B). Reprinted (adapted) with permission from (Hu et al. 2017). Copyright (2017) American Chemical Society.

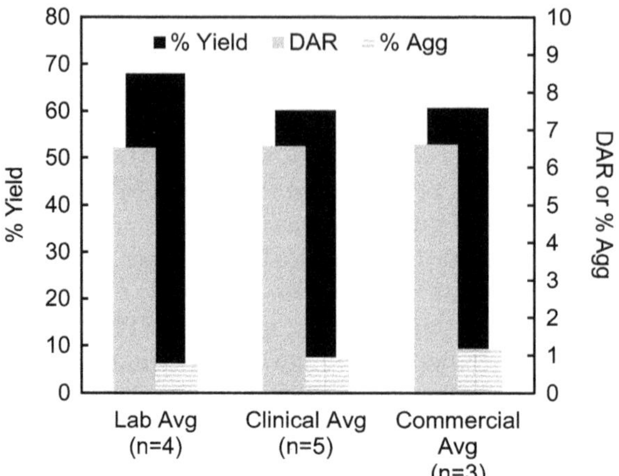

FIGURE 5.5
Conjugate quality attributes at different scales. Reprinted (adapted) with permission from (Hu et al. 2017). Copyright (2017) American Chemical Society.

5.8.4 Case Study II: Reactor Scale-Down for a Glycoconjugate Vaccine Process—Identifying the Appropriate Mixing Scalable Parameter

Glycoconjugate vaccines are chemically complex, and current manufacturing approaches start from heterogeneous mixtures of polysaccharides linked by nonselective methods to the carrier protein resulting in complex heterogeneous conjugates. The efficacy (i.e., immunogenicity) of glycoconjugates is dependent on several interconnected features of both the polysaccharide antigen and the conjugation chemistry (e.g., availability of immunogenic epitopes, sugar length, number of repeat units linked to the carrier protein, linker type, conjugate size, etc.) (Hu, Berti, and Adamo 2016b; Prasad, Kim, and Gu 2018; Chennasamudram and Vann 2014; Costantino, Rappuoli, and Berti 2011). This outlined complexity of glycoconjugates requires additional efforts during product development to decode how multiple process parameters influence their activity and to achieve manufacturing consistency.

The glycoconjugate approach was recently explored to develop a tetravalent prophylactic vaccine against *Staphylococcus aureus* infections. Two of the vaccine components were capsular polysaccharides (CP) types 5 and 8 (CP5 and CP8) (Arbeit et al. 1984) covalently conjugated to CRM_{197} carrier protein. The CP5-CRM_{197} conjugate was prepared through activation of hydroxyls on CP5 using 1,1'-carbonyl-di-(1,2,4-triazole) (CDT). Activated CP5 is then covalently bonded to the lysine's amines on CRM_{197} (Prasad, Kim, and Gu 2018). A major challenge in the development of a robust, scalable, and commercial manufacturing process for CP5-CRM_{197} conjugate was the control of conjugate size to ensure comparability to early clinical material.

Initial understanding of the activation reaction chemistry indicated fast CDT reactivity and competition between CP5 and water for available activation reagent. For competitive fast reactions, mixing is normally an important parameter to ensure product consistency is achieved. Good dispersion of CDT, to reduce high local concentration zones, was achieved through a combination of adequate mixing performance and control of reagent addition parameters such as location and rate. Appropriate mixing scaling takes additional importance in situations like this where competitive reactions take place and reagent

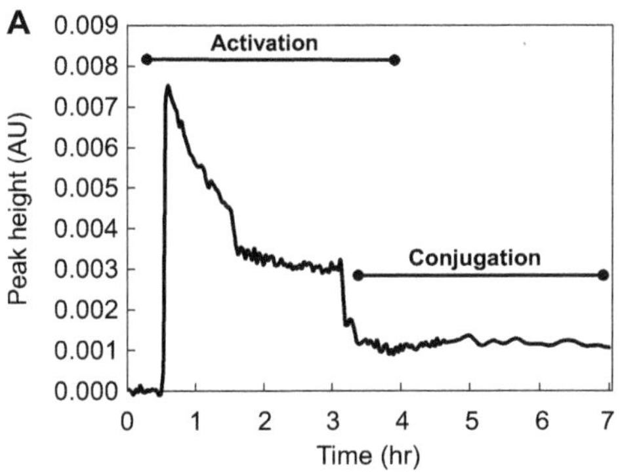
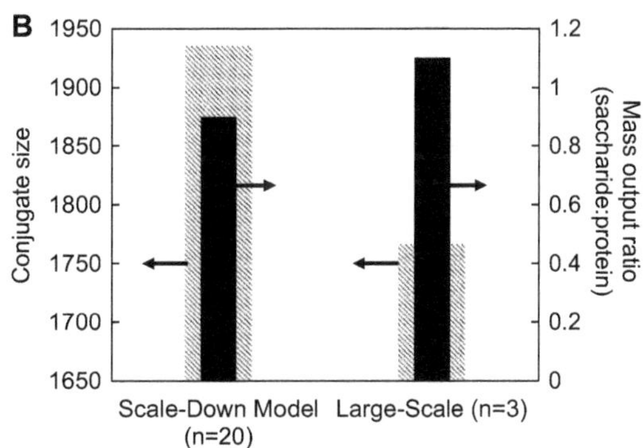

FIGURE 5.6

Conjugation scale-down model characterization: (A) CO_2 concentration over time as measured by FT-MIR and (B) comparison of conjugate attributes at different scales.

homogenization at the molecular scale needs to occur so that molecules can collide (Shah, Kostiuk, and Kresta 2012). Under poor mixing conditions, high local concentration zones of CDT can be present and favor reactivity with the highly mobile water molecules. As described in section 5.8.1, blend time is a common mixing scaling parameter for fast reactions (Grenville and Nienow 2004). Therefore, initial efforts to establish a representative small-scale model utilized blend time as the mixing scaling parameter. Blend time was determined for the large-scale reactor under standard conjugation conditions by salt-spiking studies, and attempts to design a scale-down reactor model to match the large-scale blend time were undertaken. Although blend times between scale-down and large-scale reactors were matched, conjugates prepared using the developed scale-down models resulted in quality attributes not comparable with the large-scale process. Therefore, the scale-down model was considered not suitable for process development studies.

Observations from early development and pilot-scale runs indicated that differences in reactor headspace conditions (e.g., use of closed reactor vessels vs. open vessels with nitrogen sweep) may play a significant role in final product quality attributes, driving the efficiency of the conjugation reaction. Carbon dioxide (CO_2) was known to be a by-product of the activation of CP5 with CDT prior to conjugation, and therefore, in-process analytical techniques (PAT) were established both at small and large scale to monitor CO_2 during the activation and conjugation process. Figure 5.6A shows CO_2 concentration profile collected by ReactIR mid-range infrared spectrophotometer (FT-MIR) over time. As expected, CO_2 is released in solution during activation, and over time it partitions into the reactor headspace. The gas-liquid mass transfer rate (k_La) can be used to measure the rate of CO_2 exchange and, therefore, as a scaling parameter (Brown et al. 2004; Xu et al. 2017). As a first step to establishing a new representative scale-down model, the process was scaled down by maintaining constant liquid/gas volume ratio. Then k_La was determined for small- and large-scale reactor vessels at various mixing rates to identify mixing conditions to match performance of the large-scale process (Table 5.4).

Comparability of the new scale-down model was demonstrated by assessing resulted quality attributes against large-scale process. For scale-down model qualification, the lot of CP5 polysaccharide as well as scale-independent parameters (CDT input, concentration, temperature, time, etc.) were maintained constant. Figure 5.6B compares quality attributes

TABLE 5.4

$k_L a$ Comparison of Small- and Large-Scale Reactors

Large-Scale Reactor		Small-Scale Reactor	
RPM	$k_L a$	RPM	$k_L a$
90	0.0020	50	0.0007
100	0.0018	100	0.0010
-	-	150	0.0012
-	-	200	0.0017
-	-	250	0.0020
-	-	300	0.0031

across scales demonstrating suitability of the scale-down model. The qualified scale-down model was utilized to further study the impact of mixing conditions (mixing rate, reagent addition rate, and location) on final product quality attributes. Information gained was critical to establish an appropriate control strategy for conjugate size to ensure product consistency and process robustness.

5.9 Depth Filtration

5.9.1 Description of Depth Filtration

Depth filters have become common in modern manufacturing facilities because of their low cost and single-use disposable nature. Depth filters are operated in a normal flow filtration (NFF) mode with a constant pressure applied to the feed stream or at a constant flow rate. Depth filtration is regularly used for clarification of mammalian and bacterial cell culture to provide cell-free conditioned medium or used as a prefilter for turbid feed streams loading on subsequent downstream purification steps. In the case of harvest operations, depth filters typically consist of a primary filter for removal of cell and large debris and a secondary filter designed to remove residual debris and smaller impurities (Michen et al. 2011; Schreffler et al. 2015; Khanal et al. 2019). Cells are retained within the filter matrix, and as the cells and particulates build up in the filter, the resistance to permeate flow through the filter increases. In constant pressure operation, the filtration rate decreases, while in constant flow rate operation, the feed pressure increases.

Depth filters can also exhibit shear forces at high differential pressures causing cell lysis and risking disulfide bond reduction (O'Mara et al. 2019). Cell lysis can lead to reducing environments in clarified harvest that may generate fragments and aggregates. Small-scale models can be used to understand the effect differential pressure has on cell lysis and help in the development of methods to reduce the risk of disulfide bond reduction during long hold times of clarified harvest like air overlays and pH adjustment (O'Mara et al. 2019).

Inside the depth filters there is a bed of fibers with filter aids like diatomaceous earth and, in some cases, polymeric additives that provide electrostatic and hydrophobic binding (Charlton, Relton, and Slater 1999; Yigzaw et al. 2006). Retention in depth filters is mostly based on size exclusion, but there can exist some capacity for the removal of other soluble impurities by charge and hydrophobic binding based on the media (Khanal et al.

2019; Yigzaw et al. 2006). Several groups have shown additional impurities like nucleic acids, lipids, and host cell proteins, and product-related impurities can be reduced using depth filtration during harvest operations (Yu et al. 2019; Hadpe et al. 2020). Work by Yu and coworkers has shown a scale-down model could be used to characterize the removal of low and high molecular weight species with filters containing strong hydrophobic interactions (Yu et al. 2019).

Depth filters can also be used in later downstream steps to remove turbidity and impurities that are generated by load adjustments like a low pH hold for viral inactivation or as a prefilter before virus removal filtration. Often the removal of impurities on depth filters is improved downstream when the feed stream has a lower conductivity compared to conditioned media during harvest operations.

5.9.2 General Scale-Down Principles and Parameters

Table 5.5 summarizes the relevant scale-down parameters for depth filtration systems. The performance of depth filters is influenced by the filter's internal pore size distribution, surface adsorption characteristics, operating pressure, feed flow rate, load conductivity, and temperature. The load volume and chase volume to surface area ratios are also important scale-down parameters. The feed material is often characterized by cell concentration, viability, susceptibility to shear (cell age), and turbidity.

Common methods for loading depth filters are low shear pumps such as rotary lobe pumps or placing the filter immediately after centrifugation. Typical operating pressure for mammalian cell harvest systems is 5–40 psig. Peristaltic pumps are regularly used in scale-down models to control the feed flow rate. These pumps may not appropriately mimic the large-scale production pumps regarding shear, slip, and cell damage.

5.9.3 Primary Endpoints

A successful scale-down of a depth filtration system will provide a filtered pool with similar permeate quality (turbidity, filterability, and impurity removal) and product yield as the large-scale system. The feed flow rate is generally controlled as the pressure of the feed and permeate streams are monitored throughout the process. The permeate flux should mimic the production system performance with comparable product quality in the pool.

TABLE 5.5

Scale-Down Parameters and Assessment Methods for Depth Filtration Steps

Scale-Down Parameters and Scaling Rules[a]	Assessment Methods and Techniques	
Load volume to surface area ratio	Product yield	Product concentration (UV, HPLC, activity)
Solution pH, conductivity, protein concentration, etc.	Permeate pool turbidity	UV-VIS spectrometry, nephelometry
Temperature	Generated cell lysis	Lysis measurement (lactate dehydrogenase, DNA etc.)
Feed flow rate	Permeate pool filterability	Filter-specific capacity
Feed pressure		
Wash volume to surface area ratio		

[a] Within manufacturing range unless otherwise specified.

Table 5.5 summarizes the primary endpoints to be measured and the techniques used for their measurement.

5.10　Scale-Down Models of Integrated and Continuous Biomanufacturing

A recent trend in downstream biomanufacturing has been a shift from typical batch processes to integrated and continuous biomanufacturing (ICB) operations (Cooney and Konstantinov 2014). ICB processes are not new to the biomanufacturing industry but have typically been employed for highly labile products such as enzymes and plasma fractionation. Recent advancements in technology and a need for increased manufacturing capacity have led to a renewed interest in ICB for a broader set of molecules, including monoclonal antibodies (Rathore, Kateja, and Kumar 2018; Subramanian 2017; Jungbauer 2013). Several approaches to ICB have been proposed that range from a single unit operation running continuously to a fully integrated process linking the production bioreactor through purification to formulation of drug substance.

Many ICB processes have been enabled by new or improved technologies that allow for continuous downstream operation. Harvest technologies such as continuous centrifuges, new TFF hollow fiber units, and alternating tangential flow (ATF) can be used to provide a constant flow of material to the downstream process. Several approaches have been used to make chromatography continuous from simple parallel column designs to more complex multicolumn systems such as simulated moving bed chromatography (SMB) and periodic counter current (PCC) chromatography to continuously process material (Godawat et al. 2015; Warikoo et al. 2012). Continuous low pH viral inactivation been achieved with inline mixers and a plug-flow reactor (Fiadeiro et al. 2016; Parker et al. 2018). In addition to distinctive unit operations, ICB processes generally contain other unique features like surge vessels between unit ops, feedback and feedforward controls, and increased reliance on inline monitoring.

ICB processes create several unique challenges for establishing scale-down models used in process characterization and validation. Unlike batch processes, ICB processes are dynamic systems that can contain multiple linked unit operations. This makes it inherently more difficult to establish a scale-down model, as more parts of the process need to be considered. ICB processes generally lack discrete pools for sampling and instead rely on frequent inline testing to show the system is operating as expected. In a scale-down model, the amount of sampling can impact the execution of the process as the frequency of sampling can result in enough material being removed from the system to interfere with subsequent steps. Some ICB processes also have a startup and shutdown period not found in batch processes. Scale-down models need to account for these stages of the process as they can have a significant impact on overall product quality. Many ICB processes utilize complex automation schemes that can be difficult to replicate at the small scale.

Many modern ICB processes are only just reaching the need for late-stage characterization involving small-scale models. As such, well-established approaches to scale-down models that are common for batch processes are still being established for modern ICB processes. Ultimately, the expectations for an ICB process are not different from batch processes: a process must be capable of delivering quality products.

5.11 Evaluation of Suitability of Scale-Down Systems as Models of Full-Scale Operation

For process characterization activities that are used to support marketing applications for therapeutic proteins, an evaluation of the performance of the small-scale system (also called the qualification of the small-scale model) must be conducted to ensure that results generated from the small-scale model are representative of the results that would be observed at manufacturing scale. The description of the scale-down system should include detailed information regarding potential scale-related variables, so that the qualification of the scale-down system will still hold for studies conducted years later, on equipment that may have been reassembled or may have changed slightly. Typically, a minimum of three runs are performed during qualification scale-down studies, which will allow an assessment of the reproducibility of the scale-down process and provide a more meaningful comparison to full scale. The data from the key process parameters and product quality attributes from the scale-down process should be compared to the full-scale process. Depending on the type of analysis, the comparison may be qualitative or quantitative.

Qualitative comparisons may be used for evaluation of data from complex analytical methods (product impurity or isoforms assessment by HPLC chromatograms, capillary gel electrophoresis, mass spectrometry, etc.). These evaluations should use direct, side-by-side comparisons of the two scales of operation to avoid introducing artifacts from the analysis. Other outputs are qualitative as well, such as chromatograms from a process chromatography step. While some portions of chromatograms may be evaluated using quantitative measures (elution peak asymmetry and HETP, for instance), it is more often the case that the chromatogram is used to confirm that the appropriate buffer transitions have taken place at the correct times, by recording column effluent conductivity, pH, and UV absorbance.

For quantitative attributes, statistical analyses may be used to more rigorously compare performance at the two scales. A comparison of the means for the laboratory and process systems can be conducted using equivalence testing which aims not to show differences but to conclude similarity. Confidence intervals must be set around the differences in means, which is typically 90% and depends on the number of runs and variation. A pre-defined threshold for acceptance should also be defined and be based not on process data but rather on scientific or clinical indifference. This threshold is not trivial to establish and requires careful consideration from subject matter experts to prevent an overly wide criteria that always passes or too narrow criteria that never passes. For equivalence of scales, the mean difference in measured results should be statistically smaller than the predefined threshold. Figure 5.7 shows an illustration of several cases when equivalence can and cannot be established based on the average difference and the confidence interval of that difference. Case A shows a scenario where there is a significant difference but would be considered equivalent between scales since the confidence interval falls within the acceptance threshold. Case B shows no significant difference and is equivalent between scales. Cases C and D show scenarios where there is a significant difference, and both would not be considered equivalent between scales. Case E shows no significant difference but poor precision, which prevents concluding the scales are equivalent. The risk of the Case E scenario in Figure 5.7 is not unlikely given process-scale datasets are typically limited due to fewer batches run. In this case, additional data or scientific justification may be required to conclude equivalence.

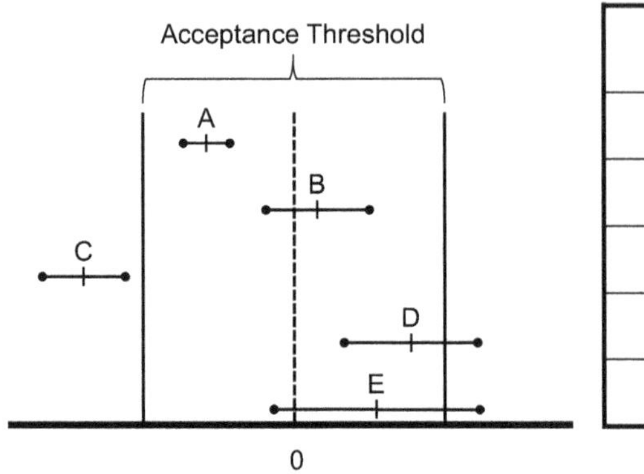

Case	Significant Difference?	Equivalence?
A	Yes	Yes
B	No	Yes
C	Yes	No
D	Yes	No
E	No	No

FIGURE 5.7
Illustration and table of equivalence testing between scales based on average difference (vertical tick marks) and the confidence interval of the difference (horizontal bars).

Often the scale-down dataset displays less variability than the process dataset, as the runs are typically conducted with identical equipment over a short time span, using the same lots of raw materials, and the product peaks are typically analyzed together. The load material used to perform the qualification runs will influence the process outputs, and a direct comparison to the full-scale process run derived from the same starting material may be useful, although this weakens the statistical evaluation by reducing the amount of information about the full-scale process. Outlier analysis may be needed when one of the scale-down qualification runs is markedly different from the other two (Box, Hunter, and Hunter 1978). When an outlier is suspected, an investigation into the root cause should be conducted; if an explanation for the deviant result can be established, the questionable run should be eliminated from the dataset and another completed. In some cases, a root cause for the inconsistency may not be identified, in which case one or more additional runs will be needed to confirm the suitability of the scale-down system's performance.

When discrepancies between scales have been established, potential causes should be identified and corrected. This investigation is an important and necessary response to the detection of significant differences between scales, as it may highlight design flaws that could improve the small-scale model. The investigation should be documented and the data from both the original and improved scale-down model included in the report to the qualification protocol. As an example, in our experience this type of investigation was triggered when a chromatographic step yield was significantly different between scales. It was discovered that the UV detector flow-cell path length was incorrect; when the correct path length was used, the difference in step yields was greatly reduced and no longer found to be statistically significant.

In some cases, there may be differences in performance observed between the scale-down and process systems (only in rare instances are the means of two datasets exactly identical!). For quantitative variables, these would have to be statistically significant in order to warrant further investigation, and the appropriate statistical test(s) should be completed to make this assessment. With qualitative variables, a significant difference arises when the output is outside of the manufacturing experience derived from a representative sample of a sufficiently large dataset to provide an accurate estimate of full-scale performance (leading to the acceptance criteria for qualitative or characterization data being within the standard range of manufacturing) (Box, Hunter, and Hunter 1978). By using multiple lots of load material for the triplicate qualification runs, it is more likely that the mean process performance of the small-scale system will approximate that of the full-scale system, but at the loss of accurate information about the run-to-run variability of the lab-scale system.

If no correction can be found, then adjustments may be made to the small-scale system to more closely match the full-scale process, even if these violate one of the scaling laws adopted for the scale-down process step design. As an example, if a scale-down centrifugation step fails to adequately clarify the process stream based on a calculation of the g-force, residence time, and settling distance of a process-scale unit, a longer centrifugation time may be tested. Also, load samples for the small-scale system that are frozen for ease of operation may develop low levels of precipitate which can be removed by filtration before use, even if the full-scale product stream is never frozen or normally filtered at that point.

Finally, if performance differences between scales cannot be corrected, a judgment may be made as to the process significance of this difference. In some cases, the differences are relatively small and may not be likely to have any influence on the more relevant process outputs. As an example, the HETP of a small-scale column chromatographic may not uncommonly have a modest increase in plate height compared to the full-scale column; while this is important for delicate separations and size-exclusion chromatography, for most bind and elute modes of chromatography, this difference may not matter. In these cases, the small-scale system can still be used for prospective validation studies. Should the difference be judged to have process significance (a major shift in product isoforms, the presence of new impurities or product isoforms, or large yield discrepancies), however, the small-scale system should not be used for process validation studies unless absolutely no other alternative exists. This may require more validation work at full scale using concurrent studies as a result.

A preliminary evaluation of the potential for success of a scale-down qualification may come from careful analysis of the scale-up of the steps following their definition from small-scale process development studies. This scale-up is simply the application of the same scaling laws used for the design scale-down, but in the opposite direction! Parallel runs conducted using samples from the full-scale process that are processed and analyzed at the same time as the full-scale batch should minimize the potential for any surprises when the qualification studies are conducted under protocol. Careful analysis of several of these parallel runs may also identify subtle, yet reproducible, scale-related effects that often require a larger dataset than the three qualification runs in order to become statistically significant.

Flexible operations using small-scale systems with various scaling factors may be supported by qualifying a range of small-scale system sizes. The smallest would be the most efficient to use for studies that involve many runs, such as multivariable robustness studies or chromatographic column reuse. Larger systems may be needed if multiple steps are used to model a full purification train, for instance. Because losses are incurred when peak pools are sampled for analysis or as retains, the scale-down factor will by necessity increase as the product progresses through the purification train. Another factor that can have the same effect is that small-scale systems may come in quantum sizes (i.e., chromatographic column diameters, or filter sizes), which will result in a maximum scale factor for the process train based on the maximum scale factor for the limiting unit operation. When considering the qualification of multiple scales of operation, a bracketing strategy should be applicable if similar equipment is used for both the largest and smallest scales. Intermediate scales of operation would therefore be covered by the successful qualification of the minimum and maximum scales.

5.12 Application of Scale-Down Systems

Scale-down models are valuable tools during process development and beyond. As mentioned earlier, one of the primary applications of scale-down models is during process development for evaluating impurity removal. However, these small-scale models can also be used for evaluation of process robustness, resin cleaning, and resin reuse and lifetime among others. Furthermore, they can also be used to support licensed commercial processes to address manufacturing investigations and post-approval process changes for comparability and to evaluate raw material changes post-licensure. All these applications were covered in detail in the previous version of this chapter, and we encourage readers to refer to that text (Godavarti et al. 2005).

Since this chapter's last publication, there has been considerable advancements in automated low-volume high-throughput screening (HTS) experiments. Efficient parallelization of HTS experiments is now easily achieved with automated UF systems, liquid handling robots, and miniature chromatography systems. These technologies have helped enable microliter purifications to screen large experimental spaces with minimal sample requirements while reducing overall costs and development times.

HTS systems are applied at all stages of process development such as sample purification (Butcher et al. 2019; Schmidt et al. 2016; Lacki and Brekkan 2011), resin screening (Zhang, Chen, and Li 2019; Chu et al. 2018; Kittelmann, Ottens, and Hubbuch 2015; Liu et al. 2015; Kiesewetter et al. 2016), and process optimization (McDonald et al. 2016; Bergander et al. 2008; Coffman, Kramarczyk, and Kelley 2008). The incorporation of automation with mini peristaltic pumps and accurate liquid pipetting make these systems invaluable tools when the number of experimental conditions, if executed in series, would require overwhelming manual labor and time. Such a case often occurs when trying to link upstream and downstream process performance through a limited number of experiments. These studies can include many upstream variables related to controlling the cell culture process that may impact the load variability and the downstream process performance and product quality. The amount of data collection now possible through HTS helps screen a wide multivariate space and gain deeper process insights and understanding.

There has been considerable work in establishing HTS experiments as small-scale models of full-scale manufacturing processes (Treier et al. 2012; Welsh et al. 2014; Petroff et al. 2016; Evans et al. 2017; Benner et al. 2019; Fernandez-Cerezo et al. 2020). Low-volume UF/DF systems and miniature chromatography columns can be used to predict larger-scale yield and product quality (Fernandez-Cerezo et al. 2020; Welsh et al. 2014). A comparison of minicolumn and lab-scale column yield, HCP concentration, and absorbance at 320 nm after Protein A purification and pH titration is shown in Figure 5.8. Panel A of Figure 5.8 shows comparable yields between the mini and lab-scale columns for three different monoclonal antibodies. Panel B shows very similar HCP concentrations between both scales for mAb 2 after pH titration from pH 4 to pH 8.1. Panel C shows similar 320 nm absorbance at both scales for mAb 2 for a range of pH values from 4 to 8.1. Absorbance at 320 nm can be roughly correlated with turbidity to identify pools that may have lower filterability for subsequent downstream steps.

Although good agreement between varying scales can be readily achieved with respect to yield and product quality, minicolumns rarely have elution pool volumes or chromatographic profiles that match process-scale columns (Welsh et al. 2014; Treier et al. 2012). At the microliter scale, larger dead-volume effects and non-ideal flow dynamics increase the complexity of establishing HTS systems as representative scale-down models (Benner et al. 2019).

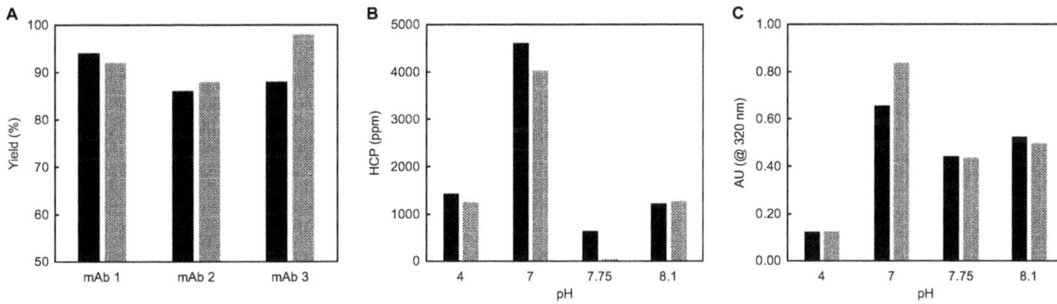

FIGURE 5.8

Comparison of Protein A performance and elution pool product quality between bench-scale columns (solid bars) and miniature columns (diagonal bars). The yield for three monoclonal antibodies is shown in panel A, the HCP concentration in the elution pool after titration to pH 4 to 8.1 is shown for mAb 2 in panel B, and absorbance at 320 nm after titration from pH 4 to 8.1 is shown for mAb 2 in panel C.

In the case of low-volume high-throughput filtration screening systems, the smallest available membranes are about 10 cm^2 and do not typically contain turbulence promoters found in larger membranes. The simpler membrane design in these small membrane cassettes prevent them from having similar pressure differentials, path lengths, and fluxes compared to process-scale membranes (Fernandez-Cerezo et al. 2020). Fernandez-Cerezo and coworkers have shown it is possible to modify the membrane flow path in these small area cassettes to increase the axial pressure drop until there is better alignment with performance at pilot-scale systems; however, the range of agreement between scales starts to deviate as the load challenge increases past 250 g/L (Fernandez-Cerezo et al. 2020). Regardless of the differences that exist, these low-volume high-through filtration systems can still be useful in understanding how buffer composition, high shear conditions, and repeated cycling impact performance and product quality.

In the case of miniature chromatography columns, good packing quality and flow non-uniformity are often the limiting factors that prevent similar performance to bench-scale and process-scale columns. Minicolumns with volumes between 50 and 600 uL are regularly run with intermittent flow as the liquid handling instruments need to apply feed material and buffers over multiple aspirating and dispensing cycles. These minicolumns also have diameters of 0.5 cm and lengths between 1 and 3 cm, which can exacerbate the effects of non-ideal flow like channeling along the wall (Knox, Laird, and Raven 1976; Farkas and Guiochon 1997) and axial dispersion effects that directly impact performance and elution pool volumes (Carta and Jungbauer 2020). With shorter column length and limitations to the flow rates that liquid handling robots can accurately deliver, it can be challenging to match the residence times used in larger scale columns. Despite these differences in operation and performance, it should be noted that exact agreement between the scale-down model and process-scale data is not needed for regulatory acceptance. To claim the scale-down model represents manufacturing-scale behavior, any scaling differences should be scientifically justified. Mechanistic models can help explain the effects of system differences and bridge the performance gap between miniature and process-scale columns (Susanto et al. 2008; Keller et al. 2015; Benner et al. 2019; Keller et al. 2017; Osberghaus et al. 2012). These models all use scale-independent isotherm parameters with an adsorption rate model that accounts for mass transfer resistance and axial dispersion effects. Benner et al. and Susanto et al. have both shown it is possible to calibrate these models with HT experiments and accurately describe the excessive peak broadening seen in miniature columns due to increased axial dispersion (Benner et al. 2019; Susanto et al. 2008).

HTS applications have mostly been in early clinical development phase. Automation and miniature unit operations help expand the experimental space while reducing time and material consumption. As these automated systems become more advanced and mechanistic understandings develop over time, high-throughput experiments will play a larger role as scale-down models to support process validation, commercial processes, and regulatory filings.

5.13 Summary

A complete process validation package is a major component of any regulatory filing. Process validation studies are performed either at full scale or using scale-down models. Scale-down models are laboratory-scale systems designed to model a full-scale unit operation used for developing a purification process, which are subsequently scaled up to production scale. Alternatively, they are used prospectively as useful tools designed to mimic large-scale unit operations.

Prior to conducting any process validation or process characterization studies, it is very important to ensure that the scale-down model system appropriately reflects the performance of the unit operation at full scale. Several parameters need to be considered in designing a scale-down model. These parameters are unique to any given unit operation. It is important to identify the critical and key parameters, scaling principles, and the appropriate endpoints for each unit operation prior to designing the scale-down model that would be relevant for comparing the two scales of operation.

For small-scale studies that are used to support product licensure applications for therapeutic proteins, the evaluation of the performance of the small-scale system should be conducted prior to initiating process characterization studies. The data from the relevant process parameters from the scale-down process should be compared, both qualitatively and quantitatively, to the full-scale processes. Qualitative comparisons may be used for evaluation of data from complex analytical methods. For quantitative variables, statistical analyses may be used to compare performance more rigorously between two scales.

Scale-down models have several applications. They have been used to evaluate removal of impurities such as nucleic acids, host cell proteins, viruses, and media additives, among others. Scale-down models are effective tools to determine useful life of chromatographic resins and to evaluate process robustness. For licensed processes, scale-down models play an important role in supporting process changes and manufacturing investigations.

Acknowledgments

The authors would like to thank Jeff Salm, Tim Iskra, Khurram Sunasara, Prashant Ganji, Xi Hu, Eric Bortell, and Brad Evans for their thoughtful comments and review of this chapter. The authors would also like to thank members of the Purification Process Development group at Pfizer, who participated in some of the studies described here.

References

Arbeit, Robert D., Walter W. Karakawa, Willie F. Vann, and John B. Robbins. 1984. "Predominance of two newly described capsular polysaccharide types among clinical isolates of Staphylococcus aureus." *Diagnostic Microbiology and Infectious Disease* 2 (2):85–91.

Banerjee, Amit. 2010. "Designing in quality: Approaches to defining the design space for a monoclonal antibody process." *BioPharm International* 23.

Barry, A., and R. Chojnacki. 1994. "Biotechnology product validation, part 8: Chromatography media and column qualification." *BioPharm* 7:43–47.

Beck, Alain, Liliane Goetsch, Charles Dumontet, and Nathalie Corvaïa. 2017. "Strategies and challenges for the next generation of antibody–drug conjugates." *Nature Reviews Drug Discovery* 16 (5):315–337.

Benner, Steven W., John P. Welsh, Michael A. Rauscher, and Jennifer M. Pollard. 2019. "Prediction of lab and manufacturing scale chromatography performance using mini-columns and mechanistic modeling." *Journal of Chromatography A* 1593:54–62. doi: 10.1016/j.chroma.2019.01.063.

Bergander, Tryggve, Kristina Nilsson-Välimaa, Katarina Öberg, and Karol M. Lacki. 2008. "High-throughput process development: Determination of dynamic binding capacity using microtiter filter plates filled with chromatography resin." *Biotechnology Progress* 24 (3):632–639. doi: 10.1021/bp0704687.

Box, G. E. P., W. G. Hunter, and J. S. Hunter. 1978. *Statistics for Experimenters*. Wiley.

Brandish, Philip E., Anthony Palmieri, Svetlana Antonenko, Maribel Beaumont, Lia Benso, Mark Cancilla, Mangeng Cheng, Laurence Fayadat-Dilman, Guo Feng, and Isabel Figueroa. 2018. "Development of anti-CD74 antibody–drug conjugates to target glucocorticoids to immune cells." *Bioconjugate Chemistry* 29 (7):2357–2369.

Brown, David A., Pip N. Jones, John C. Middleton, George Papadopoulos, and Engin B. Arik. 2004. "Experimental methods." In *Handbook of Industrial Mixing: Science and Practice*, edited by Edward L Paul, Victor A. Atiemo-Obeng and Suzanne M. Kresta, 145–256. John Wiley & Sons, Inc.

Butcher, Rebecca E., Genevieve Martin-Roussety, Rebecca A. Bradford, Andrea Tester, Catherine Owczarek, Matthew P. Hardy, Chao Guang Chen, Georgina Sansome, Louis J. Fabri, and Peter M. Schmidt. 2019. "Optimizing high throughput antibody purification by using continuous chromatography media." *Protein Expression and Purification* 159:75–82. doi: 10.1016/j.pep.2019.03.011.

Carta, Giorgio, and Alois Jungbauer. 2020. *Protein Chromatography: Process Development and Scale-Up*. John Wiley & Sons.

Cecchini, Douglas J. 2008. Applications of design space for biopharmaceutical purification processes. In *Quality by Design for Biopharmaceuticals*. John Wiley & Sons, Ltd.

Center for Biologics Evaluation and Research, FDA. 1996. Guidance for Industry for the Submission of Chemistry, Manufacturing and Controls for a Therapeutic Recombinant DNA-derived Product or Monoclonal Antibody Product for In Vivo Use.

Center for Biologics Evaluation and Research. 1997. Food and Drug Administration Docket Number FDA-1994-D-0318. Points to Consider in the Manufacture and Testing of Monoclonal Antibody Products for Human Use.

Charlton, Henry R., Julian M. Relton, and Nigel K. H. Slater. 1999. "Characterisation of a generic monoclonal antibody harvesting system for adsorption of DNA by depth filters and various membranes." *Bioseparation* 8:281–291. doi: 10.1023/A:1008142709419.

Chennasamudram, Sudha, and Willie F. Vann. 2014. "Conjugate vaccine production technology." In *Vaccine Development and Manufacturing*. John Wiley & Sons, Inc.

Chhatre, Sunil, and Nigel J. Titchener-Hooker. 2009. "Review: Microscale methods for high-throughput chromatography development in the pharmaceutical industry." *Journal of Chemical Technology & Biotechnology* 84:927–940. doi: 10.1002/jctb.2125.

Chu, Wen Ning, Qi Ci Wu, Shan Jing Yao, and Dong Qiang Lin. 2018. "High-throughput screening and optimization of mixed-mode resins for human serum albumin separation with microtiter filter plate." *Biochemical Engineering Journal* 131:47–57. doi: 10.1016/j.bej.2017.12.001.

CMC Biotech Working Group. 2009. *A–Mab: A Case Study in Bioprocess Development*. CASSS.

Coffman, Jonathan L., Jack F. Kramarczyk, and Brian D. Kelley. 2008. "High-throughput screening of chromatographic separations: I. Method development and column modeling." *Biotechnology and Bioengineering* 100:605–618. doi: 10.1002/bit.21904.

Cooney, C. L., and K. B. Konstantinov. 2014. "White paper on continuous bioprocessing." International Symposium on Continuous Manufacturing of Pharmaceuticals.

Costantino, Paolo, Rino Rappuoli, and Francesco Berti. 2011. "The design of semi-synthetic and synthetic glycoconjugate vaccines." *Expert Opinion on Drug Discovery* 6 (10):1045–1066.

European Medicines Agency. 2014. Guideline on Process Validation for the Manufacture of Biotechnology-Derived Active Substances and Data to be Provided in the Regulatory Submission-Draft.

European Medicines Evaluation Agency. 1995. CPMP Guideline in Virus Validation Studies. CPMP/BWP/268/95.

Evans, Steven T., Kevin D. Stewart, Chris Afdahl, Rohan Patel, and Kelcy J. Newell. 2017. "Optimization of a micro-scale, high throughput process development tool and the demonstration of comparable process performance and product quality with biopharmaceutical manufacturing processes." *Journal of Chromatography A* 1506:73–81. doi: 10.1016/j.chroma.2017.05.041.

Farkas, Tivadar, and Georges Guiochon. 1997. "Contribution of the radial distribution of the flow velocity to band broadening in HPLC columns." *Analytical Chemistry* 69:4592–4600. doi: 10.1021/ac970530m.

FDA/ICH. 2009a. Food and Drug Administration Docket Number FDA-2007-D-0370. Q10 Pharmaceutical Quality System.

FDA/ICH. 2009b. Food and Drug Administration Docket Number FDA-2005-D-0154. Q8(R2) Pharmaceutical Development.

FDA/ICH. 2023. Food and Drug Administration Docket Number FDA-2022-D-0705. Q9(R1) Quality Risk Management.

Fernandez-Cerezo, Lara, Michael K. Wismer, InKwan Han, and Jennifer M. Pollard. 2020. "High throughput screening of ultrafiltration and diafiltration processing of monoclonal antibodies via the ambr® crossflow system." *Biotechnology Progress* 36. doi: 10.1002/btpr.2929.

Fiadeiro, Marcus, Jill Kublbleck, Robert Fahrner, and Jeffery Salm. 2016. Enabling Continuous Low pH Viral Inactivation With Integrated In-Line Conditioning of Protein A Streams. Abstracts of Papers of the American Chemical Society, 251st National Meeting in 2016.

Fu, Cecilia H., and Kathleen M. Sakamoto. 2007. "PEG-asparaginase." *Expert Opinion on Pharmacotherapy* 8 (12):1977–1984.,

Gagnon, P., and E. Grund. 1996. "Large-scale process development for hydrophobic interaction chromatography, part 4: Controlling selectivity." *BioPharm* 9:54–64.

Godavarti, Ranga, J. Petrone, J. Robinson, R. Wright, and Brian Kelley. 2005. "Scale-down models for purification processes: Approaches and applications." *Process Validation in Manufacturing of Biopharmaceuticals*:69–142.

Godawat, Rahul, Konstantin Konstantinov, Mahsa Rohani, and Veena Warikoo. 2015. "End-to-end integrated fully continuous production of recombinant monoclonal antibodies." *Journal of Biotechnology* 213:13–19. doi: 10.1016/j.jbiotec.2015.06.393.

Grenville, Richard K., and Alvin W. Nienow. 2004. "Blending of miscicle liquids." In *Handbook of Industrial Mixing: Science and Practice*, edited by Edward L. Paul, Victor A. Atiemo-Obeng, and Suzanne M. Kresta, 507–542. John Wiley & Sons, Inc.

Gunnoo, S. B., and A. Madder. 2016. "Bioconjugation—using selective chemistry to enhance the properties of proteins and peptides as therapeutics and carriers." *Organic & Biomolecular Chemistry* 14 (34):8002–8013. doi: 10.1039/c6ob00808a.

Hadpe, Sandeep R., Vipin Mohite, Solomon Alva, and Anurag S. Rathore. 2020. "Pretreatments for enhancing clarification efficiency of depth filtration during production of monoclonal antibody therapeutics." *Biotechnology Progress*:e2996. doi: 10.1002/btpr.2996.

Hahn, Tobias, Thiemo Huuk, Vincent Heuveline, and Jürgen Hubbuch. 2015. "Simulating and optimizing preparative protein chromatography with ChromX." *Journal of Chemical Education* 92:1497–1502. doi: 10.1021/ed500854a.

Hakemeyer, Christian, Nathan McKnight, Rick St. John, Steven Meier, Melody Trexler-Schmidt, Brian Kelley, Frank Zettl, Robert Puskeiler, Annika Kleinjans, Fred Lim, and Christine Wurth. 2016. "Process characterization and design space definition." *Biologicals* 44:306–318. doi: 10.1016/j.biologicals.2016.06.004.

Hermanson, Greg T. 2013. *Bioconjugate Techniques*. Academic Press.

Hu, Qi-Ying, Francesco Berti, and Roberto Adamo. 2016a. "Towards the next generation of biomedicines by site-selective conjugation." *Chemical Society Reviews* 45 (6):1691–1719.

Hu, Qi-Ying, Francesco Berti, and Roberto Adamo. 2016b. "Towards the next generation of biomedicines by site-selective conjugation." *Chemical Society Reviews* 45:1691–1719.

Hu, Xi, Eric Bortell, Frank W. Kotch, April Xu, Bo Arve, and Stephen Freese. 2017. "Development of commercial-ready processes for antibody drug conjugates." *Organic Process Research & Development* 21 (4):601–610.

Jungbauer, Alois. 2013. "Continuous downstream processing of biopharmaceuticals." *Trends in Biotechnology* 31:479–492. doi: 10.1016/j.tibtech.2013.05.011.

Keller, William R., Steven T. Evans, Gisela Ferreira, David Robbins, and Steven M. Cramer. 2015. "Use of minicolumns for linear isotherm parameter estimation and prediction of benchtop column performance." *Journal of Chromatography A* 1418:94–102. doi: https://doi.org/10.1016/j.chroma.2015.09.038.

Keller, William R., Steven T. Evans, Gisela Ferreira, David Robbins, and Steven M. Cramer. 2017. "Understanding operational system differences for transfer of miniaturized chromatography column data using simulations." *Journal of Chromatography A* 1515:154–163. doi: 10.1016/j.chroma.2017.07.091.

Khanal, Ohnmar, Xuankuo Xu, Nripen Singh, Steven J. Traylor, Chao Huang, Sanchayita Ghose, Zheng Jian Li, and Abraham M. Lenhoff. 2019. "DNA retention on depth filters." *Journal of Membrane Science* 570–571:464–471. doi: 10.1016/j.memsci.2018.10.058.

Khongorzul, Puregmaa, Cai Jia Ling, Farhan Ullah Khan, Awais Ullah Ihsan, and Juan Zhang. 2020. "Antibody–drug conjugates: A comprehensive review." *Molecular Cancer Research* 18 (1):3–19.

Kiesewetter, André, Peter Menstell, Lars H. Peeck, and Andreas Stein. 2016. "Development of pseudo-linear gradient elution for high-throughput resin selectivity screening in RoboColumn® Format." *Biotechnology Progress* 32:1503–1519. doi: 10.1002/btpr.2363.

Kittelmann, Jörg, Marcel Ottens, and Jürgen Hubbuch. 2015. "Robust high-throughput batch screening method in 384-well format with optical in-line resin quantification." *Journal of Chromatography B: Analytical Technologies in the Biomedical and Life Sciences* 988:98–105. doi: 10.1016/j.jchromb.2015.02.028.

Knox, John H., George R. Laird, and Paul A. Raven. 1976. "Interaction of radial and axial dispersion in liquid chromatography in relation to the 'infinite diameter effect.'" *Journal of Chromatography A* 122:129–145. doi: 10.1016/S0021-9673(00)82240-2.

Kozlowski, Steven, and Patrick Swann. 2008. Considerations for biotechnology product quality by design. In *Quality by Design for Biopharmaceuticals*: John Wiley & Sons, Ltd.

Lacki, Karol M., and Eggert Brekkan. 2011. High throughput screening techniques in protein purification. In *Protein Purification: Principles, High Resolution Methods, and Applications*. 3rd ed. Wiley Blackwell.

Li, Feng, and Ram I. Mahato. 2017. *Bioconjugate Therapeutics: Current Progress and Future Perspective*. ACS Publications.

Liu, Songsong, Spyridon Gerontas, David Gruber, Richard Turner, Nigel J. Titchener-Hooker, and Lazaros G. Papageorgiou. 2015. Optimal resin selection for integrated chromatographic separations in high-throughput screening. In *Computer Aided Chemical Engineering*. Elsevier B.V.

Marshall, Elizabeth Marden, and André Bakker. 2004. "Computational fluid mixing." E. L. Paul, V. A. Atiemo-Obeng, S. M. Kresta, *Handbook of Industrial Mixing: Science and Practice*, 257–343. John Wiley & Sons, Inc.

McDonald, P., B. Tran, C. R. Williams, M. Wong, T. Zhao, B. D. Kelley, and P. Lester. 2016. "The rapid identification of elution conditions for therapeutic antibodies from cation-exchange chromatography resins using high-throughput screening." *Journal of Chromatography A* 1433:66–74. doi: 10.1016/j.chroma.2015.12.071.

Michen, Benjamin, Annegret Diatta, Johannes Fritsch, Christos Aneziris, and Thomas Graule. 2011. "Removal of colloidal particles in ceramic depth filters based on diatomaceous earth." *Separation and Purification Technology* 81:77–87. doi: 10.1016/j.seppur.2011.07.006.

Nie, Lei, Dong Gao, Haiyan Jiang, Jinxia Gou, Lei Li, Feng Hu, Tingting Guo, Haibin Wang, and Haibin Qu. 2019. "Development and qualification of a scale-down mammalian cell culture model and application in design space development by definitive screening design." *AAPS PharmSciTech* 20:246. doi: 10.1208/s12249-019-1451-7.

O'Mara, Brian, Zhong Hua Gao, Manju Kuruganti, Robert Mallett, Gautam Nayar, Laura Smith, Jeffrey D. Meyer, Jon Therriault, Cameron Miller, John Cisney, and John Fann. 2019. "Impact of depth filtration on disulfide bond reduction during downstream processing of monoclonal antibodies from CHO cell cultures." *Biotechnology and Bioengineering* 116:1669–1683. doi: 10.1002/bit.26964.

Osberghaus, A., K. Drechsel, S. Hansen, S. K. Hepbildikler, S. Nath, M. Haindl, E. von Lieres, and J. Hubbuch. 2012. "Model-integrated process development demonstrated on the optimization of a robotic cation exchange step." *Chemical Engineering Science* 76:129–139. doi: https://doi.org/10.1016/j.ces.2012.04.004.

Parenteral Drug Association. 2005. "Parenteral Drug Association's (PDA) TR 42: Process validation of protein manufacturing." *PDA Journal of Pharmaceutical Science and Technology* 59:1–28.

Parker, Stephanie A., Linus Amarikwa, Kevin Vehar, Raquel Orozco, Scott Godfrey, Jon Coffman, Parviz Shamlou, and Cameron L Bardliving. 2018. "Design of a novel continuous flow reactor for low pH viral inactivation." *Biotechnology and Bioengineering* 115 (3):606–616.

Perez-Almodovar, Ernie X., and Giorgio Carta. 2009. "IgG adsorption on a new protein A adsorbent based on macroporous hydrophilic polymers: II. Pressure–flow curves and optimization for capture." *Journal of Chromatography A* 1216 (47):8348–8354. doi: https://doi.org/10.1016/j.chroma.2009.09.033.

Petroff, Matthew G., Haiying Bao, John P. Welsh, Miranda van Beuningen-de Vaan, Jennifer M. Pollard, David J. Roush, Sunitha Kandula, Peter Machielsen, Nihal Tugcu, and Thomas O. Linden. 2016. "High throughput chromatography strategies for potential use in the formal process characterization of a monoclonal antibody." *Biotechnology and Bioengineering* 113:1273–1283. doi: 10.1002/bit.25901.

Prasad, A. Krishna, Jin-hwan Kim, and Jianxin Gu. 2018. "Design and development of glycoconjugate vaccines." In *Carbohydrate-Based Vaccines: From Concept to Clinic*, 75–100. ACS Publications.

Prashad, Amar S., Birte Nolting, Vimalkumar Patel, April Xu, Bo Arve, and Leo Letendre. 2017. "From R&D to clinical supplies." *Organic Process Research & Development* 21 (4):590–600.

Prentice, Jessica, Steven T. Evans, David Robbins, and Gisela Ferreira. 2020. "Pressure-Flow experiments, packing, and modeling for scale-up of a mixed mode chromatography column for biopharmaceutical manufacturing." *Journal of Chromatography A* 1625:461117. doi: 10.1016/j.chroma.2020.461117.

Q5A *Viral Safety Evaluation of Biotechnology Products Derived From Cell Lines of Human or Animal Origin.*

Q8(R2) *Pharmaceutical Development.*

Q9 *Quality Risk Management.*

Q10 *Pharmaceutical Quality System.*

Rathore, Anurag S., Nikhil Kateja, and Devashish Kumar. 2018. "Process integration and control in continuous bioprocessing." *Current Opinion in Chemical Engineering* 22:18–25. doi: 10.1016/j.coche.2018.08.005.

Ricart, Alejandro D. 2011. "Antibody-drug conjugates of calicheamicin derivative: Gemtuzumab ozogamicin and inotuzumab ozogamicin." *Clinical Cancer Research* 17 (20):6417–6427.

Schmidt, Peter M., Michael Abdo, Rebecca E. Butcher, Min Yin Yap, Pierre D. Scotney, Melanie L. Ramunno, Genevieve Martin-Roussety, Catherine Owczarek, Matthew P. Hardy, Chao Guang Chen, and Louis J. Fabri. 2016. "A robust robotic high-throughput antibody purification platform." *Journal of Chromatography A* 1455:9–19. doi: 10.1016/j.chroma.2016.05.076.

Schreffler, John, Matthew Bailley, Tom Klimek, Peter Agneta, W. Erick Wiltsie, Michael Felo, Pam Maisey, Xun Zuo, and Eric Routhier. 2015. "Characterization of postcapture impurity removal across an adsorptive depth filter." *Bioprocess International* 13:36–45.

Shah, Syed Imran A., Larry W. Kostiuk, and Suzanne M. Kresta. 2012. "The effects of mixing, reaction rates, and stoichiometry on yield for mixing sensitive reactions—part I: Model development." *International Journal of Chemical Engineering* 2012:750162. doi: 10.1155/2012/750162.

Smiatek, Jens, Alexander Jung, and Erich Bluhmki. 2020. "Towards a digital bioprocess replica: Computational approaches in biopharmaceutical development and manufacturing." *Trends in Biotechnology*. doi: 10.1016/j.tibtech.2020.05.008.

Smith, T. M., Wilson, E., Scott, R. G., Misczak, J. W., Bodek, J. M., and Zabriskie, D. W. 1998. "Establishment of Operating Ranges in a Purification Process for a Monoclonal Antibody." In *Validation of Biopharmaceutical Manufacturing Processes*, edited by Brian D. Kelly and R. Andrew Ramelmeier, 80–92. American Chemical Society. DOI: 10.1021/bk-1998-0698.ch007

Sofer, G., and L. Hagel. 1997. *Handbook of Process Chromatography: A Guide to Optimization, Scale-Up, and Validation*. Academic Press.

Sofer, G., and L. E. Little. 1996. "Validation: Ensuring the accuracy of scaled-down chromatography models." *BioPharm* 9:51–54.

Stephanopoulos, Nicholas, and Matthew B. Francis. 2011. "Choosing an effective protein bioconjugation strategy." *Nature Chemical Biology* 7 (12):876–884.

Stickel, J. J., and A. Fotopoulos. 2001. "Pressure-flow relationships for packed beds of compressible chromatography media at laboratory and production scale." *Biotechnology Progress* 17:744–751. doi: 10.1021/bp010060o.

Stump, Bernhard, and Jessica Steinmann. 2013. "Conjugation process development and scale-up." In *Antibody-Drug Conjugates*, 235–248. Springer.

Subramanian, Ganapathy. 2017. *Continuous Biomanufacturing: Innovative Technologies and Methods*. John Wiley and Sons.

Sunasara, Khurram, John Cundy, Sriram Srinivasan, Brad Evans, Weiqiang Sun, Scott Cook, Eric Bortell, John Farley, Daniel Griffin, and Michele Bailey Piatchek. 2018. "Bivalent rLP2086 (Trumenba®): Development of a well-characterized vaccine through commercialization." *Vaccine* 36 (22):3180–3189.

Susanto, Arthur, Esther Knieps-Grünhagen, Eric von Lieres, and Jürgen Hubbuch. 2008. "High throughput screening for the design and optimization of chromatographic processes: Assessment of model parameter determination from high throughput compatible data." *Chemical Engineering and Technology* 31:1846–1855. doi: 10.1002/ceat.200800457.

Treier, Katrin, Sigrid Hansen, Carolin Richter, Patrick Diederich, Jürgen Hubbuch, and Philip Lester. 2012. "High-throughput methods for miniaturization and automation of monoclonal antibody purification processes." *Biotechnology Progress* 28:723–732. doi: 10.1002/btpr.1533.

US Food and Drug Administration. 1987. *Guideline on General Principles of Process Validation*. US FDA.

US Food and Drug Administration. 2011. *Guidance for Industry, Process Validation: General Principles and Practices*. Center for Biologics Evaluation and Research. US Department of Health and Human Services.

Velayudhan, Ajoy. 2014. "Overview of integrated models for bioprocess engineering." *Current Opinion in Chemical Engineering* 6:83–89. doi: 10.1016/j.coche.2014.09.007.

Warikoo, Veena, Rahul Godawat, Kevin Brower, Sujit Jain, Daniel Cummings, Elizabeth Simons, Timothy Johnson, Jason Walther, Marcella Yu, Benjamin Wright, Jean Mclarty, Kenneth P. Karey, Chris Hwang, Weichang Zhou, Frank Riske, and Konstantin Konstantinov. 2012. "Integrated continuous production of recombinant therapeutic proteins." *Biotechnology and Bioengineering* 109:3018–3029. doi: 10.1002/bit.24584.

Welsh, John P., Matthew G. Petroff, Patricia Rowicki, Haiying Bao, Thomas Linden, David J. Roush, and Jennifer M. Pollard. 2014. "A practical strategy for using miniature chromatography columns in a standardized high-throughput workflow for purification development of monoclonal antibodies." *Biotechnology Progress* 30:626–635. doi: 10.1002/btpr.1905.

Xu, Ping, Colleen Clark, Todd Ryder, Colleen Sparks, Jiping Zhou, Michelle Wang, Reb Russell, and Charo Scott. 2017. "Characterization of TAP Ambr 250 disposable bioreactors, as a reliable scale-down model for biologics process development." *Biotechnology Progress* 33 (2):478–489.

Yamamoto, Shuichi, Kazahiro Nakanishi, and Ryuichi Matsuno. 1988. *Ion-Exchange Chromatography of Proteins*. CRC Press.

Yigzaw, Y., R. Piper, M. Tran, and A.A. Shukla. 2006. "Exploitation of the adsorptive properties of depth filters for host cell protein removal during monoclonal antibody purification." *Biotechnology Progress* 22:288–296. doi: 10.1021/bp050274w.

Yu, Deqiang, Mukesh Mayani, Yuanli Song, Zhizhuo Xing, Sanchayita Ghose, and Zheng Jian Li. 2019. "Control of antibody high and low molecular weight species by depth filtration-based cell culture harvesting." *Biotechnology and Bioengineering* 116:2610–2620. doi: 10.1002/bit.27081.

Zhang, Xudong, Tao Chen, and Yifeng Li. 2019. "A parallel demonstration of different resins' antibody aggregate removing capability by a case study." *Protein Expression and Purification* 153:59–69. doi: 10.1016/j.pep.2018.08.011.

6

Principles of Quality Risk Management for Validation

Tiffany Baker and Patrick Mains

6.1 Introduction

ICH Q9 defines quality risk management as a systematic process for the assessment, control, communication, and review of risks to the quality of the drug product across the product life cycle.

It is a formal process to proactively identify risks associated with pharmaceutical processes, facilities, and equipment, to assess the level of each risk, and to make decisions that will reduce each assessed risk to an acceptable level. ICH Q12, Technical and Regulatory Considerations for Pharmaceutical Process Lifecycle Management Core Guideline Guidance for Industry (ICH, 2019, p.6), takes the concept even further, describing how QRM is an enabler for the entire product life cycle. Utilizing risk management principles and tools can bridge the tacit knowledge gap between process development, commissioning and qualification, GMP manufacture, and even through decommissioning activities. QRM is no longer a "nice to have" but rather a required activity across the life cycle of a product, process, equipment, or facility by regulators around the globe.

Traditional pharmaceutical validation processes have relied on extensive testing throughout the life cycle of the process, equipment, or facility. Much of the testing is non-value added and does not improve the quality outcome. Furthermore, the "test everything" approach does little to demonstrate process and product understanding. Application of QRM tools and principles allows for the team of subject matter experts to tie quality aspects of the product being manufactured to the specific critical attributes of the manufacturing equipment through critical process parameters and to narrow the scope of process validation to focus on those elements. Application of robust quality risk management tools and principles to the validation process can also lead to right-sized scope and/or frequency of testing, allowing a firm to realize a potential reduction in the time and cost of validation exercises while still ensuring robust protection of product quality. Furthermore, performing the initial validation based on a risk-based approach will position the continuous process validation part of the life cycle to be as efficient as possible, further reducing resource load on the organization.

While QRM is a required activity per most regulatory agencies, proper utilization of the risk management tools can lead to strategic advantage with respect to deployment of personnel and other resources for validation activities and a potential for significant savings for the associated internal and external costs while maintaining quality standards of performance. It can ensure resources are focused in the right place and qualification efforts are right sized. Many companies in industry commit common errors when creating and deploying their quality risk management strategy in support of the product life cycle. This chapter will discuss how to avoid some of those common errors.

6.2 Overarching Quality Risk Management Lifecycle Strategy to Support Process Validation

ICH Q9 defines quality risk management as a systematic process for the assessment, control, communication, and review of risks to the quality of the drug product across its life cycle. It also specifically addresses the use of QRM to support validation activities in very specific ways; both "To identify the scope and extent of the verification, qualification, and validation activities" and "To distinguish between critical and non-critical process steps" (ICH, 2005). This can be accomplished through implementation of the QRM life cycle implemented in parallel with the pharmaceutical product and validation process life cycles. Quality risk management serves as the method by which a process can be assessed, understood, kept well controlled, and documented. To understand the strategy and execution of the risk management activities, though, it is critical to understand the outputs from the drug development phase and how they will be used as the foundation of the process.

The sections of this chapter will outline the QRM activities identified in Figure 6.1, as they should be performed in chronological order.

In order to walk through the strategy of lining up QRM with validation activities, it is critical to understand the strategy for setting up the risk assessments across the life cycle. The strategy outlined in this chapter can even be leveraged for benefit outside of process validation. It can be used to support the equipment commissioning and qualification effort as well, and those connections will be touched on later in this chapter. The relationships, interdependencies, and structure are best described through Figure 6.2, Pharmaceutical operations hierarchy (PDA, 2017), Figure 6.3, Flowchart of development process, and Figure 6.6, Simplified failure chain (PDA, 2017) as defined in PDA Technical Report 54–5, Quality Risk Management for the Design, Qualification, and Operation of Manufacturing Systems. Figure 6.2 demonstrates the hierarchy of operations where the production and support systems are the foundational element.

The foundation of the pyramid demonstrates that manufacturing processes are dependent on production and support systems being well controlled, therefore ultimately leading to protection of the patient at the top of the pyramid. Diving deeper into how the relationships within the pyramid are connected, starting from the top:

FIGURE 6.1
Overall sequence of process validation activities in alignment with QRM activities.

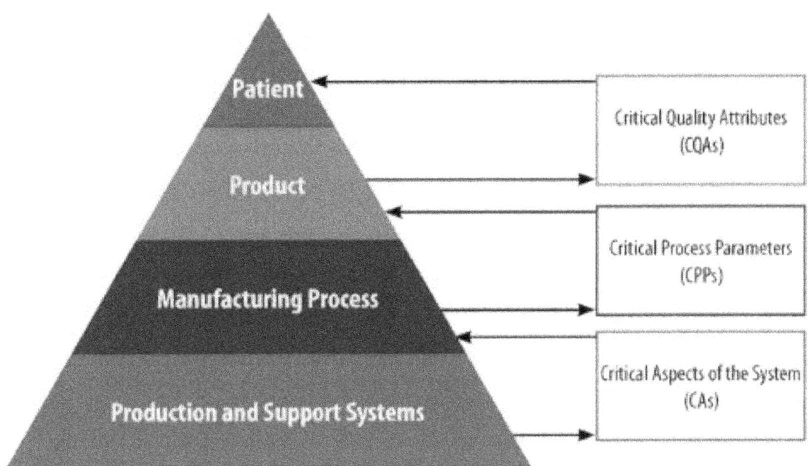

FIGURE 6.2
Pharmaceutical operations hierarchy (PDA, 2017, p. 18).

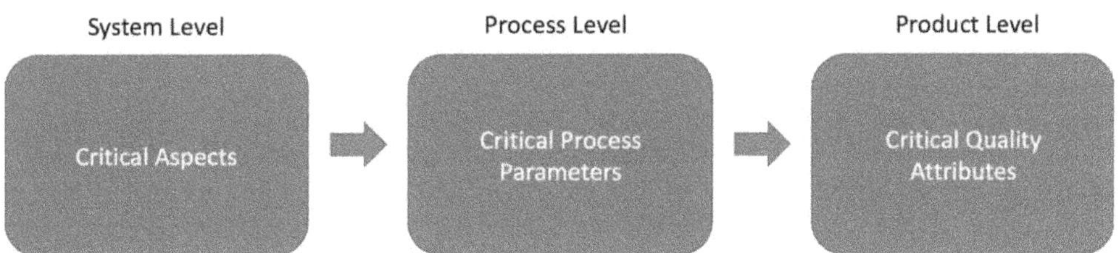

FIGURE 6.3
Flowchart of development process.

- We can ensure protection of the patient by demonstrating critical quality attributes of the product have been met. A critical quality attribute (CQA) is a physical, chemical, biological, or microbiological property or characteristic that should be within an appropriate limit, range, or distribution to ensure the desired product quality (PDA, 2017, p. 3).

- We can ensure product quality by demonstrating critical process parameters (CPPs) have been met during the manufacturing process. A critical process parameter is a process parameter whose variability has an impact on a critical quality attribute and therefore should be monitored or controlled to ensure the process produces the desired quality (PDA, 2017, p. 3).

- We can ensure a controlled manufacturing process by demonstrating that all critical aspects (CAs) of the production and support systems have operated as validated. A critical aspect is a constituent part of a system or piece of equipment that provides the ability to control one or more critical process parameters of the associated process (PDA, 2017, p. 3).

It is important to note that this pyramid supports the primary principles of quality risk management, "the evaluation of the risk to quality should be based on scientific knowledge and ultimately link to the protection of the patient."

While the pyramid shown earlier describes the hierarchy from patient to systems, it is necessary to isolate and reassemble the critical aspects on the right side of the pyramid to a sequential approach. Figure 6.3, Flowchart of development process, demonstrates the dependencies from the pyramid in Figure 6.2 for these relationships as a linear flowchart and serves as the reading frame to define the strategy for the risk assessment to be performed in support of determining scope and rigor of process validation activities.

6.3 QRM in Support of Process Development

A proactive partnership for QRM and validation activities begins at the process development and design phase of the ICH Q12 pharmaceutical product life cycle. The earlier that process development subject matter experts (SMEs) can begin work on these assessments, the more tacit knowledge will be captured and provided to future phases of the life cycle. The effort should focus on the establishment and documentation of all tacit knowledge gained during development regarding the product and proposed manufacturing process, so that a robust, commercially viable manufacturing process can be built in a thoughtful way to ensure product quality.

The goal of this early phase of the assessment will be to capture process development knowledge and combine that with the impacts that process level (CPP) failures will have on product quality (CQAs). CPPs are typically derived from other studies such as design of experiments (DOE) and design space verification, among others. This information will serve as an input to the risk assessment life cycle. The CQA/CPP assessment should assess the process, step by step, to determine which CPP failures could occur at each process step, and what the corresponding CQA impacts would be (as seen in Figure 6.4). As new information is developed, this process may need to be repeated to reassess impact. The quality target product profile (QTPP), a prospective summary of the quality characteristics of a drug product that ideally will be achieved to ensure the desired quality, taking into account safety and efficacy of the drug product (ICH, 2009, p. 16), will serve as the source document for identification of CQAs for this exercise. Several approaches can be employed to take the CQAs identified in the QTPP and make linkages between those product CQAs and the associated process CPPs. Less formal QRM tools are recommended for this phase in the life cycle, as information about the product and process is limited. A preliminary hazard analysis or simple matrix can be used to link CQAs to CPPs. In some instances these tools can even identify where studies may be helpful to better understand specific product impact from a process parameter where there may be limited understanding. It is also important to consider key process parameters (KPPs) in addition to CPPs for this effort. A key process parameter is a parameter that may impact process performance attributes (such as yield in downstream purification) but does not impact critical quality attributes (PDA, 2013, p. 21).

Many methods and tools have been developed or adapted to help manage biopharmaceutical drug Quality Risk and facilitate the performance of risk assessments. A number of these are noted in ICH Q9, and several of them may be appropriate for this type of assessment. A recommended reading list is provided at the end of the chapter that can provide source documents for how to perform risk assessments. If a hazard analysis is the chosen

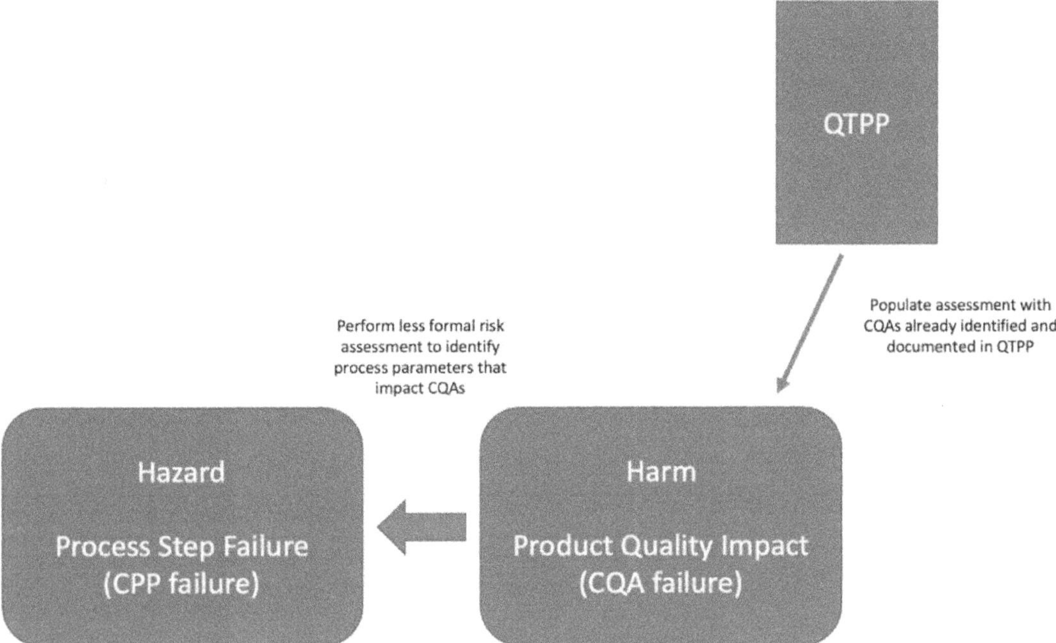

FIGURE 6.4
Simple less formal risk assessment strategy.

TABLE 6.1

Example of Hazard Analysis

Hazard (CPP)	Harm (CQA)	Impact of Hazard on Harm	Justification
Flow rate is above 30 L/min	Shear; protein damage	High	Data available to support impact of CPP on CQA

method for this assessment, it must be set up with the overarching strategy in mind. In the simple less formal risk assessment strategy (see Figure 6.4), the CPP or process-level failures are identified as the hazard within the assessment, and the subsequent CQA or product quality impacts would be captured as harms. See Table 6.1 for an example of simple hazard analysis, which will follow forward into the FMEA example later in this chapter.

Aligning this preliminary assessment with this strategy will allow simple expansion later into the process failure modes and effects analysis (FMEA) by just incorporating equipment-level failures as failure causes for each individual harm identified. This ensures early alignment with the more detailed failure chain described in Figure 6.6. We will discuss expanding the criticality assessment into the process FMEA later in this chapter. The output of the less formal risk assessment at the process development stage will serve as a direct foundation for the process FMEA, which will inform both the commissioning and qualification (C&Q) and validation strategies. See Figure 6.5 to better understand how the CQA/CPP assessment will fit into the larger QRM life cycle to support the overarching strategy.

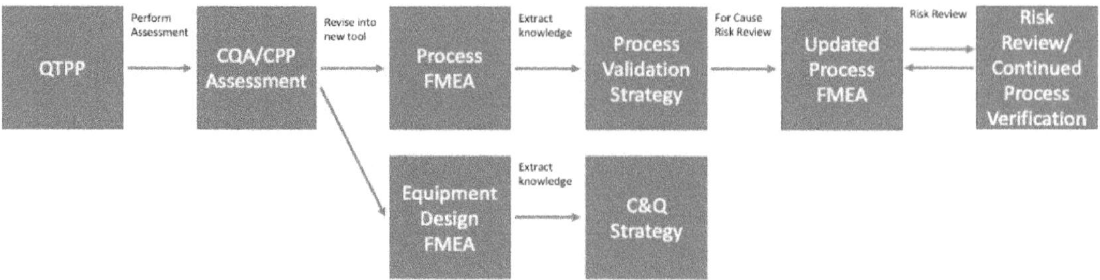

FIGURE 6.5
Risk assessment evolution from product development through process validation and verification.

A cross-functional team should be gathered to perform the CQA/CPP assessment, with responsibilities as follows:

Role	Responsibilities
Risk Assessment Owner	Initiate the risk assessment process, identify key SMEs and stakeholders, own mitigation activity resulting from the risk assessment, and communicate with all appropriate stakeholders throughout the risk assessment process.
Subject Matter Expert	Provide their subject matter expertise as an input to the risk assessment. Note: A minimum of process sciences, quality, and manufacturing should be included for this phase. Additional non-SME stakeholders required can vary by company and project.
Qualified Risk Facilitator	Ensure the risk tool methodology is adhered to throughout the assessment, assist risk assessment owner in authoring the report if needed, facilitate all working sessions and lead team to completion of risk assessment.
Senior Management	Ensure appropriate resources are allocated to the risk assessment effort.

6.4 Evolving Process Development Risk Assessments to Process Risk Assessments and Support of Process Validation

As a company transitions from a drug development phase towards a process development and validation readiness phase, the risk assessment must evolve in support of that effort. Process performance qualification is the stage that verifies the process reproducibility at commercial scale in a real-world production environment, after the equipment C&Q effort. It is the demonstration that all of the important process operations are sufficiently controlled. Transfer of the risk assessments at this phase from development teams to validation and commercial teams is an informal risk communication mechanism, and it helps to ensure tacit knowledge is maintained throughout the process. Furthermore, it ensures that the validation plan considers all critical process parameters explicitly linked to CQAs, as identified by the development teams. Both the risk assessments and the requirement trace matrix from the C&Q phase can be leveraged for process and product information when drafting the validation master plan. Figure 6.6, Simplified failure chain, demonstrates the interdependency between a CQA, its associated CPP, and its associated CA. In order to assess the risk associated with the manufacturing process as a whole, the simple hazard/

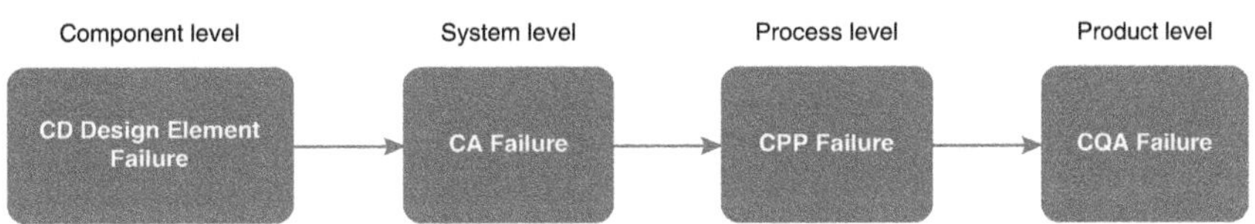

FIGURE 6.6
Simplified failure chain (PDA, 2017, p. 18).

harm structure must be transformed into a failure chain and applied to the entire manufacturing process. Figure 6.6 takes the concepts from Figure 6.4 of the interdependencies and turns them into the failure chain, upon which a risk assessment strategy can be defined. This will be the basic relationship carried through for the remainder of this chapter; it is critical for this strict linear relationship to serve as the basis for the process and design risk assessment strategy in order to produce an output that has utility to be leveraged for validation. Section 6.5 will also detail how this same strict linear relationship can be leveraged to support the C&Q strategy. Both the process and equipment design risk assessments will use this strict linear relationship as the basis for their strategy, but with different reading frames (differences in how the FMEA relationships line up with the process flow). The FMEA reading frame will be discussed further in the methodology portion of the chapter.

The process risk assessment will incorporate inputs from the process development process (the CQA/CPP assessment) and expand the failure chain (with the help of manufacturing SMEs) to include system and component level failures. Refer to Figure 6.5, Risk assessment evolution from product development through process validation and verification, for a reminder about how each assessment is evolved to support the process. A formal risk management tool should be used for the process risk assessment.

A failure modes and effects analysis is recommended, but a hazards analysis and critical control point assessment (or other formal risk tool) could also be utilized. This chapter will focus on the FMEA process-driven approach. FMEA is a systematic method for identifying, analyzing, prioritizing, and documenting potential failure modes, their effects on system, product, and process performance, and the possible causes of failure in order to prevent defects from occurring (PDA, 2017, p. 3). The hazards identified in the previously explained CQA/CPP assessment (CPP level failures) will become the failure modes in the FMEA and the harms identified (CQA impacts) will become the failure effects. CAs of the manufacturing equipment will be captured as the failure causes within the tool. The process FMEA will serve as the filtering mechanism upon which the qualification plan will be built. A second filtration assessment can be performed to determine the scope and extent of equipment qualification during the C&Q effort and will be explained in section 6.5. This step will filter out all noncritical process parameters and exclude the focus of the validation effort.

Once the CQA/CPP assessment has been translated into the process FMEA using the aforementioned strategy, each risk must be analyzed and evaluated using the defined scoring criteria. ICH Q9 provides definitions of the components of risk as well as possible scoring criteria. Ideally, scoring should be directly linked to patient impact. The patient impact assessment for each risk identified should be assessed as part of the initial CQA/CPP assessment—with each hazard directly correlating to a specific patient harm. CQA impact (which is built into the strategy) is the method by which patient impact can be assessed; product quality is the proxy for protection of patient. Once we translate the criticality assessment to FMEA, the frame of impact will shift as well, from *impact to patient* to *impact on product quality* (critical quality attributes will serve as the surrogate for protection of

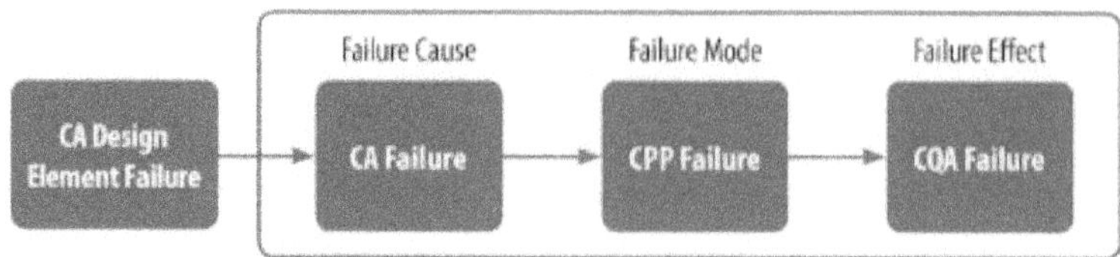

FIGURE 6.7
Simplified failure chain of the failure mode in the process FMEA (PDA, 2017, p. 18).

patient from this point forward). Rationale for the scoring criteria should be well defined, unambiguous, justified, and documented in advance. When available, scoring should be based on historical data, trending analysis, and/or nonconformances for the system. In the case of new systems, this information may not be available. In these situations, meaningful criteria should be developed and should be conservatively applied.

A trained and qualified risk facilitator must work with the risk assessment owner to define the scoring criteria and any other relevant criteria, terms, and definitions that the owner and SMEs will use to assess the previously identified risks. Ideally, this should be discussed and agreed upon by all stakeholders prior to holding risk assessment working sessions. The facilitator must ensure that the risk management process is being followed and that the scoring is being consistently applied. Another benefit to the reading frame approach is that the risk facilitator can help gain efficiencies in the process by transferring the CQA/CPP assessment relationships directly into an FMEA template. This behind-the-scenes work can save hours of working session time and reduce resource usage for SME participants.

The strategy that must be employed for the risk assessment to support the process validation strategy is captured in Figure 6.7, Simplified failure chain.

If the relationships are not aligned with the corresponding FMEA relationships within this figure, the targeted validation activities will not be apparent. If the CQA/CPP assessment was performed as described earlier in this chapter in Figure 6.6 (with CPP harms resulting in CQA impacts), the CPP failure leading to CQA impact should already be captured for the process and can just be imported as pre-work into this assessment. The additional SMEs included at this phase of the project (manufacturing process and equipment engineers) will be able to link the equipment specific critical aspects as failure causes that would lead to each CPP failure. A cross-functional team should be gathered to perform the assessment. Some roles and responsibilities are consistent with those in the CQA/CPP assessment, with an expanded group of SMEs to support the addition of system and component level knowledge. The responsibilities for the failure modes and effects analysis are as follows:

Role	Responsibilities
Risk Assessment Owner	Initiate the risk assessment process, identify key SMEs and stakeholders, own mitigation activity resulting from the risk assessment, and communicate with all appropriate stakeholders throughout the risk assessment process.
Subject Matter Expert	Provide their subject matter expertise as an input to the risk assessment. Note: Commercial Manufacturing, Engineering, Facilities, Quality Control, QA, and Validation should be included for this phase. Additional non-SME stakeholders required can vary by company and project.
Qualified Risk Facilitator	Ensure the risk tool methodology is adhered to throughout the assessment, assist risk assessment owner in authoring the report if needed, facilitate all working sessions and lead team to completion of risk assessment.
Senior Management	Ensure appropriate resources are allocated to the risk assessment effort.

TABLE 6.2

Process Validation Risk Level and Rationale

Risk Level	Rationale	Process Validation Outcome
High	Additional controls must be implemented to mitigate risk to an acceptable level.	The highest level of testing should be executed to assure performance standards are met.
Medium	Additional controls could be implemented to ensure consistent acceptable performance.	Establish appropriate targeted validation to meet suitability requirements.
Low	Robust controls and/or process knowledge is in place such that the overall risk is acceptable.	Minimal qualification with institutional gaps in knowledge or expertise*

Note: Quality review is required. A level of process qualification may be needed if an element is determined to be low risk but there is a regulatory requirement to qualify (e.g., sterilization).

The completed process risk assessment will identify the full list of critical aspects of the equipment that could impact any critical quality attribute (by way of CPPs or KPPs) during the manufacturing process. Aspects of equipment that have not been captured as having direct causal linkages with the CPPs within the risk assessment will not be considered critical and will not carry through as the focus of qualification or validation activities. These aspects will be subject to commissioning activities only. Any controls implemented in the early stages of process development should be captured as well, to ensure the control strategy of the manufacturing process is fully informed. Detection and prevention controls related to the manufacturing equipment will also be captured throughout the performance of the FMEA, which will serve as the basis for the process control strategy.

The output of the process FMEA effort will be the foundation and support for the process validation work going forward. The robust process FMEA will be the knowledge management vehicle for process and product understanding and will be revised over time through risk assessment. Testing can be focused on process parameters directly related to product CQA impact (CQA impact indicates a need for monitoring) and should be commensurate with level of risk. Low-risk items should be subject to minimal rigor, while medium- and high-risk items should be subject to a higher level of scrutiny. High-risk items should warrant more rigorous testing than medium risk—to ensure the controls are robust and reproducible, while both should be subject to some level of testing. A robust process control strategy can also be extracted directly from the process FMEA. The FMEA demonstrates which controls within the manufacturing process are directly controlling for CQAs, CPPs, or KPPs, and that accumulated list of controls should comprise the process control strategy. Utilizing the process and product relationship knowledge will allow for a targeted process control strategy and reduce issues during PPQ runs. An example of a FMEA with the associated impacts is shown below in Table 6.3.

6.5 Leveraging the Strategy to Support Commissioning and Qualification Activities

Shifting the reading frame for the risk assessment strategy can allow for gaining efficiencies in leveraging knowledge to support the equipment commissioning and qualification

TABLE 6.3

Example FMEA Line Item for a Routine BioProcess Operation

Process/System	Bioprocess Formulation
Component	Ultrafiltration Pump
Requirement	15–30 L/min

Failure Cause (CA)	Prevention Controls	Likelihood Rating	Failure Mode (CPP)	Detection Controls	Detectability Rating	Failure Cause (CQA)	Severity Rating	RPN, Overall Risk
Automation control failure	—Preventative maintenance —Calibration —Pre-use fit for use inspection	4*	Flow rate is above 30 L/min	—Inline flow meter with alarm —Pressure sensor with alarm	3*	Shear; protein damage	5*	60*, High

* RPN values are example only. Rating scales, associated criteria, and RPN thresholds will be determined by organization.

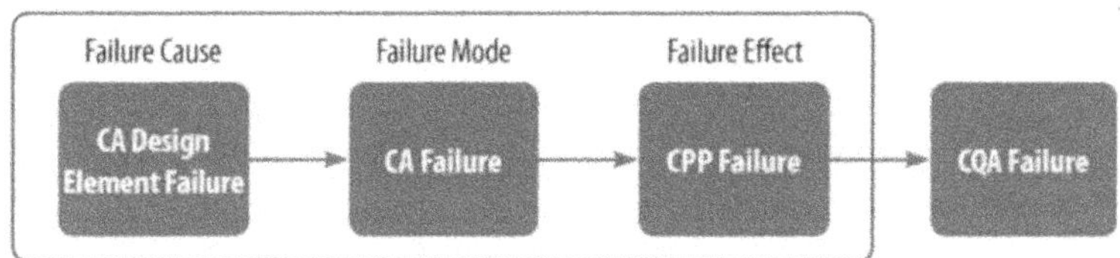

FIGURE 6.8
Failure mode chain of CAs at the component level (PDA, 2017, p. 18).

process as well. Coordinating the C&Q and process validation teams to utilize the same approach and assessment foundations can allow an organization to ensure more robust knowledge management across the organization while ensuring neither approach overlooks a critical aspect of the process or equipment. It will, by design, ensure that both processes focus on the complete list of CQAs and all associated CPPs and CAs. Each critical aspect identified will advance through the risk assessment process for further component scrutiny. Equipment that is not identified to have connections to CPPs and CQAs will not advance through this next phase of assessment, as they have not been demonstrated to have process or product quality impact. The strategy to perform the equipment design FMEAs is a building block from the strategy outlined in Figure 6.7 and can be seen in Figure 6.8. The failure cause, mode, and effect shift left one position in the failure chain so that now CPP failures are captured as failure effects, critical aspects are aligned with failure modes, and the failure causes are critical aspect design elements. This will allow the team to single out specific design elements of equipment impacting CPPs, and therefore CQAs. These design assessments serve as the identification of the qualification strategy for the equipment—if an element is identified it will be tested as it has product and process impact. If it is not identified, it will not be subject to qualification.

This is the second filtration mechanism in this process. The process FMEA was the first mechanism that allowed the removal of all aspects of the process that were not critical to quality. The second is the design FMEA that allows for exclusion of all design elements of the equipment that do not have an impact on product quality. At the end of the design assessment, the team is left with just those critical aspect design elements (CADEs) that will be qualified and focused on during the process validation. Critical aspect design elements are components, instruments, and process controls that comprise the critical aspect (e.g., temperature feedback loop) and are tested in commissioning and qualification (PDA, 2017, p. 3). Having performed both of these steps will additionally allow for the scientific justification of the reduction of activities (targeted qualification rather than exhaustive qualification) and a foundation for demonstrating product and process understanding.

The CADEs along with associated controls should be carried forward into the requirement trace matrix and incorporated into the qualification and validation strategies. All aspects of the systems must be commissioned (to ensure that they work as intended) but only the CADEs of process equipment need to be qualified (as they can be directly correlated to having product quality, CQA, impacts). The qualification of each CADE should be performed through a test specifically designed to ensure robustness of its respective controls. For example, the temperature is critical and therefore must be tested to ensure robustness, but a separate element of flow rate might not be critical and therefore can be ignored. The rigor of testing, or number of replicates for each test, should be directly related to the risk level associated with that specific CADE as determined by the design risk assessment for that piece of equipment. Table 6.2, Process Validation Risk Level and

TABLE 6.4

Example of C&Q Risk Level and Rationale

Risk Level	Rationale	Process Validation Outcome
High	Additional controls must be implemented to mitigate risk to an acceptable level.	The highest level of testing should be executed to assure performance standards are met.
Medium	Additional controls could be implemented to ensure consistent acceptable performance.	Establish appropriate targeted validation to meet suitability requirements.
Low	Robust controls and/or process knowledge is in place such that the overall risk is acceptable.	Execute commissioning only or minimal qualification for equipment with institutional gaps in knowledge or expertise.*

Note: Quality review is required. A level of qualification may be needed if an element is determined to be low risk but there is a regulatory requirement to qualify (e.g., sterilization).

Rationale, demonstrates how the level of risk corresponds to the rigor of testing. Low-risk items may only be commissioned, unless there is a recognized regulatory expectation or company commitment otherwise (e.g., as might be the case with sterilization-related processes). There should be no demonstrable connectivity to CQA impact and it should have a low likelihood of failure. Medium- and high-risk items should be tested beyond commissioning, with high-risk CADEs requiring a higher level of testing (more rigor and replicates) than medium-risk CADEs. Actual number of replicates for each test will vary by organization but should remain consistent to level of risk.

If activities performed for commissioning can demonstrate effectiveness of the CADE (and the risk assessment is performed early enough to demonstrate the impact to CQAs), commissioning activities may be leveraged as part of the qualification effort, given proper GMP documentation of those activities. In this instance, performance of the risk assessment early in the process can reduce potentially redundant work from being performed, ultimately saving project time and resources, which is demonstrated in Figure 6.9. Once all commissioning and qualification activities are completed, the systems are released for process performance qualification activities.

Pharmaceutical processes and techniques are evolving very quickly as new technologies are developed and deployed. The associated process validation and quality risk management operations must adapt to keep pace with the ever-changing process landscape. Here are two examples of such adaptation in the space:

- Single-use technologies have improved significantly in recent years. As a result, they are being widely adopted into pharmaceutical operations. One of the primary benefits it that organizations no longer need to execute cleaning validations for the single-use surfaces. However, there are frequently tradeoffs with the implementation of the new technologies. This is no exception. Organizations will need to understand the leachable and extractable profile for the new surfaces and execute studies to understand the compatibility of the products with the new surfaces. These studies should be designed using a risk-based approach to minimize surface testing while maintaining quality. Criticality assessments become critical to understanding what surfaces have direct contact with the product and therefore should be assessed though the process that has been described in this chapter.

- Individualized therapies, such as cell and gene therapies, have also become more common as the industry has developed a more robust understanding of the molecular basis for disease. This deeper understanding has led to some very exciting

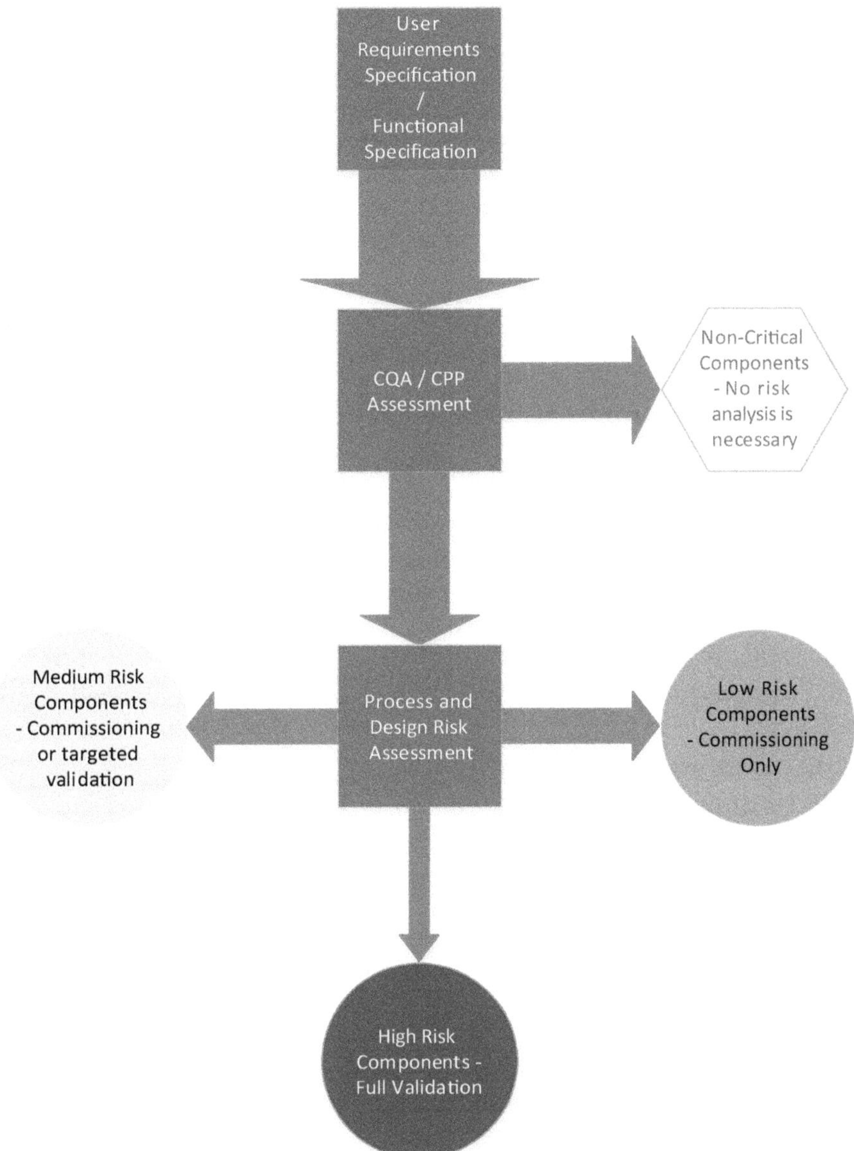

FIGURE 6.9
Example of high-level decision tree

Note: The width of the arrows is purposeful to denote magnitude/quantity. As discussed, many components will be initially assessed and filtered out until only critical components are determined to require full validation.

new therapeutic modalities. Process validation becomes more difficult for these processes because the result of the processes are not consistent compared to traditional pharmaceuticals. In traditional pharmaceuticals, the process is identical from batch to batch with the expectation that the resultant product meets a single product specification. With individualized therapies, the process is the same from batch to batch but since each dose is specially designed for an individual patient, the resultant product is different for each patient. Additionally, individualized therapies are intended to be administered much faster than traditional therapies. A traditional pharmaceutical process for production, testing, release, and distribution through to administration could be several months. Individualized therapies could be administered in days to weeks from production. This timeframe will require organizations to move away from traditional testing as a detection control and have more prevention controls that ensure that it is not possible to produce a batch that does not meet the specification. Process validation will need to adapt to this mindset in order to maintain compliance.

6.6 Risk Review to Support Continued Process Verification

Quality risk management should be a component of the periodic review process for equipment validation. Additionally, it should be considered during routine change management and deviation operations. The FMEA should be assessed upon discovery of nonconformance situations that may affect the overall validation of the system. When a nonconformance is discovered, the following considerations should be assessed:

1. Was the risk previously identified?
2. Is the equipment being operated within or outside of defined equipment parameters?
3. Was the failure mode anticipated?
4. Did the existing controls fail?
5. Are additional controls warranted?
6. What is the impact of additional controls on system/process performance?
7. Should the occurrence score be adjusted?
8. Should other systems be assessed for impact?
9. Is revalidation required?

The risk facilitator and SMEs should consider these questions and identify any other relevant concerns when a nonconformance is discovered. Changes to the risk assessment should be contemporaneously captured and approved by the relevant stakeholders. While periodic review could be considered a lagging indicator of process or equipment performance, it should be coupled with a robust real-time assessment process that will provide leading opportunities for performance assessment and correction, if needed.

A risk-based approach should be used to determine periodic review frequency. A checklist risk assessment, similar to a criticality assessment, can be used to assess systems for relative risk and assign periodic review frequency accordingly. For example, a high-risk system may need to be reviewed annually, but a low-risk system could rationally be reviewed less frequently. Rationale should be justified and documented. A robust and well-documented risk review process will provide evidence of a high level of control of the production and support systems, manufacturing processes, and ultimately the resultant product.

Risk assessments are lifecycle documents. Maintaining them as part of issue management processes and during routine scheduled review cycles will assist the owners, facilitators, and leads in making relevant decisions regarding revalidation of the system. The capacity to demonstrate system performance, stability, and control is paramount to the final decision to revalidate or not.

6.7 Common Pitfalls with Quality Risk Management Strategy Development

Some companies in industry have experienced a degree of difficulty in implementing a successful QRM strategy to support both process validation and the C&Q process. At completion of these activities, many organizations hold "lessons learned" sessions or have informal discussions with the team to better understand what went wrong. The following are some common pitfalls to avoid in setting out with a strategy like the one outlined in this chapter:

- Incomplete Expertise Represented in Risk Assessments—Ensuring SMEs have been identified to represent the entirety of all viewpoints needed for the risk assessment is critical. Not including a critical SME in the process could mean a poorly informed section of the assessment, which will result in a poorly informed process validation strategy. In order to avoid this pitfall, the SME team should be identified and vetted well in advance of working sessions.

- The Risk Assessment Is a Strict Documentation Exercise—In order to gain value from any risk assessment, it must be performed in earnest to learn information. If a team is approaching a risk assessment just as a documentation exercise, they will not be able to produce a robust assessment. The use of standard templates per piece of equipment, for example, is not recommended. Doing so would not explore the direct impacts to the specific product CQAs for which the validation is occurring, and valuable information will be overlooked.

- The Risk Assessment Is Performed Too Late in the Process—A risk assessment should be performed proactively and early on in the process in order to realize any ability to gain efficiencies. Furthermore, it is critical to perform the risk assessment early enough in the process to meaningfully mitigate identified unacceptable risks. Performing an assessment immediately prior to performance of validation activities may only allow for mitigation activities related to procedure steps or training operators, which are weak controls. The most effective mitigations are design controls (elimination of a risk rather than reduction), and that is more feasible the earlier the information about the required change is known.

- Performing Risk Assessments in Silos or as Part of Separate Efforts—The efficiencies described in this chapter can only be realized if the assessments are performed as a larger, agreed upon, end-to-end effort. Performing each of the assessments described earlier as separate efforts with no communication will put the team at a disadvantage, resulting in inconsistent risk language, inefficient deployment of resources, and overlapping assessments or missed information. The benefit of the overarching strategy is knowledge management amongst everyone involved in the process.

- *Never Invoking Risk Review*—Risk review is a required element of the quality risk management life cycle, so the first consequence of not performing risk review is noncompliance. There are many companies that have trouble successfully managing risk review, but the effects can be sharply felt in the context of this strategy. Risk review will ensure that the risk assessments (specifically the process FMEA) are always reflective of the current state of the manufacturing process level of control. Risk review is a critical part of continued process verification, allowing the organization to understand specifically which parts of the process are no longer in a state of control. Focusing specifically on those CAs or CPPs during risk review can narrow the effort required by the team.

6.8 Conclusion

A strong risk-based approach to lifecycle validation is not just necessary to meet health authority requirements but is a key component to ensuring that equipment and processes perform within defined limitations, CPPs and CQAs consistently meet requirements, and product quality meets analytical standards. Quality risk management can ensure that the product quality is at the highest possible standard with less expense of time and effort through the elimination of wasteful repetition and unnecessary operations by focusing only on the operations that are demonstrably impactful to product quality and ultimately patient safety. Ultimately, this enables the best possible outcomes for the patient by ensuring (1) a consistent supply of (2) high-quality efficacious pharmaceutical products to the market.

Recommended Reading:

For more information on quality risk management and application of specific risk tools, the authors suggest the following additional materials:

- Quality Risk Management in the FDA-Regulated Industry, 2nd Edition (2017) by Jose Rodriguez-Perez
- PDA TR 54, Implementation of Quality Risk Management For Pharmaceutical and Biotechnology Manufacturing Operations
- PDA TR 54–5, Quality Risk Management for the Design, Qualification, and Operations of Manufacturing Systems

References

ICH. (2005, November 09). *ICH Q9: Quality Risk Management*. ICH.
ICH. (2009). *ICH Q8(R2): Pharmaceutical Development*. ICH.
ICH. (2019). *ICH Q12: Technical and Regulatory Considerations for Pharmaceutical Product Lifecycle Management*. ICH.
PDA. (2013). *Process Validation: A Lifecycle Approach*. PDA.
PDA. (2017). *Technical Report Number 54–5*, Quality Risk Management for the Design, Qualification, and Operations of Manufacturing Systems. PDA.

7

Lifespan Studies for Chromatography and Filtration Media

Anurag S. Rathore and Gail Sofer

7.1 Introduction

Chromatography and filtration media often dominate the cost of raw materials for the whole process (1). As such, from a process economics point of view, it is important to be able to recycle the media to an appropriate number of cycles before replacing it with new media. The optimal number of cycles for a medium that should be targeted varies from product to product and company to company and depends on several considerations, which include media cost, approach towards and timing of the various process validation activities, cost of a batch of product, and number of batches to be run every year. The cost of the media varies in a wide spectrum, e.g., \$6500/L for rProtein A Sepharose Fast Flow® (Amersham Biosciences) to \$400/L for High S Macroprep® (BioRad). Similar variation is also seen in the various kinds of filtration media that are available in the market today.

Establishing the useful lifespan of the media, however, remains a critical issue for sponsors producing biological and biotechnological therapeutic and diagnostic products. The primary objective is quite simply the ability to consistently produce intermediates and final products that meet the defined quality and safety attributes. Achieving this goal is complicated by our limited understanding of the surface chemistry of the media and the interactions that take place between the media and the various feed components, such as host cell proteins, nucleic acids, lipids, viruses, as well as process additives. Once immobilized at the media surface, some impurities may become stabilized. Feed components are washed off to varying extents during cleaning and sanitization cycles. However, after completion of a chromatography or filtration step, a portion of these impurities can remain bound to the media and be slowly removed upon extended storage or even be carried over to the next production lot. Newer and more sensitive detection methods and techniques, such as polymerase chain reaction (PCR), can enhance our ability to understand the mechanism and nature of fouling of the media and, thus, aid in developing cleaning, sanitization, and storage procedures that preserve the function and integrity of the media and extend its lifespan. Nevertheless, validation of media lifespan requires demonstration at scale of the targeted number of reuses. These studies are often time-consuming and expensive, but they provide confidence in the continued production of consistently pure and safe biotherapeutics and reliable diagnostics.

- Regulators have stated, "Validation of the purification process should also include justification of the working conditions such as column loading capacity, column

DOI: 10.1201/9781003143130-7

regeneration and sanitization and length of use of the columns" (2). Therefore, studies designed to estimate media lifespan have slowly become a part of the "process development" of a biopharmaceutical commercial process. In January 2002, Dr. Andrew Chang of CBER presented a collection of the FDA's findings related to chromatography (3). In a review letter, a sponsor was told,

> Please provide validation data to demonstrate there is no negative impact of extended use of the . . . matrix to 150 production cycles on efficacy of cleaning and regeneration of the . . . column. Please provide data that show complete removal of viral contamination prior to reuse of the system.

In some post-approval inspections, comments included the following:

> Storage times in between runs for all of the purification columns have not been validated for entire life cycle of column. Cleaning validation study was conducted only for up to 5 uses of the column, which could be used in the purification of up to 46 lots per laboratory scale study. Cleaning validation including LAL and bioburden studies of the . . . purification columns were only validated for up to 5 uses. In addition, the cleaning validation of these columns did not include removal of process-related impurities. However, columns can be used for the following number of purification runs/years. . . . Storage in buffer not tested for attrition.

It is interesting to note that the same issues exist today, as evidenced by a recent Form 483 comment that there was "no cleaning validation for entire lifespan of purification columns." See also Chapter 1.

FDA's Therapeutic Compliance Program Guide, which serves as a guide for investigators, states,

> There should be an estimated life span for each column type, i.e., number of cycles. Laboratory studies are useful even necessary to establish life span of columns. There are situations where concurrent validation at the manufacturing scale may be more appropriate. Continued use may be based upon routine monitoring against predetermined criteria.

(4)

At a PDA/FDA conference on process validation in 2000, Dr. Barry Cherney presented CBER's current expectations on determining resin lifespan (5). He noted that the 1997 FDA Points to Consider in the Manufacture and Testing of Monoclonal Antibody Products for Human Use states that limits must be prospectively set (6). The ICH Guideline on Viral Safety states, "Over time and after repeated use, the ability of chromatography columns and other devices used in the purification scheme to clear virus may vary" (7). European regulatory authorities have also expressed concern over consistency of chromatographic performance. In one Committee for Proprietary Medicinal Products (CPMP) position statement, it was noted that elimination of host cell proteins, in most cases, makes use of chromatographic columns for which the selectivity and yield of the procedures depend not only on the quality of the material but also on the way the columns are used and reused, storage conditions, sanitization, and life span (8). Similar observations have also been made on filtration steps. One FDA Form 483 noted, "There is no integrity testing of back up filter when the primary nitrogen filter fails integrity test. There may be a lag time of up to . . . production batches before testing." It is not a big surprise that the issue of chromatography and filtration media lifespan continues to be discussed at conferences and questioned by regulatory authorities for licensure and during inspections.

In this chapter, we will discuss the various factors that influence lifespan of chromatography and filtration media and also the key operating and performance parameters that are utilized to monitor integrity of the media. Finally, we will review the different approaches that different companies have taken to successfully validate media lifespan.

7.2 Factors That Influence Chromatography Media Lifespan

Several factors are known to influence the lifespan of chromatography media, and these are listed in Table 7.1.

7.2.1 Position of Step in Purification Process

The positioning of a chromatographic step in a process has a profound influence on the expected lifespan due to the relative purity of the feed stream, which is much higher in the later stages of the process. For example, in the process of purifying albumin from plasma, the ion exchange media in the first step in which product binds is used for 600 cycles, whereas the second ion exchanger is used for 1200 cycles (9). In general, the early chromatographic steps have to face not only feed streams containing a variety of impurities that interact with the column media, but also relatively aggressive cleaning and sanitization protocols that are required to maintain the integrity of the column for the next reuse. The decision to place a chromatographic step earlier or later in a process is totally dependent on the process under consideration, and the optimal solution is frequently a result of compromise between counteracting considerations. For example, placing an affinity chromatography step later in the process will allow more reuses of the column media. However, due to the higher selectivity of affinity chromatography, using this column for capture of the product might reduce the number of steps required for purification. Finding an optimal solution in such cases requires performing small-scale studies to compare the different options.

TABLE 7.1

Factors That Influence the Lifespan of a Chromatography Column

- Position of step in purification process (capture, purification, or polishing)
- Nature of feed stream (amount and type of the various impurities)
- Mode of chromatography (bind-elute or flow-through)
- Type of chromatography media (physical and chemical stability)
- Maintenance (efficacy of cleaning, sanitization, and storage procedures)
- Column packing and attrition (packing method and physical stability of media)
- System components (design of column and supporting instrumentation)
- Quality of raw materials (nature of impurities in raw materials)
- Economics (cost of batch of media vs. validation costs)

TABLE 7.2

A CIP Protocol for a Feed Stream from Cell Culture

Feed stream:	Hybridoma cell culture
Chromatography media:	STREAMLINE rProtein A
CIP protocol:	(1) 1.0 mM NaOH and 2 M NaCl in 20% ethanol, 2 hours, 100 cm/h
	(2) 5% sodium lauroylsarcosinate, 20mM EDTA and 0.1 M NaCl in 20 mM NaH2PO4, pH 7.0, 1.5 h
	(3) 50 mM acetic acid in 20% ethanol

7.2.2 Nature of Feed stream

The nature of the feed stream is also a critical factor in determining media lifespan. Chromatographic separations performed early in the process are often complicated by the presence of a variety of components in the feed material, including host cell impurities (host cell proteins, endotoxin, nucleic acids, lipids, viruses), process-related impurities (raw materials, additives), and the various product-related impurities (10). Due to the requirement of more stringent cleaning for such steps, maximizing media lifespan requires carefully planned development of cleaning and sanitization steps early in process development. For example, in expanded bed adsorption techniques, whole broths are often applied directly from cell culture. A cleaning-in-place (CIP) protocol for such a step is shown in Table 7.2. It is obvious that this cleaning protocol is quite stringent, but it is compatible with the media, which was used for five cycles with an expectation that more could be obtained. Product recovery, purity, and breakthrough capacity remained constant over these lifespan studies (11).

Many feed streams contain colored substances that bind strongly to chromatographic media. This is particularly noticeable with *E. coli* feed streams from inclusion bodies and in plasma fractionation. Visual inspection in such cases can cause concern. Cleaning validation studies, however, can be utilized to demonstrate that these columns continue to perform consistently with no loss of capacity and no change in product purity, impurity profiles, or other performance attributes.

7.2.3 Mode of Chromatography

The mode of chromatography, bind-elute vs. flow-through, also plays an important role in determining the lifespan of the column. In contrast to the more commonly used bind-elute mode, flow-through mode is often used with the objective to bind other host cell impurities (most commonly host cell proteins and/or DNA), while letting the product of interest flow through the chromatography column. It is recommended to keep the mode of chromatography in mind while deciding on the acceptance criteria for media lifespan.

7.2.4 Type of Chromatography Media

When a chromatography column is scaled up and its diameter increases, the wall support contribution to bed stability starts decreasing. For column diameters greater than 25–30 cm, the lack of wall support may become an issue and could cause redistribution of packing particles and settling of the bed (12). This often results in formation of high-porosity regions at the column inlet and maldistribution of flow across the column. This phenomenon is more prominent for non-rigid gel materials and is often reversible within limits but almost always with a marked hysteresis. The issue of physical stability of the column bed becomes particularly significant at large scale as the column is put to more reuses during commercial manufacturing and hence the physical stability of the media must be taken into account for robust column design.

Considerations of chemical stability of the packing material include any factors that may result in deterioration of the column performance over a period of use. It may be the leaching of ligands into the mobile phase as often experienced with affinity chromatography, destruction of the matrix in the mobile phases used for column operation, regeneration, or storage (e.g., silica packings at high pH), or irreversible binding at the packing surface. These factors directly impact the column lifespan.

7.2.5 Column Maintenance

Good column maintenance is essential for maximizing media lifespan. This consists of three primary steps: cleaning, sanitizing, and storage, which have a very significant impact on column lifespan. Column lifespans are diminished when inadequate cleaning protocols are employed. Unsuitable protocols allow a continual buildup of contaminants, often leading to reduced flow rates and clogged columns. The optimal cleaning protocol depends on several factors, including compatibility of the media with cleaning solutions, nature of substances that must be removed, contact time, and temperature. A practical approach should be taken. For example, in the evaluation of cleaning calf thymus DNA from an anion exchanger, it was found that 1 M NaOH containing 1 M NaCl was effective in removing all residual DNA, as measured by the Threshold System®. However, with a monoclonal antibody sample containing a high level of DNA, the DNA was not removed even with 2 M NaOH or 3 M NaCl (13). A DNase treatment did remove the residual DNA; however, this is not an inexpensive approach for manufacturing, since the addition of DNase requires validation of its removal. A practical approach is to evaluate consistent capacity and flow properties, protein product purity, and DNA impurity levels. If no changes are observed, then the use of 1 M NaOH with 1 M NaCl should be sufficient for cleaning. Affinity media with proteinaceous ligands are usually the most difficult to clean. In the purification of Factor VIII with an immobilized monoclonal antibody column, for example, special care is taken to protect the expensive column. The feed stream and all buffers and cleaning solutions are prefiltered, and the column is kept isolated by sterile filters (14). At the other extreme, there are many ion exchangers, hydrophobic interaction, and gel filtration media that tolerate cleaning with up to 2 M NaOH and provide lifespan of up to 1200 cycles or more (15). Data on compatibility of the chromatography media with a variety of cleaning agents is generally available from vendors. However, this often serves just as a starting point, and each user must consider what is optimum for their particular application.

Sanitization of chromatography columns is required to ensure consistent performance. Microorganisms can leave behind endotoxins, enterotoxins, and other potentially harmful substances that may be quite difficult to remove. For example, negatively charged endotoxins bind strongly to anion exchangers (10). Routine monitoring for bioburden and endotoxin levels is part of a quality assurance program that ensures consistent performance of packed chromatography columns.

Storage is another critical parameter to be considered as part of the development studies. Proper storage is essential to ensure expected media lifespan. Storage solutions, temperature, and air quality may all have an impact on a firm's ability to reuse a column. When a column is stored for a considerable time period, for example when campaigning in a multi-product facility, it is advisable to periodically check the column for cleanliness and suitability for reuse. Storage solutions often have an excellent cleaning capability. There are many stories in the industry of columns that have been extensively cleaned which, after storage, have "nucleotide-like" and other substances in the rinse. As a result of this phenomenon, it is necessary to establish criteria for complete removal of storage solutions.

Other factors related to maintenance that can enhance media lifespan include proper gowning, air quality monitoring, and humidity control (16). Minimizing worker contact with the media once it is sanitized is recommended. Air quality in the purification suite is usually Class 10,000. In cold rooms, humidity control is critical, since high humidity is more likely to allow the growth of spore-forming organisms. Water quality, too, should be defined. Water for injection (WFI) is usually used for cleaning and rinsing packed columns. In some cases, however, this is not necessary—for example, in the first purification step of an *E. coli*-derived feed stream.

7.2.6 Column Packing and Attrition

Column packing and attrition also influence media lifespan. The considerations mentioned during the discussion on "physical stability" of the media in section 7.2.4 apply here as well. When in spite of rigorous cleaning and sanitizing routines deterioration in the chromatographic profile is seen, product purity changes, backpressure increases, or flow decreases, most firms remove the media from the packed column and clean it out of place with stirring. This often restores performance, but the user must requalify the column after packing, and in most cases some media is lost. New column designs that allow for automated unpacking and repacking may reduce the attrition. But newer, rigid media are sometimes more brittle and repacking may result in more fracturing of the particles. Hence, physical stability of the media should be taken into consideration while deciding on the lifespan at large scale.

7.2.7 System Components

System components can have a significant influence on the media lifespan. It is always necessary to consider not only media compatibility with cleaning and sanitizing agents, but the column hardware and system compatibility as well. If the cleaning solution causes leaching from components, for example O-rings, then the cleaning problem is only exacerbated, and batch failures may occur. Inline filters can also be problematic and should be taken out of line during cleaning and sanitization. At least one firm learned this the hard way when a filter remained in line during sanitization with NaOH. Since it was not known what might leach out of the filter, the entire column contents had to be replaced.

7.2.8 Quality of Raw Materials

Another consideration is the use of consistent, high-quality raw materials. Although the feed stream is often expected to be variable, especially with cell culture, raw material quality can be controlled. Raw material quality can have a major impact on media lifespan. For example, if laboratory-grade acetone is used as a test molecule to measure HETP, the impurities in the acetone can start to accumulate. Stories are told in the industry about cleaning agents that have been found to contain microorganisms; therefore, if detergents

TABLE 7.3

Risk Assessment of Raw Materials Used in Downstream Biotech Processes. Published from reference 18 with copyright permission from Advanstar Communications. S, O, and D denote the severity, occurrence, and detection ratings for a given raw material and the RPN score is the assigned risk priority number.

Material	Assessment	S	O	D	RPN	Risk
Chromatography resins	Material	3	3	3	27	Low
	Process	7	3	3	63	Medium
Inline filters	Material	3	3	3	21	Low
	Process	5	3	3	45	Medium
Compendial chemicals	Material	3	1	1	3	Low
	Process	3	1	1	3	Low

or similar agents are used in cleaning or sanitizing columns, a firm should ensure their quality. Buffers and salts have sometimes been found to contain high levels of undefined impurities (17). These can also accumulate and lead to a reduction in column lifespan. Loss in performance of hydroxyapatite columns due to presence of metals, such as iron, in raw materials is well known. Table 7.3 illustrates an example of risk analysis for downstream raw materials (18). Chromatography resins are generally regarded as low risk from material safety, characterization, and handling perspectives. However, the choice of the resin does have significant impact on the quality of the separation and hence the risk from the process perspective is medium. On the other hand, compendial chemicals are often well defined and understood and hence in general offer low material and process risk. Of course, chemicals that finally end up in the drug product should be assigned a higher degree of criticality.

7.3 Experimental Approaches to Determine and Validate Chromatography Media Lifespan

Determination of media lifespan can be done via concurrent validation or prospective validation (19), and these approaches are outlined in Figures 7.1 and 7.2 (20), respectively. Validation of media lifespan, however, has to be performed at manufacturing scale. The most common approach is to use small-scale data for "guidance" followed by "confirmation" and "validation" at full scale (Strategy 2) as it reduces the risk of "failure" at full scale and the resulting loss of batches if lifespan is solely determined at large scale.

While designing the experiments, it is essential to recognize that each feed stream and each process are unique. Expiry dates established for identical media used for similar end products from similar feed streams may be used to estimate media lifespan. This may be considered, for example, if a firm is producing several monoclonal antibodies from the same basic source and same culture conditions. If these data are to be used for a license application, however, one should discuss this early with the appropriate regulatory authorities to ensure they accept the concept and the data.

STRATEGY 1: CONCURRENT VALIDATION

- Perform cycling studies at large scale
- Evaluate column performance every n^{th} run
- Perform blank run every $n^{th} + 1$
- Material made in the last n cycles is quarantined until criteria are met for the final run
- **Advantages:**
  - No small-scale studies required
  - Entire dataset is pertinent as obtained at full scale
- **Disadvantages:**
  - Evaluation at large scale is more cumbersome and expensive
  - At any point, "n" lots are at risk

FIGURE 7.1
Concurrent validation strategy for determination and validation of column lifespan: performing studies only at manufacturing scale.

STRATEGY 2: PROSPECTIVE VALIDATION

- Perform cycling studies at small scale
- Evaluate column performance every n^{th} run
- Perform blank run every $n^{th} + 1$
- Once lifetime known at small scale, validate at large scale with appropriate safety factor
- **Advantages:**
  - Most work done at small scale (less expensive and faster)
  - With appropriate safety factor, no material at risk
  - Only suitable approach for evaluating clearance of hazardous impurities
- **Disadvantages:**
  - Small-scale dataset serves only as a guide and the full-scale validation is still required
  - Safety factor is critical as lifetime at small scale does not guarantee performance at large scale

FIGURE 7.2
Prospective validation strategy for determination and validation of column lifespan: performing studies both at small and manufacturing scales.

7.3.1 Small-Scale Models

A small-scale model of the chromatography step is required to be generated and then qualified before lifespan studies can be performed at small scale. The percentage of scale-down from manufacturing may be quite variable and depends on several factors, including the scale of the final unit operation. As long as the scale-down is verified to represent full scale, the degree of scale-down is not significant. In some cases, a scale-down of

TABLE 7.4

Steps for Designing Small-Scale Chromatography Models

- Create an effective cleaning procedure
- Use manufacturing feed stream
- Simulate manufacturing-scale chromatography system
- Maintain bed height and linear flow rates (contact time)
- Apply proportional sample load and maintenance solutions

2000-fold or more is acceptable. With very expensive feed streams or those which are hard to obtain plus very expensive affinity chromatography media, this level of scale-down is not uncommon.

In this section, we will briefly review some of the guidelines that should be considered while generating a scale-down model for performing lifespan studies. These are listed in Table 7.4.

First, a procedure needs to be identified or created that can be used for effective cleaning of the chromatography column. A recently published study highlighted the importance of having optimal cleaning and sanitization procedures (21). The study was performed using DEAE-Sepharose CL-6B anion-exchange media used in purification of prothrombin-complex concentrate from cryoprecipitate-depleted human plasma. The authors used a recycling procedure that includes two NaCl washes: one before and one after hydroxide washes. They found that although the majority of the material is removed during the first NaCl wash, the second wash removes material that has been solubilized by the NaOH. The cleaning and storage procedures that resulted from this investigation were found to significantly reduce carryover on the column and could be expected to have a positive effect on column lifespan.

Second, experiments should be performed using manufacturing feed stream or a feed stream that is representative of manufacturing scale. Prior to implementation of this work, the cell culture or fermentation conditions, including additives, should be well defined.

Third, attempts should be made to mimic the manufacturing-scale chromatography system as much as possible or account for the deviations by using similar or appropriate components (e.g., identical chemicals in contact with product where possible) and configurations (e.g., distance from column outlet to monitor should be appropriately scaled). Inevitable differences may occur due to differences in flow cell path lengths and diameters, tubing diameter, availability of similar wetted materials, and so on. Where deviations occur, however, they should be noted. The preferred approach is to compare the scaled-down model to the manufacturing scale in terms of the various performance criteria, which depend on the objective of the column. These performance criteria are discussed in more detail in the next section. If significant differences exist between the two scales, a careful determination of the impact of these differences should be made and, if necessary, the small-scale model should be redesigned. In any case, having a qualified scaled down model "prior" to performing small-scale cycling studies is necessary, and using a scaled-down model that is flawed would only lead to unreliable lifespan study results and a waste of time and resources. In cases where satisfactory scale-down cannot be performed, Strategy 1 outlined in Figure 7.1 can be used for determining column lifespan, provided lifespan studies do not require virus clearance evaluation or clearance of other hazardous substances.

Fourth, for most cases it is recommended to maintain bed height and linear flow velocity when scaling down. The ICH guideline on viral safety states that the validity of the scaling down should be demonstrated and that column bed height should be shown to be representative of commercial-scale manufacturing (22). However, it is often possible to scale down without maintaining the bed height, as long as residence time is maintained (23). And in some cases, it is simply not realistic to maintain the bed height. For example, in several cases firms have scaled down immobilized Protein A columns without maintaining bed height. In at least one case, the scaled down column had a bed height that was reduced two-fold from full scale. If linear flow rate cannot be maintained, retention time must be kept constant by changing the bed height. As with any such deviations from a guideline such as those produced by ICH, it is advisable to discuss the plan with regulators prior to implementation.

Fifth, apply proportional sample load to chromatography media volume. Where possible, apply the worst-case parameters to provide a safety margin. For adsorption techniques, total protein should be kept within the established specifications for manufacturing (usually expressed as grams total protein per liter of chromatography media). For gel filtration techniques, it is critical that the percentage of sample volume relative to column volume be kept within established production specifications, preferably towards the upper limit. Sample concentration should also be constant. Maintain constant ratios of wash, elution, regeneration, cleaning, and equilibration volumes relative to chromatography media volume. Buffers should be made according to manufacturing SOPs.

Table 7.5 presents an example of a successful scale down of a cation-exchange chromatography step and its use for measuring column lifespan (19). It is observed that at both scales the step yields are comparable and the key functions of the column step, i.e., reduce Chinese hamster ovary proteins (CHOP) and DNA and clear Protein A, are met at both the scales.

7.3.2 Parameters to Measure

Since the purpose of small-scale cycling studies is to determine or demonstrate lifespan for a chromatography column, a variety of functional, chemical, and physical parameters can be measured to ensure that the media will perform over the targeted number of reuses. The logical approach is to pick parameters that would be expected to affect the ability of the column in achieving its objective. In the following we list and discuss some of these operating and performance parameters that have been used in the industry to successfully determine column lifespan. It is expected that the reader will take these as suggestions and pick those that make the most sense to the application under consideration.

Table 7.6 lists some of the operating and performance parameters that are commonly used to monitor column integrity during cycling studies. These are discussed in more detail in the following text.

TABLE 7.5

Comparison of Performance of Small-Scale Model with Full-Scale Manufacturing Column (adapted from reference 18)

Scale	Cycle #	Yield, %	CHOP*, ppm	DNA, ppm	Protein A, ppm
Small scale	1	88.7	75	< 0.006	< 7.8
	51	88.3	34	0.01	< 7.8
Manufacturing scale	1	84.6	116	< 0.07	< 7.8
	50	81.1	72	< 0.03	< 7.8

* Chinese hamster ovary proteins (host-cell proteins)

TABLE 7.6

Commonly Measured Parameters for Small-Scale Models

- Chromatographic profile and related parameters
- Product yield and purity
- Clearance of impurities
- Column qualification measurements (HETP and asymmetry)
- Pressure/flow
- Media particle properties
- Product carryover (blank runs)
- End-of-life testing

7.3.2.1 Chromatographic Profile

This is the simplest observation one can make to observe changes in performance. Often, however, production chromatograms look like mountain slopes and changes are hard to decipher. Experiences collected in process development and transferred to production will enable the observer to judge the quality of the chromatogram and assess if it reflects a change in performance.

Since comparison of the chromatogram is qualitative, several attributes have been used to quantitate this analysis. These include pool volume, absorbance and/or conductivity at start of pool collection, absorbance and/or conductivity at end of pool collection, number of column volumes (CVs) from gradient start to start of pool collection, A280 and/or conductivity at peak maximum, peak area for collection, number of CVs from gradient start to peak maximum, and number of CVs from gradient start to end of pool collection. As is evident, the objective of all these parameters is to be able to spot any trends or deviations in the chromatographic profile with increasing number of cycles. Care should be taken in drawing conclusions, however, since changes in retention are often influenced by variability in the feed stream. This is particularly true for cell culture, and even more so for continuous cultures. Figure 7.3 shows a plot of conductivity of the column eluent at start collect, peak maximum, and stop collect vs. number of cycles. It is seen that the data are indicative of a consistent chromatographic profile over the 250 cycles that were investigated.

7.3.2.2 Product Yield and Purity

For most applications, quality of the final product is a key performance criterion for a chromatographic step. Product yield is also important to ensure consistent performance over column lifespan. Evaluation of both purity and impurity profiles by multiple, orthogonal methods (e.g., IEC HPLC, SDS-PAGE) measures the ability of the media to remove specific impurities relevant to this step in the process. In some cases, it may also be necessary to perform a biological assay. These determinations can be performed at an appropriate interval, depending on the targeted media lifespan.

7.3.2.3 Clearance of Impurities

In most cases, chromatographic steps are used not only to separate product-related impurities, but also to provide clearance for host cell impurities (e.g., host cell proteins, endotoxin, nucleic acids, lipids, viruses) and process-related impurities (e.g., raw materials, additives). For example, DNA and bovine IgG may need to be removed from cell culture products,

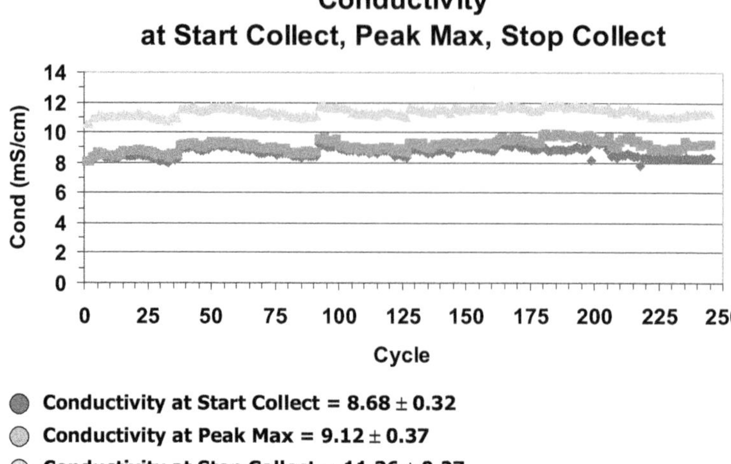

FIGURE 7.3

Conductivity at start collect, peak maximum, and end collect for an ion-exchange column as a function of the number of reuses (CM Sepharose High Performance® media, 2.6 cm diameter, 18 cm bed height). Data kindly provided by Karen Dane and Jim Seely, Amgen, Inc.

and endotoxin and DNA from *E. coli* products. It is essential to assess removal of these specific impurities over media lifespan. If, for example, a given step is designed to remove DNA and deamidated forms of a protein product, then the relevant assays should be performed to confirm consistent removal of these impurities to the specified level. This is particularly important for specific impurities that require spiking studies or those whose presence in the final product adversely affects its safety, potency, or efficacy.

One area of particular concern for cell culture products is the ability of the chromatographic media to remove viruses after repeated usage. Since viral clearance testing is quite costly, this work requires some special considerations. It is not realistic to test every few runs. For example, one firm tested the ability of Protein A Sepharose Fast Flow to remove viruses in the first, 11th, and 34th cycles. Work performed in the plasma fractionation industry indicates no loss of ability to remove viruses after more than 400 cycles (24). More recently, Brorson et al. have shown that retroviral clearance was not impacted for 100 cycles beyond the point that an immobilized Protein A media quality deteriorated (25).

7.3.2.4 Column Qualification Measurements

Periodic determinations (at an appropriate interval) of HETP and asymmetry (A_s) can be useful in pointing to any deterioration in column integrity with reuse. These measurements are particularly important for gel filtration steps but may not always be relevant, for example in "on-off" step gradient separations commonly found early in a purification process.

7.3.2.5 Pressure/Flow

A buildup of impurities can result in an increase in pressure or a decrease in flow. Changes in pressure or flow can also indicate compression or breakdown of the media, or clogging of column screens, nets, or inline filters. Pressure at certain process points can thus be a

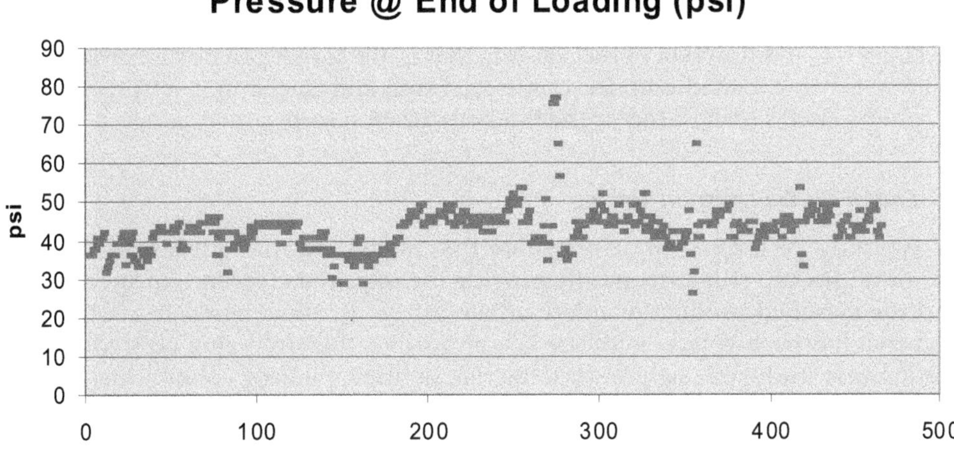

FIGURE 7.4
Pressure drop across the column at the end of the loading step for an ion-exchange column as a function of the number of reuses (S Sepharose Fast Flow® media, 2.6 cm diameter, 25 cm bed height). Data kindly provided by Dana Becker, David Dripps, and Jim Seely, Amgen, Inc.

useful indicator of the physical stability of the media and can easily be monitored for each cycle. Figure 7.4 shows a plot of pressure drop across an ion-exchanger column at the end of the loading step vs. number of column reuses. It is seen that the pressure drop slowly increases from an average of 40 psi to 45 psi after 450 reuses, indicating a slight buildup over time.

7.3.2.6 Chromatography Media Properties

Various physicochemical attributes of the media particles can be used to characterize fouling of the chromatography column during reuse. Some of these attributes are typically listed in the certificate of analysis provided by the media manufacturer. Criteria often evaluated during ion exchange reuse studies include small-ion capacity, total protein capacity, particle-size distribution, flow vs. pressure curves, and total organic carbon (TOC) removed by extreme cleaning solutions (15). A comparison of results obtained from media at the end of lifespan to the corresponding values for fresh media can yield an insight into the mechanism of media fouling, for example buildup in column pressure along with a reduction in mean particle diameter indicates breaking of media particles due to shear experienced during column repacking.

7.3.2.7 Blank Runs

Performing blank runs periodically (at an appropriate interval) is commonly used to evaluate efficacy of cleaning and potential of product carryover. In a blank run, the column is operated following the normal procedure except that the load is substituted with water or buffer. When gradient elution is used, the elution pool is collected in the same portion of the gradient where the product is typically pooled and is analyzed for any product-related or other impurity. Analytical techniques often utilized for this purpose include product-related assays (if they have the required sensitivity), SDS-PAGE, silver stain, TOC, and/or total protein assays. The target must be set appropriately, based on the quality of the feed material and function of the column, that is, the target can be higher for a capture column that is being loaded with the crude cell broth and can be more stringent for a polishing column that is being used as the final purification step.

7.3.2.8 End of Media Lifespan Testing

This is typically performed to demonstrate that media that has surpassed the targeted number of reuses can still successfully provide the required clearance of specific impurities and the specified product or intermediate purity. In some cases, it can be useful to use feed that has been spiked with the key impurities that are being cleared in the step. Performing this study may help reduce the risk in implementing column lifespans determined at small scale to manufacturing scale.

An example of measurable parameters of column performance over time is shown in Table 7.7. Drevin et al. observed a decrease in retention volume, bed height, and peak height during 300 analytical separations. These decreases were attributed to protein clogging of the top filter (26). Optimization of the washing procedure and column configuration improved column performance. This example illustrates the value of small-scale studies that allow for improvements to be made prior to scaling up for full-scale manufacturing and further extending media lifespan.

A recently published study showed the utility of performing combined reuse and characterization studies for a cation-exchange column, Macroprep High-S, used as the capture step in the manufacturing process for Neuleze, a nerve growth factor (27). The effect of multiple parameters on the % yield and clearance of impurities, such as host cell proteins, DNA, and a cell culture media component, was evaluated. Various fermentation lots, equilibration pH, elution pH, and absorbance at start of collection were investigated. This was achieved via a design of experiment (DOE) consisting of 42 experiments, and it was concluded that the media could be reused for 42 cycles. Approaches such as DOE can lead to a considerable savings in time and resources that are required to perform these studies.

7.3.3 Concurrent Validation at Pilot and/or Full Scale

In addition to the model columns that allow for large numbers of cycles to be run, analyses must be performed on production columns to determine media lifespan. This is typically done during preparation of consistency batches and concurrently in production. The operating and performance parameters that should be monitored at full scale are similar to the

TABLE 7.7

Changes in performance over 300 cycles (adapted from reference 12).

Run No.	Bed Height, cm	Cl- Capacity, mmol	Retention Vol, ml			
			Peak I	Peak II	Peak III	Peak IV
1	**10.2**	3.53 ± 0.08	48.87	74.04	84.69	100.74
150	9.5	3.53 ± 0.08	47.19	72.81	82.89	99.48
250	8.9	n.d.	46.98	71.70	82.62	99.27
300	8.9	3.54 ± 0.07	47.91	73.14	84.00	99.12

ones mentioned in section 7.3.2 and are typically chosen based on the results from small-scale studies. Small-scale end-of-lifespan testing can be performed using media from a manufacturing column that has reached the targeted lifespan. This type of study may provide an additional safety margin.

Table 7.8 shows representative data from a very large scale manufacturing facility producing Phase III clinical material on three ion exchange columns. The process was fully validated at the time these data were collected, and a comprehensive monitoring program was in place to justify continued reuse. Media replacement was based on one year's use (potentially 1000 cycles), and not based on the number of cycles. Two runs are shown in the table for each of the three ion-exchange columns used in manufacturing. Two chromatograms from each column were selected; one from the early part of the campaign and a later one. Peak profile and position were evaluated. Column yield from each run was also noted

TABLE 7.8

Data kindly provided by Holly Hutchins and Robert J. Seely, Amgen, Inc.

452 L S-Sepharose Fast Flow Column

No. cycles prior to chromatogram	41	300
Cumulative cycle no.	41	341
Most recent HETP	0.04 cm	0.04 cm
Yield (%)	79	86
Start	1.51	1.55
Peak	2.25	2.3
End	2.78	2.81
Width	1.27	1.25

378 L Q-Sepharose Fast Flow Column

No. cycles prior to chromatogram	26	208
Cumulative cycle no.	26	234
Most recent HETP	0.05 cm	0.02 cm
Yield (%)	82	80
Start	2.26	2.59
Peak	2.97	3.14
End	3.46	3.65
Width	1.20	1.06

452 L CM-Sepharose Fast Flow Column

No. cycles prior to chromatogram	25	208
Cumulative cycle no.	25	233
Most recent HETP	0.04 cm	0.04 cm
Yield (%)	70	74
Start	3.00	2.85
Peak	3.16	2.99
End	3.52	3.49
Width	0.52	0.64

along with an estimate of the number of cycles prior to the selected chromatograms. HETP tests were performed on the columns every fifth batch, and the most recent HETP value for each of the column cycles is shown in the tables. Although there is some variability, for example, in the yield for the S Sepharose Fast Flow column, the pre-established specifications for yield and retention position were met.

7.4 Experimental Approaches to Determine and Validate Filtration Media Lifespan

The key concepts that form the underlying basis for determination and validation of lifespan for chromatography media also apply for filtration media. Hence, this section will focus on aspects that are unique to filtration. The commonly used approach is to use small-scale data for "guidance" followed by "confirmation" and "validation" at full scale. While this discussion is more focused on tangential flow filtration (TFF) applications, some aspects apply to depth flow (DF) filtration applications as well.

7.4.1 Small-Scale Models

Table 7.9 reviews guidelines that could be useful when creating a scale-down model for a filtration step. Once again, this discussion focuses on the issue of media lifespan.

First, an effective cleaning and sanitization procedure is identified. It is common to try the vendor-recommended procedures as they are supported with data from the required leachables/extractables studies. Cleaning/sanitizing solutions tend to be reactive and corrosive and hence care must be taken to operate within the concentrations, temperatures, contact times, and other conditions that are covered by the vendor's package. If, for some reason, a new solution has to be used or conditions outside those recommended by the vendor have to be used, one has to plan for performing the appropriate leachables/extractables studies. These studies, however, tend to be time consuming and expensive. The issue of extractables from product-contact surfaces was recently reviewed (28).

Second, experiments should be performed using feed material that is manufactured at full scale. This is particularly true for process streams in the upstream portion of the process, since unit operations like centrifugation and homogenization are difficult if not impossible to mimic at lab scale. As a result, the feed material in the laboratory may not be representative of full scale in terms of the amount of host cell impurities and other constituents. These impurities, such as endotoxin and DNA, have a significant impact on the

TABLE 7.9

Design of Small-Scale Filtration Models

- Create an effective cleaning procedure
- Use manufacturing feed stream
- Simulate manufacturing-scale filtration system
- Use identical operating conditions

lifespan of a filter. Thus, it is best to use feed material generated at full scale or representative pilot scale.

Third, an attempt should be made to have an accurate scale-down system. Step recovery for ultrafiltration/diafiltration at lab scale are often marred by considerable losses due to high system holdup volume relative to the final pool volume. While it may not be possible to achieve the exact recovery that could be obtained at pilot or manufacturing scale, care must be taken to minimize the differences in performance of the step across the two scales. This can be achieved by ensuring that the system design reflects the manufacturing scale. Further, it is important that the membrane material and the design format of the cassette are identical to that used at manufacturing scale.

Fourth, it is common to keep membrane area per unit amount of product the same while scaling down, that is, operate at identical protein loading as compared to the large scale. It is recommended to make buffers using the appropriate SOPs and keeping the other operating conditions such as pH, ionic strength, temperature, transmembrane pressure (TMP), and cross-flow rate identical to large scale.

7.4.2 Parameters to Measure

Table 7.10 lists some of the operating and performance parameters that are commonly used to monitor filter integrity during cycling studies (29). These are discussed in more detail in the following text. Once again, the parameters that are chosen for monitoring and their specifications or control ranges depend on the intended application.

7.4.2.1 Normalized Water Permeability (NWP)

Percent recovery of NWP is perhaps the most commonly used performance parameter for monitoring the integrity of a UF (ultrafiltration)/DF membrane and should be performed after every reuse in the lifespan study. This parameter measures the permeability of the membrane using water and allows for a comparison of the integrity of the membrane pre- and post-use. Percent recovery of NWP typically declines with number of uses since every time the membrane is used, product or other species in the feed material can bind to the pores of the membrane causing decay in the permeability. It is very common to use NWP criteria for determining the number of cycles a membrane should be used, for example 75%–125% of original NWP. While the filter vendors recommend the criteria for a particular membrane product, it is recommended that cycling studies be performed by the user and a variety of performance criteria monitored. The data should then be evaluated to decide what would be an appropriate NWP criterion for the specific application under consideration.

TABLE 7.10

Commonly Measured Parameters for Small-Scale Models

- Normalized water permeability (NWP)
- Product yield and purity
- Clearance of impurities
- Filter integrity measurements
- TMP vs. flux curves
- Filter analysis
- Product carryover (blank runs)

7.4.2.2 Product Yield and Purity

Just as for chromatographic separations, product yield and purity should be monitored at an appropriate interval during the lifespan study. This is to ensure that product degradation is not induced due to repeated use of membrane. This is particularly important if one is using particularly reactive cleaning solutions, such as bleach. Minute amounts of carryover of the bleach in the system can result in a significant increase in product-related impurities in the final pool.

7.4.2.3 Clearance of Impurities

Filtration steps are often used for clearance of host cell-related as well as process-related impurities. This clearance of the appropriate impurities should be monitored during reuse studies at an appropriate interval to demonstrate that the "efficacy" of the step in performing the clearance is not marred by reuse.

7.4.2.4 Filter Integrity Measurements

These measurements are used to identify problems such as macroscopic holes in the membrane, cracks in the seals, and improperly seated modules, which can lead to product leakage and/or unsatisfactory clearance of impurities (28). A common way to do this is via an air diffusion test. When air is applied to the retentate side at a controlled pressure, it diffuses through water in the pores at a predictable rate. However, in the presence of any defects the air flows through at a significantly higher rate and, thus, fails the test value. Such measurements could easily be performed after every reuse.

Besides air diffusion, several other tests are also employed to evaluate membrane integrity. These include bubble point determination and pressure hold-decay test (30–32). It is recommended that the reader evaluate the applicability of these different tests to the application under consideration and then pick the appropriate integrity testing method.

7.4.2.5 Transmembrane Pressure (TMP) vs. Flux Curves

TMP is the average applied pressure from the feed to the filtrate side of the membrane. As TMP increases, the flux across the membrane typically increases such that the slope of the curve keeps decreasing with increasing TMP. These curves serve as a good indicator of the performance of a filtration step and are commonly used as a qualitative measurement. A carryover of product or impurities often results in decay of the TMP-flux curve. Measurements at an appropriate interval can be useful in deciding an appropriate lifespan for a membrane. It is recommended that these curves be obtained at three different cross-flow rates that span the range of manufacturer recommendations (28).

7.4.2.6 Filter Analysis

With the advent of new and more sensitive spectroscopic methods such as Fourier transform infrared (FTIR) and Raman spectroscopy, it is possible to analyze the filter surface and quantify the buildup of protein or absence of such. This kind of analysis, at least at the end of intended filter lifespan, can be done in consultation with the filter vendor and can be useful in the characterization of filter fouling.

7.4.2.7 Blank Runs

Performing blank runs periodically (at an appropriate interval) is commonly used to evaluate efficacy of cleaning and potential of product carryover. In the case of the blank run, filtration is performed using load material that does not contain any product and the resulting pool is analyzed for any product-related or other impurity. Just as for chromatographic separations, analytical techniques often utilized for this purpose include HPLC assays, SDS-PAGE, and/or total organic carbon analysis.

7.4.3 Concurrent Validation at Pilot and/or Full Scale

Once the lifespan studies have been performed at small scale, a target for number of reuses is set. Next, full-scale runs are performed to determine filter lifespan. This can be done during preparation of consistency batches and concurrently in production (33, 34). Appropriate operating and performance parameters are monitored at full scale. As mentioned earlier, blank runs should be performed at an appropriate interval at full scale to show absence of any carryover.

Table 7.11 shows performance of the scale-down model for an ultrafiltration-diafiltration (UF/DF) membrane (35). It is seen that performance of the step is comparable across scales. Small-scale reuse studies were performed using this scale-down model. Product pool pH, conductivity, % impurity, and % recovery were utilized to assess step performance. NWP and flux vs. time curves were monitored to assess cleanability of the membrane. Finally, host cell proteins (HCP), product concentration via ELISA, and DNA were evaluated for evaluating carryover by performing blank runs after every five reuses at small scale. Figure 7.5 shows data from the small-scale reuse study. It is seen that the membrane performance was acceptable based on pool pH, conductivity, % impurity, and % recovery (Figure 7.5A). Flux vs. time curves (Figure 7.5C) are seen to overlap and NWP (Figure 7.5D) is stable over a number of membrane reuses. Carryover for the blank runs (2, 7, 12, 15, and 17) is also seen as minimal with respect to host cell proteins, DNA, and product concentration (Figure 7.5B). Higher values for HCP and EIA samples for run 15 are likely due to sampling error, and this is supported by results for run 17. The small-scale data presented here gave us confidence that we would be able to validate ten reuses at commercial scale, and this was eventually performed successfully. This case study illustrates how a successful study can be planned to find and support the lifespan for a membrane step.

TABLE 7.11

Results from the Scale-Down Modeling and Qualification for a Filtration Step. Published from reference 35 with copyright permission from Advanstar Communications.

	Pool pH	Pool Conductivity (mS/cm)	% Recovery	Process Time (hr)
Scale-down model average ±3SD	6.52 ±0.08	9.92 ±0.32	147.2 ±23.6	5
Large-scale average	6.36	10.04	138	5

A

Run	%Impurity HPLC 1	Pool pH	Pool Conductivity	% Recovery
1	1.58	6.55	9.8	137
6	1.18	6.5	10	144
11	1.07			154
14	0.94			157
16	0.95	6.50	9.97	144

B

Run #	HCP (ng/ml)	Product ELISA (ng/ml)	DNA (pg/ml)
2	1.399	<0.31	< 1E+01
7	6.352	<0.31	< 1E+01
12	4.933	<0.31	< 1E+01
15	44.11	0.46	< 1E+01
17	3.634	<0.31	< 1E+01

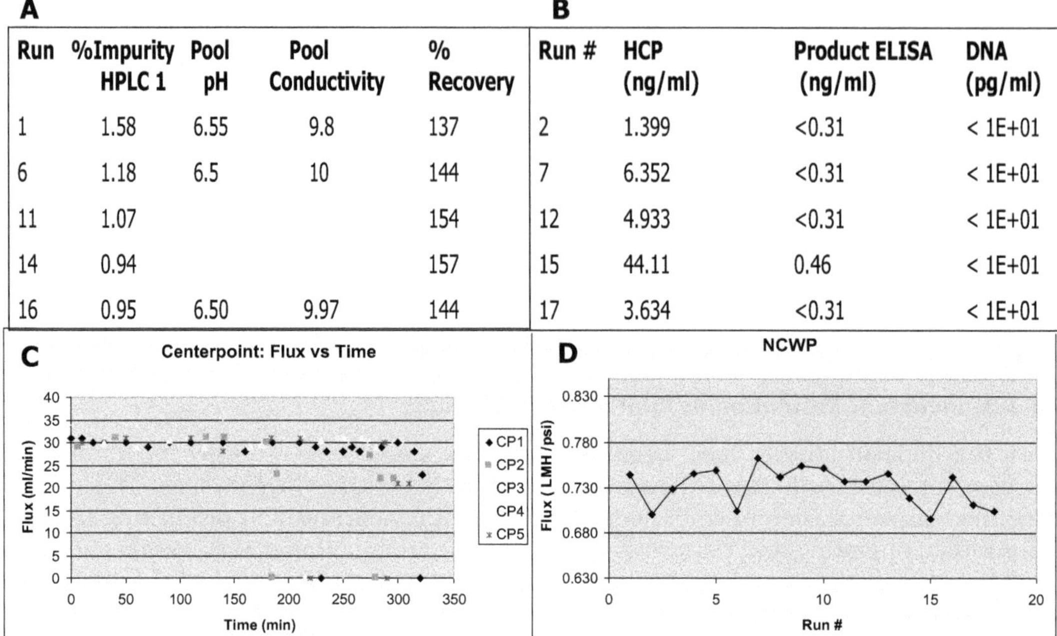

FIGURE 7.5
Data from membrane reuse studies for case study for prospective validation: (A) data from reuse studies, (B) data from blank runs for carryover studies, (C) flux, and (D) normalized water permeability. Published from reference 35 with copyright permission from Advanstar Communications.

7.5 Conclusions

In this chapter, we have discussed the various factors that influence useful lifespan of chromatography and filtration media and also the key operating and performance parameters that are utilized to monitor integrity of the media. It is clear that determining lifespan of chromatography and filtration media requires several approaches and evaluation of multiple parameters. The key output of this effort is in the form of a validation report that presents the results from qualification of the scale-down model and data from the small-scale and large-scale studies in a concise tabular form for evaluation by regulatory authorities.

It is always advantageous to build in a safety margin by performing an excess number of runs at small scale before implementing the targeted lifespan at manufacturing scale. Figure 7.6 shows a plot of the cost of chromatography media, cost of validation, and net savings as a function of number of reuses for an Amgen product (36). It is seen that while the cost of new media decreases with increasing number of media reuses, the cost of validating media lifespan increases. Thus, the number of reuses targeted is a function of the application under consideration and various factors such as cost of chromatography

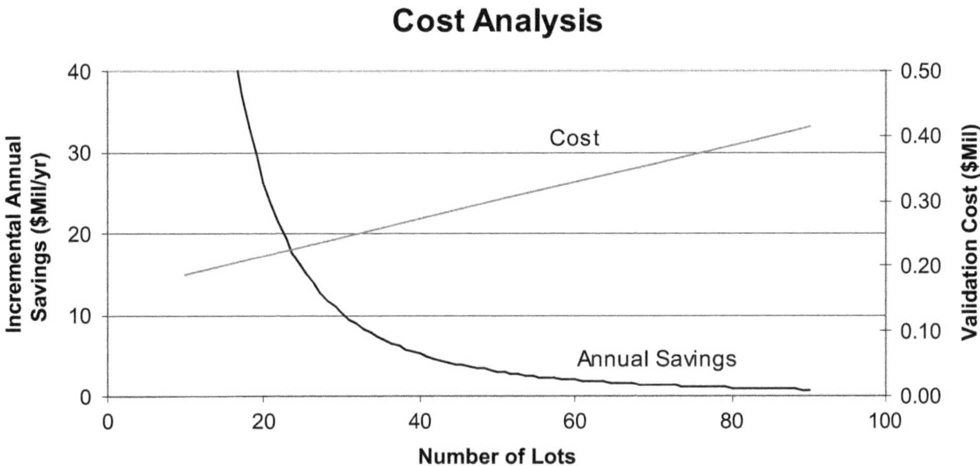

Number of Lots	Incremental Annual Savings ($M/year)	Validation Cost ($M)
10	$ 112.67	$ 0.19
30	$ 14.86	$ 0.24
50	$ 2.97	$ 0.30
70	$ 1.27	$ 0.36
90	$ 0.71	$ 0.41

FIGURE 7.6
Validation cost and net savings as a function of number of reuses for an Amgen product. Validation cost includes cost of media and validation. Data provided by Traci Taggart, Amgen, Inc. (34).

or filtration media, number of batches to be run every year, gap between different campaigns when the media will be needed to be stored, and resources available to the project. It is not uncommon for the same company to perform a different extent of activities for different products.

Indications of deterioration in chromatography and filtration media include loss of product purity, recovery, capacity, and changes in impurities' profile. More specific are monitoring of changes in retention time/volume, pressure-flow curves, HETP, and/or A_s for chromatography media and testing of the NWP and filter integrity for filters. Use of blank runs to evaluate carryover is currently the *de facto* standard for either kind of media.

In this chapter, we have attempted to present the current approach towards determination and validation of media lifespan. However, in view of the dynamism that results from tightening regulatory standards and different successful approaches that companies take, it is essential to continue to challenge and evolve one's approach and stay abreast of further developments.

References

(1) A. S. Rathore, H. Levine, P. Latham, J. Curling, and O. Kaltenbrunner, Costing Issues in Production of Biopharmaceuticals. *Biopharm*, 17(2) January (2004), 46–55.

(2) European Commission, The Rules Governing Medicinal Products in the European Union, Vol. III, Addendum 3, Guidelines on the quality, safety and efficacy of medicinal products for human use. Production and Quality Control of Medicinal Products Derived by Recombinant DNA Technology. III/3477/92, pp. 47–56.

(3) A. Chang. *Oral Presentation*. Washington, DC: Chromatography: FDA Regulator's Experience WCBP, January 2002.

(4) U.S. FDA Compliance Program Chapter 41 Inspection of Licensed Therapeutic Products, March, 1999.

(5) B. Cherney. *CBER's Expectations on Determining Resin Lifespan*. Washington, DC: FDA/PDA Process Validation Meeting, 2000.

(6) U.S. *FDA Points to Consider in the Manufacture and Testing of Monoclonal Antibody Products for Human Use*. U.S. Dept of Health and Human Services. Rockville, MD: FDA, CBER, February 1997.

(7) Viral Safety Evaluation of Biotechnology Products Derived from Cell Lines of Human or Animal Origin, 1997, ICH.

(8) EMEA, CPMP Position Statement on DNA and Host Cell Proteins (HCP) Impurities, Routine Testing Versus Validation Studies, CPMP/BWP/382/97. www.eudra.org/emea.html.

(9) Data from South African Blood Transfusion Center, Durban.

(10) A. S. Rathore, S. E. Sobacke, T. J. Kocot, D. R. Morgan, R. L. Dufield, and N. M. Mozier. Analysis for Residual Host Cell Proteins and DNA in Process Streams of a Recombinant Protein Product Expressed in E. coli Cells. *Journal of Pharmaceutical and Biomedical Analysis*, 32 (2003), 1199–1211.

(11) Data File: Expanded Bed Adsorption. Uppsala Sweden, 1996.

(12) A. S. Rathore and A. Velayudhan. An Overview of Scale-Up in Preparative Chromatography. In *Scale-up and Optimization in Preparative Chromatography*, Ed. by A. S. Rathore and A. Velayudhan. New York: Marcel Dekker, 2002, pp. 1–32.

(13) Y. Dasarathy. A Validatable Cleaning-in-Place Protocol for Total DNA Clearance from an Anion Exchange Resin. *BioPharm*, 9 (1996), 41–44.

(14) F. Feldman, S. Chandra, M. E. Hrinda, and A. B. Schreiber. Quality Assurance in Production of Plasma Proteins. In *Quality Assurance in Transfusion Medicine*. Boca Raton: CRC Press, 1993, Vol. II, pp. 259–284.

(15) R. J. Seely, H. D. Wight, H. H. Fry, S. R. Rudge, and G. F. Slaff. Validation of chromatography resin useful life. *BioPharm*, 7 (1994), 41–48.

(16) G. Sofer and L. Hagel. *Handbook of Process Chromatography*. London: Academic Press, 1997.

(17) P. Gagnon. *Purification Tools for Monoclonal Antibodies*. Tucson, AZ: Validated Biosystems, Inc.,1996.

(18) D. Low and A. S. Rathore. *Managing Raw Materials in the QbD Paradigm, Part 2: Risk Assessment and Communication. BioPharm International*, 23(12) December (2010).

(19) R. M. O'Leary, D. Feuerheim, D. Peers, Y. Xu, and G. S. Blank, Determining the Useful Lifespan of Chromatography Resins. *Biopharm*, 14(9) (2001), 10–18.

(20) A. S. Rathore. *Explore an Approach to Process Validation for an Acceptable Validation Package, Course on Process Validation*. Philadelphia, PA: Barnett International, January 2003.

(21) J. Turton and Z. Moola. Storing an Ion-Exchange Chromatography Gel in dilute Alkali During Recycling Improves Cleaning. *Biopharm*, 24–30 April 2002.

(22) ICH, Quality of Biotechnological Products: Viral Safety Evaluation of Biotechnology Products Derived from Cell Lines of Human or Animal Origin. International Conference on Harmonization, September 4 March 1997.

(23) S. Yamamoto, M. Nomura, and Y. Sano. Resolution of Proteins in Linear Gradient Elution Ion Exchange and Hydrophobic Interaction Chromatography. *Journal of Chromatography*, 409 (1987), 101–110.

(24) L. Andersson, S. E. Connor, L.-O. Lindquist, and E. A. Watson. A Validation Study for the Removal/Inactivation of Viruses During a Chromatographic Process for Albumin and IgG. International Society of Blood Transfusion, Japan, March 1996. Oral presentation.

(25) K. Brorson, J. Brown, E. Hamilton, and K.E. Stein. Identification of Protein A Media Performance Attributes That Can Be Monitored as Surrogates for Retrovirus Clearance During Extended Re-Use. *Journal of Chromatography* A, 989 (2000), 155–163.

(26) I. Drevin, L. Larsson, and B. L. Johansson. Column Performance of Q-Sepharose HP in Analytical- and Preparative-Scale Chromatography. *Journal of Chromatography*, 477 (1989), 337–344.

(27) T. N. Breece, E. Gilkerson, and C. Schmelzer. Validation of Large-Scale Chromatographic Processes. *Biopharm*, July (2002), 35–42.

(28) J. Bennan, F. Bing, H. Boone, J. Fernandez, B. Seely, H. van Denise, and D. Miller. Evaluation of Extractables from Product-Contact Surfaces. *Biopharm International*, December (2002), 22–34.

(29) Millipore Technical Brief, Protein Concentration and Diafiltration by Tangential Flow Filtration.

(30) M. W. Jornitz, J. P. Agalloco, J. E. Akers, R. E. Madsen, and T. H. Meltzer. Filter Integrity Testing in Liquid Applications, Revisited. *Pharmaceutical Technology*, October (2001), 34–50.

(31) S. Sundaram, J. D. Brantley, G. Howard, and H. Brandwein. Considerations in Using Bubble Point Type Tests as Filter Integrity Tests. *Pharmaceutical Technology*, September (2000), 90–114.

(32) A. M. Trotter, T. H. Meltzer, F. Bai, and L. Thoma. The Effects of Bacterial Cell Loading. *Pharmatical Technology*, March (2000), 72–80.

(33) G. M. Morris, J. Rozembersky, and L. Schwartz. Validation of Filtration. In *Biopharmaceutical Process Validation*, Ed. by G. Sofer and D. W. Zabriskie. New York: Marcel Dekker, 2000, pp. 213–233.

(34) Parenteral Drug Association Technical Report No. 26, 1998, Supplement Volume 52, Number S1. Sterilizing Filtration of Liquids.

(35) R. Samavedam, R. Morrison, T. Kichefski, S. Cote, and A. S. Rathore. Lifetime Studies for Membrane Reuse: Principles and Case Study. *BioPharm International*, 20(9) September (2007).

(36) T. Taggart, D. Dripps, M. Cameron, T. Kessler, J. Seely, and B. Todd. Resin Reuse Validation and Evaluation of Performance over Resin Lifespan. ACS National Meeting, March 2003.

8

Analytical Test Methods for Well-Characterized Biological and Biotechnological Products

Nadine Ritter and John McEntire

8.1 Analytical Characterization of Biomolecular Products

Historically, as technologies were developed that allowed biological or biotechnological materials to be produced in large quantities for pharmaceutical use, the statement was frequently heard in both industry and regulatory circles that "The Process is the Product." This perspective was based on the understanding that the main production agent—living organisms—produce large quantities of chemically similar material (e.g., proteins) that must undergo a variety of separation steps that can (hopefully) select the greatest yield of the highest purity of a desired molecular entity from the cellular-derived milieu. The separation processes used to sort out the one or more target proteins from other, often co-purifying, proteins are optimized, scaled, and validated to reliably achieve the same population of molecular entities from each batch of biologically produced material. The nature of these separation processes is such that even subtle changes in some steps can impart significant variations in the resulting population of proteins. It has been shown repeatedly that successfully defining and controlling the *process* can define and control the *product*, hence the rationale for the phrase.

However, analytical technologies have also been emerging that allow biological and biotechnological materials to be scrutinized in ever more sensitive and specific physiochemical detail. Increasing attention is being given to the tremendous value to be derived from adequate analytical characterization of protein products, and where appropriate, critical in-process or intermediate materials. It is understood that the majority of biologically derived products are not homogenous biomolecular species: "An inherent degree of structural heterogeneity occurs in proteins due to the biosynthetic processes used by living organisms to produce them . . . Heterogeneity can also be produced during manufacture and/or storage of the drug substance or drug product" (ICH Q6B).

During early development, in-depth, orthogonal biochemical information gained on the target molecular entity provides a better understanding of the attributes of the product that may contribute to its efficacy. It also allows an assessment of intrinsic product- or process-related impurities that could impact product safety in early clinical trials. As development proceeds and the manufacturing process generates additional product batches, analytical characterization can be used to obtain biomolecular profiles of the product and its impurities in order to evaluate process performance: "Since the heterogeneity of these

DOI: 10.1201/9781003143130-8

products defines their quality, the degree and profile of this heterogeneity should be characterized to ensure lot-to-lot consistency" (ICH Q6B).

In fact, the term commonly used is "well-characterized biological/biotechnological products" or WCBPs. Although later formally designated by the FDA as "Specified Products" (Federal Register: 14 May 1996), "well-characterized" reflects the important role of rigorous analytics in assuring safe and effective products, as has been demonstrated in numerous case studies and discussions at conferences such as the annual industry-regulatory WCBP meeting (www.casss.org).

Biomolecular characterization techniques are also used to determine the physiochemical comparability of product batches before and after a process change, to assess the success of process scale-up or scale-down, or following technology transfer of the production to a new manufacturing facility. ICH Q6B indicates,

> the manufacturer should define the pattern of heterogeneity of the desired product and demonstrate consistency with that of the lots used in preclinical and clinical studies. If a consistent pattern of product heterogeneity is demonstrated, an evaluation of the activity, efficacy, and safety (including immunogenicity) of individual forms may not be necessary.
>
> (ICH Q5E; FDA Guidance 2003; CPMP Guidance 2001)

The pattern of heterogeneity, as well as the degree of comparability, is directly related to the sensitivity and specificity of the analytical methods applied (Towns and Webber 2008). When structural differences are seen, they are expected to be investigated and a risk assessment is to be performed to determine their potential impact on product quality, safety, and efficacy (Lubieniecki, Volkin et al. 2011; Kotia and Raghani 2010).

An emerging class of "generic" biotechnology products, termed biosimilars, subsequent-entry, or follow-on biologics (CPMP Guidance 2002) expands the application of biomolecular characterization techniques to assess the physiochemical and functional comparability of protein therapeutics. Previously, comparability studies were only applicable within an innovator's own process development and commercialization activities for a given biopharmaceutical product. New regulations now allow consideration of comparability studies between innovator products and subsequent producers of the same biopharmaceutical therapeutic, which could minimize the clinical requirements for the follow-on product. For both innovator and biosimilar products, a major element of the comparability study is the selection of the analytical methods utilized in the analyses.

To be successful, many factors should be considered when selecting, optimizing, and validating analytical test methods and in using test results to establish appropriate specifications for the product. Choices are made during the development cycle regarding the types of standard and state-of-the-art technologies that may be suitable for use with the product. Current regulatory guidance documents and several biotechnology industry publications available online give considerable information on the typical analytical methods used with different types of WCBPs and other biological products and current expectations for product characterization, release, and stability testing (www.fda.gov; www.ich.org; www.bioprocessintl.com; www.biopharm-mag.com).

Practical considerations should also be factored into the selection of the methods that will be used for routine quality control testing of product batches. QC analytical methods must be robust enough to function reliably over time under varying operational conditions. Failure to fully understand the details of the analytical technology and/or failure to define the intended application of the method are prime reasons for methods that end up in QC laboratories unable to reliably perform to (unsupportable) expectations.

8.2 Physiochemical Profile of Biotechnological/Biological Products

Table 8.1 highlights elements of the structure of proteins and peptides and the associated physiochemical attributes. To develop a comprehensive profile of a product, multiple aspects of the structure require analysis. In addition, tests to determine the product's concentration and potency are conducted to assess functionality. For clarity, the differences between (1) *bioanalytical* methods, (2) *biomolecular* methods, and (3) *bioassays* should be noted.

1. *Bioanalytical Methods*: These are test methods used for the quantitative determination of drugs or metabolites in physiological samples (e.g., serum or plasma) derived from animals or humans, which include a milieu of biological components (i.e., proteins, lipids, nucleic acids) (FDA Guidance 2001). For chemical drugs, techniques such as solid phase extraction followed by GC, LC, and/or MS are used to precipitate and remove the biological matrix elements, leaving the chemical analyte in solution for subsequent analysis. For biological drugs, the target entity itself would be precipitated and removed if processed with the same techniques. Therefore, for preclinical and clinical specimens, it requires specific binding techniques such as immunological methods to quantify the biological target in the presence of such a biological matrix.

2. *Bioassays*: The FDA defines *bioassays* as functional tests used to determine the activity, potency, or biological integrity of a drug product (FDA Guidance 1996). The WHO/NIBSC developed the following definition of bioassays: "A bioassay is defined as an analytical procedure measuring a biological activity of a test substance based on a specific, functional, biological response of a test system" (Mire-Sluis, Das, and Padilla 1998). Most recently, a draft of the new USP chapters on bioassay design and development, biological assay validation, and analysis of biological assays considers the terms "bioassay" and "biological assay" to be interchangeable, and defines them as "analysis (as of a drug) to quantify the biological activity/activities of one or more components by determining its capacity for producing an expected biological activity, expressed in terms of units" (USP/NF Pharm Forum Draft Chapters on Bioassay Development and Validation, <1032>, <1033>, <1034>). Bioassays include *in vitro* methods such as cell culture assays, anti-viral assays, and infectivity assays and *in vivo* assays involving animal models (Rieder et al. 2010).

3. *Biomolecular Methods*: An additional term, biomolecular methods, encompasses the analytical technologies used to perform physiochemical characterization of biological and biotechnological products (McMillen et al. 2000). Biomolecular methods used for the analysis of protein products include various forms of HPLC, UPLC, gel electrophoresis, isoelectric focusing, sequencing, various forms of mass spectrometry, amino acid analysis, carbohydrate analysis, peptide mapping, and capillary electrophoresis, among others (Schenerman et al. 2004; Gombold et al. 2006a, 2006b).

TABLE 8.1
Physiochemical Analysis of Biomolecular Compound

Level of Structural Characterization	Analytical Information Obtained
Primary Structure	Protein sequence
	Nucleic acid sequence
	Amino acid composition
	Apparent molecular weight
	Observed molecular mass
	Post-translational modifications:
	• Phosphorylation
	• Glycosylation (monosaccharide composition)
Secondary Structure	Polypeptide chains
	Peptide fragments
	Disulfide bond linkages
	Glycosylation (oligosaccharide structure)
	Isoforms (e.g., glycoforms)
Tertiary Structure	Receptor binding
	Epitope recognition
	Cell modulator release
	Cell differentiation effect
	Replication competence
	Apoptosis
Product + Ligand Conjugate	Molar ratios of ligand: product
(e.g., PEG, chemical or biological moiety)	Ligand binding sites

8.3 Analytical Methods Used in Production Operations

Every stage of the manufacturing process is supported by established specifications. The International Conference on Harmonization (ICH) defines specifications as

> a list of tests, references to analytical procedures, and appropriate acceptance criteria which are numerical limits, ranges, or other criteria for the tests described. It establishes the set of criteria to which a drug substance, drug product, or materials at other stages of its manufacture should conform to be considered acceptable for its intended use. *Conformance to specification* means that the drug substance and drug product, when tested according to the listed analytical procedures, will meet the acceptance criteria.
>
> (ICH Q6B)

Specifications are considered contracts with the regulatory agency whereby the manufacturer agrees to make and test the licensed product as defined in the product submission and commits to using only those batches of product that pass their given quality control test methods. The total quality of a biotechnology product is recognized to be intrinsically linked to a comprehensive strategy that includes specifications, thorough product characterization, compliance with CGMP operational requirements, and validated processes and validated test methods for assessing product release and stability (ICH Q6B). It is clear that sound analytical methodology is a central tenet of this quality system.

The biopharmaceutical manufacturing process typically encompasses raw materials going into production, cell culture/fermentation conditions, the purification process, the bulk active product, the formulation of the active product, and the final drug product.

In addition, the stability of bulk and final drug product must be assessed. Each of these stages requires samples to be taken and data generated to determine if the materials are suitable for use and/or if they should be processed to the next unit operation (McEntire 1994).

Analytical methods used for the quality control testing of many pharmaceutical raw materials and excipients are typically compendial, for example as found in the US Pharmacopeia (USP 34/NF 29 2011). These methods have been validated in large-scale collaborative studies and published as official US regulatory methods. To implement a compendial method in a user laboratory, the method must be verified under conditions of actual use (21 CFR 211.194(a)(2)). The verification study should consist of testing the method with the samples in the buffer or placebo matrix to assess potential matrix interference, with limited repeatability to assure reliable performance in the user laboratory (USP <1226> Verification of Analytical Methods). Analytical quality control methods for complex and/or customized raw materials (e.g., cell culture media) are not usually pharmacopeial standard tests. These materials often require methods that assess critical product attributes such as composition, concentration, and suitable function such as growth promotion or enzymatic activity. While the vendor of the material will have some methods in place for product quality testing, they may not address the same parameters as needed by the specific applications used in biotechnology manufacturing processes. In these cases, it will be necessary to develop and validate the necessary raw materials or excipients tests. Where vendor testing is suitable to meet the user requirements, the vendor could be audited to assure the quality of their operations. Then the vendor's certificate of analysis may be accepted as the certification of each batch of material quality. The only test required upon receipt of each batch would then be an identity test method to confirm the correct material was received (21 CFR 211.84 (d)(2)).

To determine their acceptable quality, the bulk and final product must be analyzed for identity, purity, impurities, concentration, and potency (ICH Q6B). Unless the product is listed in a compendial monograph in which regulatory methods for these product attributes are given, the development of non-compendial methods will be necessary. Many historical biological products such as plasma fractionation products (i.e., human serum albumin) and vaccine products do have monograph listings in the USP/NF that must be followed for product release. In the EU, there are general monographs for certain modern biopharmaceutical products such as monoclonal antibodies (Pharm Eur. 1/2008:2031).

However, most modern biotechnology products are new molecular entities and therefore do not have monograph listings. In the absence of compendial methods, manufacturers of these products must develop and validate their own (non-compendial) analytical methods and product specifications (FDA Guidance 2000 draft). Manufacturers are also responsible for verifying that analytical methods used for product stability testing are suitably capable of detecting and, as necessary, quantifying degradation products. It should be noted that compendial methods are not necessarily verified to be stability indicating for the products listed in monographs (USP 34/NF 29 2011); the burden is on the user to confirm the appropriate methods for use in stability protocols.

During cell culture and fermentation steps, critical parameters are measured to assure adequate control of the processes. Parameters such as pH, O_2 and CO_2 levels, glucose, and other sentinel compounds are monitored to confirm they are within required limits. The test methods used to perform these measurements may be simple (e.g., pH or dissolved gas) or more complex (e.g., cell density or target protein concentration). If they are compendial, they may be used with verification, as described earlier. If they are non-compendial, they will require validation for routine use under CGMP, as described later.

Product fractionation or purification is supported with analytical methods to determine the success of a unit operation and the ability to process the material to the next step. In practice, these in-process methods are typically developed using purified forms of the target product to measure purity, concentration, or potency. Then method performance must be verified with in-process samples, since considerably greater amounts of process- and product-related impurities are present in the in-process samples than in the purified product. Also, buffer components and concentrations may be significantly different in in-process samples. For these reasons, it is usually necessary to document specificity using test buffer blanks in parallel with test samples from process development experiments to determine the effect on test method performance. In these applications, the recovery of target protein is usually calculated as a percent of starting material. When the purification process is finalized, the in-process test methods should be validated prior to use in process validation and full CGMP.

In 2003, the FDA launched a new initiative termed "Process Analytical Technology" or PAT, which is defined as

> a system for designing, analyzing, and controlling manufacturing through timely measurements (i.e., during processing) of critical quality and performance attributes of raw and in-process materials and processes with the goal of ensuring final product quality. The term "analytical" in PAT is viewed broadly to include chemical, physical, microbiological, mathematical, and risk analysis conducted in an integrated manner.
>
> (FDA Guidance 2000)

The concept marries analytical technology, process development tools (such as statistically designed experiments), mathematical modeling, and risk assessment tools (such as failure modes effects analysis) to define the most effective testing and control scheme to best assure the consistent quality of the manufactured product.

The use of PAT tools is one of many approaches included in the concepts of process design space, as described in ICH Q8 (ICH Q8(R2 2009). Establishment of a design space is based on the "a multidimensional combination and interaction of input variables and process parameters that have been demonstrated to provide assurance of quality." The ability to evaluate process design space is only as good as the data obtained from the experiments conducted on input versus output, that is, elements that produce a measurable result in the process characteristics. And the data are only as good as the type and capabilities of the analytical methods used to measure results. One element of establishing process design space is to utilize a greater number of analytical techniques to assess the impact of process changes on product characteristics. For example, in addition to methods used to measure the clearance of process-related impurities, methods for assessing product heterogeneity and product modifications (e.g., isoforms, degradants) may also be necessary to generate a meaningful design space.

8.4 Methods Used for Product Characterization, Release, and Stability Testing

A singular feature of the analysis of biotechnology products is the diversity of analytical technologies necessary to obtain the physiochemical profile. Characterization of a biopharmaceutical product is considered to be the complete description of its physical, chemical, and biological characteristics (FDA Guidance 1999). A subset of methods used for product characterization can be validated for routine product QC batch release testing. A subset of QC release methods can be suitable for use in product stability protocols. Examples of the types of analytical technologies used in the characterization, release, and stability testing of biotechnology products are shown in Tables 8.2A, 8.2B, and 8.3. These exemplify the continuum of molecular complexity among biopharmaceutical products and illustrate the corresponding set of analytical 'tools' applicable to the nature of the product. The first (Tables 8.2A and 8.2B) describes a set of methods generally utilized with monoclonal antibody therapeutics (Schenerman et al. 2004); the second (Tables 8.3) are methods applicable to viral-based products such as live attenuated vaccines or gene therapy products using viral vectors (Gombold et al. 2006a, Parts 1 and 2).

TABLE 8.2A

Most Frequently Used Lot Release Tests for Monoclonal Antibody Drug Substance and Drug Product Used with permission of the publisher.

Method	Use, Impurities or Substances Detected	ICH Q6B Category	Quality Attribute	Drug Substance	Drug Product
Protein concentration (A_{280} absorbance)	Measure protein concentration	Quantity	Dose	Yes	Yes
High-performance size-exclusion chromatography (HP-SEC)	Aggregates protein fragments	Purity	Size	Yes	Yes
Ion-exchange (IEC) or hydrophobic- interaction (HIC)chromatography, isoelectric focusing (IEF), or capillary IEF	Deamidation, protein fragments	Identity, purity	Charge	Yes	Yes
Capillary zone electrophoresis (CZE) or native gel electrophoresis	Deamidation, protein fragments	Identity purity	Charge, size	Yes	Yes
Peptide mapping	Primary structure	Identity	Structure	Yes	No
Denaturing gel or capillary electrophoresis reducing or nonreducing	Protein fragments	Purity	Size	Yes	Yes
Antigen binding assay or other appropriate bioactivity assay	Potency	Potency	Activity	Yes	Yes
Host cell proteins[a]	Residual host cell proteins	Impurities	Impurities	Yes	No
DNA[a]	Residual DNA	Impurities	Impurities	Yes	No
Process-related substances and impurities[a]	Various process-related impurities	Impurities	Impurities	Yes	No
Endotoxins (Limulus amoeboyte lysate)	Detect endotoxins	Contaminants	Impurities	Yes	Yes
Sterility	Test tor sterility	Contaminants	Impurities	Yes	Yes
pH	Measure pH	General	pH	Yes	Yes
Particulates	Impurities	Impurities	Impurities	No	Yes
Volume	Measure volume	General	Volume	No	Yes
Appearance	Evaluate color and clarity	General	Color/ clarity	No	Yes

* Used with permission of the publisher

TABLE 8.2B

Test Methods for Monoclonal Antibody Drug Substance Characterization (supplementary tests that should be performed in addition to lot-release testing) Used with permission of the publisher.

Method	Use (Impurities/Substances Detected)
ADCC, CDC, neutralization, etc.	Potency characterization
Isotyping	Verify IgG isotype
Peptide mapping with electrospray ionization mass spectrometry (ESI-MS) detection	Truncation, deamidation, oxidation, phosphorylation, substitution, alterations in oligosaccharides, incorrect sequence
Focused peptide map	Detect specific product-related impurity/substance (e.g., oxidation)
Carbohydrate composition	Determine monosaccharide and sialic acid content
Oligosaccharide profile	Determine oligosaccharides present
N-terminal and C-terminal content	Determine proportion of C-terminal lysine forms and N-terminal truncation and/or blockage
Western blotting	Heavy and/or light chain related species
Analytical ultracentrifugation	Detect and characterize aggregates
Matrix-assisted 1aser-desorption ionization time-of-flight (MALDI-TOF) mass spectrometry	Aggregates, breakdown products, verify mass
ESI-MS (intact molecule)	Aggregates, breakdown products, verify mass, non-glycosylated forms, and C-terminal variants

TABLE 8.3

Lot Release and Stability Assays Used with Live Viral Vaccines and Viral-Based Gene Therapy Products Used with permission of the publisher.

Assay	Purpose
Field flow fractionation multiangle light scattering (FFF-MALS)	Determine particle number and aggregation state
Atomic force microscopy (AFM)	Determine particle number and aggregation state
Transmission electron microscopy (TEM)	Determine particle number
Size exclusion chromatography multi-angle light scattering (SEC-MALS)	Determine particle number and aggregation state
TCID, FFA, plaque, or other assays	Determine proportion of defective particles based on the difference between total particles and infectious particles
Polymerase chain reaction (PCR)	Determine proportion of nucleic acid containing particles
Density gradient centrifugation	Determine proportion of defective particles based on relative densities of particle populations
Analytical ultracentrifugation (AUC)	Determine proportion of defective and aggregated particles based on hydrodynamic properties of particle populations
Capillary electrophoresis (CE)	Determine proportion of defective and aggregated particles based on particle mass and charge
Reversed-phase HPLC (RPHPLC)	Determine proportion of defective and aggregated particles based on hydrophobic interaction properties
Ion-exchange chromatography (IEC)	Determine proportion of defective and aggregated particles based on charge state of the particles

(Continued)

TABLE 8.3 *(Continued)*

Lot Release and Stability Assays Used with Live Viral Vaccines and Viral-Based Gene Therapy Products Used with permission of the publisher.

Assay	Purpose
Size exclusion chromatography (SEC)	Determine proportion of defective and aggregated particles based on hydrodynamic sieving properties of particle populations
SDS-PAGE (or equivalent)	Determine composition of proteins contained in preparation based on polypeptide chain sizes
Western blot	Determine composition of immunoreactive proteins contained in preparation
Process residuals (BSA, benzonase, polysorbate, etc.)	Quantify process-related impurities

Full characterization analysis is typically conducted at key points during the development cycle and after licensure. The objective of a characterization study is to provide detailed information from a wide array of techniques in order to provide a thorough understanding of the expected nature of the material. Characterization establishes the physiochemical attributes the product will, and should, have to support its safety and efficacy. When thorough characterization is performed, some "curious discoveries" have been made on the physiochemical nature of biotechnology products (Canova-Davis 1994). Errors in translation, incorporation of unusual amino acids, novel cross-links, and amino acid substitutions have all been discovered when state-of-the-art analytical methods such as peptide mapping procedures, mass spectrometry, high performance liquid chromatography (HPLC), and electrophoretic methods are used to analyze products. Of particular interest in characterization studies is the "fingerprint" of product heterogeneity and product- and process-related impurities of the product. In addition to the major quantitative analyses, qualitative assessment of SDS-PAGE and IEF banding patterns, or peptide map and mass spectrophotometric fragment patterns, can yield significant information on the capability of the production process to yield material with consistent characteristics. Also, methods that assess higher order structure are considered essential elements of product characterization and comparability studies (Wei et al. 2011). Glycoproteins require characterization of their carbohydrate moieties using a wide variety of analytical techniques (Manzi 2008, Parts 1–3).

Generally, methods used only for drug substance characterization are not validated in accordance with all ICH and FDA guidelines; however, it is expected that these methods should at a minimum be qualified to ensure they are capable of generating reliable data on the specific material being tested (Ritter et al. 2004).

8.5 QC Release Tests

Traditional chemical drugs are typically QC release tested for physical description, physical properties (e.g., pH of solution forms; moisture of solid forms; particle size), identity, assay (drug content and purity), process-related impurities (e.g., solvents), microbial limits, and, in some cases, proportion of chiral or polymorphic species (ICH Q6A). Most of these methods employ spectrophotometric, gravimetric, or chromatographic techniques. Biologically derived pharmaceuticals often require additional immunological, enzymatic, electrophoretic, colorimetric, and cell-based methods for assessing molecular characteristics and complex host and process-derived impurities. It is expected that no single analytical method will profile all biotechnology product characteristics. Each critical attribute—identity, purity, quality, potency, strength, product- and process-related impurities—can be assessed by multiple analytical procedures; each test could yield different results based on differences in method capabilities such as sensitivity and specificity. Table 8.4 is from USP General Chapter <1045> Test Procedures for Biotechnology Products (USP 32/NF 29 2011). It illustrates how a variety of biomolecular methods can be utilized to evaluate specific biotechnology product- and process-related impurities.

One class of process-related impurities requires special attention in terms of the analytical technology utilized to produce accurate measurements. Host cell proteins (HCPs) comprise a highly diverse population of process impurities, some of which have molecular properties (e.g., charge, size, polarity) that allow them to co-purify through downstream process steps along with the target biopharmaceutical protein (Champion et al. 2005). Moreover, slight variations in process unit operation conditions can alter the population of HCPs that are generated (in the case of upstream cell culture or fermentation) or cleared (in the case of downstream purification steps). The analytical method for measuring HCPs in in-process samples and purified bulk drug substance must be capable of accurately detecting a wide array of proteins produced by the host cells. Currently, the established technology for HCP testing is to use an ELISA-like immunoassay. The two most critical reagents in the method are (1) the mixture of polyclonal antibodies used to immunodetect the HCPs generated by the cell's proteome and (2) the mixture of proteomic HCPs used for the assay standard curve. In order to successfully validate an HCP ELISA method, the specificity of the anti-HCP polyclonal antibody reagent must be confirmed. That is, it must be demonstrated that the population of polyclonal antibodies present immunorecognizes a majority of the HCP present in the expression system.

Typically, the polyclonal anti-HCP antibodies are produced by immunizing animals with a mixture of HCPs (from the biopharmaceutical product's host expression system) and then collecting the hyperimmune serum that contains anti-HCP antibodies (Savino et al. 2011). Then, to verify that the antibodies produced are adequately specific for the HCP population, a Western blot study is performed. The HCP mixture is first separated on

TABLE 8.4

Analytical Methods for Biotechnology Product Impurities Adapted from USP 24 <1045> Biotechnology-Derived Articles. Used with permission of the USP.

Aggregation:	SDS-PAGE, SEC-HPLC, light scattering
Deamidation, Oxidation:	Peptide map, HPLC, IEF, MS
Proteolytic Cleavage:	Peptide map, SDS-PAGE, HPLC, IEF, MS
Amino Acid Substitutions:	AAA, Peptide map, MS, protein sequence, CE
Translation Mutations:	Peptide map, HPLC, IEF, MS, CE
Host Cell Proteins:	SDS-PAGE, Western blot, ELISA
Media Components:	SDS-PAGE, Western blot, HPLC, ELISA
Nucleic Acids:	DNA hybridization, UV, Protein binding
Affinity Antibodies:	SDS-PAGE, Western blot, ELISA
Proteases/Nucleases:	HPLC, Western blot, ELISA
Leachates/Extractables:	HPLC, MS, GC, gravimetric analysis
Process Residuals:	Karl Fisher moisture, GC, ion chromatography

2-D gels to resolve the population of proteins by charge and size. One gel is then stained to reveal the proteins present in the HCP mixture. A parallel 2-D gel of the HCP mixture is electroblotted to a membrane, which is then probed with the anti-HCP antibody preparation and reacted to visualize the immunobinding reactions. The two images should correspond as closely as possible; every HCP protein spot that is present in the stained gel should yield a corresponding immunoreactive spot in the anti-HCP Western blot. Due to differences in inherent antigenicity among the proteins from the host cell proteome and in the affinity/avidity of the polyclonal antibody species generated, as well as technical challenges in efficiently separating, electroblotting, and immunoreacting hundreds of different proteins simultaneously, it is not always possible to see 100% correlation of spots between the stained gel and Western blot.

However, there must be a very high degree of correlation in order for these reagents to be accurate enough for use in an ELISA-like method. Otherwise the signal produced in the ELISA binding reactions will only represent a subset of the possible HCPs in the test samples as well as in the HCP standard curve, and the values produced (typically in units of ppm) will underestimate the level of HCPs present. Until the specificity of the polyclonal antibodies for the HCP population is confirmed, other performance parameters of the ELISA method such as accuracy, linearity, limit of quantitation, and limit of detection cannot be validated. There are commercially available HCP ELISA test kits for several of the most common biopharmaceutical expression systems, such as *E. coli* and CHO (www.cygnustechnologies.com). With some biopharmaceutical products, 2-D gels and blots will demonstrate that the immunoreagents in these kits are adequately specific and sensitive for the user's expression system, and the ELISA method that utilizes them can be validated for its intended use. In other applications, 2-D gels/blots reveal that the commercial immunoreagents are not adequately specific, recognizing only 3/4 (or less) of the HCP population in the user's host expression system. In such cases, the user will have to develop a process-specific HCP mixture and anti-HCP polyclonal antibodies for their expression system's proteome. Then, if the specificity is confirmed by 2-D gels and blots, the reagents can be used to develop and validate a custom HCP ELISA.

For biotechnology product QC release testing, it is expected that orthogonal analytical methods will be used for key product attributes (FDA Guidance 1999). "Orthogonal" refers to methods that exploit different chemical or physical mechanisms for analysis. For example, SEC-HPLC and RP-HPLC are based on two different separation mechanisms. Similarly, IEX-HPLC and cIEF are orthogonal methods. Typically, for well-characterized biotechnology products, parameters such as purity and identity are supported with a minimum of two orthogonal QC release methods.

8.6 Stability Tests

For biopharmaceutical products, it is recognized that

> the evaluation of stability may necessitate complex analytical methodologies. Appropriate physiochemical, biochemical and immunochemical methods for the analysis of the molecular entity and the quantitative detection of degradation products should also be a part of the stability program whenever purity and molecular characteristics of the product permit use of these methodologies.

> (ICH Q5C)

Methods used in the stability protocol should detect significant changes in the quality of the product (i.e., purity, potency) with a focus on the ability of the methods to determine product degradation (FDA Guidance 1999). There are several well-known physical and chemical degradation pathways for proteins, such as aggregation, fragmentation, oxidation, and deamidation (Patal et al. 2011). However, the specific degradation pathways and kinetics (which are rarely linear) for a given protein will be unique to its primary, secondary, tertiary, and (for multimeric proteins) quaternary structure. While some pathways can be expected, such as aggregation caused by agitation of protein solutions, it cannot be assumed that any given protein will—or will not—experience multiple types of degradation over time or under conditions of stress. The analytical methods being used to assess product stability must be proven to be capable of detecting all possible degradants that could be formed if the product begins to degrade. Otherwise, there will be no data to prove the product is stable—changes could be occurring, but the methods are analytically "blind" to them.

The best way to demonstrate whether a test method is truly capable of detecting or quantifying product degradation is to conduct a forced degradation study. Forced degradation studies provide critical information on the inherent stability of the product and its degradation pathways and confirm the capabilities and suitability of the analytical methods to be used in stability testing. This one-time study on a single batch is not considered a part of the normal stability protocol. As shown in Table 8.5, it should stress the drug substance in several physical and chemical experiments, including various pH solutions, in the presence of oxygen and light, and at elevated temperature and humidity increments. For biotechnology products in solution, it should include agitation stress. For products stored frozen, multiple freeze-thaw cycles should be examined. It is recognized that stress conditions may create product degradants not formed under normal storage, shipping, and handling conditions, but the force-degraded product preparations are considered to be important as reagents used to challenge the potential stability-indicating analytical methods (FDA 1999).

Whenever significant qualitative or quantitative changes indicative of product degradation are detected during long-term, accelerated, or stress studies, consideration should be given to the potential hazards and to the need for characterization and quantitation of degradants (ICH Q5C). The goal is to ensure product safety by clearly understanding the physiochemical nature and potential adverse impact of a product's inherent degradation pathway. This is especially true for biopharmaceutical products, where degradants could create new epitopes that could trigger a neoantigenic immune response in patients, such as heat-treating product preparations to achieve viral inactivation (Smalls, Pepper, and James 2011).

Ideally, initial stress studies should be done early in product development to allow selection of the stability-indicating methods for real-time stability studies (FDA Guidance 1999). Experiments are designed to cover all potential degradation pathways of a given product, allowing the degradation conditions to proceed and removing sentinel samples at designated points and subjecting them to analysis using orthogonal analytical methods

TABLE 8.5

Purposeful (Forced) Degradation of Biotechnology Products

Chemical and Physical Treatments to Promote:

• Aggregation	• Fragmentation
• Precipitation	• Dephosphorylation
• Deamidation	• Deglycosylation
• Hydrolysis	• Oxidation
• Disulfide bond exchange	• Ligand release

(Yu 2000; Kats 2005). Accelerated and stress stability experiments can provide valuable information on the degradation pathways and kinetics of degradation (Magari 2003). But unlike chemical pharmaceutical products, expiration dates and retest periods for biological/biotechnological products cannot be determined from extrapolation of short-term results, but can only be established through real-time stability studies (ICH Q5C).

8.7 Product Potency Assays—Why the Ends Do Not Justify the Means

When discussing release and stability tests for biotechnology products, one frequently heard comment is: "Why can't a potency assay suffice as evidence of product quality and stability? Isn't it the ultimate proof of acceptable product performance?" It is true that biotechnology products may be intrinsically heterogeneous, and the specifications may include designated impurities (ICH Q6B). But it is necessary to assure continued product safety and efficacy by monitoring and maintaining the degree of heterogeneity and impurities to the levels demonstrated in preclinical and clinical trials. Ideally, the potency assay should be linked to the product's expected mechanism of action (21 CFR 610.10; ICH Q6B). For some proteins, such as monoclonal antibodies, there may be various classes of function (Xu-Rong et al. 2011).

When choosing a quality control testing scheme, the better question to ask is: "What are the parameters that best demonstrate manufacturing consistency?" These will likely encompass more than just the parameters that are known to impact clinical efficacy, because the first objective of quality control testing is to assure continued product *safety*. For biotechnology products, "since the degree of heterogeneity defines their quality, the degree and profile of this heterogeneity should be characterized to ensure lot to lot consistency" (ICH Q6B). Product potency assays alone are not capable of measuring product heterogeneity (i.e., the type and ratio of product-related substances), product-related impurities (e.g., product degradants), and process-related impurities.

However, to be suitable for the intended use of assuring product quality and stability, potency assay should be capable of measuring difference between "good" and "bad" lots of product based on the quality of the active species at time of release and over time with handling and storage. To that end, the potency assay should be sensitive to some degree of product degradation, demonstrating a decrease in potency with increased degradation. But because potency assays (and even immunoassays) are based on the measurement of a defined target activity, they are often insensitive to varying levels of impurities unless they interfere with the specific reaction being detected. Typically, potency assays are used to measure the loss of intact product activity rather than the increase of individual degradants. Also, because potency assays are often less precise than physiochemical methods, it may take a higher degree of product degradation to be accurately and reliably measured by potency than by methods such as chromatography or electrophoresis.

Some biopharmaceutical products have more than one functional domain, such as monoclonal antibodies with an epitope recognition moiety (e.g., the Fab region) and a cell-binding moiety (e.g., the Fc region). When multiple functional domains are a part of the product's critical characteristics, it is necessary to assure that both (or all) of the domains meet physical and functional specifications for quality control. The nature of the procedure(s) utilized for product potency testing range from (relatively) simple ligand binding or enzymatic procedures through cell-based assays to *in vivo* animal assays.

8.8 Comparability Assessment

There are several points at which it is advantageous to assess the comparability of product batches. At its simplest, a comparability study can be thought of as side-by-side characterization of test materials. That is, test samples from Batch A and Batch B are assayed in parallel with the types of test methods listed in Tables 8.2A, 8.2B, and 8.3 for characterization. While it is possible, it is less desirable to compare test results from samples assayed independently (i.e., on different days, months, or even years), because the effect of intrinsic test method variability can confound the ability to accurately assess the similarities or differences. This is especially true when qualitative "fingerprint" methods are used, because slight variations in method performance can significantly change subtle—but critical— product profile elements. Therefore, the best case for comparability is made when product samples are run together using validated methods. Then, even if the test methods are not fully validated for robustness over time, all of the experimental bias should at least be skewed in the same direction for all samples, allowing more confidence in the interpretation of results.

One key point for comparability assessment is to link safety and efficacy data from all phases of product development to the physiochemical characteristics of the preclinical and clinical lots. To demonstrate "developmental continuity" of product characteristics relative to the preclinical and clinical experiences, Phase I/II drug substance and drug product batches should be compared to the batches used for Phase III to show consistency of manufacturing quality, purity, and potency among them (FDA Guidance 1999). The most effective means by which to make these pre-BLA comparisons is to prospectively plan to retain samples of each preclinical and clinical batch of drug substance and drug product at each phase of development, starting with the lots used for toxicology studies. While long-term stability will not be assured, as these samples are stored before formal stability studies are conducted, the effects of degradation on the comparability of lots can be experimentally confirmed. However, if these samples are not retained, the determination of comparability will have to be based on test data collected at different points in time. Any variability in the performance of the method, including small method changes made for optimization or validation at each phase, will make the evaluation considerably more difficult.

If the characteristics of the lots, particularly in terms of impurities, change dramatically as the process is optimized, it could call into question the applicability of early toxicology and safety test results. That is, product- or process-related impurities should not be different from, or greater than, those seen in the product batches used in safety studies conducted in early development. If they are, the sponsor may have to repeat some of these earlier studies to assure that all new or elevated impurities have been experimentally tested for patient safety (Cavagnero 2003). Similarly, in drug substance and drug product stability testing, when degradation products result in heterogeneity patterns that differ from those observed in preclinical and clinical development, the significance of these alterations should be evaluated to ensure the continued safety and efficacy of the product (ICH Q6B).

After licensure, continuous improvements in the process, or technology transfer to other manufacturing facilities, can be assessed with FDA-approved Comparability Protocols. A Comparability Protocol is a well-defined, detailed, written plan for assessing the effect of specific chemistry, manufacturing, and controls (CMC) changes on the identity, strength, quality, purity, or potency of a specific drug product as these factors relate to the safety and effectiveness of the product.

A Comparability Protocol describes the changes that are covered under the protocol and specifies the tests and studies that will be performed, including analytical procedures that will be used and the acceptance criteria that will be achieved to demonstrate that the specified CMC changes do not adversely affect the product (drug substance, drug product, intermediate, or in-process material) (FDA Guidance 1996, 2003).

8.9 Development, Qualification, and Validation of a Product-Specific Non-Compendial Analytical Method

The objective of method development is to deliver a procedure that is capable of performing reliably to measure a defined attribute of a test sample. Individual firms may use different terminology for some activities (i.e., method optimization, qualification, or validation) or divide parts of the strategy among different operational groups (e.g., analytical development and quality control). The method development strategy can be outlined in terms of questions to be asked. The answers to these questions (and the data to support them) should be captured in writing in method development reports. Method development reports serve as the historical scientific record of the rationale and justification for how the method was selected and why it is considered suitable for use. Well-written, comprehensive development reports provide valuable background information to future users of the method. They support the technical content of the CMC sections of product regulatory filings (21 CFR 312.23 (a)(7)).

In addition to method development reports, it is important to document the entire life cycle of each analytical test method, from selection and development through method qualification and (where applicable) validation. But there are other significant events in the life cycle of an analytical procedure: method technology transfers (pre and post qualification or validation), method re-optimization, method requalification/revalidation, method comparability (i.e., comparing the performance capabilities of two different procedures), and method bridging (i.e., replacing an old test method with a new test method). Each of these events should be supported with a prospective experimental plan outlining the objectives of the exercise and defining what would constitute a successful outcome. For example, the general requirement for replacing an old analytical method with a new one is that the new procedure must be at least as sensitive and specific for the intended use as the old procedure (FDA Guidance 2003). After the exercise, there should be complete documentation of the experiments performed, the test materials used in the study, and all results (including raw data). Documentation should also include copies of the method SOP versions in place at the time of the study, as well as the method SOP versions generated as a result of the study (e.g., updated with new instructions after revalidation, or the method SOP from new testing site). As a part of product knowledge management (ICH Q10), these method documentation packages should be available for reference throughout the entire life cycle of the procedure.

The ICH Common Technical Document, the "generic" international regulatory dossier template for CMC quality details, indicates that a summary of the history of the analytical methods used in batch release testing and stability protocols during the preclinical and clinical development phases should be provided in the product commercial application. (ICH M4Q; ICH M4: General Questions and Answers). In addition, ICH Q10, Pharmaceutical Quality System, describes the lifecycle approach to product information whereby early phase activities such as method selection and development are incorporated into a total

Knowledge Management paradigm that remains with the product through commercialization and beyond (ICH Q10). Likewise, in the 21st-century initiatives for CGMP, the FDA acknowledges the key role of product development information in assuring that quality was "built in" to the product design, which depends heavily on the appropriate analytical methods for characterization, comparability, and QC release and stability testing (Pharmaceutical GMPs for the 21st Century 2003). For all of these reasons, method development reports will likely be of increased interest to regulatory reviewers when they are evaluating the analytical information in product license applications and annual reports.

8.10 Outline of Test Method Development Strategy

1. What is the purpose of the test? Identify the attribute of test samples that requires measurement (i.e., identity, purity, potency, concentration, or other attributes, e.g., moisture).

To do this effectively, consider what statement, or claim, about the product will be made based upon the test results:

"The $A_{280\,nm}$ protein concentration of the product batch is 5.6 mg/ml."
  "The purity of product batch 123XYZ is 98.6% by SEC-HPLC."
  "The product bands co-migrated with the reference standard on SDS-PAGE."
  "The product amino acid sequence corresponds to that of curecancerin."
  "No single impurity exceeded the limits of quantitation by SEC-HPLC."
  "Potency of the batch was 1450 mIU/mg with the chromogenic assay."
  "No degradants were detectable by RP-HPLC after 9 months at 5 °C."

Establishing the method's intended use is arguably the most critical component of test method development and validation activities (McEntire 1994). Perhaps because of the wide array of analytical techniques required for the complete analysis of biotechnology and biological products, it is often difficult to dissect the individual intended use for each type of method. But without clearly defining the method's intended use from the very beginning, the entire process of method development and validation could yield the metaphorical situation of "building the right ladder up the wrong wall."

For quantitative methods, there is a correlation between the target specifications of the intended use and the capability required of the method to support those specifications (Paul 1991). For example, for a specification of 90%–110% of a target concentration, the corresponding level of validated test method precision should be less than 5% RSD to reliably achieve accurate results using a minimum of test method replicates (i.e., n = 3). Or, turning this around, if the test method can only achieve a validated precision of 10% RSD, the specifications may only be supported for 80%–120% unless the number of test replicates is greatly increased. The requirements for method precision are inversely related to accuracy/recovery, that is, a lower recovery requires a more precise method to support the same claims than a higher recovery method would require. There are published standard probability curves (operating characteristic curves, or OC) that link test method precision capabilities and the number of samples needed to support the desired level of confidence in accuracy (Vanderweilen and Hardwidge 1982).

Certain biomolecular techniques can provide information on more than one attribute of the product, such as identity and purity, and identity and potency. If there are multiple intended uses for a single method, these must be investigated individually to assure the method can support each one adequately. For example, if SDS-PAGE is intended to support claims for purity and identity, the test method procedure should be designed to allow results to be evaluated individually for each attribute. Samples and reference standards might need to be in adjacent lanes on the gels to support the identity claim if co-migration of bands is the acceptance criteria. Sample concentration might not be critical as long as the test sample and reference standard were loaded in equivalent amounts and all bands are visible on the gel. In order to obtain purity data, lane order may be flexible but sample loading concentration critical to assure that densitometric scanning values will fall in the linear range of the test method. If an ELISA test method is intended to support claims for identity and potency, the primary antibodies must be demonstrated to be specific for the product to claim immunoidentity, whereas if potency alone were the test method claim, intrinsic background cross-reactivity might be acceptable in the presence of the proper internal controls.

2. How does the test work? Assess the nature of the method technology (immuno-reactivity, biochemical activity, quantitation of mass, resolution of polypeptides/impurities, etc.).

If necessary, refer to scientific literature, method books, manufacturer's technical literature, instrument vendor's booklets, and so on to understand exactly how the method—and importantly, the instruments used—technically function. Each technology has limitations that can significantly impact the ability of the test to meet its intended use. Edman protein sequencing and mass spectrometry can be very specific for product identity via molecular mass and/or sequence data, but neither is routinely used in a quantitative manner because the nature of the technologies renders absolute quantitation hard to achieve. SDS-PAGE with Coomassie stain and quantitative image analysis can be developed and validated for purity and impurity determination for many biotechnology products. By comparison, SDS-PAGE using silver staining methods is highly sensitive to low levels of protein and is very useful in product comparability studies, but it is very difficult to reproducibly analyze silver-stained gels using scanning densitometry to establish meaningful quantitative specifications.

If the nature of the technique is well understood and the method's intended use is clearly defined, designing the appropriate development and validation strategy should become a logical exercise to examine those aspects of the technology that are likely to affect the performance of the method for the parameter of interest. The emerging risk-based assessment strategies described in ICH Q9 could be of considerable utility in defining and prioritizing method performance parameters for a given intended use. One such exercise would be to map out the elements of the procedure, the instruments, the samples, and the reagents with a cause-and-effect diagram to identify the relationship among key elements and to uncover potential sources of variability (Ishikawa 1986). Then experiments can be designed to assess the impact of these variables on the performance of the test method. For quantitative methods, statistically designed experimental (DOE) tools can be highly valuable in assessing the effect of multiple variables simultaneously (Torbeck and Branning 1998)

3. Does the test require special reagents or materials? Identify any reagents or materials that could be critical to the reliable, robust performance of the method (e.g., antibodies, enzymes, substrates, cofactors, commercial kit materials, internal calibration standards, types of cuvettes, specific microtiter plates).

Many biomolecular methods require the use of critical reagents, such as antibodies in immunoassays or cell lines in cell-based assays, which are biologically derived components. Even common materials such as plastics and glass can interact with protein products or the method's biological reagents, and the lot-to-lot or vendor-to-vendor differences in composition can affect successful test method performance. Table 8.6 lists different types of methods and components that can contribute to variability in method performance. Any of these items can impart a significant degree of variation in test methods unless strategies are in place to prospectively address them (Ritter and Wiebe 2001). Investigate

TABLE 8.6

Potentially Critical Assay Components

Biomolecular Assay Type	Potential Batch-to-Batch Variability
Colorimetric	Unique buffer components Chromogenic reagent Commercial kit active components
Enzymatic	Unique buffer components Substrates Enzymes Cofactors Detection reagents Commercial kit active components
Chromatographic	Unique buffer components Labile mobile phase solvents Chromatography column resin Derivatization or conjugation reagents
Electrophoretic	Unique electrode buffers Gel matrix reagents Sample treatment reagents Staining reagents Commercial kit components
Immunological	Primary antibodies Secondary antibodies Conjugated antibodies Blocking reagents Detection reagents Commercial kit active components Plastic cuvettes or microtiter plates
Ligand Binding	Unique buffer components Target receptor Target ligand Detection reagents Commercial kit active components Plastic cuvettes or microtiter plates
Cell-Based Bioassay	Cell seed stock (homogeneity and viability) Cell culture (passage number and density) Media components Growth factors Antimicrobial agents Harvest reagents (e.g., trypsin) Cell reactants (e.g., induction compounds) Plastic flasks or plates

the impact on test method performance of different lots, or the range handling conditions. Develop an experimental study plan to bridge old lots of critical reagents to new ones before using the new lots in the test method. Include this study plan in the method SOP to assure future users will recognize when and how to perform reagent bridging with the test method.

> 4. What will be used as a reference standard? Determine the appropriate product-specific reference material(s) for the intended use(s) of the method.

Check for international standards if it is an existing product. Organizations such as the US Pharmacopeia (USP, www.usp.org) and the National Institute for Biological Standards and Controls (NIBSC, www.nibsc.org/) maintain certified reference standards for many currently licensed biological products and some biotechnology products. For some biologics, the FDA Center for Biologics Evaluation and Research (CBER, https://www.fda.gov/about-fda/fda-organization/center-biologics-evaluation-and-research-cber) has designated reference standards. Reference standards from a certification agency should only be used as primary standards against which in-house working standards are regularly qualified.

Note, however, that international standards for biological products are usually only certified for the calibration of potency or activity, and possibly molecular identity. They may be unsuitable for use in product assays for purity or impurities. In many cases, each biopharmaceutical firm, based on the nature of their purification process and their formulations, must generate their own purity and impurity reference standards. When the product is a new biomolecular entity for which there are no pre-existing reference standards, the innovator is responsible for establishing their own reference standards for critical product attributes, including (in some cases) key product degradants (ICH Q6B).

> 5. Can a good test run be distinguished from a bad test run independently of the test sample results? Establish appropriate system suitability controls based on the technology of the method and its intended use.

These are to be included with each run to show the run was valid. Each analytical test method should incorporate relevant system suitability measures to allow the analyst to verify that the test system is performing to expectations at the time of use (Ritter, Hayes and Dougherty 2001). Some system suitability measures are simple (i.e., calibration of a pH meter with standard solutions, or the level of precision among replicates), and others are more complex (i.e., the use of designated reference materials in structural or functional tests).

The inherent value of system suitability is that it provides a mechanism to assess the performance characteristics of the test method at the time and in the location of each use. It has been noted that defining and utilizing appropriate system suitability measures, particularly for non-chromatographic analytical test methods, is one of the most misunderstood aspects of method development and validation (Williams 1987). System suitability consists of two parts: (1) the material(s) used in the assessment and (2) the specifications associated with that material's performance (validity criteria). Typical system suitability measures for non-chromatographic methods are in the form of calibrators or controls. For quantitative methods, accuracy and precision is usually confirmed through the use of calibration standards or by the preparation of standard curves. For qualitative methods, positive and negative controls often serve as system suitability measures. In most test methods, a product-specific reference standard is included in the procedure; additional validity criteria can be designed for its use.

Each system suitability measure should have established criteria by which to determine if the materials pass or fail their performance in the test method. These are known as validity criteria—they are the specifications used to determine if a method is acceptable each time it is performed. For product-specific reference standards, validity criteria can include elution time and peak profile in chromatographic tests, migration distance and banding pattern in gel-based assays, immunoidentity in ELISAs and Western blots, concentration value in protein determination assays, activity values in potency assays, peptide fragment pattern in peptide mapping assays, and composition or fingerprint pattern in test methods for post-translational modifications.

For biomolecular methods, it can be highly valuable to utilize non-product-related test materials as additional system suitability measures. Test method reference materials, sometimes referred to as "surrogate" or "generic" test method standards, can be selected and validated as a part of development of each method. To choose the right type of surrogate test method standard, consider the physiochemical nature of the material relative to that of the product and the test method application. For example, if the biotechnology product is an IgG molecule and the test method is SEC-HPLC, a purified commercial IgG might serve as a system suitability control for column performance. If the biotechnology product is a glycoprotein and the test method is for monosaccharide composition, purified bovine fetuin may be included as a system suitability control for accuracy and precision.

Many analytical laboratories have used surrogate test method standards for a variety of biomolecular test methods. Bovine Serum Albumin (BSA) standard reference material from the National Institute of Standards and Technologies (NIST SRM 927; www.nist.gov) is often used as a system suitability control for the accuracy and precision of amino acid analysis composition and concentration (Anders et al. 2003). In a joint project between the NIST and the Association of Biomolecular Resource Facilities (ABRF, www.abrf.org), new biomolecular surrogate test method standards are currently under development (Remmer et al. 2003). Three synthetic peptides have been recently been prepared for potential use as system suitability measures in methods such as mass spectrometry, capillary electrophoresis, amino acid sequencing, amino acid analysis, and HPLC. The USP is in the process of preparing and certifying glycoprotein surrogate test method standards for use in a wide variety of biomolecular methods (USP Conference on Biological and Biotechnological Drug Substances and Products, Nov. 18–21, 2003, Arlington, VA). Also, a test method reference material was recently established by the Adenovirus Reference Material Working Group (ARWMG) for system suitability use in the analysis of adenoviral gene therapy vectors, and is now available from American Type Culture Collection (ATCC VR-1516, www.atcc.org) (Sajjadi and Callahan 2003). To best assure the quality and integrity of surrogate test method standard materials, whenever possible use material that is certified for specified physiochemical properties, such as reference materials from NIST, USP, ATCC, or the NIBSC. If test method system suitability surrogate standards are obtained from a non-certified source, document the certificates of analysis from the vendor for each batch of material.

Demonstrating that system suitability passes its validity criteria confirms that the test method run is valid and that the resulting test sample data can be confidently evaluated.

When investigating an out-of-specification event or an unexpected result, it is important to first systematically evaluate the method's system suitability measures. When any part of system suitability fails its validity criteria, the method test run should be critically reviewed to determine and correct the assignable cause. It is considered unacceptable to use the results from invalid test runs in decisions made during CGMP product manufacturing; however, invalidating test data requires a clear justification that is supported by technical evidence (FDA Guidance 2006). Building sound system suitability measures into the test method procedure often provides the empirical evidence necessary to quickly and conclusively isolate the source of the assignable cause.

6. How will the procedure be conducted? Write a draft SOP describing the steps required to perform the method.

Include details on sample preparation, preparation of standards and controls, and system suitability measures. If instruments are used, refer to the appropriate instrument SOPs in the method SOP, or give specific instructions on instrument operation in the method SOP. Specify the number of replicates for each sample/standard, show exactly how to calculate the results, and define how results are to be reported (i.e., significant digits).

When developing and validating analytical methods, it is important to understand the implication of using significant digits in defining reportable values and setting specifications. An all-too-common mistake is to allow extra digits to be reported in the test results, or worse: to incorporate unnecessary digits into the product specification values. The number of significant digits that can be accurately reported is related to the level of sensitivity. The method must be sensitive enough to measure differences that are one decimal place beyond the specification.

Also, the effect of rounding on the outcome of reportable results should be recognized when using additional significant digits. For example, in order to pass a specification of 7.0 to 8.0, assay results of 6.95 to 8.45 are acceptable. However, to pass specifications of 7.00 to 8.00, assay results of 6.995 to 8.004 are necessary. To use these results, the analytical method must be capable of generating accurate and precise values to the thousandth decimal place. The American Society of Test Methods (ASTM, www.astm.org) has published a monograph on the appropriate use of significant digits in test data (ASTM/ANSI EP29–02). The acceptable procedures for rounding values to achieve the desired number of significant digits in reportable assay values has been defined by the USP. It may be useful to review practical examples of utilizing suitable significant digits in relation to the capabilities of analytical methods, with applications of USP rounding rules (Miller and Crowther 2000, pp. 79–82).

Even the wording of the method SOP can impact the ability of different analysts to comparably reproduce the procedure. Initiators who are very familiar with the method sometimes leave out subtle details that can impact test method performance. The SOP instruction "Vortex the sample" can range operationally from gentle rotation to vigorous agitation, yielding dramatically different outcomes. Also, experienced scientists may inadvertently omit an instruction if it is assumed to be common to the technique. For example, after heating samples for SDS-PAGE, it is common practice to subject the sample to brief microcentrifugation to pool the solution droplets created in the sample vial. Failure to note this "common practice" in the SOP (and failure to specify the time and speed of centrifugation) can propagate sample-handling inconsistencies that may lead to test method problems. The SOP should be as complete and detailed as necessary to allow the method steps to be performed the same way by any analyst, including the data reduction steps. It is a good idea to allow a less-experienced analyst to run through the draft SOPs independently to see if they can complete the procedure solely based on the instructions written in the document. If not, the SOP should be revised until it provides adequate, unambiguous instructions. The best approach is always to "Clearly write the method procedure to be validated, then run the validated method procedure exactly as it is written."

7. Is the method appropriate for all types of sample(s) to be used? Consider all potential variations in test samples (concentrations, buffers, formulation constituents) that will be included in the intended use of the method.

Run the method per SOP using representative product test materials. Include analysis of buffer or formulation solutions to assess matrix effects. If necessary, optimize test method parameters to achieve suitable preliminary performance. Edit the draft SOP to reflect any changes resulting from optimization experiments.

Most analytical methods for biotechnology products are developed and optimized using samples of the drug substance, since it is often the most suitable material for these studies. But if the method is ultimately intended for use with test samples taken from in-process, conjugation, or formulation steps, it will have to be assessed for performance with those specific types of samples. As shown in Table 8.7, there are several compounds that can be used in the formulation of biotechnology products (Bontempo 1997; Frokajaer and Hovgaard 2002); McNally 1999; Perlman and Wang 1996).

Many conjugate reagents or formulation excipients interfere with the analytical methods developed for bulk product. For example, the presence of amino acids can affect compositional analysis, protein concentration assays, and sequencing results. Solubilizers such as Tween can interfere with colorimetric methods at higher concentrations. Sugars can precipitate during HPLC runs if the mobile phase becomes too polar. For both formulated and in-process samples, the concentration of drug substance may be very low ($\mu g/ml$), falling below the range of test method linearity where poor accuracy and precision may yield unreliable data. Also, unpredictable interactions between drug substance and excipients may occur. In certain cases, excipient degradation may require its own evaluation and stability testing. For these reasons, analytical methods that were validated for bulk substance may require revalidation for intermediates, conjugates, or formulated product.

8. What are the initial performance capabilities of the test method? Conduct a test method qualification or characterization study to systematically investigate the performance ranges of the method for the designated types of samples.

Prior to using a test method, studies should be conducted to investigate the working ranges of the test method for the parameters that could affect the intended use. These studies are sometimes called method "qualification" or "characterization" studies. Methods that are used only for product or process characterization and comparability studies should be

TABLE 8.7

Typical Formulation Candidates for Proteins and Peptides

› Osmotic Agents (salts)
› Chelators (EDTA, citrate)
› Cations
› Sugars (mannose, maltose, dextrose)
› Amino Acids (arginine, glycine, glutamic acid)
› Redox Agents (ascorbate, reducing sugars)
› Solubilizers (Tween, Deoxycholate)
› Stabilizers (albumin, lipids)
› Solvents (aqueous, nonaqueous)

qualified but are not required to be fully validated (Ritter et al. 2004). Methods that are used for the quality control and stability testing of Phase 1 clinical trial intermediates, drug substance, or drug product must be demonstrated to be scientifically sound, that is, "qualified", or validated using a limited experimental design prior to conducting the required late-phase full test method validation exercise (FDA Guidance 2008; ICH Q7A).

In most cases, experiments are conducted to assess all of the parameters typically included in a validation study for the type of method such as linearity, accuracy, precision, specificity, and so on. From these experiments are derived the initial performance capabilities of the test method. These results should be compared to the intended application of the test method. If it is seen that method performance is not meeting the requirements of use, optimization experiments are usually conducted until (or unless) the method becomes acceptable. If the method cannot be optimized to meet the initial intended use, either the method will have to be replaced or the acceptance specifications for method performance will have to be reassessed.

The nature and impact of test method variability, particularly with quantitative methods such as those used for purity and potency determinations, should be clearly understood for each method prior to finalizing product acceptance specifications for that method. In some cases, an analytical method can demonstrate such inherent variability that it will have to be eliminated from consideration for use with the product and replaced with a technique that can perform appropriately. If not, there will be a statistically predictable percent probability that a given test result will not fall within the product specification range simply due to test method variability (Weed 1999).

In most cases, the nature and source of test method variability can be identified via a rigorous test method development approach, with attention to even deceptively simple parameters like test sample preparation and reference standard stability. However, some sources of variation may not be detected until the test method has been used over a long period of time. Tracking and trending the performance of a new method is a valuable tool to monitor the ongoing reliability of the method for its intended use. Tracking and trending system suitability results independently of the sample results can provide a simple (but powerful) mechanism to distinguish product variability from test method variability. This information can be used to rapidly focus troubleshooting investigations to isolate and correct the root cause of change.

9. Can the method be validated to meet its intended use? Design the validation protocol using sound scientific judgment in alignment with current regulatory expectations.

There are several guidance documents on the current regulatory expectations for test method validation. For QC test methods to be used in CGMP applications, validation studies should follow the guidance given in current ICH and FDA guidelines. In addition, there are many excellent historical and current articles on test method validation, many with specific examples of validation strategies and protocols, with extensive citations for additional reading cited elsewhere in this chapter.

For those with limited experience, it is strongly recommended that each of these references be thoroughly reviewed to obtain a comprehensive understanding of the core requirements and different approaches possible when validating a test method. However, it should be noted that most are based on traditional chromatographic methodology. As shown in Tables 8.2A, 8.2B, and 8.3, biotechnology products require a broad range of methods utilizing widely different technologies. For these methods, the principle of a validation study remains the same (i.e., to demonstrate suitability for its intended use), but the experimental design to meet each parameter can differ considerably, based upon the nature of the technology. Currently, there are only a few specific publications on the validation of techniques used specifically for the quality control testing of biotechnology products (Anders et al. 2003; Allen et al. 1999; McEntire 1994; Ritter, Hayes, Dougherty 2001; Patel 2000; Findlay et al. 2000). The USP is currently revising Chapter <111> Design and Analysis of Biological Assays (http://www.pharmacopeia.cn/v29240/usp29nf24s0_c111_viewall.html), and has a new chapter in preparation on the validation of bioassays. Table 8.8 lists the required method validation parameters for different types of test methods as defined by the USP. Table 8.9 contrasts these with the required method validation

TABLE 8.8

Validation Requirements from ICH Validation of Analytical Procedures

Analytical Procedure	Identification	Testing for Impurities		Assay
Characteristics		Quantitative	Limit	Content Potency
Accuracy	−	+	−	+
Precision:				
Repeatability	−	+	−	+
Intermediate Precision	−	+ (1)	−	+ (1)
Specificity (2)	+	+	+	+
Detection Limit	−	− (3)	+	−
Quantitation Limit	−	+	−	+
Linearity	−	+	−	+
Range	−	+	−	+

- signifies that this characteristic is not normally evaluated
+ signifies that this characteristic is normally evaluated
(1) in cases where reproducibility has been performed, intermediate precision is not needed
(2) lack of specificity of one analytical procedure could be compensated by other supporting analytical procedure(s)
(3) may be needed in some cases

TABLE 8.9

USP Data Elements Required for Assay Validation

Performance Characteristic	Assay Category I	Assay Category II		Assay Category III	Assay Category IV
		Quantitative	Limit Tests		
Accuracy	Yes	Yes	*	*	No
Precision	Yes	Yes	No	Yes	No
Specificity	Yes	Yes	Yes	*	Yes
Detection Limit	No	No	Yes	*	No
Quantitation Limit	No	Yes	No	*	No
Linearity	Yes	Yes	No	*	No
Range	Yes	Yes	*	*	No

* May be required, depending on the nature of the specific test

Category I: Methods for quantitation of major components of bulk drug substances or active ingredients (including preservatives) in finished pharmaceutical products.

Category II: Methods for determination of impurities in bulk drug substances or degradation compounds in finished pharmaceutical products. These methods include quantitative assays and limit tests.

Category III: Methods for determinations of performance characteristics (e.g., dissolution, drug release).

Category IV: Identification tests.

From Table 2, USP 24, used with permission.

parameters given by the ICH. Regardless of which table is used, the parameters must be selected appropriate to the test method's intended use. The specific experiments to be conducted to achieve validation of each parameter will be based on the nature of the test method. Most of the prior references have detailed experimental designs for chromatographic assays; these experiments should be adapted for use with other, non-chromatographic biomolecular assays for biotechnology products as needed to assure the test method will meet its intended use.

It should be noted that there are two special applications of test method validation that are distinct from the test method validation applications described earlier. The USP includes a chapter describing the validation studies necessary to have an analytical method accepted into the Pharmacopeia as a regulatory compendial method (USP 27/NF 22 (2004) <1225>). Since compendial methods are intended for application in any analytical laboratory, extensive collaborative studies are needed to verify the repeatability and robustness of the method in order to establish global performance specifications. Also, validation of an analytical test method for quality control applications should not be confused with the validation of test methods applied to the detection or quantitation of biological markers in clinical samples. In addition to the parameters such as linearity, accuracy, precision, specificity, and robustness, these biomarker assays also require patient studies that correlate the activity measured *in vitro* with the *in vivo* intended application (such as the designated clinical disease or condition) (Bowsher and Smith 2002; Ishikawa 1986).

If the method is intended to monitor product stability, it should be verified for this capability (as described previously).

10. What if the method validation run(s) fail to meet performance expectations? Thoroughly investigate the root cause of the failure(s) and determine what corrective actions would prevent the same problem from occurring again.

If the assignable cause for the failure of the method to meet performance requirements is identified, implement the appropriate corrective action and repeat the affected validation run(s). If the assignable cause were related to analyst error, it would be wise to immediately address how to prevent propagating the same error in future runs of the validated test method. Sometimes SOP simply requires enhanced clarity, such as more specific instructions or a more logical organization of steps. These document adjustments can usually be justified during validation if they do not change the method itself. Note that regulatory bodies indicate laboratory error should be relatively rare (FDA Guidance 2006), so management should be alert to chronic method performance problems that are related solely to laboratory operations. Laboratory variability can usually be minimized with attention to issues such as clearly written SOPs, well-maintained instruments, a meaningful training program, and (arguably) adequate staffing to prevent chronically harried—thereby inadvertently careless—analyst performance. Regardless of how thoroughly a test method has been validated, if it is not implemented in an adequately controlled laboratory environment, it will not be able to perform reliably. On the other hand, even sound laboratory operations may not be able to compensate for poorly written method SOPs or non-robust analytical methods.

If the assignable cause for failure is not identifiable, the test method should not be considered validated. In this case, the test method should be remanded back to the method development process for further assessment of its actual suitability for the intended use(s). Experience shows that unresolved analytical test method problems that arise during method validation usually continue throughout the life cycle of test method use (Hokanson 1994; Miller and Crowther, 2000, pp. 118–130).

> 11. When should a validated test method be revalidated? When something changes that could impact its continued suitability for the intended use.

Some firms have a policy of reviewing and revalidating test methods with established frequency (e.g., every 2 years). Routinely reviewing test method SOPs against laboratory practices and test records does have the advantage of maintaining a strong connection between "creeping" performance habits and the actual written steps and can catch disconnects relatively quickly. However, periodic revalidation studies when nothing has changed might be more rigorous than most facilities can operationally support. A more effective approach is to look at method revalidation strategies from a risk-based perspective. In other words, when changes do occur, what is the level of risk that the change will impact the state of validation of the test method?

Figure 8.10 lists five categories of change that should trigger a review of the test method and possibly require a measure of revalidation: (1) changes to the product that could affect method performance (e.g., formulation excipients, product concentration), (2) changes in critical assay reagents that cannot meet prior performance requirements (e.g., gel reagents, enzymes, antibodies), (3) changes in instrumentation that cannot meet prior performance settings (e.g., automated amino acid hydrolysis systems), (4) changes in the procedure to improve robustness of the method (e.g., adding new sample preparation steps), and (5) changes in the product specifications that are beyond the test method capabilities. In most cases, the revalidation may be limited to a set of studies that "bridge" the old part of the procedure to the new part of the procedure. In cases where method performance capabilities are significantly affected and, as a result, will require changes to method specifications, a complete revalidation may be necessary. Regardless of the extent of revalidation needed, it is imperative to confirm that all analysts are sufficiently trained on the new procedure to assure successful results. It is a useful management tool to monitor trends in method performance (as described earlier) following revalidation to spot problems and make the appropriate corrections quickly.

Changes to the product that could affect method performance (e.g., formulation excipients, product concentration)

Changes in critical assay reagents that cannot meet prior specifications (e.g., gel or blot materials, enzymes)

Changes in instrumentation that cannot perform equivalent procedures (e.g., AAA hydrolysis systems)

Changes in the procedure to improve robustness of the method (e.g., new qualification procedures for critical reagents, re-optimized sample preparation steps)

Changes in the product specifications that go beyond the capabilities of the test method (e.g., specification on quantifying impurities drops from 0.5% to 0.1%; assay limit of quantitation is 0.25%)

FIGURE 8.10
Potential triggers for method revalidation.

8.11 Conclusion

Biological products are macromolecular entities that are considerably larger than most chemical products. With the exception of synthetic oligonucleotides or peptides, living cells—complex metabolic factories—produce them. The target molecules(s) must be isolated from a biochemical milieu consisting of chemical entities relatively similar to the desired product. As such, it may be difficult to completely eliminate impurities derived from the host system. The purified target may comprise several structurally heterogeneous forms, some or all of which might be active. Compared to traditional chemical drugs, biological materials are highly labile, unable to tolerate high temperatures or undue chemical or physical stress. Higher-order biological products (e.g., cells and tissues) may have a very short window of viability. It requires numerous complex analytical methods to provide an effective physiochemical profile of biotechnology products.

This chapter has presented several factors to consider when selecting the analytical methods to assess the identity, purity, impurities, concentration, potency, stability, and (in some cases) comparability of biotechnology products (Table 8.11). Since no single method can provide data on all key product parameters, orthogonal analytical methods should be used to increase confidence in the quality of the product. Methods used under GLP or

TABLE 8.11

Elements for Successful Analytical Method Development and Implementation

1. Clearly define the assay's intended use relative to the desired product attribute (e.g., identity, purity, impurities, potency, concentration, stability) and acceptance specification requirements.

2. Understand how the method technology functions to generate data on the parameter of interest. Develop appropriate system suitability measures to assess method performance independently of the test sample performance.

3. Recognize and control potential sources of method and operational variability that can impact the reproducibility of assay procedure. Incorporate system suitability measures to assure the validity of each test run and to track/trend method performance over time.

4. Confirm that the method will be scientifically sound for its intended use(s) by demonstrating its inherent performance capabilities such as accuracy, precision (intra- and inter-assay), linearity/range/LOD/LOQ, and specificity (including for product degradants, if stability indicating).

5. Verify that the assay is robust enough under the conditions of expected use to statistically support the specification requirements for the product at each phase of development and commercialization.

6. Assure that all documentation and data from each method's lifecycle events are maintained in archived files that are complete, traceable, and retrievable for use in supporting product and method knowledge management over time.

CGMP quality practices must be validated for their intended use. The strategies for qualifying and/or validating biomolecular methods should be based on the type of method, the nature of the product, and the parameter to be evaluated with the data. Laboratories that adopt validated methods (e.g., compendial methods) must experimentally verify the suitable performance of these methods in the user environment. In order to provide a complete product development record, all of these activities must be adequately documented to demonstrate how, when, and by whom they were conducted.

As has been noted, "Some data are worthless; some data are priceless. The conditions and procedures used to find data ultimately determine their value" (Torbeck and Branning 1998). All decisions regarding the control of the process and the quality of the product are based on data generated by analytical tests. If there are design flaws in the test methods, unrecognized sources of method variation, or a method chosen that cannot support the specification requirements, the data will inevitably be inadequate, inaccurate, or unreliable. So, while it is certainly critical to understand the process by which a biological or biotechnological product is produced, it is equally vital to understand the methods of analysis that are applied to the product. Otherwise it will be very difficult to distinguish between those data that are priceless and those that are worthless.

References

21 CFR 312.23 (a)(7), IND Content and Format, Chemistry, Manufacturing and Control Information.

Allen, D., R. Baffi, J. Bausch, J. Bongers, M. Costello, J. Dougherty, M. Federici, R. Garnick, S. Peterson, R. Riggins, K. Sewerin, and J. Tuis. 1999. Validation of Peptide Mapping for Protein Identity and Genetic Stability. *Biologicals* 24:2555–3275.

Anders, J., B. Parten, G. Petrie, R. Marlowe, and J. McEntire. 2003. Using Amino Acid analysis to Determine Molar Absorptivity Constants: A Validation Case Study Using B ovine Serum Albumin. *BioPharm International* February:30–37.

ASTM/ANSI EP29–02: Standard Practice for Using Significant Digits in Test Data to Determine Conformance with Specifications, 2003.

Bontempo, J. 1997. Development of Biopharmaceutical Dosage Forms. In *Drugs and the Pharmaceutical Sciences Series*, Vol. 85. Marcel Dekker, Inc., New York.

Bowsher, R., and W. Smith. 2002. Practical Approach for Clinical Drug Development: Analytical Validation of Assays for Novel Biomarkers. *AAPS Newsmagazine* June, pp. 18–25.

Canova-Davis, E. 1994. Curious Discoveries During the Characterization of Therapeutic Proteins. *ARBF News*, September.

CASSS—An International Separation Science Society. Well Characterized Biological Products. www. casss.org.

Cavagnero, J. 2003. Equivalence of Biological and Biotechnological Ingredients and Products. USP Conference on Biological and Biotechnological Drug Substances and Products, 18–21 November, Crystal City, VA.

Champion, K., H. Madden, J. Dougherty, and E. Shacter. 2005. CMC Forum Report: Defining Your Product Profile and Maintaining Control Over It. Part 2: Challenges of Monitoring Host Cell Protein Impurities. *BioProcess International* 3(9):52–57.

Code of Federal Regulations, 21 CFR 211.194(a)(2), Laboratory records.

Code of Federal Regulations, 21 CFR 211.84 (d)(2), Testing and approval or rejection of components, drug product containers, and closures.

CPMP Guidance 2001. CPMP Note for Guidance on Comparability of Medicinal Products Containing Biotechnology Derived Proteins as Drug Substance; CPMP/BWP/3207/00, September 2001.

CPMP Guidance 2002. CPMP Note for Guidance on Comparability of Medicinal Products Containing Biotechnology Derived Proteins as Drug Substance, 30 July 2002.

FDA Guidance 1996. FDA Guidance Concerning the Comparability of Human Biological Products, Including Therapeutic Biologically-Derived Products, April 1996.

FDA Guidance 1996. FDA Guidance Concerning Demonstration of Comparability of Human Biological Products, Including Therapeutic Biotechnology-derived Products.

FDA Guidance 2001. FDA Guidance for Industry: Bioanalytical Method Validation, May 2001.

FDA Guidance 2003. FDA Guidance for Industry—Comparability Protocols: Chemistry, Manufacturing and Controls Information.

FDA Guidance 2003. FDA Guidance for Industry. 2003. Comparability Protocols—Protein Drug Products and Biological Products—Chemistry, Manufacturing, and Controls Information.

FDA Guidance 1999. FDA Guidance for Industry—INDs for Phase 2 and 3 Studies of Drugs, Including Specified Therapeutic Biotechnology-Derived Products; CMC Content and Format, February 1999.

FDA Guidance 2006. FDA Guidance for Industry Investigating Out of Specification (OOS) Test Results for Pharmaceutical Production, October 2006.

FDA Guidance 2008. FDA Guidance for Industry: CGMP for Phase I Investigational Drugs.

FDA Guidance 2000. Analytical Procedures and Test Method Validation, August 2000 Draft.

FDA Guidance for Industry PAT—A Framework for Innovative Pharmaceutical Development, Manufacturing, and Quality Assurance, September 2004.

Federal Register: May 14, 1996. Volume 61, Number 94, 24227–24233. Elimination of Establishment License Application for Specified Biotechnology and Specified Synthetic Biological Products.

Findlay, J., W. Smith, J. Lee, G. Nordblom, I. Das, B. DeSilva, M. Khan, and R. Bowsher. 2000. Validation of immunoassay for bioanalysis: A pharmaceutical industry perspective. *Journal of Pharmaceutical Biomedical Analysis* 21:1249–1273.

Frokajaer, W., and L. Hovgaard, eds. 2002. *Pharmaceutical Formulation: Development of Proteins and Peptides.* Taylor and Francis Publishers, London.

Gombold, J., K. Peden, D. Gavin, Z. Wei, K. Baradaran, A. Mire-Sluis, and M. Schenerman. 2006a. Lot Release and Characterization Testing of Live-Virus-Based Vaccines and Gene Therapy Products, Part 1: Factors Influencing Assay Choices. *BioProcess International* 4:46:54.

Gombold, J., K. Peden, D. Gavin, Z. Wei, K. Baradaran, A. Mire-Sluis, and M. Schenerman. 2006b. Lot Release and Characterization Testing of Live-Virus-Based Vaccines and Gene Therapy Products, Part 2: Case Studies and Discussion. *BioProcess International* 4:56–65.

Hokanson, G. 1994. A Life-Cycle Approach to the Validation of Analytical Methods during Pharmaceutical Development. *Pharmaceutical Technology*, August:118–130.

ICH Guidance for Industry M4: The CTD—General Questions and Answers, December 2004.

ICH M4Q: International Conference on Harmonisation of Technical Requirements for Registration of Pharmaceuticals for Human Use. 2001. The CTD—Quality.

ICH Q5C: International Conference on Harmonisation of Technical Requirements for Registration of Pharmaceuticals for Human Use. 1995. Stability Testing of Biopharmaceutical Products.

ICH Q5E: International Conference on Harmonisation of Technical Requirements for Registration of Pharmaceuticals for Human Use. 2003. Comparability of Biotechnological/Biological Products Subject to Changes in Their Manufacturing Process.

ICH Q6A: International Conference on Harmonisation of Technical Requirements for Registration of Pharmaceuticals for Human Use. 1999. Specifications—Test Procedures and Acceptance Criteria for New Drug Substances and New Drug Products: Chemical Substances.

ICH Q6B: International Conference on Harmonisation of Technical Requirements for Registration of Pharmaceuticals for Human Use. 1999. Specifications: Test Methods And Acceptance Specifications For Biological/Biotechnological Products.

ICH Q7A: International Conference on Harmonisation of Technical Requirements for Registration of Pharmaceuticals for Human Use. 2001. Good Manufacturing Practice Guidance for Active Pharmaceutical Ingredients.

ICH Q8(R2): International Conference on Harmonisation of Technical Requirements for Registration of Pharmaceuticals for Human Use. 2009. Pharmaceutical Development.

ICH Q9: International Conference on Harmonisation of Technical Requirements for Registration of Pharmaceuticals for Human Use. 2006. Quality Risk Management.

ICH Q10: International Conference on Harmonisation of Technical Requirements for Registration of Pharmaceuticals for Human Use. 2008. Pharmaceutical Quality System.

Ishikawa, K. 1986. *Guide to Quality Control*. American Society for Quality Press.

Kats, M. 2005. Forced Degradation Studies: Regulatory Considerations and Implementation. *BioPharm International* July:46–52.

Kotia, R., and A. Raghani. 2010. Analysis of Monoclonal Antibody Product Heterogeneity Resulting from Alternate Cleavage Sites of Signal Peptide. *Analytical Biochemistry* 15;399(2):190–195.

Lubiniecki, A., D. Volkin, M. Federici, M. Bond, M. Nedved, L. Hendricks, P. Mehndiratta, M. Bruner, S. Burman, P. DalMonte, J. Kline, A. Ni, M. Panek, V. Pikounis, G. Powers, O. Vafa, and R. Siegel. 2011. Comparability Assessments of Process and Product Changes Made During Development of Two Different Monoclonal Antibodies. *Biologicals* 39:9–22.

Magari, R. 2003. Assessing Shelf Life Using Real Time and Accelerated Conditions—although Accelerated Tests are Needed, Real Time Tests are the Ultimate Proof. *BioPharm International*, November:28–32.

Manzi, A. 2008. Carbohydrates and Their Analysis, Part One. *BioProcess International* 6(2):54–60.

Manzi, A. 2008. Carbohydrates and Their Analysis, Part Two. *BioProcess International* 6(3):50–57.

Manzi, A. 2008. Carbohydrates and Their Analysis, Part Three. *BioProcess International* 6(6):54–65.

McEntire, J. 1994. Biotechnology Product Validation Part 5: Selection and Validation of Analytical Techniques. *BioPharm* 7:68–79.

McMillen, D., L. Bibbs, N. Denslow, K. Ivantich, C. Naeve, R. Neice, and S. Tyndall. 2000. Biotechnology Core Laboratories: An Overview. *Journal of Biomolecular Technology* 11:1–11.

McNally, E., ed. 1999. Protein Formulation and Delivery. *Drugs and the Pharmaceutical Sciences Series*, Vol. 99. Marcel Dekker, Inc., New York.

Miller, J., and J. Crowther, eds. 2000. *Analytical Chemistry in a GMP Environment: A Practical Guide*. John Wiley and Sons, New York.

Mire-Sluis, A., R. Das, and A. Padilla. 1998. WHO Cytokine Standardization: Facilitating the Development of Cytokines in Research, Diagnosis and as Therapeutic Agents. *Journal of Immunological Methods* 216:103–116.

Patal, J., R. Kothari, R. Tunga, N. Ritter, and B. Tunga. 2011. Stability Considerations for Biopharmaceuticals, Part 1: Overview of Protein and Peptide Degradation Pathways. *BioProcess International* 9(1):2–11.

Patel, U. 2000. Meeting the Challenges of Enzyme Assay Validation. *BioPharm* July:48–52.

Paul, W. 1991. USP Perspectives on Analytical Methods Validation. *Pharmatical Technology* March: 130–141.

Perlman, R., and Y. J. Wang, eds. 1996. *Formulation, Characterization, and Stability of Protein Drugs*, Vol. 9. Plenum Press, New York and London.

Pharm Eur. 1/2008:2031 Monoclonal antibodies for human use.

Pharmaceutical GMPs for the 21st Century: A Risk-Based Approach; Second Progress Report and Implementation Plan, 3 September 2003. FDA.

Remmer, H. A., N. P. Ambulos, L. F. Bonewald, J. J. Dougherty, E. Eisenstein, E. Fowler, J. Johnson, A. Khatri, M. Lively, N. Ritter, and S. Weintraub. 2003. Synthetic Peptides As Certified Analytical Standards. *Biopolymers* 71(3):354. John Wiley & Sons Inc., Hoboken, NJ.

Rieder, N., H. Gazzano-Santoro, M. Schenerman, R. Strause, C. Fuchs, A. Mire-Sluis, and L. McLeod. 2010. The Roles of Bioactivity Assays in Lot Release and Stability Testing. *BioProcess International* 8:33–42.

Ritter, N., S. Advant, J. Hennessey, H. Simmerman, J. McEntire, A. Mire-Sluis, and C. Joneckis. 2004. WCBP CMC Strategy Forum Report: What is Test Method Qualification? *BioProcess International* 2(9):2–11.

Ritter, N., T. Hayes, and J. Dougherty. 2001. Analytical Laboratory Quality, Part II—Analytical Method Validation. *Journal of Biomolecular Techniques* 12:11–15.

Ritter, N., and M. Wiebe. 2001. Validating Critical Reagents Used in cGMP Analytical Testing: Ensuring Method Integrity and Reliable Assay Performance. *BioPharm* May:12–21.

Sajjadi, N., and J. Callahan. 2003. Defining a Detailed Approach to Using the Adenovirus Reference Material (ARM). *Bioprocessing Journal* 2:83–87.

Savino, E., B. Hu, J. Sellers, A. Sobjak, N. Majewski, S. Fenton, and T. Yang. 2011. Development of an In-House, Process-Specific ELISA for Detecting HCP in a Therapeutic Antibody, Part 1. *BioProcess International* 9(3):38–49.

Schenerman, M., B. Sunday, S. Kozlowski, K. Webber, H. Gazzano-Santoro, and A. Mire-Sluis. 2004. CMC Strategy Forum Report: Analysis and Structure Characterization of Monoclonal Antibodies. *BioProcess International* 2:42–52.

Smalls, C., D. Pepper, and D. James. 2011. Protein Modification During Anti-Viral Heat Treatment Bioprocessing of Factor VIII Concentrates, Factor IX Concentrates, and Model Proteins in the Presence of Sucrose. *Biotechnology Bioengineer* 77:38–48.

Torbeck, L., and R. Branning. 1998. Designed Experiments: A Vital Role in Validation. *Pharmatical Technology* June:108–113.

Towns, J., and K. Webber. 2008. CMC Strategy Forum Report: Demonstrating Comparability for Well—Characterized Biotechnology Products—Early Phase, Late Phase, and Post approval. *BioProcess International* 6:32–43.

US Pharmacopeia 34/NF 29, 2011 <1226> Verification of Analytical Methods.

USP 27/NF 22 (2004) <1225> Assay Validation.

USP General Chapter <1045> Test Procedures for Biotechnology Products. USP 32/NF 29, 2011.

USP/NF Pharm Forum Draft Chapters on Bioassay Development and Validation, <1032>, <1033>, <1034>.

Vanderweilen, A. and E. Hardwidge. 1982. Guidelines for Assay Validation. *Pharmatical Technology* March:66–76.

Weed, D. 1999. A Statistically Integrated Approach to Analytical Method Validation. *Pharmatical Technology* October:116–129.

Wei, Z., E. Shacter, M. Schenerman, J. Dougherty, and L. McLeod. 2011. CMC Strategy Forum Report: The Role of Higher-Order Structure in Defining Biopharmaceutical Quality. *BioProcess International* 6(4):54–65.

Williams, D. 1987. Overview of Test Method Validation. *BioPharm* October:34–51.

Xu-Rong, J., A. Song, S. Bergelson, T. Arrol, B. Parekh, K. May, S. Chung, R. Strouse, A. Mire-Sluis, and M. Schenerman. 2011. Advances in the Assessment and Control of the Effector Functions of Therapeutic Antibodies. *Nature Reviews* 10:101–111.

Yu, J. 2000. Intentionally Degrading Protein Pharmaceuticals to Validate Stability-Indicating Analytical Methods. *BioPharm* 13(11):46–52.

9

Adventitious Agents:
Concerns and Testing for Biopharmaceuticals

Raymond W. Nims, Esther Presente, Gail Sofer, Carolyn Phillips, and Audrey Chang

9.1 Introduction

Adventitious agents represent undesirable contaminants of exogenous origin that are introduced during the production of biopharmaceuticals. Microbial adventitious agents include viruses, bacteria, fungi, and mycoplasma. Transmissible spongiform encephalopathy agents are also potential adventitious contaminants. Raw materials may contain adventitious agents. Adventitious agents can be introduced during establishment of cell lines, cell culture/fermentation, capture and downstream processing steps, formulation/filling, and even drug delivery. Therapeutic biotechnology products have an excellent safety record. However, the potential introduction of adventitious agents must continually be evaluated. The testing that is performed for this purpose is addressed in regulatory documents that include ICH guidelines, US Points to Consider, and European (Ph. Eur.), US (USP), and Japanese (JP) Pharmacopoeia documents. In some cases, 9 CFR (US Code of Federal Regulations) and 21 CFR 211 and 610 are applicable. Table 9.1 lists some of the regulatory documents that describe testing requirements.

Since biopharmaceuticals encompass many types of products, there is considerable variability in risks from adventitious agents. In all cases, however, the use of Good Manufacturing Practices (GMPs) (e.g., environmental controls, control of raw materials and personnel flow, and cleaning), suitable safety testing programs, and process validation (including viral and sometimes mycoplasma clearance evaluation) helps to assure patient safety and confidence in biopharmaceuticals. Some products have minimal inherent risk associated with the introduction of adventitious agents (e.g., recombinant products produced in bacteria). Other types of products, such as those used for cell and/or gene therapy, are often at the other end of the spectrum and may be associated with greater risk due to their inability to tolerate rigorous purification process conditions. Greater potential risks are often associated with the use of human cells and animal-derived raw materials. Of particular concern are materials derived from bovine and porcine sources. When viable cells are a component of the product (as in *ex vivo* transduction), there may not be sufficient time to perform the required safety testing. In such cases, product may be released prior to completion of relevant tests for adventitious agents, although this testing is still mandated in order to demonstrate that the processes are being performed under adequate controls to maintain patient safety. A risk/benefit analysis determines whether the use of these products is warranted.

DOI: 10.1201/9781003143130-9

TABLE 9.1

Regulatory Documents That Apply to Testing for Adventitious Agents

Guideline on Quality of Biotechnological/Biological Products: Derivation and Characterisation of Cell
Substrates Used in the Production of Biotechnological/Biological Products (ICH Q5D, 1997) Geneva,
Switzerland: International Conference on Harmonisation, 1997

Points to Consider in the Characterization of Cell Lines Used to Produce Biologicals (FDA/CBER, 1993)
Rockville, MD: Food and Drug Administration, Center for Biologics Evaluation and Research, 1993

Points to Consider in the Manufacture and Testing of Monoclonal Antibody Products for Human Use (FDA/
CBER, 1997) Rockville, MD: Food and Drug Administration, Center for Biologics Evaluation and Research,
1997

Guideline on Viral Safety Evaluation of Biotechnology Products Derived from Cell Lines of Human or Animal
Origin [ICH Q5A (R1), 1997] Geneva, Switzerland: International Conference on Harmonisation, 1997

Guidance for Industry: Characterization and Qualification of Cell Substrates and Other Biological Materials
Used in the Production of Viral Vaccines for Infectious Disease Indications. (FDA/CBER, 2010) Rockville, MD:
Food and Drug Administration, Center for Biologics Evaluation and Research, 2010

Guidance for Industry: Guidance for Human Somatic Cell Therapy and Gene Therapy (FDA/CBER, 1998)
Rockville, MD: Food and Drug Administration, Center for Biologics Evaluation and Research, 1998

Minimising the Risk of Transmitting Animal Spongiform Encephalopathy Agents via Human and Veterinary
Medicinal Products (CPMP, 2001). London, England: Committee for Proprietary Medicinal Products, 2001

Draft Guidance for Industry: Preventative Measures to Reduce the Possible Risk of Transmission of Creutzfeldt-
Jakob Disease (CJD) and Variant Creutzfeldt-Jakob Disease (vCJD) by Human Cells, Tissues, and Cellular
Tissue-Based Products (HCT/Ps) (FDA/CBER, 2002) Rockville, MD: Food and Drug Administration, Center
for Biologics Evaluation and Research, 2002

Organization for Economic Cooperation and Development Principles on Good Laboratory Practices, 1998:
ENV/MC/CHEM(98)17

Japan Ministry of Health and Welfare, Ordinance No. 21, 1997; Japan Pharmaceutical Affairs Bureau, Ministry
of Health and Welfare, Pharmaceutical GLP Guideline, 1995 (lyakuhin GLP kaisetsu),157–169

In this chapter, we present a description of specific adventitious agents and discuss prevention and control of risks arising from various stages of production. Throughout the
chapter, we point out risks associated with various sources.

9.2 Viruses

Newly detected viral agents continue to be a source of concern to the general public.
Examples include the West Nile virus, monkey pox, and the viruses that cause SARS.
Demonstrating freedom from adventitious viral agents enhances confidence in biopharmaceuticals, and it is a regulatory requirement for biologics. This requirement applies
for both licensed products and those destined for clinical trials. Viruses are classified
by whether they are lipid-enveloped and by size, shape, and resistance to inactivation
by physicochemical treatments. A safety testing program for adventitious viral agents
requires different assays (Table 9.2). These may include general viral screening assays such
as the 14- and 28-day *in vitro* adventitious virus screens and the *in vivo* adventitious viral
screen, as well as assays that are designed to detect specific viral agents of concern, such as
the *in vitro* bovine and *in vitro* porcine viral assays, infectivity or nucleic acid based assays
for murine minute virus (MMV) and vesivirus detection, and gel endpoint and quantitative polymerase chain reaction® (PCR) assays for other specific viral entities.

TABLE 9.2

Assay Methodologies Employed for Viral Detection

Viral Detection Assay	Specificity	Detection System	Duration	Endpoints
In vitro adventitious virus screen	Broad virus screen	Detector cell lines	14 or 28 days	Cytopathic effect, hemadsorption, hemagglutination
In vivo virus screen	Broad virus screen	Suckling, adult mice, guinea pigs, hens' eggs	28 days	Survival, egg viability
In vitro bovine virus screen	Bovine viruses	Detector cells	14 or 21 days	Cytopathic effect, hemadsorption, immunofluorescence
In vitro porcine virus screen	Porcine viruses	Detector cells	14 or 21 days	Cytopathic effect, hemadsorption, immunofluorescence
In vitro MMV screen	MMV	324K or A9 cells	14 or 21 days	Cytopathic effect, hemagglutination
PCR	Specific viruses	RNA or DNA	1–2 days	Gel visualization of amplicons
Quantitative PCR	Specific viruses	RNA or DNA	1–2 days	PCR cycle time
Next generation sequencing	Agnostic	Deep sequencing	Weeks	Sequencing reads and bioinformatics

Next generation sequencing (NGS; also termed deep sequencing or high-throughput sequencing) is a method currently used primarily during investigation of positive outcomes during viral screening assays. These methods generate reads for viral genomic material fragments and require complex bioinformatics (Onions and Kolman, 2010; Khan et al., 2017) or other approaches (Cheval et al., 2019) to interpret the results. The instrumentation platform is evolving rapidly, with smaller platform instruments with newer chemistries that produce accurate data more rapidly. This topic is discussed at length in the PDA Technical Report no. 71 (PDA, 2015). Because NGS represents a broad-spectrum detection (agnostic) method, there is considerable interest in using it as a viral detection method and eventually moving away from the *in vivo* and *in vitro* adventitious virus tests. A recent draft regulatory guidance document (European Medicines Agency, 2022) references NGS as a potential alternative method, indicating regulatory authority openness to new technologies. The industry, as a whole, has responded with much interest and discussion regarding the challenges and opportunities for implementing NGS as a lot release test for biologicals. A recent consortium publication discusses the potential replacement of the *in vivo* adventitious virus test with NGS (Barone et al., 2023), although this will, of course, still require regulatory approval.

9.2.1 Raw Materials

Typically, adventitious viral safety testing of raw materials is performed using the appropriate specific viral detection assay. Bovine-sourced materials (e.g., bovine sera, collagen) are therefore most appropriately evaluated using the 9 CFR-compliant *in vitro* bovine virus detection assay. Additionally, PCR assays for the bovine viruses of most concern (e.g., bovine viral diarrhea virus (BVDV) or bovine polyomavirus (BPyV)) may be performed on such materials. During testing, it is not uncommon for bovine serum samples to display positive results for BVDV, especially the non-cytopathic variant (Hesselink and

Makoschey, 2006). Recently, BPyV has become of some concern, especially in the European Union. Bovine serum samples test positive for this virus by PCR at a relatively high incidence, although it is not clear that the PCR results are indicative of the presence of infectious virus (Schuurman et al., 1991). Detection of infectious BPyV requires the use of a susceptible detector cell line for amplification, with a PCR endpoint for detection. Similarly, porcine-derived materials, such as trypsin, are evaluated using a 9 CFR-compliant *in vitro* porcine virus detection assay, and such testing may be augmented with PCR assays as necessary. Recently, the presence of porcine circovirus genomic material in a rotavirus vaccine has been described (Victoria et al., 2010). It is not yet clear whether the presence of the genomic material, which is thought to have been derived from porcine trypsin, is associated with infectious circovirus.

9.2.2 Cell Banks

Master (MCB) and working (WCB) cell banks are evaluated using both specific viral assays and the more general viral screening assays, depending on the level of assurance the manufacturer has on the raw materials used. At a minimum, these cell banks are evaluated using the *in vitro* and *in vivo* virus screening assays. It is quite uncommon for viruses to be detected in the cell banks produced in the biologics industry, suggesting that the various controls stipulated in the Good Manufacturing Practice requirements have had the desired effects on quality. A recent article (Chen et al., 2008) has described a case in which an equine rhinitis virus was detected during *in vitro* bovine virus assay of a cell bank. The cell bank being tested was not the source of the equine rhinitis virus, however, as the virus was found to have originated in a contaminated equine serum used as a component of a growth medium used for the detector cells of the assay. This emphasizes the need to carefully investigate each positive outcome in a viral screening assay to ensure that the virus is not a test artifact. Care must also be taken during the testing of cell banks to ensure that selection agents (e.g., methotrexate, hygromycin) that may be present in the growth media used for cell expansion do not cause excessive cytotoxicity to the detector cells used in the *in vitro* assays, as such cytotoxicity can confound interpretation of the tests.

9.2.3 Unprocessed Bulk Harvest

Manufactured biologics are expected by the Points to Consider and ICH guidelines to be tested for viral safety in a lot-by-lot manner at the bulk harvest (unpurified bulk) level. This may be accomplished using *in vitro* and, in the case of live viral vaccines, *in vivo* virus screening assays. The requirement for *in vivo* viral screening of unprocessed bulk in the case of live viral vaccines has recently changed based on the finalized (February 2010) FDA guidance document: Characterization and Qualification of Cell Substrates and Other Biological Materials Used in the Production of Viral Vaccines for Infectious Disease Indications. If the cell bank and viral vaccine seeds have been characterized with the *in vivo* viral screening assay, the *in vivo* viral screen may not be required for unprocessed bulk. These methods are intended to detect a broad range of viral contaminants. Three detector cell lines are used in the *in vitro* virus screen. Primate and human cell lines are always included, and the third detector cell line is expected to be a monolayer cell of the same or similar species as that of the substrate employed in the manufacturing process. In the *in vivo* virus screen, a variety of animal species are inoculated (suckling and adult mice, guinea pigs, and embryonated hens' eggs). The detection of viral contaminants in manufactured lots is rare, with greater risk appearing to be associated with certain types of products. For instance, cellular vaccines and monoclonal antibody products rarely have been found to contain viral contaminants. On the other hand, adenoviral vectors, which are E1-deleted and therefore expected to be replication-defective in most indicator cells used for testing, may contain small numbers of recombinant adenoviral particles that are capable of replicating in these indicator cells and of producing in the cells all the hallmarks of adenoviral infection. Recombination may also occur with retroviral vectors, leading to replication-competent retrovirus. Strictly speaking, this recombination phenomenon is a natural process and not a case of introduction of an adventitious virus during the manufacturing cycle. However, the outcome is the same, as the recombinant virus is an unwanted contaminant and, in addition, its detection in an adventitious virus evaluation may mask the presence of other potential viral contaminants. Specific testing is performed for vector preparations to detect the presence of such recombinant replication-competent virus.

The greatest risk of introducing adventitious viral contaminants during the manufacture of biologicals appears to reside with the production of recombinant proteins in systems employing Chinese hamster ovary (CHO) cells as the production substrate. It is not clear whether this is due to a higher potential of these cells to serve as viral hosts, relative to the substrates used for other types of manufacture, or to the relatively common use of such cells in terms of number of manufactured products. At any rate, several viral contaminants have been detected with some frequency in CHO cell processes: these include MMV, REO virus (REO), Cache Valley virus, and the vesivirus isolate 2117.

MMV has previously been detected in a CHO cell process (Garnick, 1996), and as a result of this experience, the 1997 Points to Consider in the Manufacture and Testing of Monoclonal Antibody Products for Human Use (Table 9.1) mandates that this virus be assayed for in bulk harvests of this type. MMV is a murine parvovirus, which represents a special challenge in that it is relatively small (~20 nm) and therefore is difficult to remove by filtration and to inactivate by gamma irradiation. It is non-enveloped and thus resistant to chemical and physical inactivation strategies. In addition, the virus is relatively hardy and the potential for survival of the virus outside of cell cultures therefore exists (Jacoby et al., 1996). In fact, the most likely route of introduction of this virus into manufacturing processes would appear to be contamination of environmental surfaces used for raw materials processing and packaging (Garnick, 1996). This virus has also been found as a contaminant in the Syrian hamster embryo cell line, BHK (Nettleton and Rweyemamu, 1980; Zoletto, 1985). The most common assays for detection of this virus are (1) cell infectivity assays using 324K or A9 cells as detector cells, which detect the fibrotropic strain of MMV [MMV(p), prototype strain]; (2) PCR (which can detect both the fibrotropic and lymphotropic or MMV[i] strains); and (3) the mouse antibody production (MAP) test, which should be able to detect both strains of MMV (Parker et al., 1970; de Souza, M. and Smith, 1989). The latter test was used by Nicklas et al. (1993) to detect MMV in a number of cell lines.

REO virus has been isolated from CHO-cell cultures on a number of occasions (Nims, 2006). This infection can be insidious, since the virus may propagate slowly in the cell substrate with minimal, if any, effect on oxygen or base demand or on the yield of recombinant protein. Members of the family Reoviridae, genus *Orthoreovirus*, viruses are spherical, 60–80 nm in diameter, and non-enveloped and possess a double protein capsid shell. The replication and assembly of these double-stranded RNA viruses occur in the cytoplasm of the host cell, where relatively large numbers of viral particles may be found packaged in crystalline arrays (Figure 9.1). It can be difficult to ascertain the animal species of origin of the various REO viruses (types 1, 2, and 3), confounding investigation of product contamination with these agents. Typically, the presence of such viruses in bulk harvest samples has been ascribed to nonhomogeneously contaminated bovine serum used during

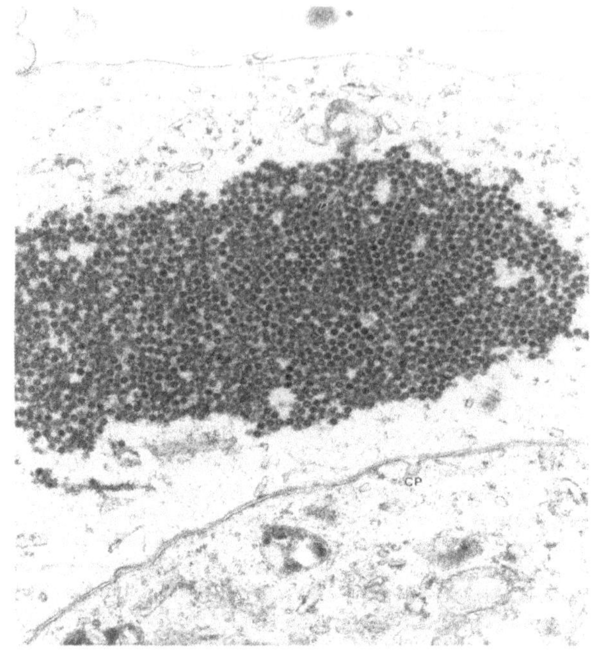

FIGURE 9.1
REO virus detected in human newborn kidney (324K) cells. 37,000X magnification, showing 65- to 75-nm spherical particles exhibiting crystalline packing pattern near cell nucleus.

the manufacturing process. The relatively low-level contamination of bovine serum lots may prevent the detection of the virus during quality control testing of the serum. The large amounts of serum incorporated into the culture medium during the manufacturing process, coupled with the relatively long culture times employed, may allow the virus to propagate to the point that the virus is detectable in the bulk harvests. The most common assays for detection of this virus are (1) the cell infectivity assay using L929 or CHO-K1 as detector cells; (2) PCR (using type-specific primers or primers designed to detect all three types); and (3) immunofluorescence staining using anti-REO antisera.

Another viral contaminant that may be introduced into manufacturing processes primarily through use of nonhomogeneously contaminated bovine serum is Cache Valley virus (Nims et al., 2008). A member of the family Bunyaviridae, genus *Bunyavirus*, Cache Valley virus is spherical and 80–120 nm in diameter with a lipid envelope containing glycoprotein spikes (Figure 9.2). The single-stranded RNA virus is known to infect livestock, being transmitted through a mosquito vector. As with the REO viruses, a low-level contamination of bovine serum lots with this virus may lead to an inability to detect the agent during quality control testing of the serum. As with REO virus, the large amounts of serum incorporated into the culture medium during the manufacturing process, coupled with the relatively long culture times employed, may allow this virus to propagate to the point that the virus is detectable in the bulk harvests. In contrast to the case for REO virus, amplification of Cache Valley virus during manufacturing runs usually leads to detectable changes in cell substrate integrity and oxygen and base demand by the cultures as cell death occurs. The virus can infect and rapidly cause cytopathic effect in a variety of

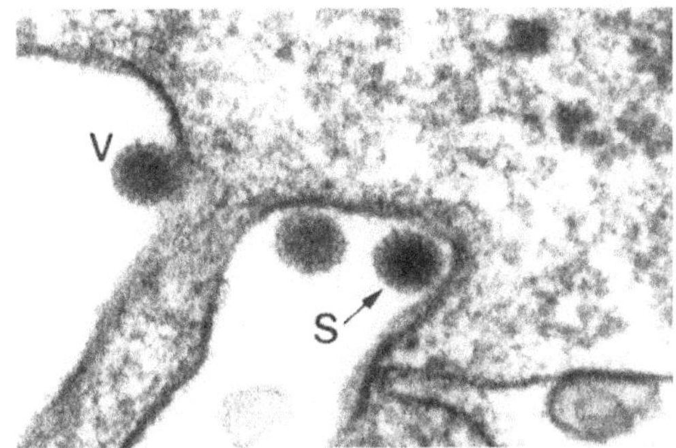

FIGURE 9.2
Cache Valley virus detected in African green monkey (Vero) cells. 100,000X magnification, showing 80- to 100-nm enveloped particles (V) with external spikes (S) budding from the membrane.

detector cells, including CHO-K1, Vero, MRC-5, and 324K. For this reason, the virus is readily detected in *in vitro* viral screens employing these detector cells. Another means of detecting this virus consists of PCR using Cache Valley- or *Bunyavirus*-specific primers.

There have been a number of occurrences of contamination of CHO-cell manufacturing of biopharmaceuticals with the vesivirus isolate 2117 (Oehmig et al., 2003; Genzyme, 2009). Vesiviruses are single-stranded RNA viruses of family Calicivirus, genus *Vesivirus*. They are non-enveloped and 30–40 nm in diameter, and the genus includes also feline calicivirus, vesicular exanthema of swine virus, rabbit vesivirus, and San Miguel sea lion virus. The susceptibility to infection by isolate 2117 appears to be limited to Chinese hamster cells, in which cells a relatively rapid lytic infection may ensue. The route of entry of the virus into biologics production processes has not been established with certainty, although the use of contaminated animal-derived materials, such as bovine sera, is considered to be the most likely source.

The four adventitious viral contaminants discussed previously are emphasized since the authors are aware of more than one instance for each virus of detection of the agent in unprocessed bulk harvest samples collected from CHO-cell manufacturing processes. As mentioned earlier in the cases of REO and Cache Valley viruses (and possibly vesivirus isolate 2117), it would appear that the relatively large volumes of bovine serum used in culture medium during scale-up of the production cell substrate may be the origin of the contamination. A low-level nonhomogeneous contamination of a large production lot of bovine serum may be undetectable using the state-of-the art methods, for which the sampled volumes are relatively small (~200 ml of serum in the case of the 9 CFR bovine virus screen, <1 ml of serum in the case of a gel endpoint or quantitative PCR assay). One approach to lowering the potential risk associated with REO virus and Cache Valley virus would appear to be gamma irradiation of the fetal bovine serum prior to use in manufacturing processes (Gauvin, 2009). Studies on the inactivation of caliciviruses indicate that UV treatment may be effective (Duizer et al., 2004; de Roda Husman et al., 2004). Gamma-irradiation at the dosages normally used does not appear to be effective for this virus, as might be expected for a virus of this relatively small size. Studies using MMV indicate that high-temperature short-time (HTST) treatment of medium containing bovine serum is effective in inactivating the parvovirus (Schleh et al., 2009) and would by implication be effective also for vesiviruses.

An isolated case of an adventitious viral contamination of a genetically engineered CHO-cell substrate with epizootic hemorrhagic disease virus (family Reoviridae, genus *Orbivirus*) has been reported (Rabenau et al., 1993).

9.2.4 Summary of Viral Safety Issues

As mentioned earlier, the detection of an adventitious viral contaminant in a manufactured biological for the most part is quite rare and has thus far been limited to certain types of products. It should be recognized, however, that a negative result in a viral safety assay does not necessarily mean that adventitious viral contaminants are not present in the bulk harvest being evaluated. This is due to limitations in the sensitivity of the various viral detection assays and to the fact that only a relatively small volume of each bulk harvest is actually sampled for testing. A negative outcome from such testing means that no adventitious viral entities were detected, but it does not guarantee that there are absolutely none present. In addition to subjecting raw materials, cell substrates, and bulk harvest samples to viral safety testing, viral clearance evaluation (purification process validation) studies are therefore performed as a means to assure patient safety. This testing is done in a proactive manner by spiking known amounts of model viruses into a scaled-down model that is validated to represent manufacturing scale. By performing viral clearance evaluation studies with model viruses selected to encompass a wide range of characteristics, confidence in final product safety is enhanced. These studies are generally performed prior to Phase 1 clinical studies and, again, if the manufacturing process is significantly changed. In addition, clearance studies involving specific viral entities may be performed in response to the detection, during routine lot release testing, of a viral contaminant. Changes in raw materials, manufacturing control parameters, and unit operations with viral clearance capability may require reevaluation of viral safety. If changes have the potential to modify viral safety, clearance studies are repeated. The ICH guideline Q5A (Table 9.1) provides information on viral safety evaluation of biotechnology products derived from cell lines of human or animal origin.

9.3 Bioburden: Bacteria and Fungi

Bacteria and fungi that contaminate a biopharmaceutical raw material, cell bank, cell culture, purification process, or product are likely to grow rapidly. Even when inactivated, these microorganisms pose a safety concern since toxins (e.g., exo- and endotoxins) may remain. A new contaminant or an increase of a specific contaminant upstream may challenge the remainder of the manufacturing process, rendering a final product that is unsuitable for use. Process clearance spiking studies are not normally performed for bioburden, since the actual personnel, environment, and water at the manufacturing site are the most likely sources of contamination. Routine testing and environmental monitoring are required to ensure that a process and product meet requisite microbial safety specifications.

9.3.1 Assays

Sterility assays are imperfect and only provide a statistical probability of non-sterility. For aseptic processes, such as the sterility assay and the aseptic fill of a final drug container, a sterility assurance level (SAL) of 10^{-3} is expected. Just imagine that you are the patient that gets that one in a thousand vial! It has been stated (Berube and Oxborrow, 1991) that "the concept of what sterile means has now become a matter of degree, i.e., there is a certain probability of sterility for each unit of product being sterilized." In some cases, bioburden testing, rather than sterility testing, is performed. Bioburden testing is intended and appropriate for non-sterile products. Bioburden testing is a type of microbial limits test. It can be quantitative and qualitative, providing a total aerobic microbial count and total yeast and molds count. It may also be used to qualitatively detect the presence of objectionable organisms. Examples of objectionable organisms are described in compendia and include *Pseudomonas aeruginosa*, *Salmonella* species, *Escherichia coli*, *Staphylococcus aureus*, *Clostridium* species, and *Mycobacteria*. Sterility testing can be thought of as a very stringent form of a microbial limits determination, in which no growth of microorganisms is permitted.

Bioburden is sometimes used for testing unprocessed bulk and typically used for setting in-process manufacturing specifications. This is allowed in the FDA's Points to Consider on therapeutic monoclonal antibodies (Table 9.1). Unless a sponsor has a sufficient database based on in-house experience, however, setting an acceptable limit for bioburden is problematic. Sterility testing is performed for master and working cell banks, for cells at the limit of *in vitro* cell age (also termed end-of-production cells), and unprocessed bulk. During the manufacturing process, in-process alert and action bioburden limits enhance confidence in the safety of the product. Sterility testing of final product is required, and purified bulk is also tested. PDA has petitioned the FDA to discontinue the regulatory requirements for purified (i.e., final) bulk sterility testing. Terminally sterilized product that is sterilized on a cycle not validated as overkill would most probably require bioburden testing of the bulk prior to fill and terminal sterilization in order to assure that the appropriate cycle SAL for the product is achieved. Overkill cycles are very robust cycles and generally do not require this consideration.

9.3.2 Cell Banks and Unprocessed Bulk

Testing at the cell banking stage is of the utmost importance for minimizing risks from adventitious agents. Testing of cell banks is described by ICH Q5D (Table 9.1), which specifies that 1% of the total number but not less than two containers of each MCB and WCB are tested for the presence of bioburden. It does not make sense to move ahead with a bank that has any detectable bioburden, and sterility assays are employed. The ICH Q5D guideline refers the reader to the European, Japanese, and US Pharmacopeias (Ph. Eur., JP, USP). However, USP < 71> describes a referee test for a pharmacopeial article purporting to be sterile. Neither Ph. Eur. 2.6 nor JP 54 refers to cell bank testing. The US FDA's Points to Consider in the Characterization of Cell Lines (Table 9.1) addresses cell bank and unprocessed bulk testing and refers to 21 CFR 610.12; however 21 CFR 610.12 does not specifically address cell bank or unprocessed bulk testing. Compliance with these methods is thus somewhat problematic, since the specific methods are not designed for testing at cell bank and unprocessed bulk stages. For example, cell banks are established by hand filling, which is not performed under conditions that can be validated from a microbiological perspective since the number of vials is too small to perform meaningful media fills.

There are two issues that are critical in the design of the sterility testing program for these early stages, namely stasis testing and pooling strategies. The USP and Ph. Eur. refer to stasis testing as *suitability* testing. Bacteriostasis and fungistasis testing provide confidence in the test results by demonstrating absence of growth inhibition by the test article.

The requirements for stasis testing at these stages is not explicitly defined in ICH Q5D, but it is expected to be demonstrated that components of the test article are evaluated for inhibitory activity (Sofer and Phillips, 2001). The suitability test (i.e., bacteriostasis and fungistasis) is necessary to assure the scientific integrity of subsequent sterility testing for a specific test article.

Pooling strategies should take into account the sensitivity of the sterility or bioburden assay. Dilution by pooling is a source of false negatives and can lead to combining "good" and "bad" materials. When cell culture duration is extended, the same issues should be addressed. Was there contamination early or late? How can pooling be justified so that testing results are not compromised?

9.3.3 In Process/Raw Materials

Raw materials have introduced microorganisms into manufacturing suites. All materials used in microbiologically controlled processes should be tested for microbial quality prior to use. Chromatographic resins, for example, generally have a microorganism specification, but that specification is a numerical limit and does not include speciation. Clearly, if any spore-forming organisms are present, an entire product batch may be compromised. Surprisingly, cleaning and sanitization reagents have also contained resistant microorganisms that have caused contamination.

In-process bioburden specifications are difficult to set. No sponsor wants a failed batch due to a low level of microorganisms during downstream processing, but how can a reasonable level be established? As with unprocessed bulk, the only way to set a limit is to gather a history of manufacturing. When a new facility or new process is established, there may be insufficient experience and data to support realistic limits. Setting limits that are too narrow can lead to out-of-specification batches. On the other hand, setting limits that are too broad is a regulatory issue in the making. Once limits are set, they must be evaluated on a periodic basis in order to assure that they remain a meaningful and effective monitoring tool.

The use of process analytical technologies (PAT) is being encouraged by the FDA (U.S. FDA, 2003a). Briefly, the concept is that rapid inline, offline, and at-line evaluation can enhance process control. There are new microbiological tools that are suitable for PAT. Most pick up metabolizing microorganisms and can eliminate the cost of continuing to process at risk. This technology is also suitable for many types of biotechnological final products such as cell and gene therapies and combination products with viable components.

9.3.4 Final Product

Final product sterility testing is addressed in many publications, including regulatory documents. However, for the small volume biotechnology product, final product testing can be problematic. Having to give up two vials of a ten-vial production lot is not easy. And as noted earlier for cell banking, the filling is often manual. Media fills, involving a larger number of vials, have been recommended by the EU for clinical studies (EMEA, 2003). Regulatory agencies have recognized that products such as cell therapies do not lend themselves to traditional sterility testing and have encouraged the use of rapid microbiological methods to release product, but this must be followed up with the traditional, required sterility assays.

9.3.5 The Positives and Negatives

A negative sterility assay is not a guarantee that there are no viable organisms. There may remain injured microorganisms that do not demonstrate reproductive growth and are, therefore, not detected by the specified assays. In some cases, such organisms may survive and be able to recover in a suitable host. Furthermore, it is impossible to test enough units to assure a 100% probability of sterility. Additionally, care must be taken to ensure the test article does not cause inhibition of microbial growth in the assay and lead to false negatives.

Even more problematic are false positives. Deciding to repeat a sterility test can be a regulatory mine field and a production and business nightmare. Common sources of false positives include equipment, supplies, materials, test media, technique, personnel, and the environment in which the sterility assays are performed. Barrier isolators employed for performance of sterility assays have been shown to provide a higher level of confidence. Isolation technology has virtually eliminated human-borne contamination and provides better control over environmental contamination. It has been observed by some that isolators reduced sterility test false positives to almost zero, and the industry-wide failure rate is estimated to be 0.01% to 0.001% (Akers, 2002).

9.4 Mycoplasma

Mycoplasmas lack a cell wall and are bound by a single plasma membrane. They are considered the smallest self-replicating prokaryote. Most mycoplasmas are parasites, exhibiting host and tissue specificities. Mycoplasmas may affect cell growth properties. They may inhibit cell metabolism, disrupt nucleic acid synthesis, produce chromosomal aberrations, produce proteases, phosphatases, and nucleases, change the antigenicity of cell membranes, mimic viral infections, and affect the yield of product. In humans, they can cause atypical pneumonia and are considered to be cofactors for some human diseases.

There are close to 100 known species of mycoplasma, but eight species constitute over 95% of the identified contaminants (McGarrity et al., 1992; Bolske, 1988). The presence of mycoplasma in a cell culture is not always obvious, and an infection may persist for an extended period of time. It has been estimated that between 5% and 35% of cell cultures worldwide are contaminated with at least one species of mycoplasma (Uphoff and Drexler, 2002). Potential sources of contamination include cell culture media components, animal sera, cell and viral stocks, and laboratory personnel. In particular, the use of bovine sera in biologics production confers the risk of introducing *A. laidlawii*, *M. arginini*, *M. mycoides*, and *M. bovis*.

It is not often that reports of mycoplasma contamination of biologics production reach the public domain. One such report described two cases involving *M. salivarium* and one involving *M. mycoides* (Dehghani et al., 2007). In a few cases, where there has been contamination of product by mycoplasma, regulatory agencies have required sponsors to perform mycoplasma clearance studies.

9.4.1 Test Methods

Mycoplasma test methods are specified in the regulatory guidelines. The ICH guideline Q5D (Table 9.1) addresses testing of cell banks for mycoplasma and states that agar and broth media procedures and the indicator cell culture procedure should be performed.

Testing from a single container is generally adequate. The FDA's 1993 Points to Consider on the Characterization of Cell Lines Used to Produce Biologicals (Table 9.1) recommends testing for both agar cultivable and non-agar cultivable mycoplasma for mammalian cells. For insect cells, both mycoplasma and spiroplasma testing must be performed. Tests are performed on virus seed and/or MCB, cell substrates, and each WCB used for manufacture of product.

During the past decade, significant harmonization of mycoplasma testing requirements has occurred. The Ph. Eur. Chapter 2.6.7 "Mycoplasmas" was revised in 1997 with the result that it became more closely aligned with the 1993 FDA Points to Consider guidance. In the fall of 2010, a USP chapter (<63>) entitled "Mycoplasma Tests" will become effective. While this chapter borrows heavily from both the Ph. Eur. and FDA Points to Consider mycoplasma test methodologies, some differences must be accounted for (Nims and Meyers, 2010). These primarily involve the criteria for assessing nutritive properties of growth media and the presence of product inhibition (inhibitory substances), terminology for incubation conditions, and specific requirements for numbers and types of positive control organisms.

For cell therapies and other products that undergo minimal processing, mycoplasma contamination of final product is a potential risk. These products must often be delivered within a matter of days, and the mycoplasma infectivity tests cannot be completed in time for product release. The use of PCR mycoplasma assays that can be performed in one day enhances the safety of such products. In a 2003 draft guidance for reviewers, FDA has noted that PCR-based mycoplasma assays are acceptable during development for product release, provided the PCR test has adequate sensitivity and specificity (U.S. FDA, 2003b). PCR is also ideally suited for screening cell lines and raw materials susceptible to mycoplasma contamination. The possibility for replacement of the existing 28-day culture method for detecting mycoplasma in lot release samples with alternative methods such as nucleic acid based testing is discussed in the latest versions (5.8 and above) of the Ph. Eur. Chapter 2.6.7 and in USP <63>. Implementation of these alternative methods involves establishing comparability with the existing culture methods (see Duguid, 2010). Criteria for establishing comparability may be found in Ph. Eur. Chapter 2.6.7.

9.5 Transmissible Spongiform Encephalopathies

Addressing regulatory requirements for prevention of transmissible spongiform encephalopathy (TSE) agents has become a big concern of biopharmaceutical manufacturers. The infectious agents are prion proteins and are extraordinarily resistant to inactivation methods. Although the infectious agent can be removed by filtration and chromatography, the reuse of filters, resins, and other contact surfaces is problematic. For biotechnological products, bovine spongiform encephalopathy (BSE) is generally the greatest concern. BSE was first identified in 1986, and one million cattle have been infected. It is likely spread by feed containing tissue from infected cattle. The infection becomes localized in high-risk tissues, such as brain, spinal cord, central nervous system, and ileum. In 2003, there were 129 human cases (vCJD), which is believed to be caused by consumption of contaminated beef. Sourcing of raw materials, such as fetal calf serum used in cell culture, is essential. Bovine materials can be obtained from countries not shown to have BSE. However, there is always a concern that BSE will be found in countries previously thought to be free of this agent.

Some regulatory authorities have gone so far as to forbid the marketing of products using bovine materials sourced from countries where other TSEs, such as elk wasting disease, are found unless the sponsor has done TSE clearance studies. As noted in a European note for guidance, the risk of transmission of TSEs is reduced by controlling the source of animals, the nature of animal tissue used in manufacture, and production processes used for preparation of the animal-derived product (CPMP, 2002a). The CPMP now requires that working viral seeds or working cell banks be rebanked if there are any unknown potential risks associated with BSE in the current banks (CPMP, 2002b).

Clearance studies with spikes of TSE agents can be performed. These studies, however, can take more than one year to complete. Testing for TSE agents is by far the most difficult of the agents discussed in this chapter. Currently, the only reliable method capable of detecting one infectious unit employs animal models, most commonly the hamster 263K PrPSc (prion protein scrapie) strain. This strain has been demonstrated to be a model for human PrPsCJD (prion protein scrapie Creutzfeldt-Jakob disease), human PrPvCJD (prion protein variant Creutzfeldt-Jakob), and sheep PrPSc (Stenland et al., 2002). There are ongoing efforts in the development of amplification methods suitable for detection of one infectious unit, but at this time they are not available. Western blots are commonly used to screen a process to determine where clearance occurs, and then, if necessary, those steps that are shown to be capable of removing scrapie agents are evaluated with the animal models (Lee et al., 2001).

9.6 Summary

For biotechnological products, absolute freedom from adventitious agents cannot be guaranteed. Often, the sensitivity and precision of infectivity assays are poor. In other cases, when the assay sensitivity is adequate, the time for performance of the assay may limit its utility in releasing product. Highly sensitive assays, such as PCR, are often very specific and can only be used to detect known agents. In spite of these limitations, to our knowledge there have been no incidences of transmittal to humans of infectious agents from purified biotechnology products. The multipronged approach used by today's biopharmaceutical sponsors mitigates potential unknown risks. Equipment design and maintenance, water and air quality, and personnel flow are key elements used to reduce risks from adventitious agents. Qualification and suitable storage of cell banks, testing of incoming raw materials, and stringent sanitization processes eliminate many risks. Compliance with Good Manufacturing Practices is also essential. A robust process that removes real or unknown risks further enhances product safety. Clearance studies using spikes with agents or model agents that represent potential risks are critical when assay sensitivity is insufficient or when routine testing is not feasible (e.g., viral and TSE clearance studies). In-process bioburden assays minimize the risk of contamination by bacteria and fungi. Final product testing for bacteria and fungi, and in some cases their byproducts such as endotoxins, further enhance the safety of biotherapeutics. Table 9.3 summarizes some of the assays and testing stages.

In the near future, we hope that more assays will be harmonized and that virus stocks can be standardized. We anticipate the use of more sensitive, inline technologies to further minimize the risks from adventitious agents. We continually find new agents as detection technology is improved. The discovery of porcine circovirus DNA in a licensed rotavirus

TABLE 9.3

Summary of Testing for Adventitious Agents to Mitigate Risk

Adventitious Agent	Test	Testing Stages
Bacteria and fungi	Sterility and bioburden assays	Cell banks; cells at limit of *in vitro* cell age; unprocessed bulk; in-process testing; final product
Mycoplasma	Mycoplasma assays	Cell banks; cells at limit of *in vitro* cell age; unprocessed bulk
TSE	Sourcing of raw materials[a]	Clearance studies[a]
Adventitious viruses	*In vitro* and *in vivo* assays; specific assays for infectivity and viral nucleic acid	Cell banks; cells at limit of *in vitro* cell age; unprocessed bulk; clearance studies

[a] Currently, the only assay of sufficient sensitivity to detect one infectious prion is the animal bioassay.

vaccine (Victoria et al., 2010) by the new NGS technique raises questions on the application of emerging technologies and the implications for their use in detection of known and unknown adventitious agents in currently licensed and developmental products. And while this may lead to concerns from sponsors, it also enables the development of better tools for control throughout a manufacturing process.

References

Akers, J. 2002. *Oral Presentation at BioReliance Workshop: Sterility Testing from Theory to Practice*, Boston, January 2002. Available from BioReliance Corp., 14920 Broschart Road, Rockville, MD, 20850.

Barone, P. W., F. J. Keumurian, C. Neufeld, et al. 2023. Historical evaluation of the in vivo adventitious virus and its potential for replacement with next generation sequencing (NGS). *Biologicals*, January 6:101661. doi: 10.1016/j.biologicals.2022.11.003

Berube, R., and G. S. Oxborrow. 1991. Methods of testing sanitizers and bacteriostatic substances. In *Disinfection, Sterilization and Preservation*, 4th ed., Block, S., Ed. Lea & Febiger, Philadelphia, 1058–1068.

Bolske, G. 1988. Survey of mycoplasma infections in cell cultures and a comparison of detection methods. *Zentbl. Bakteriol. Mikrobiol. Hyg. Ser. A* 269:331–340.

Chen, D., R. Nims, S. Dusing, et al. 2008. Root cause investigation of a viral contamination incident occurred during master cell bank (MCB) testing and characterization—a case study. *Biologicals* 36:393–402.

Cheval, J., E. Muth, G. Gonzalez, et al. 2019. Adventitious virus detection in cells by high-throughput sequencing of newly synthesized RNAs: Unambiguous differentiation of cell infection from carryover of viral nucleic acids. *mSphere* 4:e00298–19. https://doi.org/10.1128/mSphere.00298-19.

CPMP. 2002a. *Committee for Proprietary Medicinal Products/Committee for Veterinary Medicinal Products*. Note for Guidance on Minimising the Risk of Transmitting Animal Spongiform Encephalopathy Agents via Human and Veterinary Medicinal Products. Revision 2, London, December (EMEA/410/01 Rev. 2).

CPMP. 2002b. *Committee for Proprietary Medicinal Products/Committee for Veterinary Medicinal Products*. Position paper on reestablishment of working seeds and working cell banks using TSE compliant materials. London, September (EMEA/22314/02).

de Roda Husman, A. M., P. Bijkerk, W. Lodder, et al. 2004. Calicivirus inactivation by nonionizing (253.7-nanometer-wavelength [UV]) and ionizing (gamma) radiation. *Appl. Env. Microbiol.* 70:5089–5093.

de Souza, M., and A. L. Smith. 1989. Comparison of isolation in cell culture with conventional and modified mouse antibody production tests for detection of murine viruses. *J. Clin. Microbiol.* 27:185–187.

Dehghani, H., B. Rasmussen, V. Fung, et al. 2007. Case studies of mycoplasma contamination in CHO cell cultures. In *Proceedings from the PDA Workshop on Mycoplasma Contamination by Plant Peptones*. Pharmaceutical Drug Association, Bethesda, MD, pp. 53–59.

Duguid, J. 2010. Top ten validation considerations when implementing a rapid mycoplasma test. *Am. Pharm. Rev.* May/June 13(4): 26–31.

Duizer, E., P. Bijkerk, B. Rockx, A. de Groot, F. Twisk, and M. Koopmans. 2004. Inactivation of caliciviruses. *Appl. Env. Microbiol.* 70:4538–4543.

EMEA. 2003. Manufacture of investigational medicinal products. *GMP Annex13, EU,* July.

European Medicines Agency. 2022. ICH Consensus Guideline Q5A (R2) on viral safety evaluation of biotechnology products derived from cell lines of human or animal origin. Step 2b (draft for public consultation) EMA/CHMP/ICH/804363/2022.

Garnick, R. L. 1996. Experience with viral contamination in cell culture. *Dev. Biol. Stand.* 88:49–56.

Gauvin, G. 2009. Gamma irradiation of animal serum. *PDA Cell Substrates Workshop.* www.pda.org/Presentation/PDA-Cell-Substrate-Workshop/Gay-Gauvin.aspx.

Genzyme. 2009. Press release. www.reuters.com/article/pressRelease/idUS158846+25-Jun-2009+BW20090625.

Hesselink, W., and B. Makoschey. 2006. Bovine virus diarrhoea virus in bovine serum used for vaccine manufacturing. *Dev. Biol.* 126:291–292.

Jacoby, R. O., L. J. Ball-Goodrich, D. G. Besselsen, M. D. McKisic, L. K. Riley, and A. L. Smith. 1996. Rodent parvovirus infections. *Lab. Anim. Sci.* 46:370–380.

Khan, A. S., S. H. S. Ng, O. Vandeputte, et al. 2017. A multicenter study to evaluate the performance of high-throughput sequencing for virus detection. *mSphere* 2:e00307–317. https://doi.org/10.1128/mSphere.00307-17.

Lee, D. C., C. J. Stenland, J. L. Miller, et al. 2001. A direct relationship between the partitioning of the pathogenic prion protein and transmissible spongiform encephalopathy infectivity during the purification of plasma proteins. *Transfusion* 41:449–455.

McGarrity, G. J., H. Katani, and G. H. Butler. 1992. Mycoplasmas and tissue culture cells. In *Mycoplasmas - Molecular Biology and Pathogenesis*, Maniloff, J., McElhaney, R. N., Finch, L. R., and Baseman, J. B., Eds. American Society for Microbiology, Washington, DC, pp. 445–454.

Nettleton, P. F., and M. M. Rweyemamu. 1980. The association of calf serum with the contamination of BHK21 clone 13 suspension cells by a parvovirus serologically related to the minute virus of mice (MVM). *Arch. Virol.* 64:359–374.

Nicklas, W., V. Kraft, and B. Meyer. 1993. Contamination of transplantable tumors, cell lines, and monoclonal antibodies with rodent viruses. *Lab. Anim. Sci.* 43:296–300.

Nims, R. W. 2006. Detection of adventitious viruses in biologicals—a rare occurrence. *Dev. Biol.* 123:153–164.

Nims, R. W., S. K. Dusing, W.-T. Hsieh, et al. 2008. Detection of Cache Valley virus in biologics manufactured in CHO cells. *BioPharm Int.* 21:89–94.

Nims, R. W., and E. Meyers. 2010. USP <63> Mycoplasma tests: A new regulation for mycoplasm testing. *BioPharm Int.* 23(8):54–59.

Oehmig, A., M. Buttner, F. Weiland, W. Werz, K. Bergemann, and E. Pfaff. 2003. Identification of a calicivirus isolate of unknown origin. *J. Gen. Virol.* 84:2837–2845.

Onions, D., and J. Kolman. 2010. Massively parallel sequencing, a new method for detecting adventitious agents. *Biologicals* 38(3):377–380.

Parenteral Drug Association. 2015. Technical Report No. 71. Emerging Methods for Viral Detection. Parenteral Drug Association, Inc. ISBN: 978-0-939459-77-9. www.pda.org/bookstore/product-detail/2831-tr-71-virus-detection

Parker, J. C., S. S. Cross, M. J. Collins, Jr., and W. P. Rowe. 1970. Minute virus of mice. 1. Procedures for quantitation and detection. *J. Natl. Cancer Inst.* 45:297–303.

Rabenau, H., V. Ohlinger, J. Anderson, et al. 1993. Contamination of genetically engineered CHO-cells by epizootic haemorrhagic disease virus (EHDV). *Biologicals* 21:207–214.

Schleh, M., P. Romanowski, P. Bhebe, et al. 2009. Susceptibility of mouse minute virus to inactivation by heat in two cell culture media types. *Biotechnol. Prog.* 25:854–860.

Schuurman, R., B. van Steenis, A. van Strien, J. van der Noordaa, and C. Sol. 1991. Frequent detection of bovine polyomavirus in commercial batches of calf serum by using the polymerase chain reaction. *J. Gen. Virol.* 72:2739–2745.

Sofer, G., and C. Phillips. 2001. Sterility testing for master cell banks, working cell banks, and unprocessed bulk. *BioPharm Int.* 14:36–38.

Stenland, C. J., D. C. Lee, P. Brown, S. R. Petteway, Jr., and R. Rubenstein. 2002. Partitioning of human and sheep forms of the pathogenic prion protein during the purification of therapeutic proteins from human plasma. *Transfusion* 42:1497–1500.

U.S. FDA. 2003a. Guidance for industry PAT—A framework for innovative pharmaceutical manufacturing and quality assurance, draft guidance, August. www.fda.gov/cder/guidance/5815dft. htm.

U.S. FDA. 2003b. Draft Guidance for Reviewers. Instructions and Template for CMC Reviewers of Human Somatic Cell Therapy Investigational New Drug Applications (INDs), August. Rockville, MD.

Uphoff, C. C., and H. G. Drexler. 2002. Comparative PCR analysis for detection of mycoplasma infections in continuous cell lines. *In Vitro Cell Devel. Biol. Anim.* 38:79–85.

Victoria, J. G., C. Wang, M. S. Jones, et al. 2010. Viral nucleic acids in live-attenuated vaccines: Detection of minority variants and an adventitious virus. *J. Virol.* 84:6033–6040.

Zoletto, R. 1985. Parvovirus serologically related to the minute virus of mice (MVM) as contaminant of BHK 21 cl. 13 suspension cells. *Dev. Biol. Stand.* 60:179–183.

10

Biotech Facility Design for Validation

Phil DeSantis

10.1 Introduction

The design, construction, and startup of biopharmaceutical (biotech) facilities governed by current good manufacturing practice (CGMP) regulations requires a different approach from what might be applied for other manufacturing facility projects. This is because such facilities must comply with specific requirements of the applicable global regulations. These are fully described by Subpart C of 21CFR211, the United States CGMP regulations, which begins:

§211.42 Design and construction features.

    (a)   Any building or buildings used in the manufacture, processing, packing, or holding of a drug product shall be of suitable size, construction and location to facilitate cleaning, maintenance, and proper operations.

The PIC/s Guide to Good Manufacturing Practice is representative of the regulations applicable to other jurisdictions, including the European Union, and has similar language:

> Chapter 3 *Premises and Equipment—Principle*
> *Premises and equipment must be located, designed, constructed, adapted and maintained to suit the operations to be carried out. Their layout and design must aim to minimise the risk of errors and permit effective cleaning and maintenance in order to avoid cross-contamination, build-up of dust or dirt and, in general, any adverse effect on the quality of products.*

The regulations go on to provide more guidance, much of which will be covered by this chapter. Based on the author's experience, it is of little value to develop a safe and efficacious product with a reliable process and then expect to produce it in a poorly designed or substandard facility.

Further, to meet the business goal of the manufacturer, that is to deliver the product to the patient, validation of the included processes must be successfully completed. Of course, this requires that the facility and manufacturing equipment that is determined to directly affect product quality must be qualified. This chapter will emphasize the importance of both good design and good project management practices to achieve the goals of a CGMP project.

Before proceeding, however, it is wise to define and understand what the goals of a project are. A project is a series of interrelated activities of limited scope and duration with a defined outcome. In general, project success is defined by return on investment. A capital project, such as those discussed in this chapter, results in some durable (i.e., long-lasting) assets and may be judged by the financial return on those assets. Capital project success is most often driven by budget and schedule, that is, what the level of investment is and how long it will be before that investment returns value. Engineers and builders are trained to drive projects to completion in the least amount of time and with the lowest cost, while still reaching the defined outcome. It is important to note that a third driver is necessary to achieve the outcome. The assets must perform as expected. In short, the third driver may be described as "quality" (including quality of the facility and the products that it will produce).

There is an old adage that has appeared frequently in the literature and is repeated periodically online: "Budget, schedule, quality: pick two." If interpreted figuratively, this may be taken to mean that a project must be adequately funded and given enough time to be planned and executed in order to achieve its desired outcomes. This is true enough. On the other hand, biopharm users have often taken this adage to mean that a tight budget and aggressive schedule always result in poor quality. This, of course, is nonsense. Many CGMP projects are undertaken with the all-important "time to market" consideration as a schedule driver. Responsible firms understand that aggressive scheduling may cost more but that rigorous budgets may still be applied. In the end, research has shown that when best project management practices are applied, all three goals are accomplished: on time, under budget, and with the expected quality of the facility and its output (1). This chapter will look at how these seemingly competing objectives can actually work in synergy. Section 10.2 looks at projects in general and section 10.3 discusses the relationship between project activities, deliverables, and validation, where validation is an essential element of quality.

10.2 Project Management

We should be clear in our discussion of facilities projects that while we speak of success as measured by return on assets, that real return is dependent upon the performance of the facility and equipment in support of product manufacture. This means acceptable product delivered to the patient without recalls, rejects, reworks, and a minimum number of deviations, investigations, delays, and interruptions. Certainly, facilities and equipment do not cause all of these problems; however, poor design can contribute to a significant proportion of them.

One of the most important factors contributing to overall project success is the application of a rigorous "stage-gated" project management system. Stage-gating is the application of defined review stages over the course of project development where scope, budgets, and schedules are reviewed and authorization for further work is given. Final budget and schedule are not firmly established until the final approval of some defined stage (2).

In building a project management structure it is valuable to clearly define the various necessary project activities, milestones, and deliverables as well as the project team members and other stakeholders. Of course, this may be done using a series of documents, plans, procedures, schedules, spreadsheets, and so on. In fact, each of these items plays an important part in a project management structure. It is valuable to organize these into a user-friendly graphic interface similar to the GMP Project Roadmap™ shown later (3). *Although this chapter is not a primer on project management, it is important to understand the stages of a project and how each supports and leads to validation.*

10.2.1 A GMP Project Roadmap

The GMP Project Roadmap™ is a proprietary configurable representation of an effective capital project management system for GMP projects. Shaded nodes indicate activities most closely related to validation.

The GMP Project Roadmap™ shown is a generic version provided for illustration and guidance of further discussion on project management and validation. (It is an example and not intended to be critiqued in detail. Project management guidelines are configurable and may be unique to each company and even to individual projects.) Figure 10.1 indicates the "front-end" or planning end of a project. Figure 10.2 relates to project execution. Of course, in the real world there is sometimes overlap among activities, but this chapter will emphasize the importance of front-end loading to ensure project success, especially as it relates to CGMP compliance and validation. This principle is strongly supported by many expert organizations studying project management (2). A graphic interface like the GMP Project Roadmap™ may be supported by active links to various guidelines and templates to assist the team in accomplishing the tasks described in each node of the diagram. Some of this guidance will be briefly described in the chapter.

10.2.2 Chartering

This phase initiates the project and establishes the business case that the project is meant to satisfy. The business case may be commercial, such as new product introduction and capacity increase. The business case may also be driven by regulatory compliance, for example to improve environmental controls within a facility or even to replace a substandard facility with one that is more compliant. The business case should also define the other major project drivers, such as time to market, cost restrictions/anticipated budget, and desired return on investment. Major assumptions or restrictions should be considered. Projects without a clear business plan are usually on a rough road toward successful completion.

10.2.3 Concept

Once a business case is established and possible technical solutions are determined, the next phase of design results in a Concept, often summarized in a Conceptual Design Report. This usually includes the following (this is neither an exhaustive nor a required list and varies based on project scope and complexity):

- Site Surveys and Site Master Plan
- Soil Samples
- Process Control Strategy
- Project User Requirements
- Conceptual Layouts
- Conceptual Equipment Arrangement Drawings
- Conceptual Major Equipment List
- Process Flow Diagrams (PFDs)
- Utility Flow Diagrams (UFDs)
- Heat and Material Balances

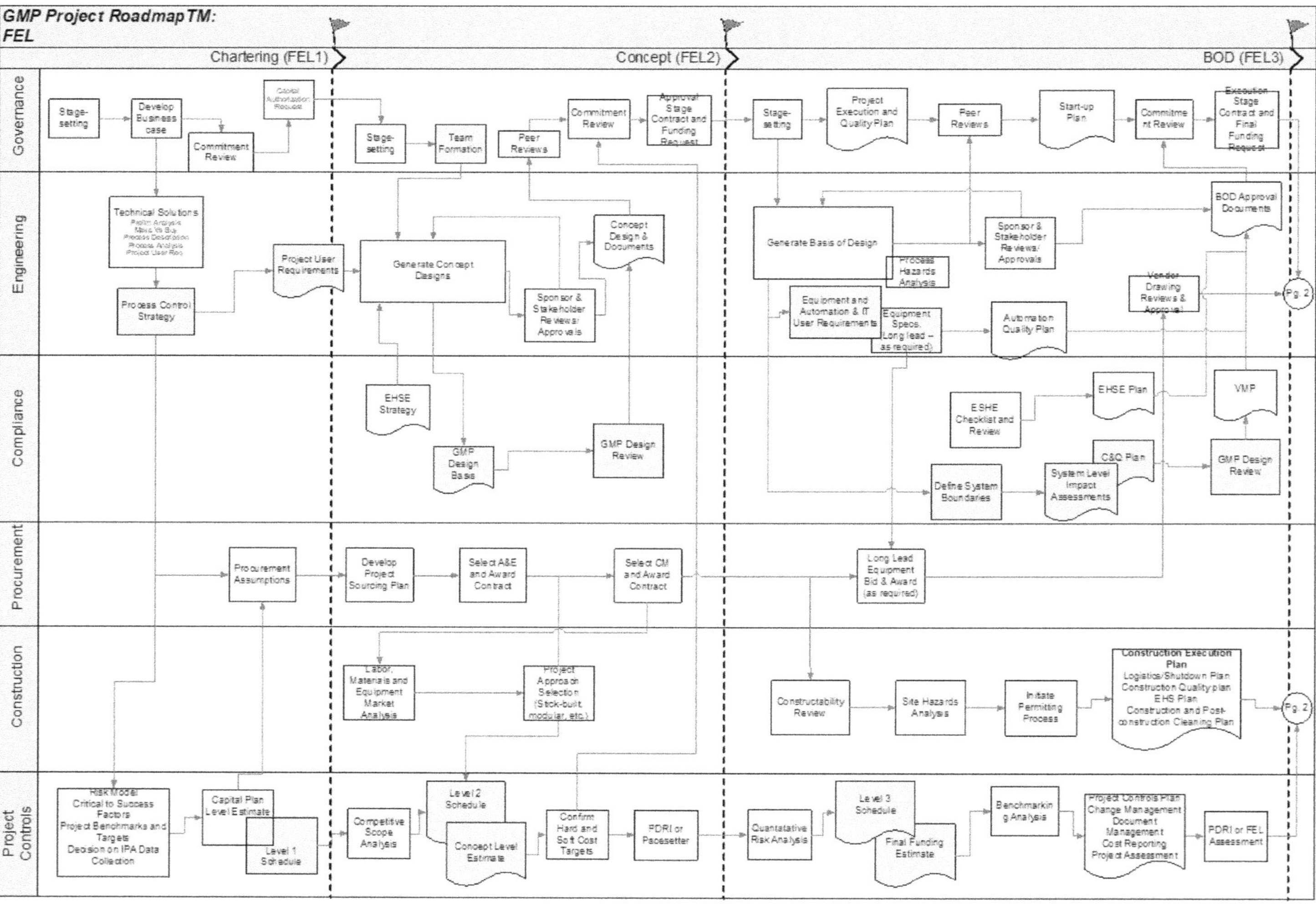

FIGURE 10.1
GMP Project Roadmap©—front-end loading.

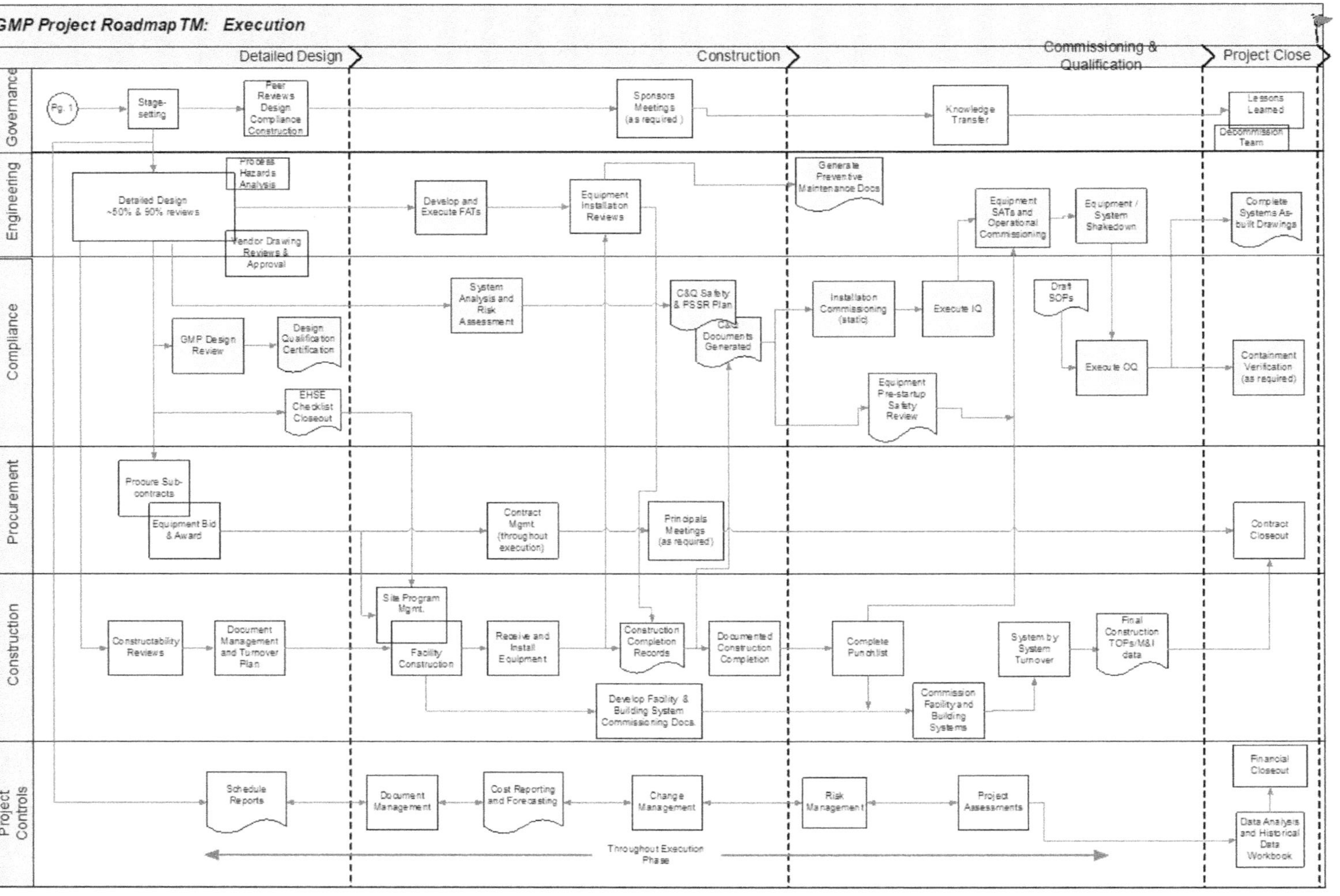

FIGURE 10.2
GMP Project Roadmap©—execution.

- Initial Process Hazards Analysis
- People, materials, product, waste flow diagrams
- Area Classification and Pressurization Scheme
- HVAC Concept
- Containment Strategy
- Identification of Critical Utilities and Support Systems
- Validation Strategy
- Maintenance Strategy
- Design Safety Philosophy Document (DSPD)
- Environmental Impact Review
- Mechanical/Equipment Room Conceptual Configuration
- Material Handling Strategy
- Gowning Requirements
- Conceptual GMP Design Review Report (Optional)
- Conceptual Project Schedule (Level 1)
- Conceptual Project Cost Estimate

10.2.4 Basis of Design (BOD)

The Basis of Design (BOD) phase is sometimes called Basic Design, Preliminary Engineering, or Design Development. This phase develops the Concept into more detailed drawings and documents that become the basis for the construction drawings and specifications needed to actually build the facility. BOD drawings are schematic, that is, they define system components, their relationship and connectivity, but are not spatially precise.

The following lists typical BOD content, sometimes compiled into a Basis of Design Report and collection of drawings. There may be some overlap with Conceptual Design and the list varies depending on project scope and complexity. For small to modest project scope, it may be decided to combine the Concept and BOD phases.

- General Arrangement (GA) Drawings
- Preliminary Discipline Specific Drawings
- Area Classification drawing
- Piping and Instrumentation Diagrams (P&IDs)
- Utility Schematics
- HVAC Schematics
- Equipment List
- Motor Load List
- Electrical Single Lines/Schematics
- Line List
- Tie-in List
- Instrument Index
- Room User Requirements/Room Cards

- Automation Philosophy and Control System Architectural Diagram
- Building Automation Functional Specification
- Building Management System (BMS) Architecture Diagram
- Process Hazards Analysis
- Safety Risk Analysis: Health Hazard Criteria
- Regulations, Codes, and Standards Review
- Constructability Review
- Design Safety Plan
- Maintenance Strategy
- Procurement Strategy
- Risk Management Analysis—Cost and Schedule
- Environmental Review
- Document Control and Filing Index
- GMP Design Review
- Project Cost Estimate—Control Budget
- Project Schedule (Level 2)
- System-Level Impact Assessments
- Commissioning and Qualification Plan
- Validation Plan

10.2.5 Project Execution—Detailed Design

The documents developed in the Detailed Design phase are specific as to dimension, detail, and orientation and are used by contractors to build the facility, its incorporated systems, and manufacturing equipment. These documents are developed by architects, engineers, and equipment specialists, most often third-party service providers and equipment vendors. This phase is also where detailed commissioning and qualification documents are prepared based on the design information. Detailed drawings are issued for construction (IFC) and are among the documents typically developed during detailed design, as follows.

- GMP Design Review—Critical Systems
- Design Qualification Certification
- Detailed Project Schedule (through Equipment Qualification)
- Architectural Finishes, Details, Specifications
- Mechanical/Equipment Room Layouts
- Civil Details and Installation Specification
- Reflective Ceiling Plan
- Planning/Construction Permits
- HVAC Duct Routing Plans—Issued for Construction (IFC)
- Equipment Specifications
- Construction and Installation Specifications
- Cable Tray Routing Plans—IFC

- Isometrics—IFC
- Pipework Special Design Details
- Pipework Specification
- Instrumentation Specification—IFC
- Loop Drawings—IFC
- Process Control System (PCS) Design Specifications
- Building Automation Design Specification
- Construction Safety Plan
- Maintenance Plans
- Value Engineering Options/Report
- Submittal Log

10.2.6 Project Execution—Construction

During this phase of execution facilities are built, equipment installed, and construction turnover dossiers finalized. Turnover dossiers may include:

- Shop drawings
- Construction quality inspection reports
- Pressure test reports
- Slope checks
- Motor inspection and test reports
- Grounding checks
- Weld inspection reports
- Concrete test reports
- Equipment manuals and vendor drawings
- Factory Acceptance Test (FAT) reports
- Construction completion reports/acceptance

10.2.7 Project Execution—Commissioning and Qualification (C&Q)

The final phase of the capital project is commissioning and qualification. Process performance qualification (PPQ) runs are subsequent to the capital portion of the project. These are confirmatory runs using actual processes and production materials and are discussed elsewhere in this book.

10.3 Managing for Validation

Each of the nodes indicated on the GMP Project Roadmap™ is important to project delivery. Rather than review each node and activity, we will focus on those that have the most direct effect on and/or relationship to validation. It is true, however, that the three major factors leading to a qualified facility and validated processes are the same ones that contribute most to overall project success (4):

- Team organization
- Front-end loading
- Project controls

This chapter will review each of these in light of how they relate to validation. The discussion will take a very broad view of validation to include its necessary components.

10.3.1 Team Organization

When building a project team for any manufacturing project it is necessary to start with Project Management and a Project Owner, usually representing a user group (e.g., Production). Core team additions are technical/process experts, engineers/designers, and construction managers. Some of these may be qualified third parties (contractors). Part-time stakeholders who may not be fully devoted to the project include regulatory professionals, maintenance personnel, and R&D. GMP projects are unique in that core project teams must include a commissioning/qualification lead, a validation lead, and a Quality unit representative.

It is essential that these last three are necessary parts of the core project team. For any project of significant size, these roles may require full-time focus or at least be considered a main responsibility. The bigger or more complex the project, the more focused these roles must become. They play an essential role in reviewing and approving all project decisions that have an effect on product quality.

10.3.1.1 Commissioning and Qualification (C&Q) Lead

As the GMP Project Roadmap™ indicates, the close of a GMP capital project runs through facility and equipment qualification. This typically includes all commissioning and qualification activities that are funded by and managed under the capital project. All manufacturing equipment is qualified along with facility and utility systems that have a direct effect upon product quality, such as process water purification and distribution. Early planning for commissioning and qualification supported by an expert in these disciplines is necessary.

10.3.1.2 Validation Lead

Process validation is not often managed under the core project management team. PPQ lots are often designated for commercial sale and, therefore, are not part of the capital cost. Additionally, process validation requires special expertise often from outside the User or Engineering organizations. Nonetheless, a GMP project is not actually complete until commercial (or clinical) distribution can begin. It is important that the organization plan for the date when actual return on the capital investment can begin in order to determine product success. In addition, process validation expertise is important in reviewing and approving designs and key GMP documents included in the GMP Project Roadmap™.

10.3.1.3 Quality Representative

Often overlooked in GMP projects is the importance of a dedicated Quality unit team member. Experience teaches that designs and plans that have been developed without

Quality involvement had to be changed when Quality entered the picture. Quality must be a core team member, even full-time for larger projects, from team formation through PPQ. It is also important that the designated Quality representative have approval authority for GMP activities and documents. It is not effective in most cases for a junior-level team member to participate in reviews and discussions yet defer approvals to a more senior member who is not committed to the project. This is so important that some companies have actually written this Quality project role into an SOP or Quality Standard, making it a GMP violation to structure a GMP project team without Quality on board.

10.3.2 Front-End Loading (FEL)

The "planning" stages of a project are often referred to as the "front end" and have been proven to be critical to the ultimate project success. All projects should begin with a business case, followed by one or two stages of design that gets a project to a decision point.

This holds true for validation as well as other measures of success. Team building is one of the initial points of FEL but earlier was emphasized separately for its critical importance. The team ensures that all the required workflows and documents described later are performed, reviewed, and approved as required. Here we focus on those project elements shaded in the diagrams to indicate influence on GMP compliance and validation.

10.3.2.1 Process Control Strategy

The FDA Guidance for Industry *Process Validation: General Principles and Practices* emphasizes the importance of process knowledge and understanding as the first phase of process validation (5). The result of that understanding should be a process control strategy that outlines everything that is important to ensure a reliable and repeatable process. Included should be material considerations, equipment, and process considerations including sequences and parametric ranges. This should be available in the initiating phase of the project (FEL 1) or as early as possible in Conceptual Design. This strategy is a necessary element of process validation.

10.3.2.2 Project User Requirements

Project User Requirements are driven first by the Process Control Strategy and then expanded to include all related process, compliance, and business drivers for the facility and project. As the name implies, the User group is the primary source of requirements, but in the capital project sense the term "user" may be expanded to include all project stakeholders. Note that this chapter does not use the common term User Requirements *Specification* (URS). The author prefers to reserve the term "specification" for its literal use, avoiding possible confusion with subsequent engineering documents.

It really does not matter who drafts User Requirements, but two important factors need to be considered. First is input from core stakeholders, including the user (after all, these are *user* requirements), technical/process experts, and, because this is a GMP facility requiring validation, Quality. The second important note is that these "requirements" are not technical solutions. They should be limited to what is needed rather than how to accomplish it. For example, if a clean room is required, its classification may be stated (e.g., Grade C or ISO 7), but technical details such as filter grades and number of air changes should be left to the engineers providing the solutions in the subsequent design.

Besides providing a foundation for the initiation of design, another valuable use of User Requirements is to provide a foundation for facility and equipment qualification (henceforth EQ). Each requirement listed may be assessed to determine why it is included. A recommended approach is to designate all requirements that have a direct effect on product quality as QCR (quality-critical requirement or some equivalent acronym). Other requirements may be designated as business, safety, environmental, or whatever other category the team determines to be valuable. By identifying quality-critical requirements, the team provides a source to trace design specification as well as qualification acceptance criteria. For these reasons, User Requirements should be approved by the core project team members, including Quality.

10.3.2.3 GMP Design Basis

It is wise to define what regulations and compliance issues are applicable early in the project. This may be to identify which GMP regulations must be followed as well as regulatory guidelines (e.g., FDA Guidance for Industry). Companies may also have internal Quality standards, design guidelines, and/or procedures (e.g., SOPs) that impact the project. These may be summarized in a document before or at the initiation of Conceptual Design. Alternatively, these requirements may be built into the project charter or a similar document.

10.3.2.4 GMP Design Review (Concept)

For mid- to large-size projects the Conceptual Design will be completed as described earlier and a "stage-gate" will be reached. In many companies, this stage-gate is a financial one wherein the project requests funding authorization to proceed. On the GMP side, this is an ideal time to perform a first GMP Design Review. This review is a cooperative effort among core team members and other invitees that reviews the Concept against a predefined set of criteria, often guided by a checklist. Questions may be raised and unanswered questions followed up by the design team. They should be guided by the User Requirements, the GMP regulations, and other documents specified in the GMP Design Basis. A checklist format has proven to be very effective (an example is provided in the Appendix). A roundtable format has also proven to be most effective but is not the only way this may be done. Once all issues have been resolved (some of which may be deferred to the next design stage), a summary report may be issued for approval by the team. This can be considered a step toward Design Qualification.

10.3.2.5 Equipment User Requirements and Specifications

As FEL proceeds into Basis of Design, individual equipment systems may require dedicated User Requirements. These can be driven by the Process Control Strategy, company or international standards (i.e., ISO), or other needs. These may be developed and formatted similarly to the Project User Requirements and will be subject to the same approvals. Technical or engineering specifications follow and become sources of information for commissioning and EQ protocols.

10.3.2.6 System Boundaries and Impact Assessments

As BOD proceeds through development of system schematics (P&IDs), it is a good practice to set system boundaries. Since many systems are interconnected to other systems (e.g., utilities or other process steps), it is important to know to which system each component shown on a P&ID belongs. This can be done by marking up or color-coding the drawing. This allows for the system to be looked at as a unique entity and a system-level assessment of quality risk completed. Determination of significant quality impact will mean that a system requires EQ and not just commissioning.

10.3.2.7 GMP Design Review (Basis of Design)

The GMP Design Review at the completion of BOD (FEL 3) is probably the most important and one that should not be overlooked. This is because most GMP compliance issues on a project level should have been addressed in the design at this point; additionally, because it is early enough in the project to correct errors or make changes without incurring the cost of implementing field changes after construction has started. There may be more items to review as Detailed Design proceeds, but this is the one that covers the greatest scope. Again, all issues and questions need to be resolved before a GMP Review report can be approved.

10.3.2.8 Validation Planning

Every GMP project needs to plan validation. Depending on the scope of the project this may include a Commissioning and Qualification Plan, a Process Validation Plan, a Computer Validation Plan, or any combination along with other related plans. Commissioning and Qualification are usually considered part of the capital project (because their costs are capitalized) and the total project cost and schedule are not closed out until EQ is completed. Studies have shown that the success of a capital project (judged by the triple criteria of cost, schedule, and quality) are best assured when C&Q planning is completed as part of FEL, that is, by the completion of BOD (6).

Because the facility does not begin paying back until PPQ is completed and commercial product is released, it is also important to have a validation plan in place that covers all areas whether or not they are managed under the facility capital project. The earlier these plans are in place, the greater the probability that the work will be completed on time and as expected.

10.3.2.9 Project Controls (Schedule and Estimate)

Although Project Controls are not usually considered GMP activities, they are shaded on the FEL diagram. All activities take time and require resources. Project Controls must consider that validation activities extend the project schedule and also have a cost. More often than not, the capital portion of the project ends with equipment qualification. However, a well-planned project takes into account the costs in material and labor, as well as the time required to perform and approve PPQ lots that carry the project to the point of financial return. For PPQ lots that will ultimately be distributed for commercial or clinical use costs are expensed rather than capitalized. Nevertheless, it is wise to consider PPQ and its ancillary activities like analytical and cleaning validation into the overall program schedule. This should include interim milestones (e.g., protocol preparation) as is done for EQ.

10.3.3 Project Execution

The work of validation continues in parallel to the work of the facility project through the detailed design and construction phases. For the capital portion of the project validation culminates in commissioning and qualification, but for the overall project where specific processes have been identified validation must proceed through PPQ. For clinical manufacturing facilities that are designed for process development there may be no specifically identified process and so the project schedule will end with EQ.

Some firms, if not most, when considering PPQ will account for cost and schedule separately from the capital portion of the project. This does not mean they are ignored or managed less rigorously. To avoid disorganization, the team organization discussed in section 10.3.1 should identify an overall program manager, separate from the facility project manager, who oversees the PPQ-related activities. These include technology transfer (e.g., from process development to operations), analytical validation, and material logistics for PPQ lots.

10.3.3.1 *Final Design Review and Design Qualification (DQ)*

As detailed design proceeds, specific process and critical building systems are specified and may require targeted GMP Design Reviews. This is the case when these systems are unique or custom designed. Detailed formal GMP Design Review is usually not required for standard commercial process equipment such as autoclaves and mixers. These designs have been proven through extensive industrial use, and the specificity of each system will be confirmed through normal engineering review as shown in the Execution diagram. In addition, these designs are further confirmed through the EQ process.

Where a system is entirely unique to the project, for example Purified Water or Water for injection (WFI) distribution, a formal system GMP Design Review may be instituted. However, even in these cases, the BOD documentation is often complete enough to allow the GMP Design Review to be completed at that stage.

Once the GMP Design Review has been completed and approved by key stakeholders, including Quality, the team may issue a certification that the design has been qualified to meet User Requirements and compliance to applicable regulations. This may be designated as Design Qualification as defined in European Union and PIC/s GMP Guidelines (4, 7).

10.3.3.2 *Factory Acceptance Tests*

The Project Execution Plan or Procurement Plan may define if and where factory acceptance testing (FAT) will be required. FAT is strictly a matter of contract between the project and equipment vendors and is not a regulatory requirement. It may be a valuable exercise, however, especially where automation is included with the system. FAT is highlighted in the diagram because it is allowable to document some of the FAT test results to be leveraged into EQ protocols. This must be predetermined, that is, tests that will be leveraged predefined, and documentation must follow the same standards as set for EQ.

10.3.3.3 *Preventative Maintenance*

In order for systems to be qualified, maintenance procedures must be in place. For individual systems this includes preventative maintenance procedures, including schedules for periodic, run-time, or predictive maintenance. Maintenance is an essential element of validation and facilitates the continuing process verification described in the FDA Guidance (5).

10.3.3.4 *Commissioning and Qualification (C&Q)*

Commissioning plans and EQ protocols may be developed during the Detailed Design phase of the project as specific system information and vendor documentation is received and approved by Engineering. It is important that detailed and approved design documents be used as reference. These need to be compared to User Requirements to identify the specific system characteristics and functionality that satisfy the User Requirements identified as essential to product quality.

10.3.3.5 *Containment Verification*

For facilities encompassing highly active, biohazard, and/or toxic materials, an additional step to verify containment may be required. Depending on the containment design, this may be included in the qualification of environmental controls, but for more sophisticated containment devices (e.g., glove boxes/isolators) this may require a separate protocol. Although this may be primarily a safety or health concern, it may also prevent cross-contamination of products and therefore is a validation consideration.

10.3.3.6 As-Built Drawings

For qualified process equipment and facility systems schematic drawings, usually P&IDs, must be verified and maintained as part of the EQ records. This is usually done by marking up or "red-lining" these drawings according to a specific approved procedure. The red-lined drawings are retained in the validation files and are used as reference to develop certified as-built schematics that are used by the facility for future operations and as basis for future changes.

10.4 Facility Design

The foregoing sections emphasize the importance of planning, organization, and communication throughout a facility project to facilitate and ensure validation. Experience has shown that when projects fail, it is often the lack of attention paid to validation that is the root cause. However, even when validation receives the appropriate attention it is insufficient when placed into substandard facilities and equipment or facilities that employ poorly designed utility systems. Good facility and equipment design is necessary in order to ensure that facility, equipment, and the processes they house can be validated and that they will continue to perform reliably and consistently. This includes facility layout, equipment, environment, and critical utilities.

The GMP regulations provide only general guidance on facility design as in the excerpts provided in the Introduction and the following sections of the regulations. The possible exception to this general approach is EU GMP Annex 1 for sterile products, which has several more specific guidelines. None of these regulations, however, are definitive with regard to design criteria. Fortunately, after several decades of facility design under GMP, manufacturers and their supporting contractor engineers and architects have become expert in understanding proven design concepts. There are numerous publications compiling these concepts, among which are a series of ISPE Baseline® Guides for facility and equipment design (8). These and other publications provide significant details and alternative approaches to comply with the general requirements of the regulations and, in doing so, ensuring continuously validated processes (9).

The following are a digest of points to be considered when designing a biopharmaceutical facility for GMP compliance and validation. These points were originally compiled as part of a GMP Facilities Design Guide developed by Schering-Plough Global Engineering Services around 2009. These are not requirements but are intended to satisfy the GMP regulations.

10.4.1 General Design Considerations

10.4.1.1   The production facility will be designed, constructed, and maintained in a manner to minimize the impact of the external environment.

10.4.1.2   The premises will be designed and constructed such that they prevent entry of insects and other animals and the migration of extraneous material from the outside into the manufacturing areas and from one area to another.

10.4.1.3   Buildings and facilities will have adequate space for the orderly placement of equipment and materials to prevent cross-contamination and/or mix-ups between different components, drug product containers, closures, labeling, in-process materials, and drug products. The layout should also provide for appropriate access for routine operation, maintenance, and repairs.

10.4.1.4   Facility design will permit appropriate flow of materials, equipment, and personnel such that the risk of mix-ups or contamination is minimized. Unidirectional flow is recommended. If this is not possible, specific procedural controls are necessary. Special consideration shall be given to multi-product facilities, as extra precautions are necessary to prevent cross-contamination. The facility should provide for separate material and equipment flows until the material or product is contained in a closed system. Personnel, material, and product flows will be reviewed during GMP Design Reviews.

10.4.1.5   Corridor dimensions and door openings shall take into account the movement of large equipment and materials. Appropriate wall protection (e.g., bumpers, railings) should be utilized where necessary to protect against damage to equipment, walls, and doors. Mobile equipment should also be designed to minimize damage to the facilities.

10.4.1.6   Operations will be located and performed within specifically defined areas of adequate size. There shall be separate or defined areas as necessary to prevent contamination or mix-ups during the course of the following procedures:

- Receipt and holding of materials, components, and labeling pending approval for use.
- Holding rejected materials, components, and labeling before disposition.
- Storage of released components, in-process materials, and drug products after release and labeling.
- Manufacturing, packaging, and labeling operations.
- Quarantine storage prior to release of drug products or APIs.
- Control and laboratory operations.
- Sampling of intermediates, APIs, and finished or unpackaged drug products.
- Personnel dress/gowning, hygiene, and food consumption.

10.4.1.7   Controlled (Non-classified) and Classified areas are to be segregated from, and kept at a positive pressure to General non-production areas.

10.4.1.8   As a general rule the pressure cascade should be from the highest to lowest classification, but special consideration may be necessary (for example in airlocks) to contain or limit exposure to biohazards, toxic, or high-potency materials, drug substances and drug products. (See HVAC Design Requirements.)

10.4.1.9   In areas where potent or toxic materials are processed outside of glove boxes or isolators, consideration should be given to enhanced cleanability. These areas should have a separate HVAC system.

10.4.1.10   Room finishes and materials selection should be appropriate to the activities performed in the respective areas. Personnel safety is an important consideration. Floors and surfaces should be non-slip even in wet conditions.

Consideration will also be given to the following characteristics: cleanability, durability, functionality (e.g., low humidity, anti-static), and maintenance (ease of repair or renovation).

10.4.1.11   In Controlled and Classified areas, the walls, floors, ceilings, and so on should be constructed so that the primary surfaces are accessible for cleaning. Exposed horizontal surfaces in Controlled areas are generally acceptable if they are accessible for cleaning. They should, however, be sloped to allow free draining.

Horizontal surface should be avoided as much as possible in Classified areas. Windows, doors, and fixtures should be flush with primary surfaces.

10.4.1.12   As much as possible, process drainage should be through the wall into a non-controlled area (mechanical Support area).

Floor drains are allowed in Controlled and Classified areas only when required (e.g., in a washroom). They must be of adequate size and should be designed for clean service (i.e., food and pharmaceutical) with a removable trap that may be easily cleaned and sanitized. In all cases process drains must be provided with an air break or a suitable device to prevent backflow from the drain to the process vessel.

Drains in Controlled and Classified areas (Grades C & D) should be capped when not in use. Drains in Grade A and B areas are not allowed unless specifically approved.

10.4.1.13   Premises shall be designed in a manner to prevent unauthorized access to production and support areas. Additional controls may be required for limited access areas such as sterile manufacturing, label control, controlled drug storage, and quarantine areas.

10.4.1.14   In most circumstances, airlocks with interlocking doors should be installed when accessing Classified areas or between Classified areas of different grades. A notable exception is when there is a local Grade A operating environment established within a room that is established and maintained at Grade B.

10.4.1.15   Another allowable exception shall be a locally controlled environment operating at Grade C (e.g., a weigh booth) within a Grade D environment.

10.4.1.16   In gowning areas and airlocks, architectural finishes and materials should comply with the requirements of the higher classification into which the area opens.

10.4.1.17   Doors, windows, walls, ceilings, and floors will be intact with no holes or cracks evident.

10.4.1.18   Doors giving direct access to the exterior from production areas are to be used for emergency purposes only and will be alarmed. These doors will be adequately sealed.

10.4.1.19   In-process storage areas are General areas if the product stored in these areas is not exposed. The facility should generally comply with the requirements specified for General areas. There should be suitable electronic systems or other means to identify (e.g., labeling, signs, bar coding, etc.) and segregate materials.

10.4.1.20   Sampling and weighing rooms should provide a level of protection equivalent to, or greater than, the level of protection required by the product at that stage of its processing. (e.g., materials that will be processed in a Grade C Classified area must be sampled and weighed in a Grade C Classified area or equivalent locally controlled environment, e.g., a weigh booth).

10.4.1.21 When providing utilities or instrumentation service to Controlled or Classified areas, there should be mechanical rooms independent of the processing room. These areas should be designed to minimize the presence of piping and instrumentation in the processing areas. These areas should also be designed to minimize the need for service personnel to enter the Controlled or Classified area for routine maintenance/calibration and to minimize the need for processing personnel to enter the mechanical room or chase during routine processing operations.

10.4.1.22 Adequate, clean locker rooms shall be provided for personnel entering the manufacturing facility from outside. These shall be physically separated from Controlled or Classified manufacturing areas by at least one intermediate area (e.g., an anteroom or a corridor). Personnel should enter from the "street" or public side and exit into the operations area with the appropriate dedicated plant garments.

Locker rooms shall be designed to maintain adequate separation between street clothes and plant uniforms. This may include separate lockers or sections for storage of plant uniforms. Also, locker rooms should be suitably marked or separated to indicate the limit where street clothes and shoes may be worn. Stepover benches provide suitable barriers but are not required.

10.4.1.23 Break rooms, bathrooms, and mechanical Support areas should meet operating needs and building code requirements. These areas should be separated physically by at least one intermediate area (room, vestibule, corridor, etc.) from manufacturing areas and by at least two intermediate areas (room, corridor, gowning room, airlock, etc.) when accessing Controlled or Classified manufacturing areas (Note: The intervening zone may or may not be the same classification as the manufacturing zone). There should be no breakrooms or rest rooms within Classified or Controlled (Non-classified) areas.

10.4.1.24 Adequate maintenance workshop space and spare parts storage facilities should be available or provided, which should have direct access to utility areas, including those requiring gowning.

10.4.1.25 Systems for sewage, trash, and refuse disposal will be designed and maintained in a sanitary manner. Waste flow should not contaminate the process it is derived from or be a source of cross-contamination in other areas. Drains shall be of adequate size and shall be provided with an air break or other mechanical device to prevent back-siphonage. Where pass-throughs are utilized, separate waste and material pass-throughs should be provided.

10.4.1.26 Any permanent equipment parts in contact with the floor, walls, or ceiling should be sealed or installed so that access for cleaning is assured.

10.4.1.27 Wiring leading to systems and fixtures in Controlled and Classified areas should be enclosed in appropriate conduit to enhance cleanability. Grounding wires should be designed using an insulated wire instead of bare wire.

10.4.2 Warehousing/Shipping Areas

10.4.2.1 Finishes and materials used in the warehouse area should emphasize durability to traffic and stress. They should be clean and dry, as appropriate, and maintained within temperature and humidity limits that are consistent with material requirements. Storage areas should be designed to ensure safe storage conditions. All areas should be signed and labeled.

10.4.2.2   Shipping and receiving areas should be designed with ample space to promote a logical and orderly movement of materials and to ensure that forklift traffic does not impede upon personnel traffic during normal operation. Shipping and receiving areas should be separated to minimize opportunities for improper staging or materials/product mix-up.

10.4.2.3   Shipping and receiving bays should protect materials and product from the weather. Considerations should be made for minimizing the impact of outside environmental conditions, such as the use of automated doors, docking systems, and other barriers.

10.4.2.4   A transition point from the warehouse environment to manufacturing areas should be clearly marked and may require a vestibule or airlock to maintain the integrity of a controlled area. Shipping and receiving areas should not be designed with direct access to Controlled or Classified areas, whereas access to General areas may require only a marked transition zone.

10.4.2.5   A separate area shall be provided for sampling and incoming inspection activities. The physical and environmental conditions of this area should be consistent with the area in which the materials will be later processed.

10.4.3 Packaging Areas

10.4.3.1   Primary packaging areas, where exposed product and primary packaging components are handled (including product-contact dosing aids), should be designed to the same level of control as the associated manufacturing areas.

10.4.3.2   Primary packaging areas should be physically separated from secondary areas as well as from adjacent primary packaging (filling) areas via floor to ceiling partition or walls.

10.4.3.3   Physical or spatial separation must exist between secondary packaging areas/lines to prevent mix-ups and cross-contamination. Partial-height partitioning is acceptable. Unidirectional design layout is recommended.

10.4.3.4   Adequate space must be provided for storage and staging of components, in-process materials, and finished goods. Limited access storage areas need to exist, as required, for labeling and coding.

10.4.3.5   Packaging areas and packaging lines shall be adequately and clearly marked.

10.4.4 Laboratory/Quality Control Areas

10.4.4.1   Laboratory design must ensure sample integrity and the validity of the test results. Most analytical laboratories may be designated as Support areas. However, where environmental conditions may affect analyses (e.g., humidity, viable particulate), laboratories may be Controlled or Classified.

10.4.4.2   Laboratory areas should normally be separated from production areas. However, those used for in-process controls can be located within production areas provided the process operations do not adversely affect the accuracy of the laboratory results and the laboratory activities do not adversely affect the production process or materials. Laboratories within the operating area and not separated by an intermediate space (e.g., anteroom, corridor) should have the same level of architectural design and area classification as the areas they support. For in-process laboratory areas supporting multiple products, separation from the production area is required.

10.4.4.3 Laboratories for biological testing will be separated from the production areas. Additional requirements may exist in such laboratories for handling these particular substances. These laboratories are Classified and, depending upon materials handled, may require additional hazard classification (e.g., NIH Biocontainment guidelines).

10.4.4.4 Quality control laboratories should be designed to suit the operations to be carried out in them. Sufficient space should be given to avoid mix-up of samples and cross-contamination.

10.4.4.5 There should be adequate storage space for the receipt, handling, and storage of samples, standards, and records.

10.4.5 Ancillary Areas

10.4.5.1 Whenever possible, support staff and general office space should be located outside of the manufacturing areas. If necessary, offices within the production area should have cleanable finishes consistent with the adjacent surroundings and/or provide a barrier to the operating area.

10.4.5.2 Maintenance workshops shall have ample space and be physically separated from the production areas. Space and suitable storage facilities should be allocated for equipment change parts and spare parts. General personnel corridors should be routed outside of the production areas. Adequate space should also be provided in the manufacturing areas for the storage of cleanroom maintenance toolboxes and the like.

10.4.6 API and Biotechnology Manufacturing Areas

10.4.6.1 There should be facilities to maintain master cell banks in segregated areas under storage conditions designed to maintain viability and prevent contamination.

10.4.6.2 Primary biological manufacturing areas (bioreaction) should be designed to prevent contamination of culture and reaction products. Although largely closed operations, these areas may involve multiple manipulations (e.g., sampling). Therefore, they should be Classified, typically Grade C or better.

10.4.6.3 Appropriate precautions should be taken to prevent potential viral contamination from pre-viral to post-viral removal/inactivation steps. Therefore, open operations prior to the viral removal/inactivation steps should be performed in areas that are separate from other operating activities and have separate air handling units.

10.4.6.4 Vents and drains from areas containing live biological organisms, including viruses, shall be provided with sufficient safeguards to prevent contamination of the surrounding environment. This may include HEPA filtration of vent streams and containment and/or deactivation of liquid waste.

10.4.7 Non-Sterile Drug Product Production Areas

Although most products produced by biotechnology are delivered as sterile dosage forms, this section is included for completeness.

10.4.7.1   All non-sterile drug product manufacturing operations where raw or in-process materials, product, and product contact components are exposed should be in Controlled areas. Entirely enclosed operations may be in General areas. Corridors within non-sterile manufacturing operations may be Controlled or General, depending upon whether product will be exposed.

10.4.7.2   The facility shall provide for specific gowning locations, dependent upon the area classification, where personnel can properly dress prior to entering the area. The size of the gowning area should be appropriate for the number of users. Intermediate changing areas may be needed within the facility to enter into cleaner areas. Finishes should match those of the area being serviced. Clean garment storage and used garment collection bins must be provided at each gowning area. Alternately, a laundry pass-through system for disposal may be utilized.

10.4.7.3   A suitable transition zone or airlock shall be used to separate areas of differing air classification at the points where personnel, materials, or equipment enter and exit such areas. Transition zones from Support to General areas may be a simple door (e.g., from lab to corridor). Transition zones from warehouse to General corridor may be a normally closed door with a clearly marked floor area for lay down of pallets or containers. Transition zones into Controlled environments should comprise a vestibule with opposing doors. Airlocks should be employed when potential airborne cross-contamination is of particular concern (e.g., classified environments, processing of powders). Airlocks shall have interlocked doors to prevent more than one door from being open at the any time (except in emergency).

10.4.7.4   Process systems should be designed to minimize drug product exposure to the environment. Closed or contained systems are preferred. Where possible, maintenance access to the equipment should be from non-controlled areas.

10.4.8 Sterile Manufacturing Areas

Although many biotechnology APIs are isolated as non-sterile drug substances and later formulated and filled aseptically, the control of microbial contamination is usually critical even in the non-sterile steps. Design principles common to sterile product manufacturing are usually applied to all stages of biotech.

10.4.8.1   The sterile facility design shall utilize a core concept whereby the aseptic environments are isolated from other less controlled areas. Critical operating areas (e.g., aseptic filling, lyophilizer loading) within the sterile core should be visible from the lesser-controlled areas to permit observation by authorized personnel.

10.4.8.2   Access control features should be incorporated into the design of Classified areas. There should be separate areas within the clean area for various operations of component preparation, product compounding/preparation, and filling. Technical space (e.g., for autoclaves and lyophilizers) should not be directly accessible from Classified areas.

10.4.8.3   Compounding of product prior to filter sterilization and aseptic processing shall be in a Grade C area. Compounding of product prior to filling and terminal sterilization shall be in at least Grade D.

10.4.8.4   Preparation and holding of filling and packaging components and other materials prior to sterilization shall be in Grade D.

10.4.8.5   Aseptic operations, including filling, sampling, and connections, or where sterile product or components are exposed, shall be performed under Grade A conditions, with a surrounding Grade B area. U(see exception in section 10.4.8.7 below). Filling of non-sterile product prior to terminal sterilization shall also be in a Grade A area, with at least a Grade C background.

10.4.8.6   Where aseptic operations are performed, the preferred design will utilize full isolators or restricted access barrier systems (RABS) suitable for "advanced aseptic processing." Design details and preferences should be discussed with the project team.

10.4.8.7   Where isolators are used for aseptic operations, the background environment shall be at least Grade C. RABS requires a Grade B background.

10.4.8.8   Capping of filled and stoppered vials may be as an aseptic process carried out within the sterile core (A/B classification or full isolator) using sterilized caps. In this case, the capping should be segregated from the filling operation to minimize exposure of the product to particulate from the capping operation. Segregation within the sterile core may be a separate room, a full barrier isolator, or a spatial separation incorporating a RABS to limit particulate generation in the surrounding room.

10.4.8.9   Alternatively, capping may be as a clean operation outside the sterile core. In the latter case, the stoppering operation shall be supplied with Grade A quality air within at least a Grade D background until the cap has been applied and crimped. In this case, a RABS should be employed to limit product exposure to the surrounding environment. Also, in this configuration, there should be a means to detect and reject unstoppered or incompletely stoppered vials prior to capping.

10.4.8.10   Unless specific environmental conditions are required for product stability or the inspection of packaging integrity, secondary packaging operations may be within General areas.

10.4.8.11   Access corridors to gowning rooms for Grades A/B and C should be Grade D or better. Gowning rooms should be designed as airlocks and used to provide physical separation of the different stages of gowning (pre-gowning, sterile gowning). They should be supplied with HEPA-filtered air. The final stage of the changing room should be the same grade as the area into which it leads.

10.4.8.12   Entry to the sterile manufacturing suite (Grades A/B and C) shall be through separate airlocks for personnel versus materials/equipment. The airlocks shall have mechanically interlocked doors to prevent more than one door from being open at the same time. Horizontal sliding doors are prohibited.

10.4.8.13   Personnel, equipment, and production waste materials may exit through a common airlock. Where there is no material airlock, exit from Grade C areas may be through the gown room. Grade A/B areas require a separate de-gown room.

10.4.8.14   Grade A/B gown rooms require a two-stage (pre-gowning, gowning) arrangement utilizing adjacent but separate rooms. The initial room may be Grade C or D and should include a handwashing facility. The second (gowning) room must be Grade B or better and incorporate a clear barrier (i.e., stepover bench) to demarcate the ungowned from the gowned side.

10.4.8.15   Sinks and drains are prohibited in the aseptic areas (Grade A/B). Floor or wall drains (where necessary) in other Classified areas should be capped. Wet vacuums with HEPA-filtered exhaust are the preferred method for cleaning spills.

10.4.8.16   Equipment and component sterilizers (autoclaves, ovens, etc.) and lyophilizers shall be built into room architectural wall, floor, and ceiling assemblies in order to maintain continuously smooth, clean, and sanitizable conditions. Where possible, maintenance access to this equipment should be from Non-classified areas.

10.4.8.17   Operating and control panels serving the aseptic area should be located in adjacent mechanical rooms. Alternatively, they may be constructed integral to process rooms such that operator access is on the clean side and maintenance access is from the Non-classified mechanical rooms.

10.4.8.18   Individual manufacturing processes (i.e., multiple products or similar product, different batches) within a common room should have adequate physical separation (e.g., isolators, barriers, mass air flow devices) to prevent cross contamination of product.

10.4.8.19   Grade A and B areas should include provisions for continuous remote particulate monitoring, although this is not strictly required.

10.5 HVAC Design

Environments or areas are defined as Classified and Non-classified. Applications and design criteria for each of these environments are outlined as follows. The directional airflow cascade scheme should be from "more clean" to "less clean" area. Special design for containment may be necessary where highly active or toxic materials are handled or where biocontainment guidelines apply.

The following tables are guidelines. These may be modified based upon specific product or operational requirements.

10.5.1 Classified Areas

10.5.1.1   Classified areas include processing operations for sterile API, biotech manufacturing, high-potency products, inhalation drugs, and all aseptic operations. Table 10.1 provides design philosophy and basic requirements for Classified environments:

TABLE 10.1

HVAC Requirements—Classified Areas

CLASS	Grade A (unidirectional)	Grade B	Grade C	Grade D
Fed. Std. 209e (retired) (operational)	Class 100	Class 10,000	Class 100,000	not specified
ISO 14644 at rest operational	Class 4.8 Class 4.8	Class 5 Class 7	Class 7 Class 8	Class 8 not defined
PARTICLE LIMITS	See Table 10.2.			
APPLICATION	Local zone of high-risk operations, i.e., aseptic filling zone, stopper bowls, open ampoules or vials, aseptic connections, loading lyophilizer, primary closure process.	Background to Grade A aseptic operations, filling of inhalation products, purification of biotech products (open processes).	Less critical stages of aseptic processing, preparation of solution prior to filter sterilization, filling of terminally sterilized products, biotech purification (closed processes).	Less critical stages of aseptic processing, including preparation/holding of components prior to sterilization, compounding of terminally sterilized products, access corridors for aseptic and biotech operations.
AIR Filtration	Unidirectional air flow with 99.97% efficiency HEPA or better	Ceiling-mounted HEPA filtration, low-wall return-air; plenum design preferred.	Ceiling-mounted HEPA filtration, low level return-air.	Central or ceiling-mounted HEPA filters. Low-level returns preferred.
DIRECTIONAL AIRFLOW	Highest positive pressure.	Positive to Grade C, 0.04 to 0.06" WC (10–15 Pa) between areas.	Negative to Grade A/B (Class 100), Positive to Grade D, 0.04 to 0.06" WC (10–15 Pa) between areas.	Positive to Non-classified area, 0.04 to 0.06" WC (10–15 Pa) between areas.
DESIGN TEMPERATURE	68 °F ± 2 °F (20 °C ± 1 °C)	68 °F ± 2 °F (20 °C ± 1 °C)	70 °F ± 2 °F (21 °C ± 1 °C)	70 °F ± 2 °F (21 °C ± 1 °C)
DESIGN RH %	50% ± 5%	50% ± 5%	50% ± 5%	55% ± 5%
CHANGES PER HOUR AND RECOVERY GUIDE	Linear velocity adequate to maintain unidirectional flow in the critical zone. Typical 90 ± 10 ft/min at one foot from filter face.	The number of air changes must be adequate to guarantee the grade (classification) performance requirements. The design basis should consider the room size, equipment arrangement, personnel required, air flow patterns and process operations, with a 15- to 20-minute recovery time.	The number of air changes must be adequate to guarantee grade (classification) performance requirements, and should be related to the size of the room, the equipment, air flow patterns, process operations and personnel present in the room, 15- to 20-min. recovery.	The number of air changes must be adequate to guarantee the grade (classification) performance requirements and should be related to the room size, operations, air flow patterns and personnel flows, with a 15- to 20-minute recovery time under at-rest conditions. HEPA filtration is required.

TABLE 10.2

Total Particles per Cubic Meter

CLASS	At Rest		In Operation	
	0.5 µM	5 µM	0.5 µM	5 µM
A	3520	20	3520	20
B	3520	29	352,000	2900
C	352,000	2900	3,520,000	29,000
D	3,520,000	29,000	not defined	

10.5.1.2   Gown room and airlocks between areas of different classification should be designed to the higher class. Area classifications are designated according to EU Annex 1 for more universal application. Firms may choose their own nomenclature.

10.5.2 Non-Classified Areas

10.5.2.1   Non-classified areas are identified as Controlled, General, or Support. Although most biotech products are delivered as sterile dosage forms and require Classified environments, description of Controlled areas is included for completeness. Controlled areas are designed to achieve Grade D conditions for the purpose of maintaining control and ease of cleaning. Operational requirements for these areas shall be designated by the manufacturing organization. The actual process operations may be open or closed. General operations within these areas are those where manufacturing activities do not expose the product or product contact surfaces to the environment. (Exposure of vessel interiors in a General area is allowed only during cleaning.) The areas may be for storage, staging, analytical laboratories, amenities, and other non-production activities. All spaces need to be designed such that the environment does not impact the quality, purity, and integrity of the products or of the test results associated with product manufacture or quality release. (Note that certain laboratory activity, e.g., microbiological, may require a Classified environment.)

10.5.2.2   Note that the Controlled designation is typically not used for sterile manufacturing operations where microbial control is critical. These areas should be Classified.

10.5.2.3   Table 10.3 is based upon typical industrial quality standards.

Gowning rooms, transition zones, and airlocks between areas of different classification should be designed to the higher class.

10.6 Process Water and Pure Steam Systems

10.6.1 General Requirements

10.6.1.1   Process water and Pure Steam (used directly in a manufacturing process or sterilization) are critical systems, as they directly impact product quality. These systems must be designed, constructed, commissioned, and qualified to provide water and steam that meets defined User Requirements.

TABLE 10.3

HVAC Requirements–Non-Classified Areas

CLASS	Controlled	General	Support Offices, Labs	Support Warehouse, Mechanical
Fed. Std. 209e (operational)	Not applicable			
ISO 14644	Class 8	Not applicable		
PARTICLE LIMITS	0.5 µM at rest: 3,520,000/M^3	Not applicable		
APPLICATION	Non-sterile drug product manufacture, primary filling and packaging, final API isolation steps.	API and intermediate manufacture, closed biotech reactions, closed drug product manufacturing processes.	Offices, chemical/analytical labs, locker rooms, amenities.	Warehousing, day staging, shipping/receiving, mechanical rooms, shops, utility areas.
FILTRATION	95% ASHRAE or HEPA, ceiling mount or central bank.	Standard industrial or commercial practice to provide comfort conditions.		Not applicable; space heaters may be employed.
DIRECTIONAL AIRFLOW	Positive to adjacent General or Support areas, sufficient to maintain directional flow of air under normal operating conditions.	May be positive or negative to adjacent non-manufacturing areas depending upon subject operations.	Neutral or negative to surrounding Support areas, depending upon operations supported.	Neutral or positive to outside environment.
DESIGN TEMPERATURE	72 °F ± 2 °F	72 °F ± 4 °F Conditioned for human comfort.	72 °F ± 4 °F Conditioned for human comfort.	72 °F ± 4 °F Heated and ventilated only to maintain conditions below design temp.
DESIGN RH %[1]	55% ± 5%	55% ± 5%	55%–60% Winter humidification optional.	No RH control.
AIR CHANGES PER HOUR	Varies depending upon room size, flow factors; should be adequate to achieve Grade D/ISO 8.	Based upon calculated heat load.		As required to maintain temperature uniformity.

10.6.1.2   Process water systems and use points must be clearly marked. Exposed permanent piping should be avoided as much as possible in production areas, with only use points exposed in the room, where possible. If water lines must be routed in the process room, they should not be run over areas where product may be exposed.

10.6.2 Water Pre-Treatment

10.6.2.1   Water pre-treatment is a series of unit operations required to purify the feed water to a level that ensures that the final treatment step will consistently produce water or steam meeting chemical and microbiological specifications.

10.6.2.2   Pre-treatment system design must be based upon sufficient data on incoming source water quality to account for seasonal variations (typically at least one year of data, reported quarterly).

10.6.2.3   When more than one source is used, equivalent data are required for each source of feed water.

10.6.2.4   The feed water data should provide evidence that the feed water consistently complies with World Health Organization (WHO) and local drinking water standards. If it does not, additional treatment steps (e.g., chlorination) may be prescribed.

10.6.2.5   Supplemental testing should be conducted beyond what may be reported by normal source water testing. As a minimum, this should include tests for

- Total hardness
- Total solids
- Alkalinity
- Silica (total and colloidal)
- Total organic content

10.6.2.6   The use of ultraviolet (UV) light is allowed in pre-treatment systems to help limit microbial growth. However, UV light is not a recommended practice as the primary means to reduce high microbial counts in source water. In these cases, chlorine or other effective chemical pre-treatment is preferred.

10.6.2.7   Inline filters may be used when particle removal is of concern at any unit operation within the pre-treatment system. Filter efficiency should be selected based on the expected size particle to be removed. The use of microbially retentive filters in pre-treatment systems is discouraged and is not allowed in recirculating lines.

10.6.2.8   Sample points should be provided to evaluate the performance of each unit operation. When parallel units are specified (e.g., dual multimedia filters, softeners, or carbon beds), sample points should be located after each unit.

10.6.2.9   The location of sample points should be accessible for sampling at working level and should also be safe and secure. Drain funnels for sample points should be provided.

10.6.3 Water Systems Final Purification

10.6.3.1   For non-compendial (e.g., deionized) process waters, final treatment may be reverse osmosis, deionization, or a combination of the two.

10.6.3.2   The preferred final purification methodology for Purified Water USP/EP or for feed water to a WFI still is reverse osmosis (RO) followed by continuous electronic deionization (CEDI or EDI).

10.6.3.3   The RO membranes and EDI unit should be specified to be hot water sanitizable.

10.6.3.4   When using ion exchange technology, EDI is the preferred technology,

10.6.3.5   Material of construction for each unit operation should be designed based on the requirements for sanitization. When a system is designed for hot water sanitization, the internal components should be constructed of stainless steel. No PVC internal components are allowed.

10.6.3.6   WFI may be generated using multi-effect or vapor compression distillation technology. Other methods may be acceptable (e.g., reverse-osmosis in combination with ultrafiltration or other technologies) and must be specifically approved.

10.6.3.7   WFI distillation units should be constructed of 316/316L stainless steel and materials suitable for the high temperature and pressure conditions. Note that other materials suitable for sanitary service may be considered (e.g., for vapor compression stills) upon approval. Construction shall be of sanitary design, as appropriate, with minimum dead-space-diaphragm valves at sample locations.

10.6.4 Water Systems Storage

10.6.4.1   Water storage capacity should be defined to provide adequate available water during periods of use, with allowance made for adequate supply during sanitization of the water treatment system (minimum sanitization allowance shall be 4 hours per day; usually done on an off shift). Water use and capacity analysis should be included as part of the Basis of Design.

10.6.4.2   Storage tank shall be constructed of 316/316L stainless steel (316L for interior welded surfaces), ASME rated for pressure (min. 3.5 bar/50 psia) and full vacuum. All welds shall be polished smooth, and the interior finish shall be Ra 0.5 μm (Ra 20 μ-inch) or better and electropolished. Vendor shall provide stainless steel mill certification, surface profilometer report, and vessel ASME (or equivalent) stamp. Where WFI-quality water is stored at a high temperature (>70 °C), polymeric storage vessels (e.g., PVDF) are not recommended.

10.6.4.3   Tanks should be insulated for heat retention. Exterior skin over insulation may be 304 SS (i.e., 316 SS is not required). It shall be finished to a polished appearance approximately 150 grit (US).

Tank shall be heated via an external heat exchanger, either on the main circulating line, on an independent heating loop or both. Steam jackets are not preferred.

10.6.4.4   Tank should include

- 45 cm (18″) manway with sight-glass and light
- sanitary clamp-type fittings, to be located at the top or bottom only

- temperature probe
- pressure relief device
- optional: microbially retentive, hydrophobic vent filter, to be traced and insulated or steam-jacketed
- spray ball at the water return line, designed to wet all exposed surfaces

10.6.5 Water System Distribution

10.6.5.1 For hot (>70 °C) distribution tubing shall be 316L stainless welded sanitary tubing, with sanitary clamp-type fittings. O-rings to be Kalrez, EPDM, or steam-resistant Viton, all designed to withstand sanitization temperatures. Tubing and fittings interior finish to be Ra 0.5 µm (Ra 20 µ-inches) or better and electropolished. Exterior surfaces shall be finished to a polished appearance approximately 150 grit (US). Vendors to provide mill certification or other suitable certification of materials (including O-rings) and an isometric of the complete system with the identification of all weld points.

10.6.5.2 Fittings to be limited only as required for instrumentation and other required connections to system components. All clamps in the field to be accessible for maintenance. Direct connections to process systems or equipment should be designed to prevent backflow into the system. Valves at point of use (POU) shall be zero dead leg diaphragm type.

10.6.5.3 Flow in distribution loops should be designed for a minimum velocity of 1.5 m/s (5 ft/s) and/or to maintain turbulent flow at normal full flow conditions as calculated by a minimum Reynolds number of 20,000 with all user points closed. System design should be to maintain a return flow of approximately 0.5 m/s (1.5 ft/s) when user points (typical diversity factor) are open.

10.6.5.4 Distribution systems should be designed to minimize number and length of dead legs. In no case should any dead leg be greater than two pipe diameters from the interior wall of the adjoining main. All pipe runs to be self-draining with a minimum slope of 1:100 to low points.

10.6.5.5 Preferred design temperature for Purified Water distribution loops is ambient (20–22 °C). Depending upon specified use, they may also be designed as hot (>70 °C) or cold (4–8 °C). Trim heat exchangers should be employed to maintain ambient circulating loops below 22 °C. All Purified Water loops should be provided with means (e.g., heat exchanger) to provide for hot water sanitization at 80 °C or higher. Alternatively, ozone sanitized systems will be considered.

10.6.5.6 WFI loops may be hot loops designed to be self-sanitizing. Hot loops typically operate at 80 °C, with 70 °C minimum temperature. Ambient or cold WFI loops are allowed if justified by process demands. Any loops operating at a temperature of less than 70 °C should have a daily scheduled sanitization procedure at a temperature of at least 80 °C.

10.6.5.7 For ambient WFI-grade loops, storage tanks shall be maintained hot (80 °C) while the ambient return flow is directed to the inlet of the circulating pump and not back to the tank. This so-called "chase your tail loop" is replenished with hot water from the storage tank and tempered using an inline heat exchanger of sanitary design. The loop may be periodically sanitized by suspending cooling at the heat exchanger and displacing the tempered water with hot from the storage tank for a defined period.

10.6.5.8 There shall be no microbially retentive filters in recirculating lines. Where sterile water is required, these may be installed at point of use.

10.6.6 Stainless Steel Tubing Fabrication

10.6.6.1 Specifications for high purity water and steam systems should include the requirements for system fabrication. The fabrication specification should address weld quality requirements, accepted welding techniques, welder's qualification, welding procedures, welding processes, test coupons, inspection requirements, and documentation.

10.6.6.2 The fabrication procedure should include limits established to accept or reject welds and other installation tests.

10.6.6.3 Documentation requirements should include welding logs, isometric drawings, and installation reports (cleaning, pressure testing, and passivation).

10.6.6.4 The fabrication procedure should address the inspection requirements and methodology. Third-party (independent from installer) inspection is preferred. Boroscope is considered the minimum required inspection method. X-ray can be used for pipes that are not accessible.

10.6.6.5 Use of chemical cleaning and passivation of all product contact stainless steel welded parts is required. If a passivation procedure is conducted, a detailed report must be provided. The report must include information such as acid/base type and concentration used, circulation times and temperature, final rinse water quality used, and endpoint determination.

10.6.7 Valves

10.6.7.1 Valves are to be diaphragm type of sanitary design, and diaphragms are to be encapsulated Teflon, EPDM, or Teflon. O-rings shall be Kalrez, EPDM, steam-resistant Viton, designed to withstand sanitization temperatures.

10.6.7.2 Zero dead leg valves to be used at POU and wherever else possible.

10.6.7.3 Sample valves should be sanitary design; diaphragm or Millipore ESP valves are preferred.

10.6.7.4 All automatic valve user points to be fail closed, unless fail open is the design requirement necessary.

10.6.8 Heat Exchangers

10.6.8.1 Heat exchangers must be designed to be fully drainable and suitable for sanitary operations. Preferred heat exchangers are double tube sheet types. Plate and frame heat exchangers may be allowed providing that they are of a fully welded construction. Differential pressure between the heating/cooling fluid and the water should be provided, with higher pressure on the product side. Differential pressure monitoring, as well as pressure relief, should be provided on all heat exchangers.

10.6.8.2 Spool pieces should be provided to enable bypassing of heat exchangers for degreasing and passivation.

10.6.9 Pumps

10.6.9.1 Pumps should be centrifugal and single seal sanitary design. Pump construction should be of 316/316L stainless steel (water contact side), with suitable impeller construction and EPDM or encapsulated Teflon gaskets. The pump impeller should be one size smaller than the maximum specified for the casing.

10.6.9.2 Pumps should be connected using sanitary clamp connections for ease of service.

10.6.9.3 Pumps should be specified for use at no greater than 70% of capacity. All pumps should be capable of sanitizing by steam with temperature monitoring to be installed at all pumps. Pump curves are required for all pumps.

10.6.9.4 Standby pumps should not be piped (connected) into the distribution system to avoid the potential of dead legs.

10.6.10 Pure Steam Systems

10.6.10.1 Pure Steam is defined as steam that is produced by a steam generator which, when condensed, meets requirements for USP/EP WFI. At sites that manufacture product sold in the European Union, Pure Steam used to sterilize porous loads should also meet the requirements of European EN285 for noncondensable gases, dryness, and superheat (4).

10.6.10.2 The term Pure Steam is preferred. Use of the term "Clean Steam" is considered to be equivalent and subject to the requirements of this guide.

10.6.10.3 Chemical Free Steam (CFS) is steam produced from pre-treated water with no volatile boiler additives, which may be used for humidification or selected process operations not involving sterilization or processing of sterile products. Systems generating CFS are not required to comply with the other requirements of this section.

10.6.10.4 Pure Steam shall be generated by distillation in a Pure Steam generator unit specifically designed for the purpose. Pure Steam may also be obtained from the first effect of a multi-effect still.

10.6.10.5 Feed water to the generation unit shall meet the requirements of the manufacturer to ensure that condensed Pure Steam will meet the USP/EP WFI monograph.

10.6.10.6 Pure Steam (PS) distribution systems shall be constructed of 316L stainless steel pipe or tubing. Lines shall be sloped to thermostatic type condensate traps. The traps shall allow the system to gravity drain. The distribution system shall be of sanitary welded construction. All fittings shall be sanitary type. Diaphragms valves are not recommended. Sanitary ball valves may be used.

10.6.10.7 Pure Steam or its condensate should not be returned for reuse. Condensate from Pure Steam should be quenched and discharged to drain.

10.6.10.8   Pure Steam sample points shall be provided with a cooling water location nearby for connection to the steam-sampling device. Each use point shall be provided with a valve for sampling of steam using a portable condenser.

10.6.11 Water and Pure Steam Instrumentation and Controls

10.6.11.1   Water and Pure Steam systems shall be provided with automated control systems adequate to ensure repeatable and reliable operation. Where appropriate, these systems may be interfaced with a higher-level control or monitoring system or linked back to a central control system.

10.6.11.2   At a minimum, instrumentation and control systems should include monitoring and/or testing of the following:

- Temperature, pressure, and conductivity monitoring for each component of the pre-treatment system, as applicable.
- Regeneration/backwash of pre-treatment unit operations.
- Sanitization sequence for unit operations after primary treatment.
- Alarms and interlocks for the integration of the pre-treatment, final treatment, and storage and distribution systems, as required.
- "Dump on start" function for WFI stills, based on conductivity of the distillate.
- Storage tank—"dump on high conductivity" as a quality function.
- Temperature control and recording for the storage and distribution system.
- Monitoring of conductivity (online) at the supply to storage tank and on the return from the distribution loop.
- Monitoring of total organic carbon (TOC) at the supply to storage tank and on the return line from the distribution loop. Monitoring may be done online, with the application of a suitable SOP.

10.7 Process Equipment, Piping, and Instrumentation

10.7.1 Equipment Requirements

10.7.1.1   Equipment used in processing of biopharmaceutical drug products and APIs should be of appropriate design, of adequate size, and suitably located for its intended use, qualification, cleaning, sanitization/sterilization (where appropriate), and maintenance.

10.7.1.2   The requirements for process equipment, piping, and instrumentation for a manufacturing system will be based upon User Requirements (UR). This document will be specific to the system being designed and will include quality, business, and other (e.g., safety) requirements from the user perspective. It is not intended to be an engineering or technical specification and will normally not include engineering details unless they are truly required by the user (e.g., to match the process as described in a regulatory filing).

10.7.1.3   An engineering specification will be prepared by an engineer or firm designated by the Project Execution Plan. This will describe technical details in order to meet the User Requirements and is the main document used to communicate with vendors, designers, and/or contractors. This may also be called a technical specification.

10.7.1.4   For systems involving complex automation, especially where a high degree of customization is required, a Functional Requirements Specification (FRS) may be required. In these cases, it is permissible (and even recommended) to separate the mechanical (engineering) specification from the automation (FRS). In most cases of skid-mounted equipment with integral standard automation packages, a single engineering specification will suffice.

10.7.1.5   Design details (e.g., datasheets, isometrics, P&IDs, electrical diagrams, general arrangement drawings, etc.) are developed by an engineer or firm designated by the Project Execution Plan. For specialized production equipment, these details are usually supplied by the vendor.

10.7.1.6   Purchase agreements should affirm the user's right to audit and inspect vendor's fabrication facilities, including quality systems. Vendors shall prepare/present a project quality plan or applicable quality manual for the equipment to be delivered.

10.7.1.7   Equipment must be compatible with the product and process for which the equipment is intended. Product contact materials must be resistant to corrosion and must not contribute to product contamination, either by chemical interaction or by promoting microbial growth. Drug product and biotech manufacture require sanitary design.

10.7.1.8   The equipment and control panels associated with the manufacturing system in Controlled areas should be designed with sloped surfaces when feasible to provide ease of cleaning. Panels in Classified areas should be flush mounted to walls.

10.7.1.9   Equipment access should be provided to allow for ease of cleaning, qualification, inspection, validation, processing, sampling, maintenance, and calibration activities. Special access devices, such as lifters and manways, should be identified in the specification, when appropriate.

10.7.1.10 The lubricants used must be compatible with the product and process. Food-grade lubricants and/or generally recognized as safe (GRAS) substances must be used for drug product, bioprocessing, and final purification and isolation for APIs. Refer to 21CFR178.3570 on *Lubricants with Incidental Food Contact* for a listing of FDA-approved lubricants (10). Equivalent international standards may also be applied.

10.7.1.11 Seals with potential product contact should be suitable to the application and designed to withstand process conditions (e.g., sterilization). The seals should be of minimal wearing/shedding material. Seal lubricating fluid (that could potentially have product contact) must be identified and treated as a potential product contact fluid or material.

10.7.1.12 When equipment is designed for steam sterilization it should be ASME rated for pressure (min. 3.5 bar/50 psia) and full vacuum.

10.7.2 Fixed Equipment—Materials of Construction

10.7.2.1 Product contact materials of construction and surface finishes must be suitable for the product, for the process, and with other materials used in cleaning, sanitization, or sterilization requirements. Product contact surfaces should have a smooth surface free of pits, cracks, crevices, folds, projections, and other imperfections.

10.7.2.2 External materials of construction and surface finishes should be compatible with the surrounding area and/or room classification. They should present a clean and/or polished appearance, be smooth, cleanable, and not subject to deterioration or corrosion. Exterior painted surfaces are not considered suitable for Classified areas and are not preferred in Controlled areas.

10.7.2.3 If painting is necessary on non-product contact surfaces, it should be shielded (as applicable) to separate the product from the painted surface. Chemically resistant, non-chipping, and non-wearing exterior paint systems must be documented (e.g., baked-on powder coat finish or two-part epoxy paint are examples of acceptable painting systems).

10.7.2.4 Vendors are required to provide project-specific certification of product contact materials, including mill certificates where appropriate. In no case should reference to a model number be considered adequate certification. Where mill certificates are not available, the vendor must provide written certification with specific reference to the purchased equipment identified by serial number, if available, or to the bill of materials for a commodity item (e.g., valves).

10.7.2.5 Clean-in-place (CIP) equipment should be considered to facilitate cleaning of product contact areas.

10.7.2.6 Product contact materials should not require a separate finish (e.g., no paint on product contact surfaces). If a coating is required, the coating for product contact surfaces must be resistant to cracking or chipping. When required for corrosion resistance or other surface properties, surface treatments must be specified (e.g., anodizing, passivation).

10.7.2.7 Preferred material for permanent/fixed vessel/equipment for drug product processes is 316/316L stainless steel (316L for interior welded surfaces). All welds shall be polished smooth, and the interior finish shall be Ra 0.5 μm (Ra 20 μ-inch) or better. Where hard-to-clean materials (e.g., protein residues) are processed, electropolished interior finishes may be required. Vendor shall provide stainless steel mill certification, surface profilometer report, and vessel ASME (or equivalent) stamp.

Note: Vendor standard finishes up to Ra 0.8 μm (32 μ-inch) may be accepted when particularly difficult cleaning issues are not expected (e.g., dedicated equipment, soluble actives).

10.7.2.8 Jackets and exterior skin may be 304 SS and shall be finished to a polished appearance, approximately 150 grit (US).

10.7.2.9 Product contact valves design should be appropriate for the intended use and for cleaning (e.g., sanitary valves for drug product processing and bioprocessing). Product contact valve surface finish should be consistent with the surface finish of the attached process equipment. All product contact surfaces of the valve should be cleanable.

10.7.3 Single-Use Equipment

10.7.3.1 The use of pre-cleaned, pre-sterilized single-use systems has expanded greatly in the biopharmaceutical industry in recent years. These system now encompass all major unit operations including bioreaction, separation, clarification, purification, sterilization, and even filling (11). Single-use systems have the advantage of reducing requirements for in-house sterilization and cleaning, resulting in reductions in utility cost. They can also reduce space and HVAC requirements by providing closed systems that may be sited in a "ballroom" layout rather than a fixed configuration required by stainless steel.

10.7.3.2 Materials for single-use equipment must be determined and certified to be compatible with the products and processes involved.

10.7.3.3 Allowance for waste decontamination and removal must be targeted at single-use materials.

10.7.4 Process Piping

10.7.4.1 Process piping materials and internal contact surface finish should be consistent with the associated process equipment.

10.7.4.2 Product contact piping, tubing, and fittings used in sterile drug product manufacturing and bioprocessing must be sanitary in design in compliance with ASME BPE Guide (latest revision) (12). Acceptable designs include sanitary clamp (e.g., Tri-Clamp®) and European Hygienic fittings. Threaded fittings (e.g., IPT, NPT) and flat flanged fittings must not be used in these applications.

10.7.4.3 The process equipment and piping systems should be designed to be self-draining. Process piping systems sterilized in place should be designed to allow for removal of trapped air and condensate.

10.7.4.4 All process piping points-of-use, sampling ports, and access ports used (e.g., electrical, material handling) should be appropriately identified.

10.7.4.5 All product contact gasket and O-ring materials should be suitably resistant to processes (e.g., able to withstand chemical reaction), cleaning solutions, and sterilization. The gasket inspection or replacement intervals should be specified based on the life expectancy of the material and extended use, as appropriate.

10.7.4.6 Preferred O-ring material for sanitary service are encapsulated Teflon, EPDM, steam-resistant Viton, and Kalrez 6221/6230. All elastomers in product contact must comply with 21CFR177, sections 2400(d) and 2600 or equivalent international standard.

10.7.5 Process Instrumentation

10.7.5.1 Instrumentation should be chosen for accuracy and reliability over the entire process range. Factors that affect reliability or accuracy of devices include process conditions, ambient conditions, and location of instruments.

10.7.5.2 Instrument sensors should be installed in locations that provide ease of removal for calibration and validation purposes and are not easily snagged. If fitted in areas where they are prone to damage, instrument sensors should be fitted with protective caps.

10.7.5.3   Product contact materials shall be compatible with the process as well as cleaning and sanitizing materials and conditions. Exterior materials of construction should be similar to the associated process equipment.

10.7.6 Process Automation

10.7.6.1   Identification of all measured or controlled parameters, their range, and the degree of control required must be specified. Critical process and environmental parameters that directly affect product quality will also be identified. Measuring instruments must be accurate enough to reliably measure critical parameters. Preferred accuracy is one order of magnitude better than required operating range (i.e., for temperature control within ± 2.0 °C, sensor should be accurate to ± 0.2 °C)

10.7.6.2   Various control methodologies are acceptable, depending upon the project or site automation architecture and the application. These include stand-alone programmable logic controllers (PLC), PLC with supervisory control and data acquisition (PLC-SCADA), direct digital control (DDC), single-loop control, distributed control systems (DCS), and building management systems (BMS). As a minimum, the control system should:

- measure, control, and record critical process parameters.
- measure and record critical quality attributes, as defined in User Requirements.
- provide specified interlocks, permissives, and alarms. Provide a record of alarms.
- be capable of generating accurate and complete copies of GMP records in electronic form suitable for inspection and review.
- control sequences of operations; allow for selection of multiple recipes or processes as required.
- provide fail-safe shutdown, with memory protection.
- provide logical and physical security to prevent unauthorized access to the system and its data.
- have controls to ensure electronic record integrity and security.
- have an audit trail to provide a record of user actions.

10.7.6.3   All automated systems are subject to a lifecycle development approach, beginning with User Requirements.

10.7.6.4   Process automation systems must comply with regulations compiled in US Code of Federal Regulations 21CFR Part 11 and EU-GMP Annex 11.

10.8 Compressed Air and Process Gases

10.8.1   The system materials of construction shall be consistent with a quality specification and include all of the site distribution and storage systems leading to the use point. They should prevent accumulation of moisture/oil and particle generation. Materials such as copper (with brazed connections) and stainless steel have been proven suitable for the operation. Stainless steel should be used downstream from any final filter into the production area. Valves and fittings should be specified to match the piping material finish and rated to operating pressure requirements. Sample points should be provided.

10.8.2   Where technically feasible, the use of non-oil lubricated air compressors is preferred for product-contact applications.

10.8.3   All compressed gases that are used in contact with the product require the use of equipment designed to remove particles as well as to remove oil or water that may be present in the gas. Additionally, POU filtration is recommended for sterile gas requirements.

10.9 Microenvironments

Microenvironment equipment includes closed units such as stability chambers (environmental chambers), incubators, isolation units, biosafety cabinets, glove boxes, and cold boxes (e.g., refrigerators, freezers), as well as partially open equipment such as laminar flow hoods, fume hoods, and down flow hoods. All microenvironments are designed to provide a specified set of environmental conditions different from or more tightly controlled than the surrounding room conditions.

10.9.1 General Design Requirements

10.9.1.1   The operating unit must specify environmental operating set points (or target value) and ranges for the microenvironment. These parameters are to be documented as part of the User Requirements or other suitable document. Environmental parameters that may be specified include:

- temperature, relative humidity
- level of light or oxygen
- airborne particle count (total and/or viable)

10.9.1.2   Appropriate equipment and instrumentation must be in place to ensure the specified environmental conditions are reliably met and monitored. A disaster recovery plan must be in place for the equipment on site and the associated microenvironment. This will include (where appropriate) emergency/backup power and/or redundancy of equipment in specified circumstances.

10.9.1.3   The design of all new microenvironment equipment should take into account the ergonomics of all operator interfaces. This should consider working height, reach, visibility, lighting, ventilation, cleanability, and change-out of filters.

10.9.2 Partially Open Equipment Design

10.9.2.1   Air ingress (direction, "laminarity," velocity or volume, and intake vs. exhaust) must be specified, as appropriate.

10.9.2.2 The interaction with the room air in which the unit is located must be considered, and appropriate air balancing must be provided.

10.9.2.3 The air filtration requirements must be specified, including appropriate filter change-out provisions.

10.9.2.4 Materials of construction must provide for ease of cleaning/sanitization/decontamination. There should be appropriate vision panels to facilitate visibility of operations. It is preferred that non-vision surfaces that are exposed to the microenvironment be constructed of stainless steel. However, other materials may be used if they are suitably compatible with the use. Exposed painted surfaces are generally not permitted.

10.9.2.5 Instrumentation should include pressure differential and airflow. As appropriate, sash height interlocks should be installed that alarm in event of an opening that exceeds the qualified height.

10.9.3 Cold Boxes, Incubators, and Stability Chambers

10.9.3.1 Temperature, relative humidity, light, and oxygen concentration set point(s) and operating range(s) must be specified, as appropriate.

10.9.3.2 Water/steam quality used for humidification must be specified. Potable water and/or plant steam may be suitable if product is not contacted.

10.9.3.3 Interaction with the room/external environment in which it is located must be considered. As appropriate, temperature and relative humidity differentials must be accounted for.

10.9.3.4 Materials of construction must be cleanable and suitable for the environment. Viewing panels may be installed in the door or wall to facilitate viewing of contents. The viewing panel must be designed such that condensation on the viewing surface is minimized.

10.9.3.5 The equipment must be designed to facilitate ease of maintenance. Non-invasive preventative maintenance is preferred (e.g., to change fans, filters, and lighting). Door gaskets should be designed to permit expedient changing.

10.9.3.6 Instrumentation should include an appropriate number of monitoring probes and justified based on the size of the equipment. The monitoring probe(s) should be separate from the controlling probe. There must be appropriate parameter recording capabilities, preferably continuous. There must be appropriate parameter alarming capabilities (facilities and/or procedural), including audible and visual alarms in the local area.

10.9.3.7 Security and access control must be considered. This is particularly important when controlled drugs are contained in the equipment.

10.9.4 Isolators and Glove Boxes

10.9.4.1 Particle concentration (total and/or non-viable), temperature, relative humidity, and oxygen concentration target values and operating ranges as well as working light levels must be specified, as appropriate to the use of the unit.

10.9.4.2	Air filtration requirements, including recirculation vs. fresh air make-up/ exhaust, must be specified. The ventilation/exhaust system must serve only one isolator/glove box (multiple connected chambers may constitute a single unit) and must include independent inlet and exhaust air systems.

10.9.4.3	Where required (i.e., aseptic isolator), HEPA-filtered and/or unidirectional air should be provided. In these circumstances, a HEPA prefilter should be provided for the supply air mainly to provide backup in the event of a failure of one of the filters. When the isolator will be aseptic, the air-handling unit (AHU) and filtration system must be capable of providing Grade A conditions.

10.9.4.4	There should be HEPA exhaust filters to prevent contamination of the microenvironment in the event of a backflow or to ensure containment when processing toxic or hazardous materials. All HEPA filter systems should be designed to permit integrity testing in place. HEPA filters should be resistant to cleaning solvents used to clean the isolator.

10.9.4.5	Filter maintenance is critical in these units, the "bag in–bag out" design for filter change-out has proven suitable and is recommended when toxic or high-potency materials are handled.

10.9.4.6	In aseptic isolators, a minimum positive pressure differential of 10 Pa (0.04″ WC) under all operating conditions and a minimum differential of 25 Pa. (0.10″ WC) during product processing is recommended. If the requirements justify a negative pressure (e.g., due to operator safety when handling sterile but toxic materials), then consideration should be given to enclosing the unit in a positive pressure envelope (the room itself can constitute the envelope).

10.9.4.7	Isolators with an active pass-through for product flow (e.g., a "mouse hole") must maintain an air velocity of 0.5 to 0.7 meters per second (1.6 to 2.3 feet per second) from the isolator to the outside environment.

10.9.4.8	Isolator gloves/sleeves and half-suits must be rugged and provide for appropriate dexterity, as well as designed to facilitate ease of changing and frequent integrity testing with minimal impact on the contained environment. Both one-piece and two-piece glove/sleeve construction have proven adequate. Materials of construction must be carefully considered and justified based on process and material contact needs. (e.g., Neoprene or Hypalon). Gloves must also be compatible with cleaning agents and not allow pass-through of solvent. It should be possible to integrity test sleeves and gloves.

10.9.4.9	Half suits (when used) should be double walled and provide individual air supply (preferably controlled by the operator) to permit operator comfort. It is preferred that this air be HEPA filtered; however, the specific quality should be established based on process needs. Material of construction should also be evaluated based on process needs (e.g., nylon-lined PVC).

10.9.4.10	Material transfer apparatus (e.g., airlock, rapid transfer port—RTP) must be sized considering the material being transferred, and the port must be designed to transfer material without integrity loss of the microenvironment. Examples of RTP designs that have been successfully used include the Double Porte De Transfert Entanche (DPTE), the CQ Trans Plus ®, Automatic Transfer Valve, The Passport ™, and High Containment Port.

10.9.4.11   The gaskets must be resistant to sporicidal agents (for aseptic isolators) and designed for ease of change-out. Consideration must be given to passage of air between the room and microenvironment (or, as appropriate, between the microenvironment and the room).

10.9.4.12   If an airlock-style transfer port is installed, there must be an interlock (mechanical or automated) to prevent opening of both doors simultaneously. There must be facilities or operational procedures that ensure that operations cease if the interlock fails. Isolation unit design must provide for minimal use of "conventional" doors during non-product exposure operations (these doors are a potential source of contamination) and no use of "conventional" doors during product exposure operations.

10.9.4.13   Materials of construction must be resistant to the operating environment (e.g., heavy solvent load or corrosive conditions), as well as the cleaning/sterilization process (e.g., peroxide used for chemical sterilization). The material of construction must also be resistant to adsorption of chemical sterilants (e.g., polycarbonate has a tendency to adsorb hydrogen peroxide). These considerations also apply to the supporting AHU, the transfer ports, and any other apparatus installed within the microenvironment.

10.9.4.14   If the microenvironment will be operated at a negative pressure, then a flexible material of construction is not recommended. The surface should be smooth with minimal crevices and, as appropriate, rounded corners. The chamber should be equipped with "low point drains" to facilitate proper cleaning. There must be minimal sharp edges inside the equipment to prevent glove tears.

10.9.4.15   Isolators/glove boxes should be located in rooms/areas that facilitate suitably clean and orderly execution of operations. They should be located in a room or area that restricts access to only personnel related to isolator operations. The unit should be located to prevent condensation on the internal walls (e.g., room supply air grille should not impinge directly on the walls of the unit, temperature differentials between room and operating environment, etc.) The quality of air in the surrounding area must be established based on operating needs and worker comfort. Aseptic isolators used to manufacture sterile products must be located in an area designed to at least a Grade C environment. (Note: Area may be designated Grade D in actual use.)

10.9.4.16   If an aseptic isolator will be decontaminated using a chemical agent, then the agent must be sporicidal. Peracetic acid, hydrogen peroxide, and chlorine dioxide are frequently used and may be used without qualification of the material itself (use of the material in the equipment, however, must be qualified). Other agents may also be used if suitably qualified. If the agent is purchased, then it must be received, tested, released, and stored in a manner similar to other GMP materials. If the agent will be generated on site, then the generation equipment must be qualified in a manner similar to other GMP equipment.

10.9.4.17   The equipment should be designed to enable multiple modes of operation (e.g., routine operation, sterilization, cleaning, drying, flushing, etc.) There should be ample provision for compressed gas, steam, chemical decontamination agents, electrical, and so on to match process needs, and the connections should match operating requirements (e.g., sanitary fittings). There should be minimal horizontal surfaces to prevent accumulation of material. The internal design should enable access to all surfaces for cleaning without major dismantling. When temperature or humidity-sensitive sanitization/sterilization processes are used, the equipment must be equipped with appropriate baffling, assist fans, and control systems to ensure uniform distribution as such.

10.9.4.18   Special consideration must be given to exhaust filter design. The potential for chemical contamination during the chemical decontamination cycle and subsequent blowback into the operating environment must be considered. Filters should guard all vacuum ports. In the case of an aseptic isolator, a microbial retentive filter must be used.

10.9.4.19   Clean-in-place systems are preferred and must consider the type of product as well as complexity of the unit. CIP-assist (e.g., difficult to clean certain areas) has also been used successfully. Under all circumstances, the cleaning cycle must consider disposal of the cleaning agent, thus considering use of cleaning agents at high volume at low pressure against low volume at high pressure. The equipment should also provide for an appropriate drying cycle and the use of a drying gas (air or nitrogen) under appropriate circumstances. All surfaces inside the isolator should be free and the isolator should have a drain to remove the CIP agent.

10.9.4.20   Instrumentation should include an appropriate number of monitoring and controlling probes and be justified based on the planned processes (e.g., operational, decontamination, drying, etc.) as well as size of the equipment. The monitoring probe(s) must be separate from the controlling probe. There should be appropriate parameter-recording capabilities (preferably continuous). It is preferred that aseptic isolators be equipped with continuous non-viable particle monitoring instrumentation, and it is preferred that the probe be located at the supply air discharge to the microenvironment.

10.10 Conclusion

Validation is dependent on facility and equipment design that will ensure product quality by providing for the proper environments, avoiding mix-ups, preventing cross-contamination, and ensuring process control. Understanding and applying good design is not enough. Beyond that, the management of a biopharmaceutical facility project must take into account the importance of validation as a necessary and critical milestone. The project team must include key members who are focused on validation and all other aspects of product quality. These members must be involved in all relevant reviews and decisions. This integrated approach has proven to be a significant driver toward overall project success.

Note

1 Humidity may be allowed below the stated range, for example in winter, if process operations allow.

References

1. Independent Project Analysis, *Capital Excellence for Pharmaceutical Projects*, Presentation to Schering-Plough, Nov. 2005
2. European Standard EN285, *Sterilization- Steam Sterilizers -Large Sterilizers*, European Standard EN285, 2015
3. DeSantis, P., *The GMP Project Roadmap*TM, 2021, http://desantisassociates.com/services/project-management/
4. Eudralex, *Good Manufacturing Practices (GMP Guidelines)*, Eudralex, Vol. 4, 2023
5. US Food and Drug Administration, Guidance for Industry, *Process Validation: General Principles and Practices*, US Food and Drug Administration, 2011
6. Independent Project Analysis, *Commissioning and Qualification Best Practices*, Conference Notes, Nov. 2003
7. PIC/s, *GMP Guide for Good Manufacturing Practice for Medicinal Products*, PIC/s, 2023
8. International Society for Pharmaceutical Engineering:
   a) ISPE Baseline® Guide Volume 1: *Active Pharmaceutical Ingredients*, 2007
   b) ISPE Baseline® Guide Volume 2: *Oral Solid Dosage Forms*, 3rd Ed., 2016
   c) ISPE Baseline® Guide Volume 3: *Sterile Dosage Forms*, 3rd Ed., 2018
   d) ISPE Baseline® Guide Volume 4: *Water and Steam System*, 3rd Ed., 2019
   e) ISPE Baseline® Guide Volume 5: *Commissioning and Qualification*, 2nd Ed., 2019
   f) ISPE Baseline® Guide Volume 6: *Biopharmaceutical Manufacturing Facilities*, 2nd Ed., 2013
9. International Standards Organization, ISO 14644, *Cleanrooms and Associated Controlled Environments*
10. US Code of Federal Regulations, 21 CFR part 178.3570, *Lubricants with Incidental Food Contact*
11. Trotter, A. M. and Pendlebury, D., *Single-use Technologies and Systems, Handbook for Validation in Pharmaceutical Processes*, 4th edition, Agalloco, et al., editors, CRC Press, 2022
12. American Society of Mechanical Engineers, *Bioprocessing Equipment*, American Society of Mechanical Engineers, 2022
13. DeSantis, P., *A Framework For Quality Risk Management Of Facilities And Equipment*, Pharmaceutical Online, Jan. 2018, www.pharmaceuticalonline.com/doc/a-framework-for-quality-risk-management-of-facilities-and-equipment-0001
14. DeSantis, P., *Facilities And Equipment Risk Management: A Quality Systems Approach*, Pharmaceutical Online, Dec. 2017, www.pharmaceuticalonline.com/doc/facilities-and-equipment-risk-management-a-quality-systems-approach-0001
15. DeSantis, P., *Facility Design for Validation, Handbook for Validation in Pharmaceutical Processes*, 4th edition, Agalloco, et al., editors, CRC Press, 2022
16. Jornitz, M. and Backstrom, S., *Modular Facilities—Meeting the Need for Flexibility, Handbook for Validation in Pharmaceutical Processes*, 4th edition, Agalloco, et al., editors, CRC Press, 2022
17. Schering-Plough Corporation, *Global Engineering Services Internal Document*, GMP Facilities Design Guide, 2009
18. US Code of Federal Regulations, 21 CFR part 211, *Current Good Manufacturing Practice for Finished Pharmaceuticals*

Appendix—GMP Design Review Template

The following template is provided for GMP Design Review as described in section 10.3.2. The body of the document may be derived from the design guidance provided in sections 10.4–10.9, with initial examples provided as follows. The overall format of the document may be constructed like a validation protocol or report following the user's conventional format. It should include:

- Title page
- Approvals
- List of GMP Review participants, including affiliation (may include third parties)
- List of documents reviewed
- Matrix/questionnaire (example following)

Requirement	Observation	Notes/ Actions

1   Facility Design Requirements

1.1   General Considerations

1. Corridor dimensions and door openings shall take into account the movement of large equipment and materials. Appropriate wall protection (e.g., bumpers, railings) should be utilized where necessary to protect against damage to equipment, walls, and doors. Mobile equipment should also be designed to minimize damage to the facilities.

2. Operations will be located and performed within specifically defined areas of adequate size. There shall be separate or defined areas as are necessary to prevent contamination or mix-ups during the course of the following procedures:
   - Receipt and holding of materials, components, and labeling pending approval for use
   - Holding rejected materials, components, and labeling before disposition.
   - Storage of released components, in-process materials, and drug products after release and labeling.
   - Manufacturing, packaging, and labeling operations.
   - Quarantine storage prior to release of drug products or APIs.
   - Control and laboratory operations.
   - Sampling of intermediates, APIs, and finished or unpackaged drug products.
   - Personnel dress/gowning, hygiene, and food consumption.

3. In areas where potent or toxic materials are processed, consideration should be given to enhanced cleanability. These areas should have a separate HVAC system.

4. In Controlled and Classified areas, the walls, floors, ceilings, etc. should be constructed so that the primary surfaces are accessible for cleaning. Exposed horizontal surfaces in Controlled areas are generally acceptable if they are accessible for cleaning. They should, however, be sloped to allow free draining.

5. Horizontal surface should be avoided as much as possible in Classified areas. Windows, doors, and fixtures should be flush with primary surfaces.

11

Process Validation at Contract Manufacturing Organizations: Approaches, Incentives, Benefits, and Risks

Maria Wik and Scott Rudge

11.1 Introduction

Pharmaceutical companies must have robust quality systems in place to ensure drug products meet required quality standards and are manufactured to meet health authority requirements. This applies to innovators, drug sponsors, and contract manufacturing organizations (CMOs) alike. CMOs are seen by regulatory agencies as an extension of the company that contracts with them.

Process Validation is a critical aspect of the overall quality system, and effective Process Validation contributes significantly to assuring drug quality. *Process Validation* is defined as the collection and evaluation of data, from the process design stage through commercial production, which establishes scientific evidence that a process is capable of consistently delivering quality product (1). Process Validation involves a series of activities taking place over the life cycle of the product and process. The US Food and Drug Administration (FDA) describes process validation activities in three stages (1) as shown in Figure 11.1.

Each stage and its key objective is described in Table 11.1.

The three stages of Process Validation are to be applied to pharmaceutical manufacturing whether the stage is performed by the drug sponsor, by a contract manufacturing organization, or a combination. Stage 1 is generally executed by a development organization (in-house or contracted). Stages 2 and 3 are generally executed by a manufacturing organization (in-house or CMO) with all stages having input and leadership provided by a stand-alone or integrated Validation organization, and all stages having appropriate Quality oversight.

While a manufacturing organization may have different priorities from the development organization, whether in-house or contracted, contract manufacturers and sponsor organizations have specific incentives that should be considered when developing an ultimately successful validation program. In this chapter, the different incentives, approaches, benefits, and risks encountered in Process Validation planning and execution, specifically when working with CMOs, will be explored.

DOI: 10.1201/9781003143130-11

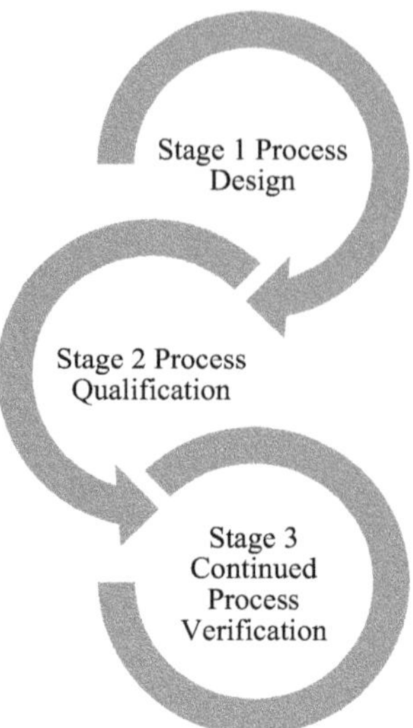

FIGURE 11.1
Process Validation stages and life cycle.

TABLE 11.1

Process Validation Stages and Key Objectives

Stage	Description	Key objective
1	Process Design	The commercial manufacturing process is defined based on knowledge gained during development and scale-up activities.
2	Process Qualification	The Process Design is evaluated to determine if the process is capable of reproducible commercial manufacturing.
3	Continued Process Verification	Ongoing assurance is gained during routine production that the process remains in a state of control.

11.2 Stage 1 Validation: Process Design

11.2.1 Definition

Process Design is the activity of defining the commercial manufacturing process that will be reflected in planned master production and control records. The goal of this stage is to design a process suitable for routine commercial manufacturing that can consistently deliver a product that meets its quality attributes. Generally, early process design experiments do not need to be performed under the cGMP conditions required for drugs intended for commercial distribution that are manufactured during Stages 2 and 3. Experiments should, however, be conducted in accordance with sound scientific methods and principles, including good documentation practices (1).

Stage 1 consists of two substages:

- Building and capturing process knowledge and understanding
- Establishing the strategy for process control

11.2.2 Industry Approach

Stage 1 studies are generally performed at small-scale laboratories using qualified scale-down models (2). Process information from earlier product development activities, platform technologies, and engineering first principles are leveraged in the process design stage. A risk tool (3), such as a preliminary hazard analysis (PHA), risk map (4), or failure modes and effects analysis (FMEA) is performed to identify important process parameters using process development and manufacturing experience to date (5). A quality target product profile (QTPP) and cross-functional team input on rank order of critical quality attributes (CQA) are key inputs to the FMEA. If the manufacturing site is selected when the FMEA is performed, it is beneficial to include subject matter experts (SMEs) that are familiar with the manufacturing facility and equipment control capabilities during the FMEA. Similarly, Analytical Development/Quality Control personnel that have insight into analytical method testing capabilities should be consulted as well.

Based on the FMEA, process parameters (process inputs) are prioritized for studying in single-variable experiments, fractional factorial design of experiments (DOEs), response surface DOEs, and worst-case experiments. Appropriate analytical methods based on a CQA list need to be in place to perform testing during the characterization experiments.

Based on the FMEA and subsequent experiments, process parameters are classified as critical and noncritical (key and non-key), and their control ranges are defined. Based on the understanding of how the process parameters impact performance parameters (process output/performance) and CQAs, the control strategy is developed. The control strategy becomes the basis for master batch records.

Care should be taken when conducting the FMEA to correctly classify occurrence or probability of likely failure modes. The pH and conductivity of a solution should be highly controlled if matched salts and acids/bases are used to establish the pH and the balance used to weigh salts has the accuracy required by the USP. Transient excursions from normal operating ranges (NORs) or proven acceptable ranges (PARs) are more likely, such as a slug of cleaning solution being trapped in a dead leg or by an ill-considered valve sequence that creates a transient spike in pH far beyond the limit it would be useful to study. A manufacturer that consistently makes and uses buffers that are out of pH and conductivity specification has operating issues that a DOE-based process characterization study will not correct.

Equally unlikely is the pH or temperature of a fermentation or cell culture being incorrect by a constant offset throughout the duration of the step. Again, transient spikes are much more common but do not lend themselves to efficient study by statistical methods, as spikes at different times during the unit operation may have very different impacts. The path to increased quality in biomanufacturing is probably not through the widening of parameter specifications and the justification of deviations but through the hardening of calibration standards, process controls, and manufacturing procedures, in order to stay easily within a narrow range.

The risks contained in the process FMEA should be updated at the end of process characterization and residual risks above the risk tolerance threshold transferred to a corporate risk register. The continued steps in quality risk management of risk control, risk communication, and risk review should be applied to these residual risks.

11.2.3 Approaches, Incentives, Benefits, and Risks

The Process Design activities are typically owned and led by personnel in a development organization, however cross-functional input is critical. Functional areas and SMEs that should be represented during Process Design planning, execution, and closure are listed

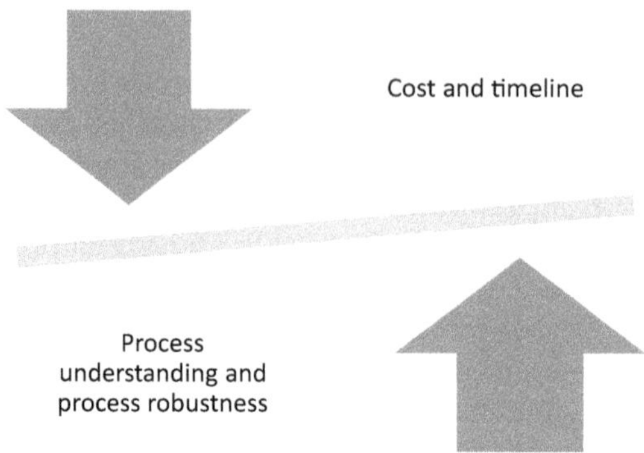

FIGURE 11.2
Balancing incentives during Stage 1.

TABLE 11.2

Functional Areas and Role during Stage 1

Functional Area/SME	Role during Stage 1
Development	Lead the effort. Transfer knowledge from Research and Discovery groups to understanding of molecule function and design.
Manufacturing	Provide input on facility and equipment capabilities and limitations to ensure process control ranges studied can robustly be met during Stages 2 and 3.
Analytical Development/ Quality Control	Provide input on analytical method testing capabilities and limitations to ensure methods are suitable for use and sample matrices ultimately required to be qualified are included in analytical method validations.
Regulatory	Provide insight on overall regulatory submission strategy.
Validation	Provide Validation leadership and ensure necessary Validation documents are planned for.
Quality	Provide Quality oversight appropriate for stage and ensure readiness for Stage 2.
Statistics	Provide input on design of experiments, number of experiments, and data evaluation as needed.

in Table 11.2 along with the key input they should provide during this stage to ultimately ensure successful Process Validation.

This cross-functional approach is necessary regardless of how the project is executed. Project execution and delivery can be based on a number of combinations of options listed below:

- Process developed in-house by sponsor, Process Design by sponsor, and manufacturing by sponsor
- Process developed in-house by sponsor, Process Design by sponsor, and manufacturing by CMO
- Process developed by a contract development organization (CDO), Process Design by CDO, and manufacturing by CMO (same organization as CDO or different)

During Stage 1, the various key stakeholders will bring different viewpoints, priorities, and key objectives to the table. Each function or organization is driven by different incentives.

Clearly, the ultimate common goal is successful overall Process Validation to bring a new product to the patient. However, it is important for the team to understand the different incentives from the various players to adequately assess, address, and execute the Process Design phase well.

Manufacturing organization incentives:

- Study wide enough process parameter ranges to avoid as many deviations as possible (cost avoidance for deviation investigations during Stage 2 and Stage 3).
- Study wide enough process parameter ranges to reduce risk of batch rejection during Stage 2 and Stage 3 (significant cost avoidance for rejected batches).

Development organization incentives:

- Gain in-depth process understanding.
- Maintain design and function of product as originally intended.
- Deliver Process Design within a timeline that meets overall commercialization timeline. Process Design stage can become a rate-limiting step, especially for break-through designation or fast-track programs. The Development organization will need to balance the need for in-depth process understanding and timeline.

It is important to transparently discuss these different incentives and find a balance. With the Validation lifecycle concept, there is also an opportunity to continuously improve and adjust ranges throughout the product life cycle.

The use of Process Design to avoid deviations in the manufacture of the product must be examined skeptically. Although the intent of detailed process understanding is to ensure knowledge of cause-and-effect relationships between process parameters and quality attributes, the method of knowledge acquisition (through statistical design of experiments) quite often leads to scattered findings, which can be interpreted falsely as "lack of relationship". Generally, range-finding experiments are conducted to ensure that there is no effect on product quality across a range for one or more specific parameters. However, the rejection of the null hypothesis in these experiments is misleading. The rejection of the null hypothesis means that a linear or sometimes binomial cause-and-effect relationship cannot be certain with 95% confidence in the range investigated. This does not mean such a relationship does not exist. An experiment designed to find if a causal relationship exists between parameter X and quality attribute Y cannot prove that no effect exists. In many cases, there are unit operations models in existence that predict an effect will be found, but the extent of the effect is dependent on the particular biology and/or chemistry of the specific process. Therefore, experiments that take the models into account should be undertaken to understand the magnitude of the effect, rather than statistical experiments designed to conclude the effect can be ignored. These can be done in the form of one factor at a time studies, or studies that include the engineering model as the statistical basis for analysis.

In some cases, the elements of the product optimized in research and development during lead discovery and optimization are lost in the search for a deviation-free manufacturing process. Lead selection for biologics typically involves designing a molecule for compatibility with the target organ or tissue, optimizing glycosylation (for example) and isoelectric point for formulation. Quite often, the features of a biomolecule are engineered into the host cell line that will produce it. This functionality needs to be confirmed in

Process Design through frequent communication with the discovery scientists. Making the link between product design and process design is an essential feature of Quality by Design that is often overlooked.

The confirmation of these molecular design attributes is difficult with statistical experiments. This confirmation must be obtained with detailed analytical procedures and knowledge of the up-regulation and down-regulation of specific genes during the manufacturing process. With expanded analytical capabilities in gene detection and protein characterization, these factors can be known as conditions progress through cell culture or fermentation. The product developer has every incentive to understand these attributes and to know how they are affected by manufacturing conditions.

11.3 Stage 2 Validation: Process Qualification

11.3.1 Definition

During the Process Qualification (PQ) stage of Process Validation, the Process Design is evaluated to determine if it is capable of reproducible commercial manufacture. During Stage 2, cGMP-compliant procedures must be followed (1).

Stage 2 consists of two substages:

- Design of Facility and Qualification of Utilities and Equipment
- Process Performance Qualification (PPQ)

Process Performance Qualification combines the qualified facility, utilities, equipment, and trained personnel with the commercial manufacturing process, control procedures, and components to produce commercial batches. Success in this stage confirms Process Design and demonstrates performance of the commercial manufacturing process.

11.3.2 Industry Approach

Design and qualification of facility, utilities, and equipment can oftentimes happen in parallel with the Process Design stage. In many cases it has already been completed if an existing facility and equipment will be used. The more that is known about the Process Design during the facility, utility, and equipment qualification activities, the better. This will ensure that the process can be operated in alignment with the design and qualification of the facility, utilities, and equipment. If the process is going into a facility that has already been qualified, an assessment is needed to ensure that the existing qualification is adequate.

Ideally, validation studies that are not process specific can be decoupled and not executed during the PPQ substage to reduce complexity during execution of the PPQ. These include:

- Mixing validation
- Media and buffer hold times
- Cleaning validation (when bracketing approaches are used)
- Bioburden and endotoxin control

Also, process validation studies that are not scale dependent may be decoupled from the PPQ batches. These studies include:

- In-process hold times
- Extended generation studies
- Resin and membrane cleaning
- Resin and membrane reuse

Depending on how the facility and company operates, however, this may not be possible. Parallel execution can increase the risk due to the increased complexity (number of samples, additional processing that is not part of routine manufacturing, etc.). The simplest Process Performance Qualification is the best Process Performance Qualification.

A case can be made to conduct all these studies separately from Process Performance Qualification batches. The facility-specific studies should be the responsibility of the manufacturing organization. These can be generic and apply to numerous products. Specifically, studies demonstrating that the facility has control over exogenous contaminations should be conducted by the manufacturer. There is rarely a process-specific cause for exogenous contamination. The contract manufacturer has some incentive to require its customers to pay for customized studies on operations under facility control. Customers should resist these studies. Contract manufacturers will be forced to absorb those costs. However, studies conducted on a generic level can be used as a competitive cost advantage to attract more business.

To accomplish the decoupling, for stainless steel facilities generic and worst-case soils can be identified and cleaning validation studies performed outside of manufacturing production windows if the facility can allocate time for this purpose. Coupon studies can also be leveraged to further decouple or reduce the amount of large-scale testing.

Similarly for mixing and media/buffer/in-process pool hold times, representative media or solutions can be identified and studies can be carried out outside of manufacturing production windows. For hold times, small-scale studies can be done during the Process Design stage to determine chemical stability. The at-scale work, ideally using representative media/solutions outside of the manufacturing production window, can be limited to establishing hold times to demonstrated microbial control.

Use of single-use technology further simplifies the validation, in that cleaning validation is not required and hold time validation can be completed outside of production altogether. Representative materials and environments need to be considered when executing hold times studies but can typically be done outside of production areas and at smaller scale.

Process-specific studies that are not scale dependent can also be conducted outside of the manufacturing area. This can reduce cost as additional steps not normally taken during routine manufacturing that extend time in the plant can be avoided. These process steps include holding of intermediates for stability testing, running blank gradients for resin cleaning studies, and incorporating additional expansion steps for extended generation studies. Manufacturers have been able to derive extra revenue from these studies, driving up the cost of PPQ batches. Taking these studies out of manufacturing and back into the laboratory aids in simplifying the PPQ exercise. Presuming the scale-down model is adequate, there is no scientific need to repeat these studies at large scale, although many do.

Membrane and resin-reuse and -storage studies can also be executed with a combination of small-scale studies and to avoid needing to run blank gradients at scale, which creates the need for additional processing steps. The process can be designed such that suitable transitions coming out of storage can be used to obtain a sample and perform necessary testing traditionally performed on samples from blank gradients.

Another approach to leverage to simplify PPQ execution is by using bracketing approaches when complex variables, stock-keeping units (SKUs), or protein concentrations are involved.

During the PPQ execution, which traditionally requires a minimum of three consecutive batches (2) if filing globally, the focus on flawless execution becomes a critical component. The hand-off of process understanding gained during Stage 1 must be robust from the development personnel that have led the efforts during Stage 1 to the manufacturing and quality control staff that takes over the execution of Stage 2 activities. Effective communication and careful planning is instrumental. The more familiar manufacturing personnel are with the process, the better the chance of success during PPQ execution. Direct prior experience via clinical manufacturing or engineering runs can be very beneficial. Development personnel should also spend time on the floor to support the knowledge transfer. No samples can be missed, and the focus during the PPQ execution is data generation as much as it is product generation. All data required for subsequent filing must come from the PPQ execution and ideally in as few batches as possible.

Reducing the number of studies conducted on the manufacturing floor reduces the samples taken and the data generated to a minimum. This benefits everyone but the accountants who like to profit from full-scale manufacturing studies. A focus on fundamental understanding of the process in Stage 1 can help limit the studies required in Stage 2, as the Stage 1 studies focus more on process fundamentals and less on statistical approaches.

11.3.3 Approaches, Incentives, Benefits, and Risks

Process Performance Qualification activities are typically owned and led by personnel in the manufacturing organization. However, cross-functional input is critical. Functional areas and SMEs that should be represented during Process Qualification planning, execution, and closure are listed in Table 11.3 along with their key inputs needed during this stage to ensure ultimate successful Process Performance Qualification.

TABLE 11.3

Functional Areas and Role during Stage 2

Functional Area/SME	Role during Stage 2
Development	Train manufacturing personnel, provide person-in-plant support, ensure process runs as designed.
Manufacturing	Execute PPQ batches, obtain all samples, documentation of all data required.
Analytical Development/ Quality Control	Testing of all samples. Typically, a higher volume of testing that needs to be planned for during execution of Stage 2.
Regulatory	Provide insight on overall regulatory submission strategy.
Validation	Provide Validation leadership, ensure validation procedures are adhered to, and ensure necessary validation documents are generated.
Quality	Provide quality oversight appropriate for stage and ensure readiness for Stage 3.
Statistics	Support data evaluation to start preparing for Stage 3 process monitoring.

Manufacturing organization incentives:

- Simple execution of PPQ. Perform as few side studies and non-routine processing as possible.
- Processing step times and hold times established during PPQ to allow for future manufacturing flexibility and to level load resources.

Development organization incentives:

- Gain additional process understanding (especially if there has been no, or limited, prior at-scale experience). This can require additional samples, processing steps, and parallel studies.

These opposing needs between the manufacturing and development organizations should be carefully balanced to ensure successful delivery of the PPQ data package.

Special considerations when PPQ is executed at a CMO:

- Process familiarity: Sponsor (or another CMO) likely will have slightly different process platform technologies. A balance between Process Design requirements and facility/equipment capabilities and staff familiarity with technology will need to be struck to ensure success. Again, the sooner the manufacturing organization that will execute the PPQ can get involved with the Process Design activities, the better this can be managed. Ideally, the manufacturing organization's support laboratory will conduct the Process Design studies so that equipment capabilities are taken into account.
- Document and nomenclature familiarity: CMO and sponsor (or other CMO) likely will have slight differences in nomenclature (for instance naming of samples, sampling time points, etc.), and details like this cannot be overlooked. Sponsor and different CMOs will also likely have differences in document templates and, from a flawless PPQ execution perspective, it is best to use templates that the manufacturing organization executing the PPQ is most familiar with, and this could mean that the sponsor as a critical reviewer of documents will need to spend some time getting trained on and familiar with that CMO's documents.

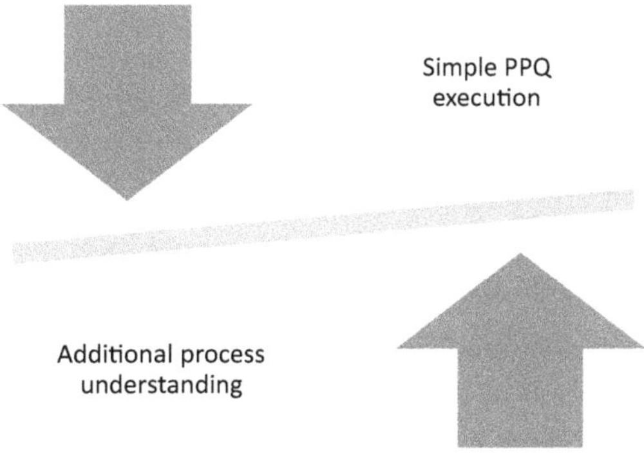

FIGURE 11.3
Balancing incentives during Stage 2.

- Real-time data review: It is critical that systems are set up to allow for rapid and transparent data sharing between CMO and sponsor so that there are no delays in understanding any potential process issues encountered.

Having robust Quality Agreements and Master Service Agreements in place that address the different incentives and special considerations is important to help ensure successful overall Process Validation.

11.4 Stage 3 Validation: Continued Process Verification

11.4.1 Definition

The goal of the third validation stage is continual assurance that the process remains in a state of control (the validated state) during commercial manufacturing. Evaluating the performance of the process identifies problems and determines whether action must be taken to anticipate, correct, and prevent problems so that the process remains in control (1).

Continued Process Verification (CPV) is a program rather than an event. This should be primarily a manufacturing responsibility, but the development organization has input as well. This stage requires routine monitoring of the manufacturing process to identify any trends affecting product quality attributes. Monitoring under CPV includes deviations, complaints, process changes, the use of statistical tools such as control charts, and Nelson rule violation evaluations. CPV helps detect potential product quality issues based on historical evaluation of product-related quality attributes and process parameters.

11.4.2 Industry Approach

Additional parameters or attributes not considered critical to quality or otherwise not specified in the control strategy may be included in the CPV program to enhance process learning and support investigations to identify root causes and sources of unexpected variability. A CPV signal is designed to identify potential new variation or unexpected patterns in the data. Signals are defined as classic statistical rules to evaluate data for outliers, shifts, and drifts in performance (6). Because these conditions are within the control strategy, signals should not automatically be considered formal process deviation investigations, but signals can be escalated to become a formal deviation within the Quality Management System. CPV can also identify opportunities to improve process performance and/or optimize process controls. The cross-functional team that has been involved throughout Stages 1 and 2 will meet at regular intervals to review the process performance and determine actions.

11.4.3 Approaches, Incentives, Benefits, and Risks

CPV activities are commonly owned and led by personnel in a manufacturing, development, or validation organization based on the overall organizational structure. Commonly there is both support provided by statistics SMEs and cross-functional review input.

There are fewer competing priorities between functions and/or sponsor and CMO during this stage. Underlying differences in priorities can be to the extent that sponsor vs.

CMO wants to drive process improvements dependent on the nature of the improvement. If process failure rates are high, both sponsor and CMO will likely want to drive the improvements. Process simplifications that could lead to fewer resources being needed and/or more manufacturing and resources flexibility could be more appealing to the CMO organization than to the sponsor. Process yield improvements could be more of an incentive for the sponsor than the CMO dependent on compensation structure of the business contract.

11.5 Conclusions

To ensure an overall successful Process Validation, it is imperative to take an integrated team approach. Expertise from a variety of functions (development, manufacturing, statistical SMEs, quality, validation) should be involved throughout all three stages. In this chapter, we have reviewed the industry best practices and incentives for the three stages of Process Validation from the point of view of both the sponsor and CMO.

The need for handshakes and transparency of different viewpoints is true for both internal Process Validation activities executed exclusively in-house as well as for projects where there is a sponsor and a CMO involved. In the latter, it is critical to have effective Quality Agreements and Master Service Agreements in place that drive transparency and alignment.

References

1. Guidance for Industry, Process Validation: General Principles and Practices, U.S. Department of Health and Human Services, Food and Drug Administration, January 2011, Revision 1.
2. Guideline on Process Validation for the Manufacture of Biotechnology-Derived Active Substances and Data to Be Provided in the Regulatory Submission, European Medicines Agency, 28 April 2016, EMA/CHMP/BWP/187338/2014.
3. International Conference on Harmonization, Q9 Quality Risk Management, Rockville, MD: US Department of Health and Human Services, Food and Drug Administration, November 2005.
4. B. Andreasen, J. Blasi, H. Fabritz, J. Feldthusen, N. Guldager and G. Moelgaard, "Risk-MaPP, ICH Q9, ASTM 2500 in Action: Project Advantages of Practical Quality Risk Management Approaches", *Pharm Eng.*, 1–7 (2011)
5. Biopharmaceutical Manufacturing Process Validation and Quality Risk Management, Pharmaceutical Engineering, May 2016.
6. Douglas C. Montgomery, *Introduction to Statistical Quality Control* (5th ed.), Hoboken, NJ: John Wiley & Sons, 2005, ISBN 978-0-471-65631-9, OCLC 56729567.

12

Validation of a Filtration Step

Jennifer Campbell, Anissa Souidi, Guillaume Lesage, Renato Lorenzi, and Pascale Richert

12.1 Filtration Validation Overview

Process validation is defined as the collection and evaluation of data, from the process design stage through commercial production, which establishes scientific evidence that a process is capable of consistently delivering quality product. Process validation involves a series of activities taking place over the life cycle of the product and process (1). The lifecycle concept links product and process development, qualification of the commercial manufacturing process, and maintenance of the process in a state of control during routine commercial production. It is an assurance that a process is robust and reproducible and will consistently produce a product that meets specifications. Sources of variation must be identified, and these variations must be controlled and monitored. Final testing of the product is not sufficient to assure quality (2–4).

Filters have functions relating to flow rate, throughput, sterilizability, organism or particle retention, extractable levels, particle shedding, product stability, compatibility, toxicity, non-pyrogenicity, and thermal and pressure tolerance. The manufacturer is best equipped to assess some of these functions, and most manufacturers document claims of sterilizability, lack of toxicity and non-pyrogenicity, maintenance of integrity under pressure, extractable levels, particle shedding, organism retention, and air and liquid flow rates as a function of pressure. The user may accept the validation claims from the filter manufacturer. However, the responsibility for validation rests with the user (5).

If a sterilizing grade filter is used to sterile filter the product in a process, then the filter must be validated to be sterilizing grade. However, the same filter could be used to filter a process intermediate to remove particulate and reduce bioburden, but it is not intended to sterilize the process intermediate. In this case, the filter does not have to be validated to be sterilizing grade. Instead, it must be validated to remove an adequate amount of particulate and bioburden to ensure optimal performance of the downstream operations. Only the claims made regarding the performance of a filter device must be validated. A method must be developed to validate the claim, and this constitutes validation protocols.

12.1.1 Phase of Validation

As a drug moves through the different phases of manufacturing, emphasis on validation increases. Phase I is performed at the laboratory scale. Limited process data exist and the drug is in clinical testing. At this point assays may not be well developed and the process

DOI: 10.1201/9781003143130-12

is not well defined. The final formulation may not be set. Key concerns are filter membrane selection, chemical compatibility, product and/or preservative binding, and assay validation. Membrane compatibility screening determines if filter materials are compatible with the process fluid to be sterilized. Any effects of the filter on the product formulation need to be described, such as adsorption of preservatives, active drug substances, and extractables.

Phase II is typically performed at pilot scale. The process is more defined and assay development is progressing. Further scale-up may be required, and during scale-up process operating parameters such as differential pressure and flow rate, temperature and filtration time must be evaluated. Product yield should be evaluated. In this phase it is strongly recommended to have the bacteria retention performance evaluated especially for challenging/difficult-to-validate products such as formulations associated to liposomes or emulsions. Integrity testing of critical filters should be documented.

Phase III is typically performed at large manufacturing scale. Process validation is a requirement, as the product will be released for manufacturing if its biologics licence application (BLA) is approved.

The validation criteria will define the acceptable ranges of critical parameters in the process, such as pressure, flow rate, temperature, and processing time. This chapter will focus on the validation of the critical parameters of filtration unit operations, as well as the cleaning requirements and sampling plan for reuse filters such as tangential flow devices.

12.1.2 Scales of Validation

The process scales at which validation occurs may differ. Certain operations may be validated using scale-down studies. This is advantageous as a cost savings and sometimes as a safety consideration. In the case of viral spiking or microbial challenge studies, scale-down experiments are often performed for safety reasons. When performing scale-down studies, it is important to mimic the process-scale conditions as closely as possible (see Figure 12.1). Despite differences in volumes, the scale-down studies should emulate the holding times, mixing times, and transfer times of the manufacturing-scale process. Maintaining these times reduces differences in product quality between the two scales. For example, if the manufacturing process incorporates an overnight hold step at 4 °C, this step must be performed in the scale-down study. However, certain aspects cannot be duplicated, such as heat transfer rates, surface-to-volume ratios, and pumping rates.

Properties not affected by the scale of operation, such as compatibility, extractables, and cleanliness, can be tested at small scale. Product assays measuring the effect of the filtration on the product can be performed at small scale. Bacterial retention and viral retention testing can be performed at small scale as long as the full-scale process is modeled properly. Full-scale validation should demonstrate that the scale of operation does not alter the product quality or filtration process. The filter validation can be integrated with the process validation (6).

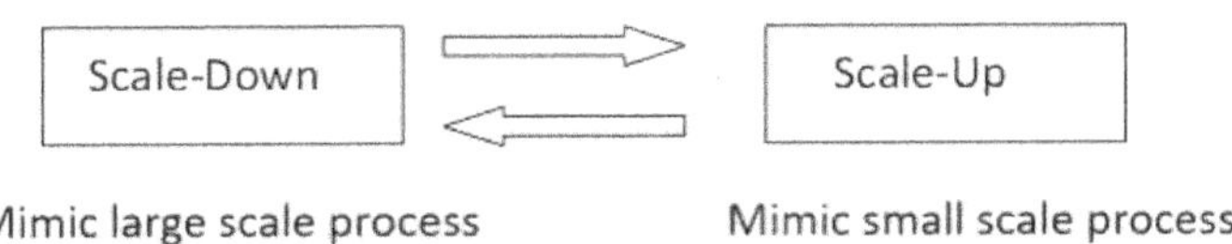

FIGURE 12.1
Scale-up and scale-down.

12.2 Sterilizing Grade Filter Validation

The sterilizing grade filter validation program derives from the pharmacopeia parenteral preparation monography that underlines the drug product critical attributes that include sterility, particles levels, critical excipients concentration, batch homogeneity, drug identity, purity, bioactivity, and stability.

This validation program focuses primarily on filter retention performance with bacteria in the drug product and on integrity testing of the filter that both address the sterility assurance aspect. Other considerations derive from the large internal product contact surface area of a membrane filter with the expectation that it is not reactive, additive, or absorptive to such an extent that it would affect the quality of the filtered product.

These elements are captured in Table 12.1. However, validation of the filter for its intended use is the drug manufacturer's responsibility.

12.2.1 Regulatory Environment

Following the initial requirement for filter validation in 1987 with the FDA guidance for sterile drug product produced by aseptic processing, most regulatory authorities around the world are now calling for the same level of qualification. The prominent references when validating a critical filtration step used in sterile drug product manufacturing are listed in the references section (7–10, 11).

12.2.2 Bacteria Retention Test

The sterilizing grade filter primary function is to sterilize the fluid by the capture of microorganisms by size exclusion. The membrane retention performance is evaluated during a process simulation at laboratory scale combining the actual sterilizing membrane exposed to high levels of a diminutive bacteria and the drug product for highest levels of processing parameters that include duration, pressure differentials, flow rates, loading (volume/cm^2), and temperature.

TABLE 12.1

Responsibilities and Contributions of the Filter Manufacturer and the Filter User in Sterilizing Filtration Validation

Filter Validation Activities	Filter User	Filter Manufacturer
Bacteria retention in drug product with simulation of worst-case processing conditions	✓	Service
Chemical compatibility verification	✓	Documentation + Service
Extractables and leachables documentation	✓	BioPhorum Operating Goup/USP 665
— Drug product interaction	✓	available extractable data dossiers +
— Patient safety evaluation	✓	Leachable service—Service
Line flushing design		
E&L removal monitored by TOC & conductivity	✓	✓
Transmission/adsorption studies of API and excipients	✓	Process simulation and sampling
Filtration system sterilization validation	Autoclave, steam in place	Gamma irradiation
Filter integrity test limits with product	✓	Service
Particulate matter compliance	✓	-

A sterilizing grade filter is a filter that reproducibly removes test microorganisms from the process stream, producing a sterile filtrate. The ASTM® F838 standard describes the filter manufacturer protocol for the retention test of 10^7 CFU/cm^2 of *Brevundimonas diminuta* ATCC® 19146. In the biopharmaceutical industry, a 0.2 or 0.22 µm pore size rated filter is typically used as a sterilizing grade filter.

Typically, the challenge bacteria for retention validation is *B. diminuta* and it is cultured according to ASTM® F838 standards to ensure organism viability, diminutive size, and monodispersion. Nevertheless, process bioburden should be evaluated in order to establish if a smaller bacteria than *B. diminuta* exists and in such case be used as a challenge bacteria.

The viability of the test organism is evaluated in the drug product in a preliminary test. If viable, then the challenge organism is resuspended into the product for the process simulation. This preferred method enables to evaluate the bacteria and product interaction and any product impact on the capture mechanism.

Nevertheless, some drug products may contain components that are inhibitory to the challenge organism or some processes conditions are bactericidal to the challenge organism. If this is the case, the bacterial challenge using the inhibitory drug product is not valid. Several modified methods may be used for the bacteria retention validation in the case of bactericidal effect, as detailed next:

> The filter can be preconditioned with the drug product and then challenged over a reduced yet significant period of time where the bacteria are viable.

Another approach is to precondition the filter with the drug product, then flush it to remove the drug product and challenge it with the appropriate organism in a modified solution (drug product without the bactericidal component or with a modified pH) or with a product surrogate. In any case, the filter is exposed to the drug product to prove that the product does impact the membrane and change the pore sizes.

Lastly, the filter can be preconditioned with a drug product at the process conditions and then challenged with bacteria in milder process conditions such as with modified temperature.

It is recognized that there are many options to design a process simulation for the bacteria retention test. The following is a generic schematic of the setup used for the bacteria retention test.

Figure 12.2 shows the bacteria retention test setup schematic: a pressure-dispensing vessel with drug product contaminated with *B. diminuta* delivers the fluid through a manifold to three test filters and one size control filter. The filtrates are collected into individual bottles and their entire volume is assessed for sterility by membrane filtration.

Success criteria for the bacteria retention test of the filtration system are:

1. Challenge level: the minimum challenge level of 10^7 CFU/cm^2 of effective filter area is demonstrated from viable microorganism enumeration over the test duration and filtered volume data.
2. Test organism is shown to pass through a 0.45 µm membrane.
3. The challenge is conducted under conditions that simulate the worst-case processing conditions.
4. Filtrate sterility assessment method by membrane filtration is qualified to recover small numbers of the test organism.

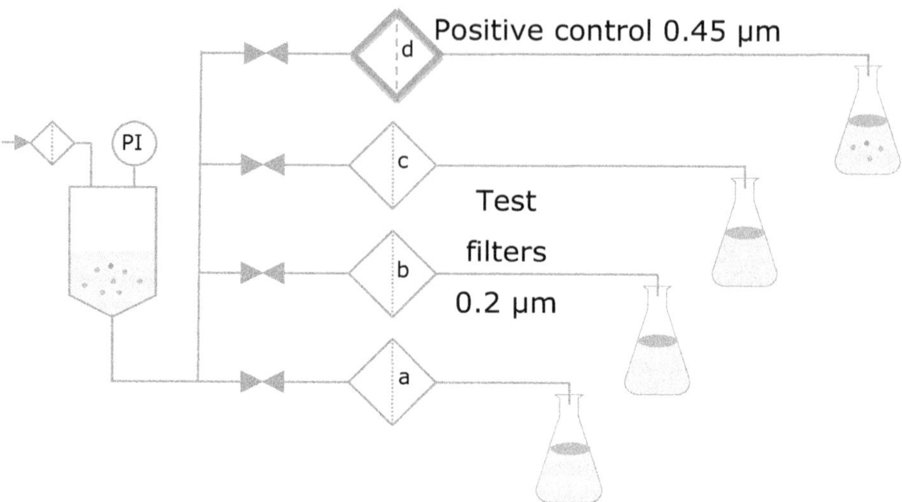

FIGURE 12.2
Bacteria retention test setup schematic.

5. One test filter shall have a pre-use physical integrity test value at or near the filter manufacturer specification.

6. Retention capability: all test filters retain all the challenge test organism. The entire filtrate has been assessed for sterility.

The pre-use post-sterilization integrity test (PUPSIT) is incorporated into the simulation for processes seeking EU GMP compliance. Worst-case conditions are now more reasonably set so the simulation resembles the actual process. In particular, the widespread intermittent filtration processes are more formally simulated with active filtration cycles and idle times that are distributed from the start until the end of the test. The implementation of pressure monitoring in final fill and in lab-scale experiments enables a closer and more realistic process simulation that shall be verified by pressure profile superposition.

12.2.3 Filter Integrity Test

The sterilizing grade filter used in critical filtration is to be tested post-use for integrity by a physical method correlated to the retention capability of the filter to ensure that the filtration system remained properly assembled during processing and to ensure the absence of defect. The integrity test performed with compressed gas on the wetted filter is a one or two points verification of the gas flow-rate level against a reference profile of the integral filter model.

The diffusion test or forward flow test is the verification that the gas flow rate on the horizontal portion of the curve remains below a maximum associated to a filter model and size integrity. The bubble point test is the verification that the inflection point on the gas flow-rate profile where the water is expelled from the largest pores occurs at a pressure higher than a minimum associated to a membrane pore size.

The combination of the two tests provides the highest level of information on the filtration system characteristics and integrity status. Depending on the filter and membrane type and on the test conditions, one or the other test may be impractical, leading to performing only one of the two tests.

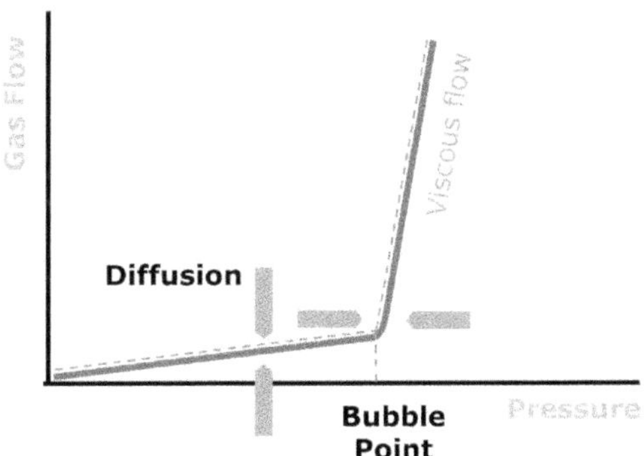

FIGURE 12.3
Gas flow rate vs. pressure profile.

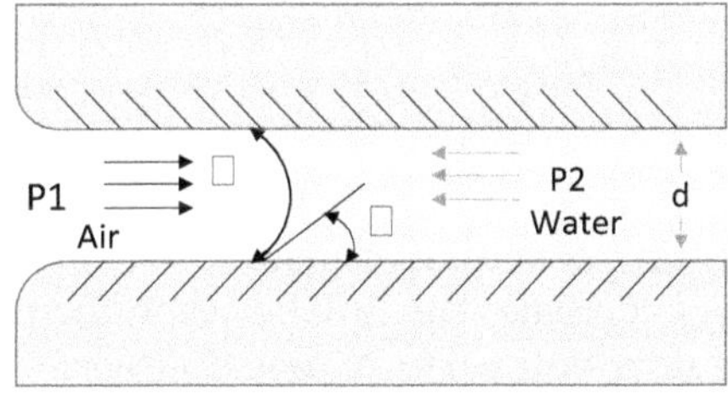

FIGURE 12.4
Forces within a wetted membrane pore.

12.2.3.1 Bubble Point Test

The bubble point test has a direct correlation with bacterial retention, whereas the diffusion test has an indirect correlation with retention. The theory behind the bubble point integrity test is that the pores of the filter membrane retain liquid due to the surface tension of the liquid and the capillary forces of the pores (see Figure 12.4).

P1 represents the upstream air pressure; *P2* represents the capillary forces equilibrating P1; θ is the angle of contact; σ is the surface tension, and *d* is the diameter of the pore.

Smaller pores retain liquid more strongly than larger pores. If pressurized gas is used to displace the liquid from the pores, the largest pores clear of liquid at a lower gas pressure than the smaller pores. In practice, with an automatic test instrument or manually with a pressure regulator the gas pressure is slowly raised on the upstream side of the wetted filter until liquid is displaced from the largest set of pores, allowing bulk gas flow through the filter.

There is a direct correlation between the size of the pores and the pressure required to release the liquid wetting the pores (12). This can be displayed graphically as the relationship between the microbial count value and the bubble point value (see Figure 12.5).

The bubble point (*BP*) pressure is expressed as:

$$BP = \frac{4 \cdot k \cdot \gamma \cdot \cos\theta}{d}$$

(12.1)

where
k = shape correction factor
γ = surface tension
θ = contact angle
d = pore diameter

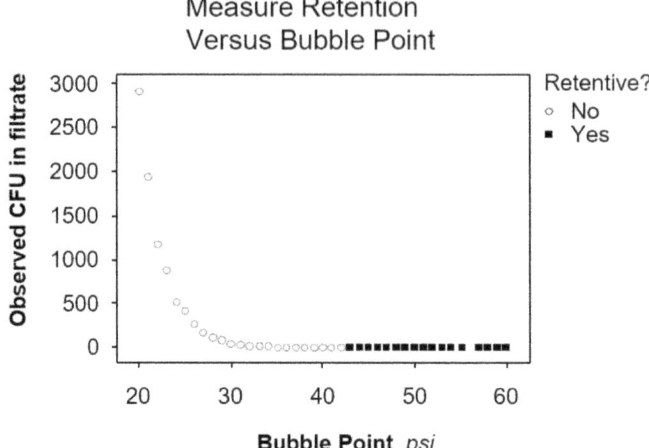

FIGURE 12.5
Relationship of bubble point values to microbial retention.

The bubble point integrity test is particularly sensitive to capture potential defects in small area—less than 0.2 m²—filters. It creates the link between the production-scale filter integrity test with the membrane retention performance that was validated on small discs and tested with the bubble point. Finally, it is insensitive to temperature variation and a sharp monitor of the fluid composition as it is extremely sensitive to any surface tension and angle of contact variation. The bubble point test requires the membrane to be fully wetted; therefore, the manufacturer wetting instructions for volume and pressure must be carefully followed.

12.2.3.2 Diffusion Test

The diffusion test is based upon the diffusivity of gas into the liquid wetting the filter membrane pores. The amount of gas diffusion into the wetting liquid is a factor of the solubility of the gas in the wetting fluid, the path length of the membrane, the membrane porosity, the gas pressure, and the total membrane area.

$$Diffusion = \frac{K \cdot (P1 - P2) \cdot A \cdot \rho}{L} \tag{12.2}$$

Where:

        K = diffusivity/solubility coefficient
$(P1–P2)$ = pressure differential across the membrane
        ρ = membrane porosity
        L = effective path length
        A = membrane area

As increasing gas pressure is applied to the upstream side of the filter, the amount of gas that dissolves in the wetted pores and travels across the membrane increases.

The diffusion test is an excellent and easy integrity test since it can detect potential defects in small and larger filtration systems. Also with the lower pressure profile needed for the test, the now common single-use filtration system assemblies are exposed to a lower

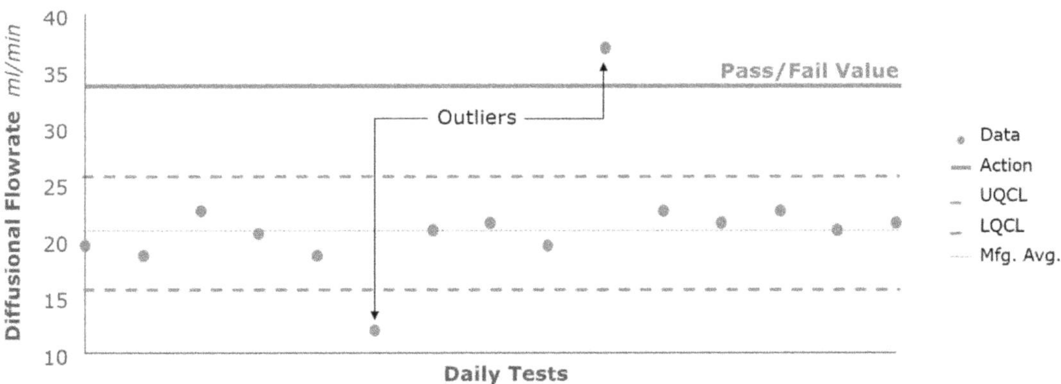

FIGURE 12.6
Tracking of integrity testing values.

mechanical stress, the amount of gas to be managed downstream is reduced, and the risk of product foaming is minimized in case of PUPSIT for the mitigation of protein aggregation risk with air liquid interfaces. The sensitivity to temperature variation during the test and to the level of plugging of the membrane call for a monitoring of the results to ensure optimum test conditions.

Values above the diffusion specification indicate a gross defect in the filter or a faulty connection in the filter housing. Values below typical may indicate a plugged filter, a temperature variation artifact (heat-up or cool-down), or an obstruction. Diffusion values should be trended to identify and treat outliers with respect to upper and lower quality control limits (see Figure 12.6).

Regardless of the type of integrity test selected, it is preferred by regulatory agencies to test the filter *in situ*, which means not only the filter is tested for integrity, but the housing and connections as well. A post-use integrity test is required by regulatory agencies. Pre-use integrity testing is a good practice and favorably received as part of the validation plan.

12.2.3.3 Integrity Test Validation

The wetting fluid for the sterilizing filter integrity test must be selected from either the filter manufacturer standard fluid (water, IPA 70/30) or the process fluid to be filtered (e.g., buffer or the drug product). Since filter integrity test (FIT) parameters are specific to the filter and membrane type, the wetting fluid, test gas, and temperature, it is important to verify the test robustness for stability and reproducibility within the actual test conditions. Indeed, the FIT is a critical element of the filtrate sterility assurance, and FIT limits must be selected responsibly from quality datasets.

12.2.3.3.1 Product-Specific FIT Limits

FIT with the product avoids flushing of the membrane after the product filtration, saving time and flushing fluid. It is very useful when the product is incompatible or immiscible in water or surface active components adsorb to the membrane during the filtration. If testing is required at the start of process (e.g., PUPSIT), flushing is not required and product dilution is avoided.

When product-specific FIT limits are sought for a particular product/filter combination, an initial laboratory study is typically conducted using typically 47 mm disk filters or

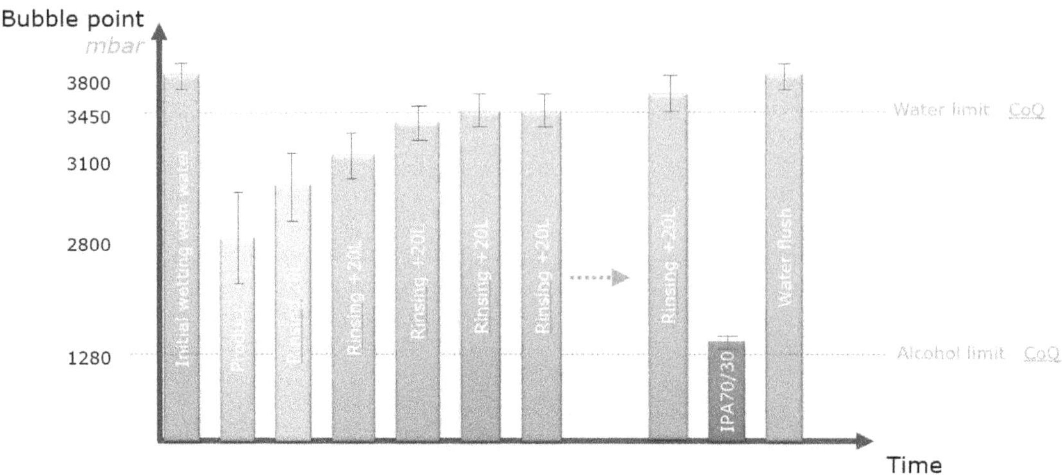

FIGURE 12.7
Example of bubble point evolution and variability across the filter unit life cycle.

bench-scale devices to relate product-wetted value to the water-wetted values for the same filters (12). The FIT limits are then derived by multiplying the filter specifications with the standard fluid by the bubble point ratio or the diffusion ratio between the product and the standard fluid.

Because there may be variability associated with the phenomenon and the measure, bubble point and diffusion are measured with a minimum of one lot of product and three test filters and wetting, and measures are repeated until stabilization. FIT limits for critical product are often tested over three lots of fluid and membranes to document the consistency in FIT results and the robustness of the determined limits.

If the product tests are inconsistent, it can be assumed the product lots are too variable to validate a product bubble point test, for example, and the diffusion test may be proposed instead. Alternatively, a fixed-volume water flush may be recommended to lower the FIT variability to acceptable levels.

Figure 12.7 is an example of bubble point evolution and variability across a filter unit service life when wetted successively with water, product, several water flush volumes, and one alcohol reference test. If the bubble point variability is too important with a product, the filtration system shall be flushed with a fixed volume and a new limit with an acceptable standard deviation shall be implemented. Alternatively, a diffusion test at low test pressure will be resilient to the product interference and would be a robust FIT option.

Most aqueous solutions suppress the water bubble point due to a change in the surface tension of the fluid. When surface active component adsorption on the membrane is suspected, it is important to simulate the filtration process (volume/cm^2) so the FIT measure conditions are close to the actual process conditions.

Meanwhile, it is not always possible for the laboratory study to emulate exactly the production-scale filters and conditions as found in the actual process. Per PDA Tech Report No. 26 (12), "the scale down study is only the first part of the validation; the second part is obtaining additional ongoing product attribute data." The product-specific FIT parameter is then confirmed under normal processing conditions, as part of the PQ, on three consecutive filtration runs. If the PQ results deviate from the laboratory study, then the laboratory FIT limit study is discarded in favor of an in-process FIT limit study using filter

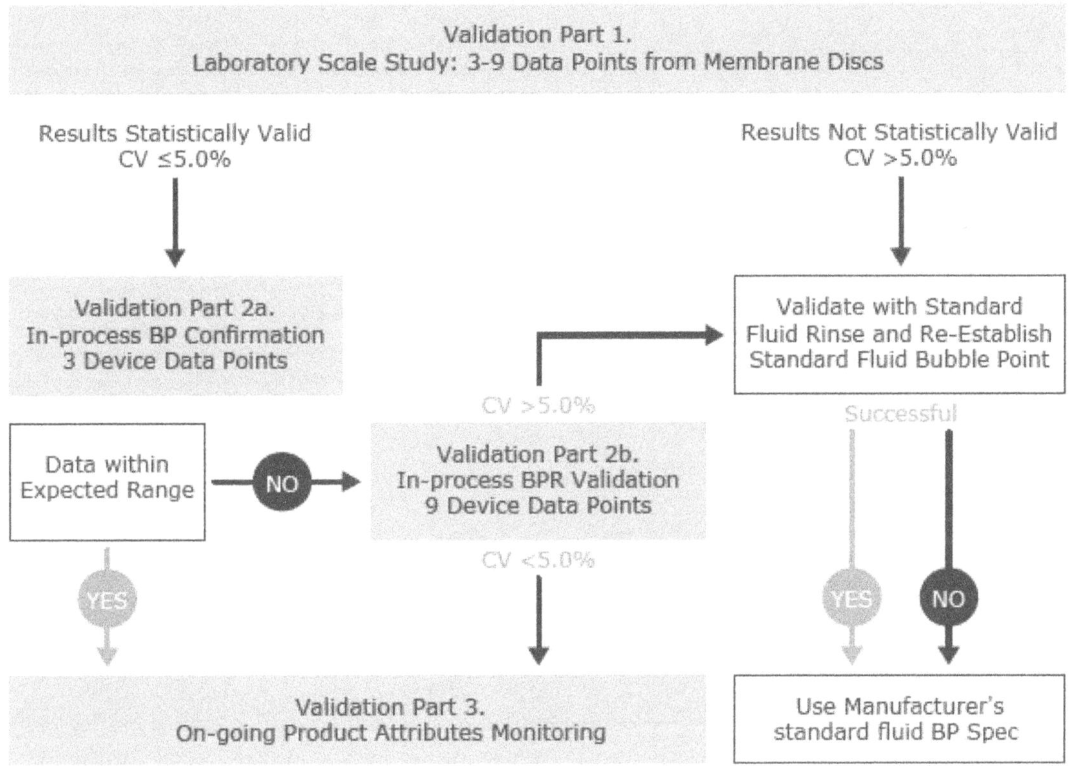

FIGURE 12.8
Approach for validation of product bubble point testing.

devices in the process stream under normal processing conditions. For example, acceptable bubble point ratio variability is achieved by a coefficient of variation CV ≤ 5.0% (13). See Figure 12.8.

12.2.3.3.2 FIT Failure Troubleshooting

Protocols should be in place in the event of a filter integrity testing failure. PDA Technical Report 26—§7.7 Decision tree for the analysis and the troubleshooting of FIT failure—is an excellent basis for developing end-user SOP. For example, protocols should include checking the sealing of the filter in the housing and all connections in the system. Rewetting can be performed with increased volumes and increased wetting pressure.

The decision tree in Figure 12.9 can be used to aid in the troubleshooting of integrity testing procedures.

Note that written procedures detailing the sequence of actions to be taken in the event of a FIT failure are required by inspectors. If additional wetting is to be performed, the wetting volumes and pressures should be specified. If an alcohol wetting or soak is to be performed, the volumes and times, along with the alcohol flushing protocol, must be specified.

Finally, FIT robustness shall be reviewed annually for every product and filter combinations. Where frequent FIT failure and retest are detected, the wetting procedure shall be reviewed and the FIT limits and test type shall be reevaluated until a "first time every time" FIT pass status is reached, as it is reasonably achievable in a GMP process.

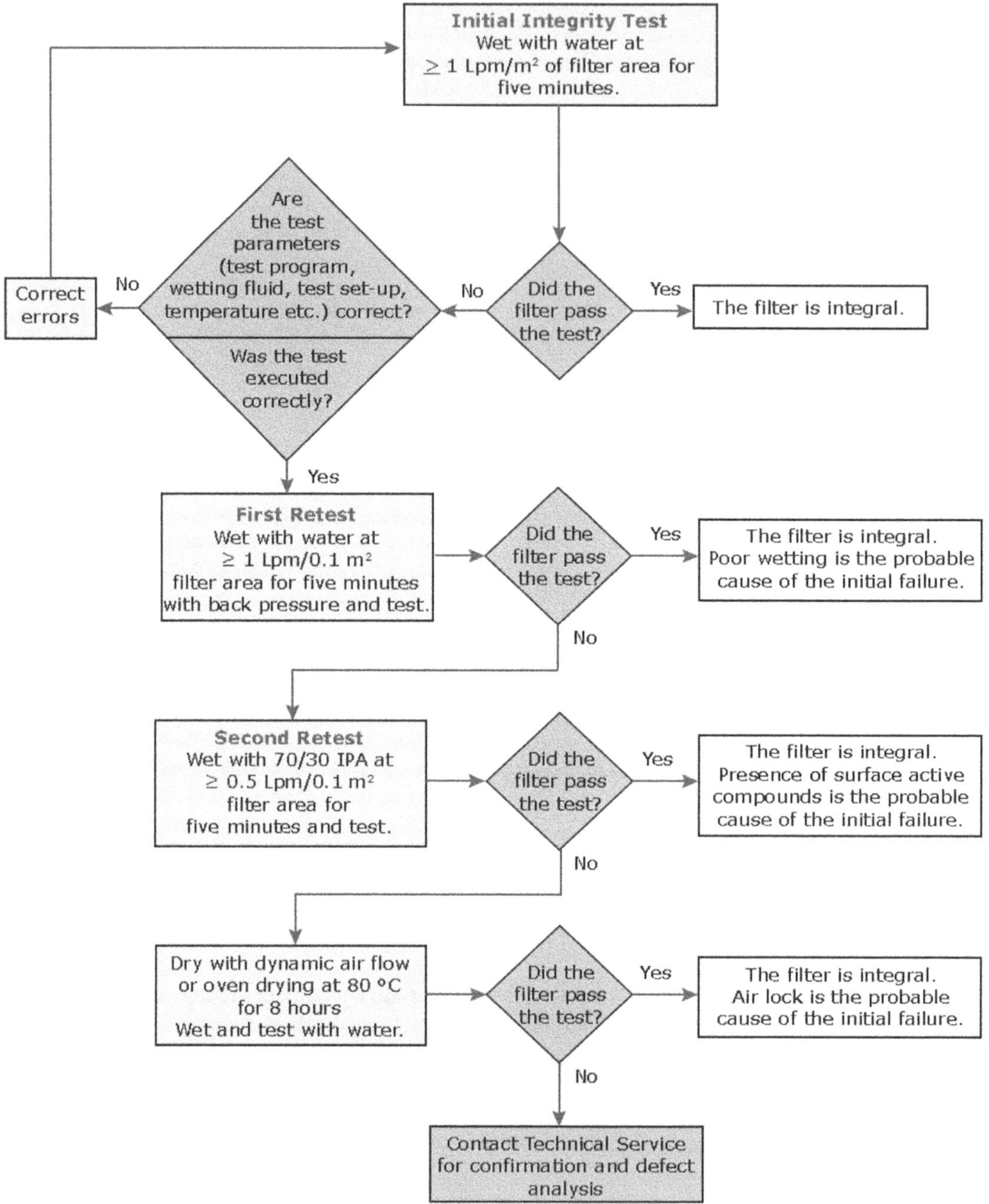

FIGURE 12.9
Integrity testing troubleshooting decision tree.

12.2.4 Filter Interaction with the Process Fluid

A filtration system should neither add anything to the fluid being processed (extractables and leachables, E&L) nor remove anything from the fluid being processed (adsorption). In reality, trace extractants and adsorption are likely. The purpose of validation, then, is to quantify the effects of extractables and adsorption by empirical studies.

12.2.4.1 Chemical Compatibility

The initial item of a filter selection is the verification of the membrane and filter components' chemical compatibility with the fluid and processing conditions. Often overlooked because filters are designed with material of construction inert and compatible with aqueous solutions of the biopharmaceutical industry, the exercise is nonetheless important and formal.

Considering the various degradation mode of polymers that include melting, dissolution, swelling, environmental stress cracking, oxidation, hydrolysis, photodegradation, and the extremely thin membrane wall thickness of about 0.02 µm, the compatibility profile of membranes may be different from the commonly available compatibility tables based on solid plastic specimens.

The chemical compatibility shall be evaluated by literature review or by actual testing with a combination of the following tests performed before and after exposure to a product in process conditions:

- Visual aspect variation—discoloration/dimension
- Mass variation
- Bubble point variation
- Flow rate variation
- Eventually: thickness, tensile strength, surface FTIR of filtration system components

Low level of variation would document the membrane inertness and compatibility with the fluid and processing conditions.

12.2.4.2 Adsorption/Transmission Study

The nominal transmission of the drug ingredients through a filtration system is carefully studied at the beginning of the filtration process and around process interruptions: in those instances, diluted solutes may interact with the membrane porous structure. This adsorption or binding study is to determine whether a given filter adsorbs components from a drug product. Adsorption can cause loss of API or excipient, induce biomolecule conformational changes, and reduce activity or stability. If adsorptive interaction or conformational changes are discovered, the drug manufacturer must determine if the interaction affects drug safety and efficacy. If it does, that filter is not acceptable for use in the manufacturing process. If it doesn't, it must be determined if it is possible to compensate for the effect of the filter (e.g., pre-flushing the filter membrane with a preservative to tie up the binding sites).

Product adsorption is assessed by product assays pre and post filtration. A drug ingredient-binding dynamic may be affected by flow rate, API concentration, excipient concentrations, pH, ionic strength, and temperature of the solution. For this reason, it is important to conduct adsorption assays on product filled in the actual process.

Preliminary adsorption studies are most easily performed on 47 mm membrane discs by scaling down the process volume and flow rate based on the membrane area. Samples are taken from the feed, from sequential fractions, and from the pooled filtrate to analyze for component concentration. From the dataset is derived the needed volume for the saturation of the filter binding capacity and the flush volume requirement for an acceptable content concentration.

Adsorption studies are critical for final filling applications. The results of the adsorption studies will determine the initial flush volume of product to discard. Interruptions in

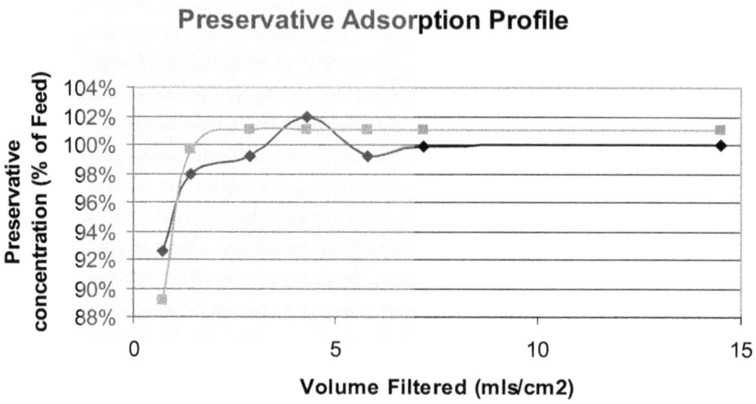

Preservative Adsorption Profile

FIGURE 12.10
Adsorption profile of preservatives in a drug product.

processing such as line stoppage increase exposure time of the drug product to potentially adsorptive medium and should be considered in the validation process.

In contrast, adsorption studies are important but somewhat less critical for tank-to-tank filtration: often in this case the quantity of material adsorbed to the filter is undetectable when assaying the concentration in the filtered bulk.

12.2.4.3 Extractables and Leachables Product Impact Evaluation

To evaluate filtration system E&L impact on proteinaceous drug product formulation, stability studies shall be designed to enhance the fluid and filter interaction. Forced degradation studies may be undertaken by either leaving the drug product bathing a sterilized filtration system for an exaggerated duration or spiking the drug product with aliquot of initial flush volume of the sterilized filtration system. The product is then characterized with assays, including analytics for protein degradation by chemical instability (oxidation, deamidation, etc.) or physical instability (denaturation, adsorption, aggregation, etc.). These analytics include measures of surface tension, visible and subvisible particle evaluation, dynamic light scattering (DLS), nephelometry, size exclusion chromatography (SEC), differential scanning calorimetry (DSC), capillary electrophoresis, HPLC GC-MS (14). Such worst-case study shall provide reference information for the elaboration of subsequent stability evaluation.

Those sophisticated forced degradation studies are typically not applied to small molecule formulations. However, acceptable product stability must be validated within nominal utilization parameters of the processing equipment.

12.2.4.4 Extractables and Leachables Patient Safety Evaluation

Filters must be validated to show that they do not add extractables and leachables to the drug product being filtered to the extent that will alter the drug product safety or efficacy (15, 16).

With the widespread availability of modern analytical equipment, the E&L evaluation for filtration system was significantly upgraded following the release by the Biophorum operation group protocol (17, 18) and with the elaboration of the USP <665> chapter on plastic components and systems used in pharmaceutical manufacturing.

Since then, filter and single-use system manufacturers have published thorough information that is compliant with the guideline.

This documentation, alongside with risk assessment guidelines (19), facilitates dramatically the evaluation of a filtration system for the nature and amount of E&L components in model fluids, hence accelerating the patient safety assessment by a toxicologist for a specific drug posology (20).

"Extractables" refers to compounds that can be extracted from plastic or elastomeric materials in solvents of different physicochemical properties under aggressive conditions; while "leachables" refers to compounds that leach from the plastic or elastomeric materials into actual drug product under normal use conditions. In sterile filtration, the proximity of extractables and leachables lies in the fact that the sterilization by moist heat or gamma irradiation is typically the major contributor to the release of polymer residue. This extreme preconditioning is common to both extractables and leachables evaluation protocols. The extractables are inherent in the filter manufacturing and use process and are present to some degree in all filter devices. Extractables may include filter materials of construction, wetting agents' residual solvents, antioxidants, unreacted monomers, lubricants, stabilizers, molding and lubricant agents, and surfactants. Process variables such as temperature, contact time with solvents, sterilization methods, and flushing procedures all have an impact on extractable levels. Extractable levels will increase with increased contact time, increased temperature, and more rigorous sterilization methods. Flushing is intended to remove extractables to an acceptable level as described in the next section.

Although filter manufacturers ensure that the materials of construction meet biosafety standards, such as ISO 10993 and USP <88> and <87>, these evaluations are conducted with extraction solvents, conditions, and evaluation protocols focused on acute toxicity that are far from real equipment processing conditions.

In contrast, BioPhorum-harmonized extractable documentation will include extraction of pre-sterilized process-scale filters in the following solvents: 50% ethanol, 0.5 N NaOH, 0.1 M H_3PO_4, and water for injection for up to 7 days at 40 °C. For a drug product modeling, the most representative solvent combination includes solvents with similar polarity and pH characteristics. Those studies provide quantitative and qualitative information on the E&L level in the first filtrate volume of a single-use filtration system.

Extractables are analyzed using a wide range of analytical techniques for identification and quantitation as listed in the Biophorum updated protocol (17). These include HS-GC-MS, GC-MS, LC-UV-MS ESI, LC-UV-MS APCI, ICP-MS, TOC, pH, NVR that are characterizing volatile, semi-, and non-volatile organic compounds, and elemental impurities.

The first step in an E&L evaluation is to identify all product-contact materials. From BioPhorum Operating Goup (BPOG) E&L documentation, data are selected and compiled to reflect drug product- and process-specific conditions resulting in a list of compounds and amounts (18).

The second step is to perform for each compound a patient safety risk assessment, comparing its E&L level to its published permitted daily exposure (PDE) level (20). Multiple factors should be taken into consideration, including but not limited to the following:

- Chemical compatibility between material and drug product
- Drug product modeling by representative model solvent
- Contact time
- Contact temperature
- Surface area-to-volume ratio

- Proximity of material to final product
- Extractable substance toxicity profile
- Dosage form
- Route of administration
- Target population (adults/children)
- Posology

Based on the patient safety evaluation outcome, the application knowledge, and the risk assessment guidelines, it is possible to introduce eventually mitigation factors such as system pre-flushing and initial volume discarding.

Finally, if the patient safety evaluation based on BPOG data identifies a potential risk for the patient, then a deeper evaluation shall be undertaken with a leachable study measuring extracted compounds in the actual drug product for the actual processing conditions. These data are again evaluated for patient safety evaluation against PDE.

12.2.5 Filtration System Flushing

In final fill, a filtration system is flushed for reasons including the ones listed next:

- Particle flushing
- TOC/conductivity/extractable flushing
- pH stabilization
- Air removal
- Membrane preconditioning to minimize protein aggregation
- Excipient and API nominal filtrate concentration
- Robust filter integrity test

The extractable and leachable substances flushing from a sterilized filtration system is typically monitored by TOC or conductivity. The chemical species removed on the initial flush are fundamentally the extractable substances released by the sterilization process. Their elimination follows a pattern documented by filter supplier for a particular membrane, filter size, and sterilization method in validation guides. A profile example is shown in Figure 12.11.

The information derived from this graphic is only moderately interesting for designing a flushing procedure. It is possible to transform this flushing information into a more useful profile with the following approach that can be applied to a variety of polymeric processing equipment: the TOC raw data in ppb is transformed into an amount in mg C by the multiplication with the aliquot volume. Then the cumulative profile is built from a spreadsheet as illustrated next:

This TOC in mg C cumulative flush profile uses the same information as in the previous figure, and now it readily provides insightful information: e.g., a 5 L flush to waste would discard 93% of initial E&L (13/14 mg) and would leave 1 mg in the header tank of the filling machine.

This graphical representation has many advantages as it documents the total amount of organic matter released by a filtration system and it intuitively shows the impact of discarded volume on E&L elimination as well as the remaining quantity in the filtrate.

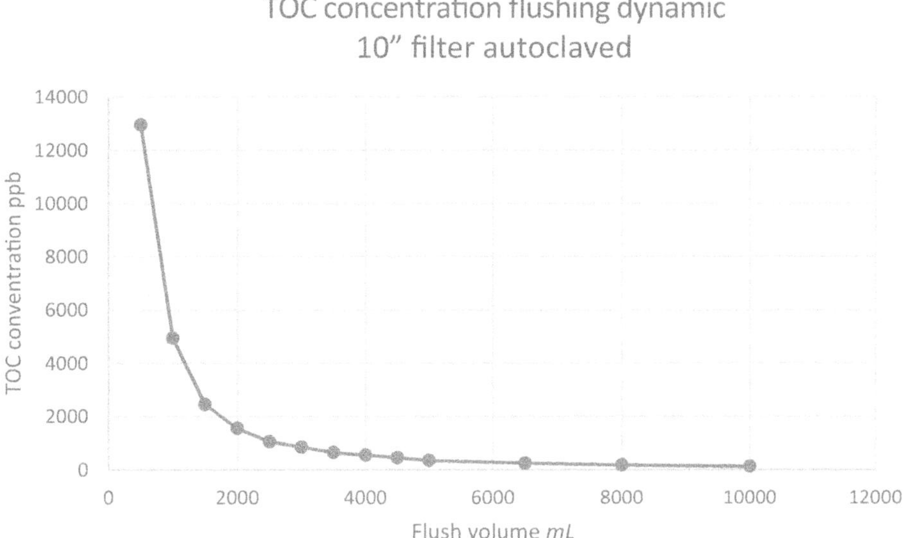

FIGURE 12.11
TOC in ppb flush profile in water for injection of a 10" filter sterilized by autoclave.

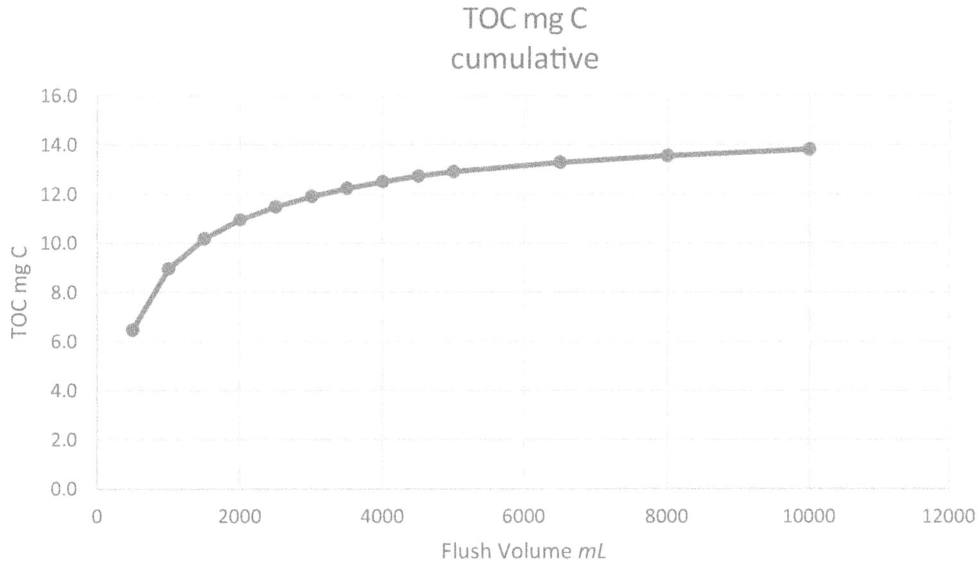

FIGURE 12.12
TOC in mg C cumulative flush profile.

The information derived from this profile is in line with the levels from the BPOG extractable data reports available from filter suppliers where TOC is evaluated after 1 and 7 days. For aqueous products, the initial flush truly removes the majority of the sterilization extractables, and this information is the scientific ground for E&L mitigation programs.

12.2.6 Filtration System Sterilization

Filtration line sterilization is validated according to the standards associated to the sterilization method. Steam sterilization by means of steam in place and autoclave are qualified for proper air removal and steam penetration with thermocouples and biological indicators according to recommendations from ISO 13408–5 and ISO 17665–1. Qualification data are generated, evaluated, and maintained within the end-user organization.

A single-use filtration system sterilized by gamma irradiation is validated for bioburden level and sensitivity to low irradiation dose following ISO 11137–1 standard. The end-user organization typically obtains from the supplier a validation summary, including a quarterly bioburden evaluation with its low-dose sensitivity audit report and a certificated of processing from the sterilization contractor indicating the delivered dose in kGy as measured by the dosimeters for a particular lot.

Production-scale filtration systems are rarely sterilized by ethylene oxide in the biopharmaceutical industry. Sterilization process validation would, however, adhere to the standard ISO 11135.

12.2.7 Fit for Duty and Pressure Monitoring

The filter is typically exposed to harsh sterilization conditions during steam in place and to a lesser extent during autoclaving. Subsequent processing steps, including wetting, integrity test, blowdown, and product filtration with potential pulsing from a peristaltic pump or from an intermittent valve opening, may stress mechanically the filter element. A gamma-sterilized filtration system will undergo a similar stress profile during processing. To date, moist heat sterilization was often seen as the harshest step in a filter life cycle because of the polymer's weaker profile at elevated temperatures and the potential and accidental presence of a high level of condensate punctually. Nevertheless, past the sterilization phase, the entire processing sequence must be evaluated to qualify the filter resistance to the process. Hence, engineering runs will verify the filtration system "fit for duty" by evaluating the filter integrity post-use.

Pressure monitoring enables the documentation of any eventual filter plugging or accidental pressure spikes, the trending of filtration performance over multiple batches, and finally the confirmation of acceptable levels during processing. The filtration equipment and process characterization by pressure mapping as proposed by ISO 13408–2 is the final element of the validation: since filtration systems have become more complex and SOP includes numerous steps as described earlier, the pressure monitoring is the only tool to document nominal pressure profiles in normal conditions and eventual artifacts—e.g., backpressure due to a closed valve or a blinded phobic filter—during degraded conditions of processing and integrity testing.

12.3 Clarification/Prefiltration Filters

Filter manufacturers qualify their clarification and prefiltration media and devices with claims of biological safety USP 88, USP 87, 21 CFR toxicity compliance, hydraulic and thermal stress resistance, flow rate and pressure drop, TOC, conductivity, extractables, and non-fiber shedding. It is the drug manufacturer's responsibility to review this data, audit the filter manufacturer, and validate the prefilter in their specific process. The user should demonstrate drug product consistency and activity/stability and evaluate product yield, process time, chemical compatibility, adsorption, and extractables with their drug product solution.

Prefilter compatibility, extractables, and binding studies are performed in the same manner as described in the Sterilizing Grade Filter Validation section. The extractables data from prefilters are taken into account for the patient safety evaluation performed along with the sterilizing grade filter (21).

12.3.1 Filter Capacity

Filter capacity must be evaluated with process fluid in order to determine the most adequate filter type and sizing in order to protect the following membrane filter from clogging. Testing procedures should be application specific and may require different flow conditions (constant pressure vs. constant flow, single-pass vs. multi-pass). This testing can be performed at small scale. Acceptable methods include Vmax™ testing, constant flow trials, and breakthrough trials.

12.3.2 Toxicity and Fiber Shedding

Prefilters used in the final process filter train are subject to the same toxicity requirements as are the sterilizing grade filters (22). Although it is preferable to have a fiber shedding claim for all the filters in the drug manufacturing process, prefilters do not need to have a fiber shedding claim as long as there is a non-fiber shedding filter downstream. 21 CFR, part 211.72 states, "If use of a fiber-releasing filter is necessary, an additional non-fiber-releasing filter of 0.22 micron maximum mean porosity (0.45 micron if the manufacturing conditions so dictate) shall subsequently be used to reduce the content of particulates in the injectable drug product."

12.3.3 Processing Considerations

The validation of the clarification step should include testing the subsequent operation at the range of outcomes from the filtration step. This allows for establishment of the normal operating ranges for the filtration step and gives enough process knowledge to determine what the minimum specification for clarity is to ensure a successful subsequent operation. The filtrate clarity may be expressed in nephelometric turbidity unit (NTU) or may be appreciated by another indicator such as the capacity in Vmax of the subsequent filtration step.

The clarification step may be followed by a unit operation that requires a particulate-free filtrate from the clarifying filter. In the case of a subsequent chromatography step, the clarifying filtration system must generate a particulate-free filtrate that will not impair the service life of the chromatography media. A range of operating pressures, flow rates, and volumetric loading (volume/membrane area) should be tested to set specifications that meet the criteria for particulate removal. A clarifying filter may give a filtrate of acceptable clarity at a low flow rate and operating pressure, but not at a higher flow rate and/or pressure. The pressures and flow rates tested should generate a filtrate free of host cell proteins and cellular debris. A clarity specification is necessary to ensure the required purity of the feed stream moving forward in the purification process.

If the feed stream is subject to forming colloids or aggregates as the processing time increases, this must be considered in the validation. Also, previously frozen feed streams can present difficulties in filtration. If a frozen feed stream is thawed and held for an extended period of time prior to filtration, the validation of that filtration step must be performed under the same conditions simulating hold time. This is an instance where scaling of the process may require revalidation, as the amount of time for a small volume of feed stream to thaw is much less than the thawing time for a large volume. The difference in thawing time could contribute to product quality differences due to the degradation by proteases and other enzymes present in the feed stream.

12.4 Viral Clearance Filter

12.4.1 Definitions and Regulatory Requirements

Viral contamination in both upstream and downstream operations are a real risk and have broadly affected the plasma, vaccine, and recombinant protein industry in the past. Today, these industries have much better safety records due to the firm measures that the industry and regulators have taken to lessen viral safety risks.

Viral clearance filters are specifically designed to retain viruses using a size-exclusion mechanism.

In order to understand the regulatory requirements, it is first necessary to define the different categories of viruses according to how they contaminate the product and to define viral clearance.

Viruses can be grouped into three classifications according to how they contaminate the product (23):

- Endogenous: Viruses whose genome is part of the germ line of the species of origin of the cell line and can be produced in culture by cell lines from these species. In plasma-derived products, endogenous viruses are natural contaminants of donor blood. Mammalian cell culture systems are susceptible to endogenous retrovirus-like particle (RVLP) contamination. RVLPs are present in virtually all mammalian cell lines and cannot be screened out of the master cell bank because they integrate their genome into the host cell genome.

- Non-endogenous: Viruses from external sources present in the master cell bank.

- *Adventitious*: Contaminant viruses that are unintentionally introduced contaminant viruses. Contamination with adventitious virus may occur through several routes such as the addition of contaminated raw materials and extraneous contamination.

Viruses can also be categorized according to their relevance as model test viruses:

- Relevant virus: The identified virus, or of the same species as the virus known to or likely to contaminate the cell substrate or any other materials or reagents used in the production process.

- Specific model virus: A virus which is closely related to the known or suspected virus (same genus or family), having similar physical and chemical properties to those of the observed or suspected virus.

- *Nonspecific model virus*: A virus used for characterization of viral clearance of the process when the purpose is to characterize the capacity of the manufacturing process to remove and/or inactivate viruses in general, that is, to characterize the robustness of the purification process.

12.4.2 Viral Clearance Validation Study

Viral clearance is defined as the elimination of viral contamination by removal of viral particles or inactivation of viral infectivity and can be classified as specific or general (23):

- Specific-virus clearance: "to provide evidence that the production process will effectively inactivate/remove viruses which are either known to contaminate the starting material or which could conceivably do so."
- *General-virus clearance*: "to provide indirect evidence that the production process might inactivate/remove novel or unpredictable virus contamination." This relates to robustness of the unit operation.

Regulatory guidance for viral clearance promotes three complementary approaches to control viral contamination. First, cell lines and other raw materials are controlled and qualified for use through virus screening measures. Second, specific steps in the manufacture of the product are included in the process line and validated to clear virus. Third, the final product is screened for the presence of virus (24).

The ICH makes the following recommendations in Topic Q5A (23):

1. *Test the process source material*: Test cell lines, raw materials, media components, and so on for viruses.

2. *Test the process*: Assess the capability of the manufacturing process to clear viruses.

3. *Test the product*: Test the product at appropriate points in the process for the absence of contaminating infectious viruses.

The overall process needs a product sterility assurance limit (SAL) of 6, meaning that in 10^6 doses, no more than one dose will be *contaminated*. For biotech products that carry endogenous viruses, RVLPs are found at titers of 10^6–10^9/ml. In order to reduce the titer to 10^{-6}/ml in the final product, industry standard requires a 12–15 log reduction value (LRV) in the process. It is difficult to measure the effectiveness of viral clearance beyond 6 LRV due to limits of detection of the assay, therefore at least two effective steps should be used in the process to achieve 12–15 LRV. These effective steps should be orthogonal, meaning they utilize different viral clearance methods, such as size exclusion and adsorption (25). If a process relies on one methodology for viral clearance, it is more likely to fail. If two adsorption chromatography steps are used to clear virus, a shift in the pH of the feed stream may reduce the binding capacity of the chromatography media, resulting in breakthrough of virus from both steps. If orthogonal methods are implemented, the adsorption step is coupled with a size-exclusion or inactivation step, neither of which will be affected by the pH shift. More recent trends also include testing at process extremes such as flow rate, protein concentration, and pressure (24).

Industry standard is to demonstrate 6 LRV for adventitious viruses. This requires at least one effective step. An "effective" step is defined as providing a minimum of 4 LRV and is scalable and robust. An "ineffective" step provides an LRV of 1 or less. A "moderately effective" step falls between the two. Robustness refers to two concepts, the first being that the viral clearance step is capable of clearing a wide range of viruses and the second being that the step can withstand perturbations of process variables. Minor changes in the feed stream should not affect the performance of the viral clearance step (26): low pH inactivation, heat inactivation (HTST), radiation, solvent-detergent, and chaotropes can be used for virus inactivation, while virus removal is typically obtained by chromatography and membrane filtration (26). Although inactivation methods are considered robust, they are likely to impact the product quality and their effectiveness can be limited by a broad range of chemical, physical, and biological resistance properties. Virus removal using chromatography also has limited operating windows of efficient viral clearance and is affected by the chemical properties of the virus, the resin type, and the operating conditions such as ionic strength, pH, temperature, and flow rate. The virus filtration represents a robust, size-based viral clearance that is scalable, is easy to use, and provides high product recovery with minimal product damage (27, 28).

12.4.3 Virus Spiking Studies

The capacity of a purification process to inactivate or remove viruses is demonstrated using model viruses during virus spiking studies, also called viral clearance studies. Most viral clearance validation studies incorporate a Phase I study. For plasma products, HIV is endemic in the blood donor population and a range of model viruses is used, with a full panel of viruses submitted with the BLA. For CHO-derived biotech products, xenotropic murine leukemia virus (xMuLV) is the RVLP model virus while murine leukemia virus (MMV) is known to contaminate the rodent production cell cultures and is typically used to represent a worst-case adventitious viral vector due to its small size (20 to 25 nm) and its high resistance to chemical inactivation (29, 30). Relevant viruses are those identified viruses of the same species likely to contaminate the cell substrate. Specific models are viruses of the same family of the suspected or identified virus. They are available in high titers, safer than the virus of concern, and practical if the actual virus is not culturable. Relevant and specific model viruses are used to demonstrate the capability of the process to remove viruses present in the master cell bank. Nonspecific models are viruses with different properties than the suspected or identified virus. This category of viruses is used to demonstrate the robustness of the process to remove/inactivate viruses with wide physicochemical properties (31).

Table 12.2 shows a list of the common viruses used in viral clearance studies and their properties.

TABLE 12.2

Typical Viruses Used in Viral Clearance Study

Virus Name	Family	Size (nm)	Envelope	Shape	Chem. Resist.
Murine minute virus (MMV)	Parvoviridae	18–26	No	Icosahedral	Very high
Porcine parvovirus (PPV)	Parvoviridae	18–26	No	Icosahedral	Very high
Poliovirus (Pol)	Picornaviridae	22–30	No	Icosahedral	Medium
Simian vacuolating virus 40 (SV40)	Polyomaviridae	40–50	No	Icosahedral	Very high
Bovine virus diarrhea (BVDV)	Flaviviridae	50–70	Yes	Pleo/Spher	Low
Reovirus type 3 (Reo3)	Reoviridae	60–80	No	Spherical	Medium
Murine leukemia virus (MuLV)	Retroviridae	80–110	Yes	Spherical	Low
Pseudorabies virus (PRV)	Herpesviridae	120–200	Yes	Spherical	Medium

Viral clearance validation studies are typically performed at specialized contract testing organizations (CTOs) that are "equipped for virological work and performed by staff with virological expertise in conjunction with production personnel involved in designing and preparing a scaled-down version of the purification process."(23) The validation is performed with viral spiking studies using a qualified scaled-down model of the filtration process, in which a heavily virus-spiked solution is filtered through a test filter to characterize the retention capability of the filter. It is performed for known endogenous viruses in the master cell bank (MCB) to provide assurance for clearance of such viruses and for unknown contaminants to characterize the robustness of the removal/inactivation step. Viral clearance validation is performed at small scale for multiple reasons. Viruses are spiked into the feed stream, and thus enormous quantities of virus and drug product would be required at the process scale. The required quantities are not practical. The safety of manufacturing personnel is also a concern. The scaled-down model uses the same membrane and operating conditions (flow rates and pressures) as are in the manufacturing scale as well as a feed from a representative process with the same buffer, concentration, and production history. It should include all critical parameters, maintain relative values, and represent the production procedure. The parameters include:

- Same challenge solution (product)
  - pH
  - Protein concentration of the input/output
  - Ionic strength
- Same process parameters
  - TMP or flow rate
  - Process time
  - Process interruptions
  - Temperature
  - Relative volume/filter area

It is also important to establish worst-case conditions for viral clearance validation in order to perform studies that respect the regulatory requirements (32).

A minimum quantity of virus should be used so as not to change the characteristics of the test product. The industry has shifted to reduce the quantity of virus spike added to the feed. A 0.1%–1% v/v ratio of virus suspension to challenge solution is typical. Contract testing laboratories are now producing cleaner virus preps with higher titers, allowing for smaller volumetric spiking loads. This is more representative of the feed and allows for less filter fouling from the virus spike. Virus should be non-aggregated to avoid overstating physical removal. TEMs of the virus prep should reflect monodispersed intact particles with no cellular debris. Hold controls should be included to monitor infectivity of the virus during the process conditions. The product should be evaluated for toxicity to the indicator cells used in the assays. Duplicate studies should be done to provide reproducible results.

The reduction factor is the $\log_{10}$ of the ratio between the total virus load before clearance and the total virus load after clearance.

$$Reduction\ factor = log_{10}\left[\frac{(V1\,x\,T1)}{(V2\,x\,T2)}\right] \tag{12.3}$$

Where:
$V1$ = volume of spiked feed stock that has been used to challenge the clearance step
$T1$ = virus concentration of spiked feed stock prior to the clearance step
$V2$ = volume of material that has been treated by the clearance step
$T2$ = virus concentration of material after the clearance step

The current guidance states, "sufficient sample volumes should be tested to ensure that there is a high probability of detecting virus in the sample if present" (33). If no virus is detected in the sample, the theoretical virus concentration in the sample must be calculated before the LRV is determined. For samples representing a percentage of the total volume, it is possible that the volume not sampled contains virus. The probability (p) of this is dependent on the total volume of filtrate (V), the sample volume of filtrate (v), and the number of viruses in the total volume (n). The following equation quantifies the probability:

$$p = \left| \frac{(V-v)}{V} \right|^n \tag{12.4}$$

If the sample volume is a very small percentage of the total volume, the probability is expressed as the Poisson distribution:

$$p = e^{-cv} \tag{12.5}$$

Where:
c = virus concentration. If 95% confidence limits are applied
$p = 0.05$.

In solving for c, the equation becomes:

$$c = \frac{3}{v} \tag{12.6}$$

This value for c is then used as the value for $T2$ in the first equation, and the reduction factor is calculated as previously.

   Serial dilutions should be performed when the expected titer of the undiluted sample is outside the statistically significant range. Multiple replicates of the final dilution should be assayed to increase the accuracy and precision of the assay. Assay methods can be quantitative or qualitative. Quantitative methods include 50% tissue culture infectious dose ($TCID_{50}$) assays where the cell culture is scored either infected or not. The titer is measured by the proportion of the culture infected. Another method is to measure the genomic equivalent units (geu) using polymerase chain reaction (PCR). The genome is amplified by PCR until the limit of detection of the assay is reached. The number of cycles to reach the limit of detection is indicative of the amount of genomic material present in the original sample.

   Quantitative assays include plaque assays in which one plaque corresponds to a single infectious unit. For negative results, when only a portion is assayed, there is sampling error involved. The amount of virus to achieve a positive result should be calculated using the Poisson distribution and taken into consideration when determining LRV. The FDA recommends 95% confidence limits when calculating LRV.

Probability of detection is:

$$p = e^{-cv} p = e^{-cv} \ or \ c = \frac{\ln(p)}{-v} \ c = \frac{\ln(p)}{-v} \tag{12.7}$$

Where:
c = is the concentration
ln= is the natural logarithm
p = is the % confidence
v = is the volume of the material assayed

Example: 100 mL of xMuLV with initial titer of 1.5×10^6 TCID$_{50}$/mL suspension is filtered through a viral clearance scale-down device. There is no virus detected in the filtrate. If 4 mL of the filtrate was assayed, the LRV is calculated as follows:

Initial virus load: 1.5×106 TCID50/mL $\times$ 100 mL $= 1.5 \times 108$ TCID50

Final viral load is less than the limit of detection.
For 95% confidence: $c = \ln(0.05)/-4 = 0.75$
Final virus load is less than 100 mL $\times$ 0.75 or < 75
LRV is calculated as: $\log 1.5 \times 10^8 - \log 75 = 8.2 - 1.9 = 6.3$ LRV > 6.3
The equipment setup and strategy for a standard viral spiking study is illustrated in Figure 12.13 and the steps are outlined as follows:

1. Attach small-scale devices to the tank reservoir.
2. If using an adsorptive prefilter, attach a dip tube at the prefilter outlet to collect the filtrate and avoid dropping effect.
3. Attach downstream of filter devices to tared collection vessels.
4. Spike product with virus and gently mix. Note the exact volume.
5. Prefilter challenge material with a 0.2 or 0.45 µm membrane (low protein binding).
6. Remove a sample of the spiked feed to determine the initial virus concentration and a hold control sample.
7. Gently transfer spiked feed into pressure vessel (in BL2 cabinet).
8. Pressurize vessel at the appropriate pressure.
9. Start filtration and record the weight over the time.
10. When desired amount of filtrate has been collected, shut downstream valve off and remove filtrate vessel.
11. Determine the exact filtrate volume.
12. Perform a buffer recovery step if applicable.
13. Sample feed and compare to initial to determine if virus inactivation is occurring.
14. Perform a post-use integrity test.

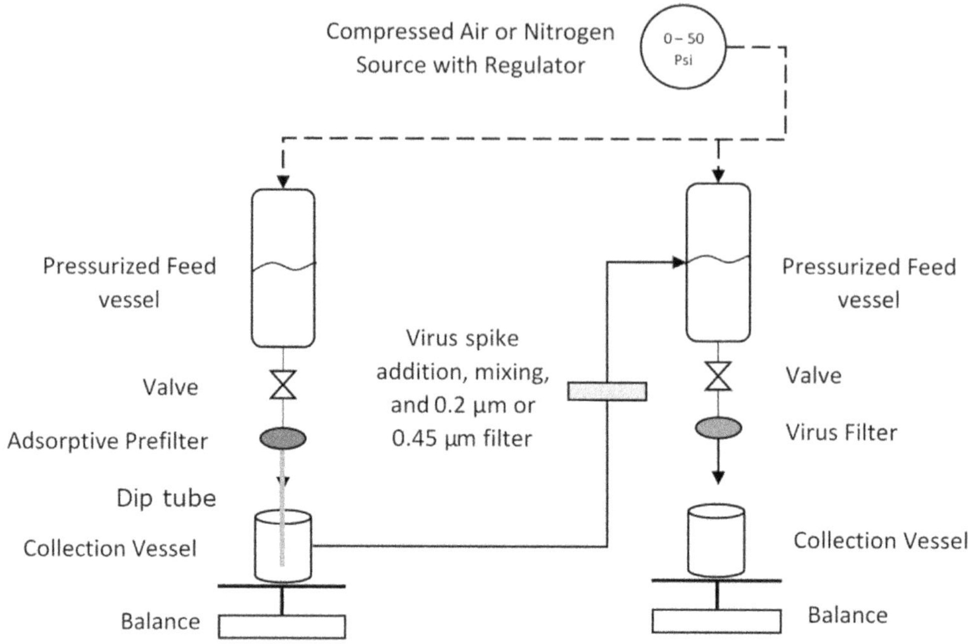

FIGURE 12.13
Standard viral spiking strategy.

12.4.4 Inline Virus Spiking Studies

The use of adsorptive prefilters in series with viral clearance filters has been shown to enhance the viral clearance filters performance and robustness by retaining aggregates in feed streams (34, 35). During conventional virus spiking studies, the inline prefilter is separated from the viral clearance filters for batch prefiltration followed by batch viral clearance filtration. This makes the scaled-down model not fully representative of the inline manufacturing process and, for a number of proteins, leads to aggregates formation in the prefiltered feed that significantly reduces the viral clearance filter capacity comparing to manufacturing (36).

The inline spiking method has been developed to overcome these challenges by using inline prefiltration with direct measurement of virus filter removal capabilities leading to shorter runs and less handling.

The equipment and setup used for the inline virus spiking method are illustrated in Figure 12.14. A dual syringe pump controls the virus injection and sampling by adding and removing virus at the same rate. The virus is mixed using an inline mixer to ensure an accurate sample. The rate of virus injection/removal is adjusted manually to ensure the same ratio of feed to virus is being challenged across the virus filter (Figure 12.14).

The spike titer required in the injection syringe was calculated using the rearranged steady-state mass balance:

$$C_{sample} = C_{syringe} \ x \ \frac{Q_{syringe}}{\left(Q_{syringe} + Q_{filter} \right)} \tag{12.8}$$

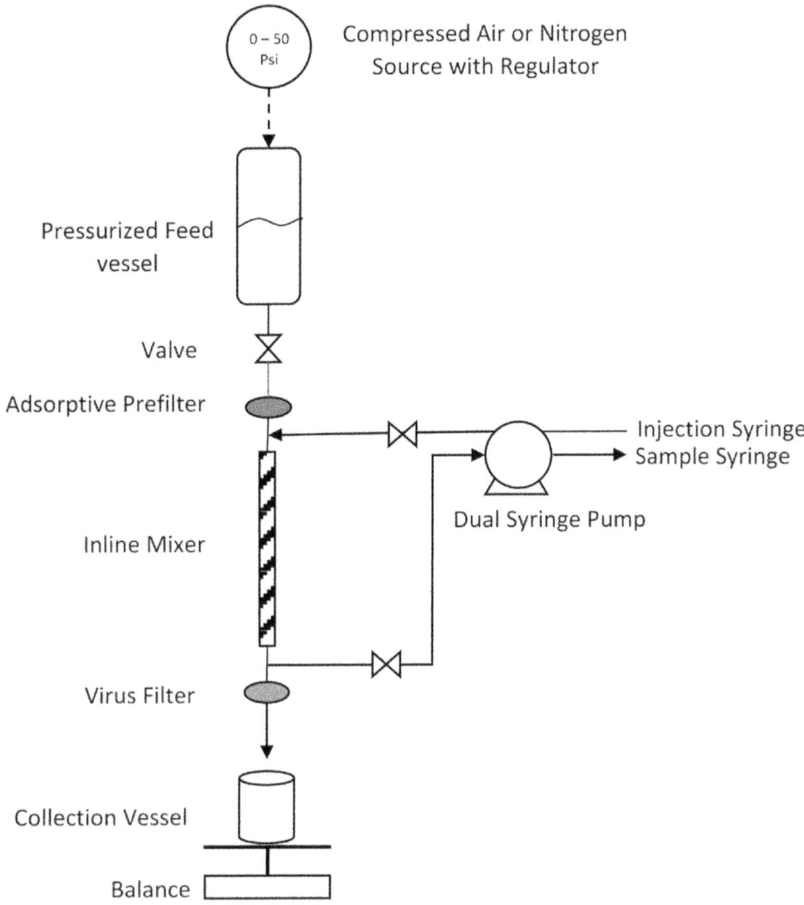

FIGURE 12.14
Inline spiking system.

Where:
C_{sample} = concentration of virus in the sample syringe (and the virus filter feed)
$C_{syringe}$ = concentration in the spike syringe
$Q_{syringe}$ = spike syringe volume flow rate
Q_{filter} = virus filter feed volume flow rate

Some alternatives can be used to minimize air–liquid interferences causing aggregates formation during virus spiking studies, such as replacing the vacuum filter with a different filter like a syringe filter at constant pressure with a dip tube and using a syphoning method for pouring feed stream between vessels.

12.4.5 Integrity Testing of Viral Clearance Filters

Viral clearance filters are available in normal flow filtration (NFF) and tangential flow filtration (TFF) configurations, all devices being single use, regardless of the configuration. Regardless of the type of device selected, it must be integrity tested. NFF devices can be integrity tested using the diffusion test as described in section 12.2.3.2. The user should comply with the manufacturer's recommendations and specifications for the wetting fluid, test gas, and test pressures. The CorrTest™ integrity test of TFF viral clearance membranes is patented by Millipore Corporation. The CorrTest integrity test uses a pair of immiscible fluids, one of which is employed as a wetting fluid and the other as an intrusion fluid. The result is a ratio of two membrane permeabilities measured at preselected operated conditions. Customer integrity tests are combined with vendor tests on devices and membrane to provide assurance of consistent and reliable virus retention. For scaled-down devices, the performance of some integrity tests leads to inaccurate results because of the very small signal obtained. As alternatives, a pressure hold test can be used in which no pressure decay is measured over the time of the test, or a bubble test in which the visual detection of airflow through the filter indicates a filter defect (32).

12.4.6 Other Considerations

Effluent from the filter must test negative for USP oxidizable substances after the appropriate flush volume. Manufacturers ensure that the materials of construction meet USP requirements for Class VI Biological Tests for Plastic and are nontoxic per USP General Mouse Safety Test (USP XXIII, pp 1699–1703. Section 88 Biological Reactivity Tests, *in vivo*). Endotoxin limits are confirmed to be < 0.5 EU/ml using the LAL test. Filters must be non-fiber releasing as defined in 21CFR 210.3, as this might affect the viral clearance filters' performance.

Gaps and failures linked to virus breakthrough can be observed for small viral clearance filters under extreme running conditions like extreme flux decay (37) and virus overloading (38).

Process interruptions can also induce virus breakthrough (39, 40), and it is very important to take these pauses into consideration during virus spiking studies validation.

12.5 Validation of Tangential Flow Filters

12.5.1 Overview

Tangential flow filtration is widely used in biological and biopharmaceutical purification processes, both upstream and downstream (41). Figure 12.15 shows an example of TFF used in the typical monoclonal antibody process.

Examples of upstream applications for TFF are

- mammalian cell culture clarification and perfusion
- mammalian and bacterial cell harvest
- bacterial cell lysate clarification

Examples of downstream applications are

- product concentration and formulation
- buffer exchange

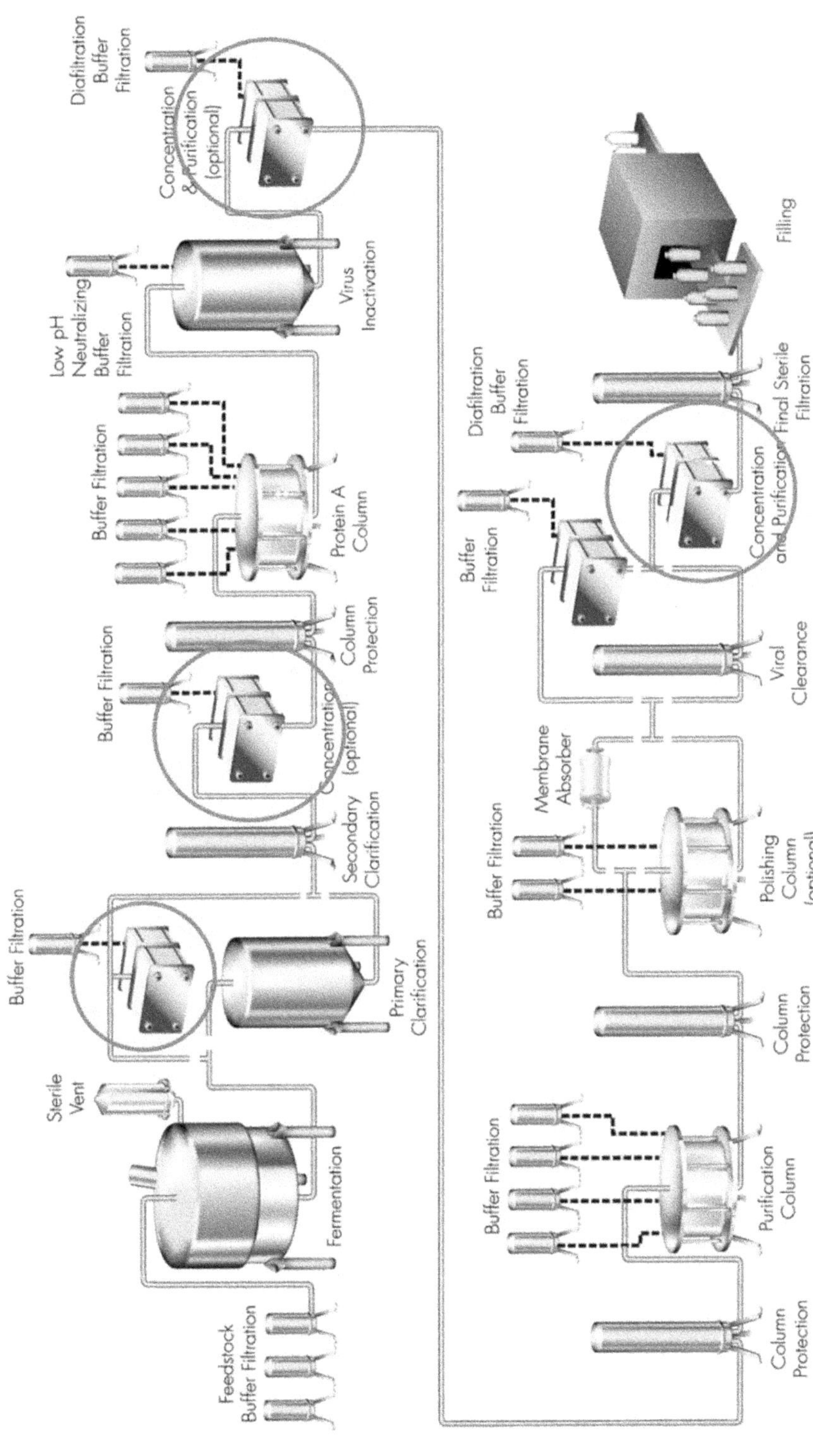

FIGURE 12.15
Generic monoclonal antibody manufacturing process.

- fractionation or separation of soluble molecules based on their molecular size
- small molecule depyrogenation

Although each of these applications have different performance outcomes, each TFF operation must demonstrate performance consistency. It means the TFF process should be designed to meet predefined product quality and process attributes over the claimed number of runs, as well as ruggedness of the process over the predefined range of operating conditions.

Process attributes include yield, costs, and process time. These process attributes do not affect patients but only the needs to maintain adequate drug supply and maintain adequate profitability. Nevertheless, yields or process time changes should be an indication of some process modifications, and impact on product quality must be evaluated. These process attributes are of interest to regulators since they could potentially affect patients through altering product safety and/or efficacy and purity.

A standard TFF ultrafiltration (UF) process consists of several steps. Table 12.3 provides the most important process parameters for each of these steps.

Variability of process parameters that could have a potential impact on final product quality and on process attributes has to be studied.

An example of potential impact of process parameters on final product quality is described in Table 12.4.

From a quality systems perspective, risk assessments help determine which systems, materials, methods, or processes parameters have a potential to impact product quality.

Quantifying risk helps provide rationale to define the level of efforts and the amount of validation testing to perform. Risk increases over the areas of the production workflow. As product gets closer to delivery to patient, the risk increases. This increase in risk is known as the distance along the product stream.

So, the efforts for validation could be different for TFF applications based on their position in the process stream.

As an example, the TFF UF step for concentration and formulation purpose should have more stringent requirements able to minimize the formation of aggregates, the presence of bioburden or endotoxin, or components carried over from the previous batch, module storage, or extractables compared to a clarification step placed immediately after the bioreactor and upfront multiple purification steps.

TABLE 12.3

Typical Process Steps

Process Step	Process Parameter
Installation	Module type, membrane cutoff, filtration area, torque, hydraulic compression
Flush	Volume of retentate and permeate water in L/m^2
Integrity test	Test pressure and air flow specification
Equilibration	Volume of retentate and permeate buffer in L/m^2
Concentration/Diafiltration	Feed flow rate, TMP, loading, concentration endpoints, number of diavolumes, temperature
Recovery	Volume of retentate diafiltration buffer in L/m^2
Cleaning	Water flush as above, cleaning agent type, temperature, concentration, contact time, feed flow rate, loading in L/m^2, NWP measurement conditions
Sanitization	Sanitization agent type, temperature, concentration, time, feed flow rate, loading in L/m^2, environment

NWP = normalized water permeability; TMP = transmembrane pressure

TABLE 12.4

Potential Impact of TFF Process Parameters on Product and Process Attributes

Process Step	Process Parameter	Typical Manufacturing Range	Potential Impact on Product or Process Attributes
Installation	Membrane cutoff	Based on application	Maximum achievable concentration
	Torque force/compression	Defined by membrane supplier	Yields, final product concentration
	Area (m²)	Based on application	Process time
Flush	Retentate and permeate water flush (L/m²)	30–60	Purification level, contaminants residuals
	Integrity test	Defined by membrane supplier	Yields, final product concentration
Equilibration	Retentate and permeate buffer (L/m²)	2–10	Product precipitation, degradation, yields
Concentration diafiltration process	Feed flow rate (L/min/m²)	1–10	Fouling, polarization, purification, product shear stress, process time, module reuse (cost of goods)
	TMP (psi)	10–45	Process time, maximum achievable concentration
	Diavolumes (N)	3–10	Purification level, residuals, process time
	Loading (L/m²)	20–40	Process time
	Temperature (°C)	4–30	Product degradation, process time, integrity test results
Recovery	Recovery volume (L/m²)	2–4(MWV) minimum working volume	Yields, final product concentration
Cleaning	Cleaning agent type, concentration, contact time, temperature, feed flow rate	Application specific/ validation defined	Cost of goods, process time

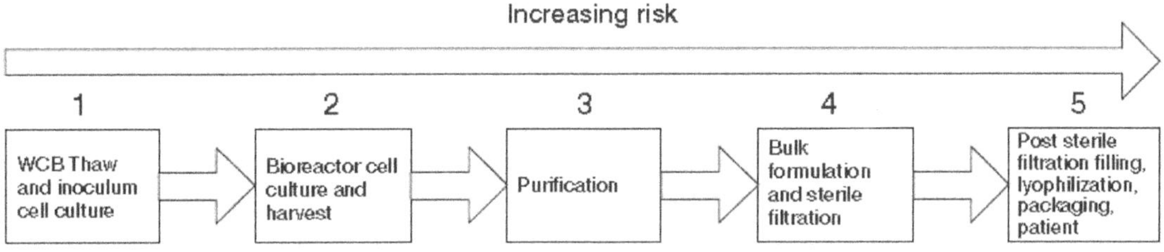

FIGURE 12.16

Risk in biological process. From: "A 3-D Risk Assessment Model" *Journal of Validation Technology* [Autumn 2008] pp. 70–76 (42).

Process developing studies, using representative small-scale models, should be conducting to establish the process requirements able to produce a product with predefined quality attributes. Selection of TFF membranes, operating parameters, and characterization studies should be performed as part of process development to gain information and increase process knowledge.

Characterization studies have to be run to define a range of (critical or key) operating parameters for the manufacturing plant, within which experimental data show no impact on product quality. Set points, as well as normal operating ranges (NOR) for transmembrane pressure, feed flow (or cross flow), minimum and maximum volumes (or loading as L/m^2 or kg/m^2), differential pressures, flow rates, temperatures, and processing times should be established at process development scale and confirmed at manufacturing level accordingly to the process parameters' criticality knowledge coming from characterization studies.

The same studies should provide design information for the new system and should be completed before full-scale production begins.

The installation qualification (IQ) and operational qualification (OQ) portions of the validation plan should be completed prior to full-scale production. PQ may be initiated concurrent with process validation.

The validation of a TFF unit operation involves common elements like the ones already listed in the first paragraph of this section, plus specific claims valid for TFF application related to membrane proprieties (retention/sieving, integrity), mode of operation (batch, fed batch, single pass), and TFF module reuse (validation of cleaning and sanitization procedures and reuse).

Some of the most important elements are hereafter taken into consideration.

12.5.2 Key Elements in TFF Validation

12.5.2.1 Chemical Compatibility

For all applications, the compatibility of the membranes as well as all wetted components of the system must be determined with each process fluid. It is imperative that none of the wetted surfaces, including the filters, corrode, extract, swell, or weaken with exposure to any of the process fluids. Process fluids include the feed stream, buffers (commonly used in the membrane conditioning and diafiltration steps), flushing, cleaning, sanitizing, storage, and depyrogenating agents. Membrane retention and selectivity must not change over the course of use of the devices. The lifetime of the expendable device must be determined. The cleaning and sanitization of the membranes must be validated.

12.5.2.2 Extractable and Leachables

It is required that equipment and materials used in drug manufacturing should not alter the potency, purity, and safety of the drug product. The membrane and the TFF system must not add chemical contamination to the product. This is of particular importance if the TFF unit operation is near the end of the purification process (i.e., formulation TFF UF).

The drug manufacturer has the final responsibility to assure that TFF modules are fit for use for the particular process. Extractables and leachables data could be retrieved from the TFF module supplier, but there are no acceptable criteria for extractables and leachables for TFF modules. Risk assessment should be applied to identify the highest risk component based on TFF modules' material of construction, process design (diafiltration for purification or product recovery), cleaning regimes, application and position of the step in the manufacturing workflow, upstream or downstream, or final step (43).

Most TFF filter manufacturers ensure that the materials of construction meet USP requirements for Class VI Biological Tests for Plastics and are nontoxic per USP General Mouse Safety Test (44). Effluent from the filter must test negative for USP oxidizable substances after the appropriate flush volume.

Users of TFF membranes must also demonstrate that the preservative solution and the cleaning and storage solutions are effectively and completely removed from the device by the recommended flushing procedure. The manufacturers' flushing guidelines should be used to establish the appropriate cross-flow, pressure, and flushing volumes to remove preservative or storage solution. Users have to validate the flushing procedure, flushing volumes, and/or contact time at manufacturing level accordingly to the specific equipment design. The assay showing clearance of the preservative or storage solution should be easy to use. The validation of an assay that can be run in the purification suite will save processing time, rather than collecting samples and submitting them for QC testing.

12.5.2.3 Retention

Retention is a measure of the amount of a solute that does not pass through the membrane relative to the amount of that solute in the feed stream. Process development studies should select the membrane able to provide the target acceptable retention for product and other components in the feed stock. Retention of the drug product should be consistent throughout the lifetime use of a membrane. For UF applications, the membrane should exhibit the highest retention for the product of interest, to ensure adequate recovery of the drug product, and the lowest retention for contaminants and/or undesirable components. Increased retention can be due to foulants decreasing the pore size of the membrane or a polarization layer of solutes impeding the passage of other solutes to the membrane surface. Decreased retention can be due to harsh cleaning regimens altering the membrane pores. The molecular weight cutoff of membranes can change with exposure to a chemical that degrades the membrane matrix.

Variability in retention has to be carefully evaluated during process development for applications where the target is the purification or separation of species with different MW. Lot-to-lot variability of both TFF modules and product stream has to be taken into account and specific studies have to be run.

TFF devices should be integrity tested *in situ* to show integrity of the filter devices and the system. Acceptable integrity tests include diffusion testing as described in section 12.2.3.2 and the pressure hold test. Because TFF devices have an inlet (feed) and an outlet (retentate) port on the upstream side of the filter membrane, either port may be used to introduce gas pressure to the upstream side of the membrane. If the feed port is used to introduce gas pressure, the outlet port (retentate port) must be completely closed to integrity test the device, and vice versa. The membranes must be fully wetted prior to the test. Systems should be stabilized for an appropriate amount of time to ensure even pressure distribution. A laboratory-scale TFF system may require only 2–3 minutes stabilization time, whereas a manufacturing-scale TFF system may require 10 minutes or more. Diffusion measurements are taken at the permeate port, which represents the downstream side of the membrane. The pressure hold test is performed by applying a known pressure to the upstream side of the device with the second upstream port closed and the permeate port open and measuring the pressure decay over a specified time. It is strongly recommended to test TFF membranes for integrity before and after use.

12.5.2.4 Process Flux and Total Process Time

Process development experiments should determine the average flux at the process temperature and pressure for having the process done in the target time frame. Consistency and reproducibility of process flux should be monitored run to run. A decreasing flux value indicates a difference in this rate of passage; the difference being due to some form of resistance. This resistance is the result of fouling, which is the physical or chemical binding of solutes to the membrane surfaces. In a reuse process, membranes are cleaned in order to remove these foulants and restore process flux. A decreased process flux value indicates inadequate removal of foulants during the cleaning regimen. Figure 12.17 shows a tangential flow filter with an adsorbed layer of foulants.

A decreased flux means that additional processing time is required to meet the concentration and/or purity specifications of the process intermediate. See Figure 12.18 for an example of trending process flux over multiple process runs.

12.5.2.5 Pressure Profiles

As well as monitoring process flux, pressure profiles should also be monitored. At a constant feed flow rate, the same solution will generate equivalent pressure profiles. A change in pressure profile at a constant feed flow rate is indicative of a change in fluid characteristic like viscosity or channel geometry. In a robust process, changes in the feed stream should be minor. Changes in viscosity could happen in high protein concentration applications and microfiltration applications, since cell densities may vary in a cell culture clarification.

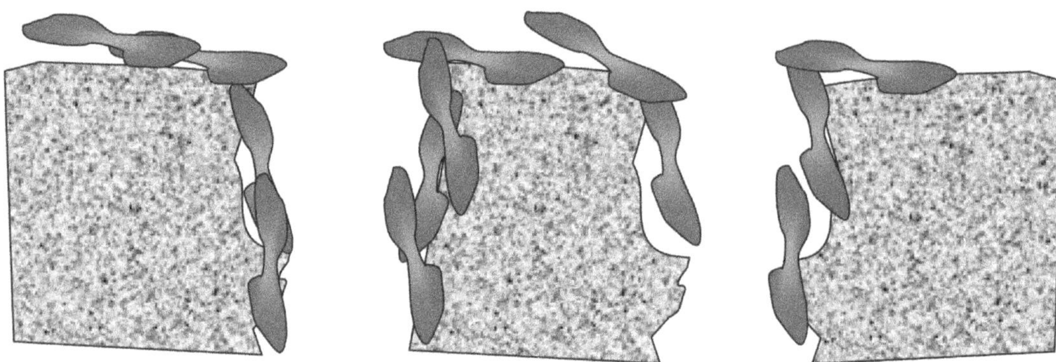

FIGURE 12.17
Solutes fouling the pores of a membrane.

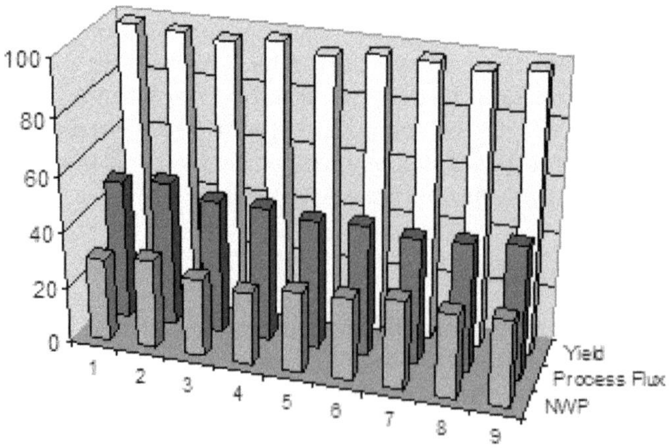

FIGURE 12.18
Yield, process flux, and clean water flux monitored over multiple process runs.

12.5.2.6 Yield

Product yield should be consistent over membrane reuse. Changes in yield may be indicative of changes in product retention. In clarification applications, a decreased yield of product in the filtrate of an micro filtration (MF) step may be attributed to the polarization layer impeding the passage of product or product loss due to aggregation, denaturation, or precipitation.

A decreased yield of retained product in a common TFF UF concentration step may be attributed to unwanted passage through the membrane where the molecular weight cutoff has increased, or again losses due to aggregation, denaturation, or precipitation. Product assays should be developed to ensure that the product is recovered from the system in an active non-denatured and non-aggregated form. See Figure 12.18 for an example of trending drug product yield over multiple process runs.

Product yield is related not only to membrane performances; product recovery procedure validity cannot be evaluated at small scale as it is strictly dependent on the system design (pipe sections, recovery strategy, minimum internal working volume, etc.) and has to be validated at manufacturing level. Post-process (concentration/purification/clarification) product recovery strategies need to be set up at manufacturing level in order to guarantee the target yields (45). The product recovery step of a TFF process has to be designed to limit hold-up volume losses without incurring protein degradation or excessive product dilution. The amount of buffer volume used for product recovery has a potential impact on a product attribute such as product concentration.

12.5.2.7 Bioburden and Endotoxin

The membrane and the TFF system must not add microbial contamination to the product. This is of high importance if the TFF unit operation is near the end of the purification process. Maximum bioburden and endotoxin level must be specified and documented. Bioburden control and endotoxin removal are in most cases performed after processing concurrently with cleaning with caustic agents, typically 0.5–1.0 N NaOH. Some UF TFF membranes are susceptible to degradation by these cleaning agents. In this case the membranes can be bypassed when the system is cleaned. Manufacturers are responsible for determining whether the endotoxin test is necessary and, if it is, what the endotoxin limit should be (46).

Ultrafiltration TFF is also used to remove endotoxins from feed streams whose components are low molecular weight molecules, like LVPs, contrast media solutions, cell culture media, and buffer solutions (57). The LRV of endotoxin can be validated at the laboratory (small-scale) level. Generally, a 10 kDa cutoff membrane is used for depyrogenation applications. An LRV higher than 3 is generally accepted by industry for single endotoxin removal step by TFF UF.

12.5.3 Bacterial Cell Harvest and Lysate Clarification

Special considerations surround the validation of a TFF system for microfiltration due to exposure to cells. Whether bacterial, yeast, or mammalian, the cleaning of the TFF system must be adequate to show removal of all cells between runs.

The processing of bacterial cell products involves two potential MF steps:

a. cell harvesting step, in which the cells are concentrated, and the spent media removed

b. lysate clarification step, in which the cell debris is removed from the product

Important considerations in process validation of bacterial cell harvests are the potential for changes in the feed stream. Bacterial cells grow rapidly, and there may be a range of cell concentrations over which the harvest must take place. Also, the pH and viscosity of the feed stream may vary between fermentations. Bacterial cell walls make them relatively insensitive to lysis by shear. The bigger validation issue surrounding bacterial cell harvest is the containment of recombinant organisms and equipment decontamination. If a claim is made that the filtrate from the TFF system should be free of recombinant organisms, then protocols should specify testing of the filtrate for such.

Bacterial cell lysate clarification presents a challenge in that cell wall debris, organelles, and hundreds of host cell proteins are released upon cell lysis. In a lysate clarification, product is being passed through the membrane and cell debris is retained. The large amounts of proteinaceous material present in the bacterial lysate may complicate identification of contaminants. This should be taken into account during assay development and qualification.

12.5.4 Mammalian Cell Clarification

The clarification of product from mammalian cells is relatively simple since these cells are capable of secreting products and can be clarified from the product.

In mammalian cell separations the objective is to keep the cells intact, thereby preserving a cleaner feed stream. Traditionally, in mammalian cell culture the amount of biomass in a culture is expressed as a cell density. Another important parameter is cell viability. Cell viability is a measure of the proportion of live, healthy cells within a population. A marker such as lactate dehydrogenase can be used to measure cell lysis, or alternatively trypan blue can be used to count viable vs. non-viable cells. Cell lysis can affect the quality of the product by releasing proteases and other enzymes that may damage or alter the product. Variability in the feed stream quality may impact further downstream unit operation. During the cell separation by TFF, the pressures and shear (measured as fluid velocity or flow rate in a channel with constant dimensions) must be kept to a minimum to avoid cell lysis. Feed flow rate and TMP are critical parameters for which an optimal range must be identified at development level and validated at manufacturing.

Among different fermentation processes (batch, fed batch, perfusion), a batch process will have a higher cell viability and lower cell count than a perfusion process. In a perfusion process, which could last for several days or weeks, it is critical to sterilize the TFF unit and maintain sterility of the perfusion system over the entire process time. Procedures for cleaning and sterilizing the TFF system and modules should be validated. The cell culture should be monitored over time to ensure that no contamination is occurring and that product expression and cell viability are not affected by the process parameters.

It is common in the industry to steam MF systems and devices used for vaccine production, although the devices do not have to be sterile. In applications where steaming is not desired, or the devices cannot be steamed, they are often cleaned *in situ* and then removed for steaming of the system. Devices must be cleaned prior to steaming, or the steaming cycle will fix the existing residues on to the membranes.

12.5.5 Protein Concentration and Diafiltration

One of the most common TFF UF applications in biopharmaceutical manufacturing process is final product formulation. In final formulation, the protein product is concentrated to the desired final concentration and buffer exchanged by diafiltration (DF) with the desired formulation excipients while removing residual small-molecule buffer components from previous purification operations.

As a protein is being concentrated in a UF process, a polarization layer builds up on the surface of the membrane. This polarization layer results from the transmembrane pressure driving solutes in the feed stream against the membrane, where they are retained.

If this polarization layer is not controlled and becomes excessive, it can impede the passage of small solutes that were intended to pass through the membrane (47). The polarization layer becomes a second layer of resistance to filtration above the membrane itself. This is of concern if it interferes with the intended separation.

Polarization is a reversible phenomenon and can be changed by changing process parameters like feed flow and TMP. Optimization is run at PD level during characterization studies addressed to identify a suitable range of process parameters. At manufacturing level, the user has to confirm that the selected range of process parameters is able to produce a product with the target quality attributes. The user has also to demonstrate that process parameters are controlled and remain inside the qualified range. Product must be tested for quality as well as yield.

The diafiltration process is used for two main reasons:

1. to produce a product that meets the specification for a maximum level of the contaminant being removed from the drug product in purification applications

2. in formulation applications to exchange buffers, with the desired formulation excipients while removing residual small-molecule buffer components from previous purification operations

The DF process should not lead to any changes in the composition of the finished drug product. Stability and solubility of the drug product in the starting buffer and the final DF buffer should be determined during process development at the laboratory scale. In the validation of the DF process, samples should be taken from the starting solution and throughout the DF to show removal of contaminants.

Validation of the DF operation should also show that the change in buffer does not cause effects that alter the permeability of the membrane or product characteristics (aggregation, denaturation, fragmentation, or precipitation). Rejection for small molecular weight contaminants in a UF process is calculated according to the following formula. The specification for the maximum allowable contaminant concentration in the drug product after the DF step must be measurable by a reproducible and accurate assay. Consideration should be given to the limit of detection of the assays for the contaminants being removed.

$$Contaminants\ Remaining\ \% = [e^{(R-1)(\frac{V_d}{V_s})}] x\, 100 \tag{12.9}$$

Where:
V_d = diafiltration volume
V_s = system volume
R = retention of contaminant

Figure 12.19 shows the graphic relationship of number of diavolumes and purification level by considering three different retention value for contaminants.

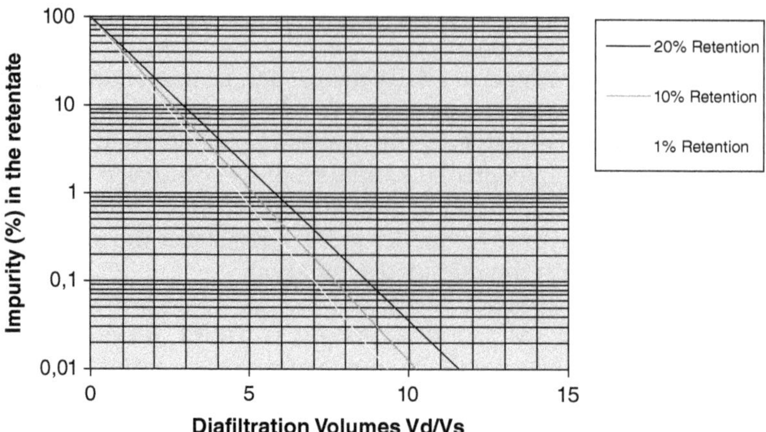

FIGURE 12.19
Diavolumes vs. solute clearance with different membrane retention levels.

12.5.6 Validation of a TFF System Cleaning Protocol

12.5.6.1 General Considerations about Cleaning of TFF System

Cleaning and sanitization of process equipment should be validated. Section 211.67(a) of part 21 of CFR (48) states:

> Equipment and utensils shall be cleaned, maintained, and, as appropriate for the nature of the drug, sanitized and/or sterilized at appropriate intervals to prevent malfunctions or contamination that would alter the safety, identity, strength, quality, or purity of the drug product beyond the official or other established requirements.

Because of the complex structure of the membrane and its limited chemical resistance, the development and validation of cleaning procedures of TFF systems represents a real challenge. The broad range of applications encountered in biopharmaceutical manufacturing exposes TFF modules and production equipment to many types of soils and foulants, including proteins, lipids, carbohydrates, fatty acids, inorganic salts, and more, with a range of chemical and physical characteristics. These, along with the TFF membrane properties, determine the nature of the interaction between the process fluid and the membrane, such as the foulant binding affinity (van der Waals forces, electrostatic and/or hydrophobic interaction, etc.). Process operating parameters, specifically feed (or cross) flow rate and transmembrane pressure (TMP), control the extent of concentration polarization, which when poorly controlled may aggregate and precipitate components at the membrane surface, potentially leading to irreversible fouling and membrane pore occlusion. An understanding of the possible fluid/membrane interactions and analysis of the operating conditions should be run at development-level stage and can provide indications about the appropriate cleaning regiment at large scale to restore the membrane as close to its original state as possible.

While restoration of the membrane to its pre-use state is the most important aspect, in the expanded scope of cleaning the complete TFF system there are many more components, including vessels, pumps, piping, and instrumentation, that require consideration during both the system design phase and the cleaning cycle development. Frequently, the system and membranes are cleaned concurrently, which presents savings of time and cleaning chemicals, though in some instances, during product changeover or when more aggressive chemicals are used for the skid, the system may be cleaned separately from the membranes. When cleaning systems and devices it is critical that the cleaning agents and rinse solutions fill all the interior surfaces of the device and system. All wetted components of the system must have fluid contact, including but not limited to instruments, piping, pump heads, and any dead legs. Tanks should be given special consideration. Closed tanks may require spray ball cleaning. In the case of fed-batch concentration, the large tank not permanently attached to the system must not be forgotten, nor should the piping or pump that connected it to the tank on the system. These components may be cleaned separately or attached to the system for cleaning.

Cleaning may be manual, semi-automated, or fully automated. Manual cleaning may require ongoing verification of operator training and performance. The SOPs will require regular review, and operators should be trained on the process at intervals compliant with the company's policies. Thus, it is more labor intensive but may involve less capital outlay for the manual system.

For a fully automated system, SOPs should be in place describing the preparation of the cleaning solutions. This will include the concentrations and temperatures of the solutions. An automated system will need to be programmed for the volume and duration of cleaning and rinse cycles and the pressures and flow rates of these solutions. The software controlling the automated system will require validation. Although automated cleaning procedures may be more consistent, in certain applications manual cleaning may be more thorough. If washing machines are used to clean small pieces of equipment, the machines must be validated. Despite which method of cleaning is utilized, the end result of the cleaning will be validated through sampling and assaying for residue of product, contaminants, or cleaning agents.

Cleaning program development starts with process development; due to long-term studies, high cost, and lack of reproducibility with a large-scale manufacturing unit, concurrent validation for TFF equipment and membrane cleaning, storage, and reuse is acceptable and justified.

The following paragraphs are mainly addressed to evaluate cleaning approach for the TFF modules, even if most of the touched points are valid for hardware too.

12.5.6.2 Cleaning Validation Protocol

A cleaning protocol must be developed, documented, and validated. Procedure has to be consistently applied and effectiveness has to be regularly verified.

Elements of the Cleaning Validation Protocol are similar to that of other sanitary equipment with additional specific testing for TFF technology like normalized water permeability test (NWP) and integrity testing (IT).

The Cleaning Validation Protocol should include:

- Validation objective
- Responsibilities for performing and approving the validation study
- Description of the equipment to be used
- Cleaning procedures to be used for each product, each manufacturing system, or each piece of equipment
- Interval between end of production and beginning of cleaning procedures ("dirty hold time")
- Period between validation and revalidation

- Sampling procedures, sampling locations, and any routine monitoring requirements
- Analytical methods, including the limit of detection and method quantitation
- Acceptance criteria

The cleaning protocol is very often developed using the filter manufacturer's recommendations for compatible cleaning solutions, contact times, and temperatures. Many filter manufacturers also provide flushing guidelines for removal of preservatives and common cleaning agents.

The objectives of a cleaning/sanitizing protocol for a TFF system are multifold. When new modules are installed, flushing procedures should be able to remove the preservative solution the modules are stored in after manufacturing and before shipment. Post product processing cleaning serves to remove residual proteins and contaminants and restore membrane permeability. That's a guarantee for consistent batch-to-batch process flux and process time, minimizing the risk of product carryover. Proper cleaning should also prevent or eliminate microbiological contamination, should enhance the long-term performance of the device, and can extend the lifetime of the device. Validated cleaning, at the end, provides repeatable product and process quality attributes.

12.5.6.3 Chemical Compatibility and Cleaning Agent Selection

The development of the cleaning protocol requires first an investigation into the compatibility of the filtration device as well as all the wetted components of the system with the cleaning agents.

TFF membrane modules have limitations in terms of pH, temperature, and pressure, specific to the membrane polymer and filter support and cassette materials of construction, which influence the selection of cleaning agents and conditions. Membrane retention and selectivity must not change over the course of use of the devices. The lifetime of the expendable device must be determined.

Cleaning agent selection is based on specific criteria that should consider:

- Suitability to remove product residues
- Compatibility with the equipment material of construction (MOC)
- Ease and sensitivity of assay method
- Ease of removal and verification of removal
- Low toxicity

Example of chemicals used in TFF cleaning protocols are reported in Table 12.5.

Bioburden reduction or elimination (sanitization) is very often run concurrently with cleaning. Effectiveness of cleaning procedure for sanitizing has to be evaluated and, in case of needs, an extra dedicated step has to be added. Other suitable and most common chemicals dedicated to TFF membrane and equipment sanitization are hydrogen peroxide and peracetic acid. Compatibility of such materials with membrane and system component has to be verified prior their use.

Post-use and post-cleaning integrity testing is run to verify that no adverse effect of cleaning procedure on membrane structure occurs. The trend of IT value versus time or number of batches has to be analyzed and can be used for defining the working life of the modules.

TABLE 12.5

Example of Chemicals Used in TFF Cleaning Protocols

Typical Cleaning Agents	Cleaning Agent Activity
Alkaline chemical (NaOH)	Proteolytic attack, solubilization of lipid
Oxidizer chemical (NaOCl, >pH 7)	Oxidation and proteolysis
Acidic chemical (phosphoric acid)	Protein hydrolysis and DNA solubilization
Detergent formulation (surfactants)	Solubilization and emulsification
Enzymes	Hydrolysis and solubilization
Solvents (water, alcohols)	Solubilization

12.5.6.4 Cleaning Procedure

The post-use cleaning process for a TFF system is made of a series of steps. It starts, before the "dirty hold time" expires, with a water rinse, or better with a dedicated buffer, to remove loose soils. Cleaning solution(s) wash (with rinses in between in case of multiple cleaning solutions) is then run. A post-cleaning flush is run to remove any residuals of cleaning/sanitizing agent. A final water rinse is typically dedicated for a sampling step. At the end, modules and membranes are stored in a preservative solution to keep them in a state of cleanliness until used (clean hold time).

When developing SOPs for cleaning, factors to consider include the evaluation of the maximum length of time in which the equipment can sit between processing and cleaning. Oftentimes, the focus is on the product and speeding it along in the downstream process, and equipment is left to sit before cleaning while operators concentrate on the next unit operation. If residuals are allowed to dry on the equipment surfaces, a cleaning regimen that was adequate immediately after processing may not be able to clean dried residue. The maximum "dirty hold time" and clean hold time (storage time after cleaning and before the next use) must be determined and validated.

The effectiveness of a cleaning procedure is a function of different parameters and their combination. As with cleaning any surface, the temperature and concentration of the cleaning agent, as well as the flow rate and exposure time, affect the cleaning.

It's very often identified as a time, action, concentration, temperature (TACT) approach.

Cleaning cycles have to be tailored to the type of soil observed. For some applications it may be necessary to clean with two separate cleaning reagents, requiring a flush between the cleanings.

Throughout the lifetime use of the membrane, the required volume of cleaning solution and water for flushing or number of cleaning/rinsing cycles should be batch-to-batch consistent. An increase of the number of cleaning/rinsing cycles (that means increase of cleaning volume, reagent concentration, exposure time, or temperature) indicates a process far from fully controlled. The test until clean approach (or elsewhere called "clean until clean" that means repeated recleaning and retesting) is not well accepted by regulatory agencies anymore. Constant retesting and resampling can show that the cleaning process is not validated since these retests actually document the presence of unacceptable residue and contaminants from an ineffective cleaning process.

12.5.6.5 Measuring Cleaning Effectiveness for TFF Modules

An effective cleaning protocol removes residual protein, foulants, and other contaminants from the membrane surface and cassette feed channels and restores the membrane so that performance returns to a predictable, consistent level. Key elements to monitor when assessing cleaning effectiveness include:

- Normalized water permeability
- Process reproducibility
- Flush water residuals
- Bioburden and endotoxin
- Physical analysis of contamination through destructive techniques

12.5.6.6 NWP

Normalized water permeability is an established method for determining the cleanliness of a cassette after cleaning. In combination with process flux reproducibility, they directly indicate both the effectiveness and the consistency of the cleaning procedures. NWP is a measurement of the passage of clean water through a membrane under standard pressure and temperature conditions. The NWP is simply the filtrate flow rate (L/hr) time a temperature correction factor divided by the area of the membrane (m^2) and the differential pressure (TMP).

A consistent decrease in water flux after processing and cleaning indicates fouling of the membrane and implies that the cleaning step did not remove all of the foulants. A new membrane will show a decrease in water flux after the first use; typically, no more than a 10% decrease for cellulosic membranes and a 15%–20% decrease for polyethersulfone membranes. This is normal and indicative of the clean membrane becoming conditioned with solutes. After the first cleaning, the water flux value should be consistent, typically within 10%.

Post-use fouled membranes typically have NWP values that are significantly less than 50% of the membrane's original NWP specification because of adsorption of materials such as proteins on the membrane surface. After cleaning, the NWP values are compared to initial (pre-process) levels and may be analyzed for trends over time (Figures 12.18 and 12.20).

It is important to measure clean water flux using an appropriate water source, such as microfiltered deionized water, reverse-osmosis permeate, water for injection, or 18-megohm water. Tap water contains organic and inorganic solutes that can foul membranes, resulting in unreproducible measurements (49).

12.5.6.7 Process Reproducibility

The goal of a cleaning cycle is to ensure consistent process performance on a product stream. Reproducible process flux, process time, and predictable product yield are directly related to membrane cleaning. Long-term monitoring of the process and cleaning performance provides a means to proactively address potential operating problems and enhances overall process consistency. Stable process pressures and flow conditions indicate absence of plugging and suitability of device design. Consistent water and process fluxes coupled with consistent yields are strong statistics in a membrane reuse validation study (see section 12.5.6.12).

12.5.6.8 Flush Water Residuals

In commercial applications, the TFF devices and system are flushed to displace preservative, storage, and cleaning solutions and to remove residual process materials. Sampling

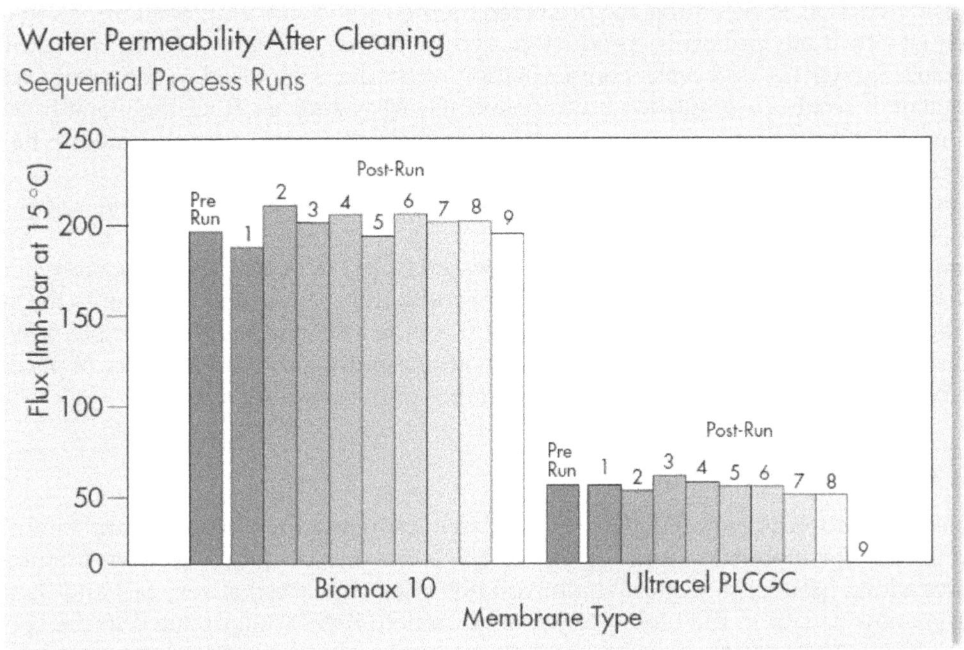

FIGURE 12.20
Example of batch-to-batch post-cleaning water permeability

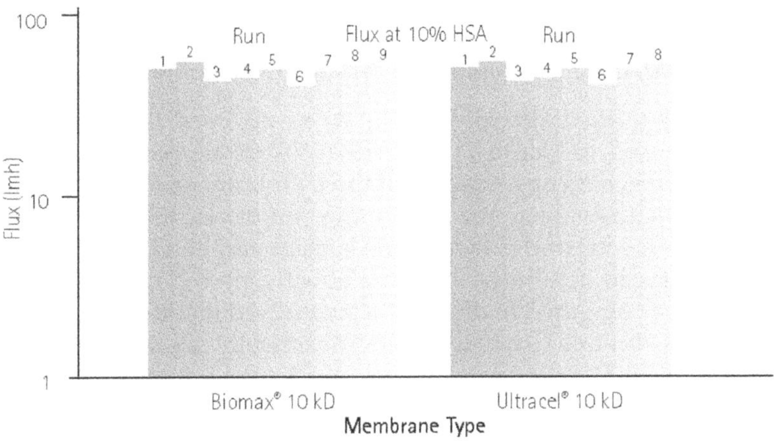

FIGURE 12.21
Batch-to-batch process reproducibility example.

methodologies for closed systems like TFF is rinse sampling. Analysis can be run in- or offline. The flush cycle performed immediately prior to the process feed is termed the "critical flush" and is the most important cycle to identify trace of contaminants. A change in the observed residuals in the critical flush step also indicates a deviation in the cleaning step.

TOC measurements of the permeate and retentate flush streams provide a reliable and highly sensitive means of detecting organic contamination in a cassette (50). Because of its sensitivity, TOC is becoming the preferred method for identifying contamination from inadequate cleaning protocols, product carryover between batches (MAC = maximum acceptable carryover), and cross-contamination of products exposed to the same process equipment. If residue is identified by a nonspecific assay such as TOC, the worst-case scenario must be used and all residue identified in the process must be assumed to be that which is most toxic.

Typical TFF system critical flush volumes of 20 L/m^2 of membrane are used and should yield TOC values below a 1.0 ppm level (51).

Removal of cleaning agent is sometimes assessed by pH or conductivity measurement. If the cleaning agent can be measured by pH or conductivity, these can validate as the assay to demonstrate removal of the cleaning agent from the TFF system. For example, if WFI is being used to rinse 0.5 N NaOH from a system after cleaning, the conductivity or pH of the rinse WFI could be used as reference for clearance of the cleaning agent.

12.5.6.9 *Bioburden and Endotoxin*

The cleaning procedure should also be effective in reducing the bioburden and endotoxin levels. Limits for bioburden and endotoxin levels have to be set. Bioburden monitoring requires a long time for detection. Endotoxin detection (LAL test) is very fast and accurate and somehow linked to the bioburden contamination. Both are correlated to the specific TFF application and working environment.

Validated methods for the LAL assay are available from the FDA (52). It should be noted that the LAL test is sensitive and fast, but certain buffers may interfere with the LAL assay since it is enzyme based, and this should be accounted for in the assay validation. The rabbit test can also be used to test for endotoxin, as it detects pyrogenic substances.

12.5.6.10 *Physical Analysis of Contamination through Destructive Techniques*

Use of specific assays (NIR, FTIR, HPLC, ELISA) as opposed to nonspecific (TOC, pH, conductivity) can be used to identify the residue. Visual assessment and Fourier transform infrared (FTIR) spectroscopy analysis of the membrane surface may be used during the development phase to evaluate the membrane and device after cleaning. After many process cycles, the feed screen and membrane surface can be examined for residuals by autopsying the cassette to determine cleaning effectiveness. The surface of the cleaned membrane may also be compared to the FTIR of an unused membrane sample. This procedure is a destructive technique that may be used to assist in validating a cleaning protocol.

12.5.6.11 *Residue Sampling Methods and Limits Acceptance Criteria*

ICH Q7A section 5.25 states, "Acceptance criteria for residues and the choice of cleaning procedures and cleaning agents should be defined and justified." Section 12.74 states, "Residue limits should be practical, achievable, verifiable and based on the most deleterious residue" (53).

A validated protocol must be in place for the sampling and testing of residues and for the determination of the maximum acceptable residue level. Determination of residues after the cleaning and rinse step can be performed multiple ways, however per section 12.73, "The sampling methods used should be capable of quantitatively measuring levels of residues remaining on the equipment surfaces after cleaning" (53). Destructive methods involve dissection of the device followed by visual or FTIR analysis. Obviously, this is not an option in a reuse application, so non-destructive methods are employed. Non-destructive methods include direct surface sampling such as swab testing, rinse sampling, coupon sampling, solvent sampling, product sampling, and carryover testing. In TFF applications, the most used sampling method is rinsate sampling during the critical water flush procedure (51).

There are many points to consider in setting residue limits. Limits must be practical with respect to the cleaning situation and verifiable with respect to the assay. An analytical methodology must exist for the specific product, and the rationale behind the chosen limit must be scientifically sound. Limits for cleaning validation of TFF modules and equipment are based on considerations about the nature of the residues and the stage along the manufacturing process (example, upstream vs. purification vs. formulation). Acceptance criteria used for TFF cleaning validation in cell culture, purification, and formulation/filling processes may be different. A cleaning validation program based on risk assessment with justifiable and achievable acceptance criteria has been proposed (54).

Limits for cleaning validation generally include measures about the molecule of interest (and sub-products or degradants), cleaning agent (e.g., detergents, solvents), and bioburden and endotoxin levels.

The toxicity of the active matter dictates the specification for residuals, with the level of detection of the assay commonly used. A warning here is that with advancements in technology, assays are continually becoming more sensitive. Setting a residual limit at the level of detection of the assay may prove troublesome if the assay becomes capable of detecting a lower level of residual. Selecting an industry standard value (such as less than 1 ppm) avoids this complication and will be acceptable for most applications.

Table 12.6 provides an example of typical residuals acceptance limit for a TFF system in a biotechnological process.

PDA Technical Report Number 29 recommends the following formula for determination of *MAC* = maximum allowable carryover (55):

$$MAC = \frac{TD \times BS \times SF}{LDD} \tag{12.10}$$

Where:
MAC = maximum allowable carryover
TD = single therapeutic dose
BS = batch size of next product
SF = safety factor
LDD = largest daily dose of next product to be manufactured using same equipment

Typical safety factors for injectable products are 1/1000th—1/10,000th of a daily dose. The safety factor selected will be dependent upon the potency and toxicity of the compound, as well as the analytical limit of detection. Other considerations are the solubility of the product and contaminants in the cleaning reagents and the cleanability of the system. If the equipment is used to manufacture multiple products, then the effects of cross-contamination must be considered and limits calculated using the worst-case scenario. The worst-case scenario assumes that all residue present will be the most toxic contaminant present.

TABLE 12.6

Example of Typical Residuals Acceptance Limit for a TFF System in a Biotechnological Process

Parameter	Proposed limit	Comment
TOC	10 ppm total, 1 ppm per residue	All TOC is assumed to be product as worst case in formulation applications
Product	*MAC* generally 0.001–0.0001x max therapeutic dose	Limit valid for TFF UF formulation
Endotoxin	Typically < 0.5 EU/mL	Endotoxin can rise to 25–50 EU/mL in clarification applications processes with a Gram-negative host
Bioburden	10 CFU/100 mL rinse minimum	WFI limit derived. Higher level accepted if a subsequent sterilization cycles is foreseen
Conductivity	1–5 μS/cm	Valid for inorganic cleaning agent like NaOH or phosphoric acid

12.5.6.12 Membrane Reuse

Tangential flow filtration devices have the benefit of reuse. This is a cost savings to the user, but the savings can be offset by the cost of validating a cleaning protocol and a reuse protocol. Many factors are involved in the decision to opt for single use or reuse. The size of the batch and the number of batches per year will usually be the determining factor.

For a clarification application, normal flow or tangential flow can be used. A commonly used rule of thumb is to use NFF for batch sizes less than 1000 L or if less than ten batches are produced per year. TFF becomes more economical with larger batch sizes or frequent processing, in which case the larger initial outlay for the TFF hardware and reuse validation costs are offset by the cost savings of reusing the membranes (56).

Operating costs will include the filter membranes, cleaning reagents, and WFI. Fixed costs include the hardware and validation costs. Because the validation of TFF systems includes cleaning and reuse, there are higher fixed costs associated with TFF. NFF has higher operating costs, since the filters are single use.

Membrane reuse validation studies starts at the laboratory scale concurrent with process development and cleaning studies. Scale-down studies provide an indication of the starting number of TFF modules reuse. Prospective reuse validation is not practical due to the high costs and time needed to fulfill the requirement of multiple use (some time more than 100–150 reuses). Concurrent validation at manufacturing level as product lots are processed is acceptable. Maximum number of uses should be validated under a membrane lifetime protocol as long as performance parameters are monitored.

Maximum storage and cleaning time may be relevant to confirm exposure does not affect process or product consistency.

Membrane reuse validation should demonstrate (43):

- Batch-to-batch reproducibility in terms of process and product quality attributes
- Removal of soil, foulants, and contaminants in order to prevent lot-to-lot carryover
- Containment of bioburden and endotoxin levels within acceptable limits

A lifetime study approach considers monitoring some process and product parameters; NWP, integrity test, and process parameters (including TMP, flux, membrane retention, and process yields) are monitored every batch and should be trended over time. Product attributes (CQAs like aggregates, level of contaminants, osmolarity, etc.), as well as flush water residuals (MAC, bioburden, endotoxin), need to be monitored with a different

TABLE 12.7

Recommended Frequency for Parameters Control along the Reuse Validation Study

Parameter	Frequency
NWP	Every batch
Integrity	
Process parameters	
Product quality attribute	Frequent[*]
Flush water residuals	
Physical analysis	During PD or cleaning validation

[*] Based on historical data, company experience, and risk assessment

frequency. Frequency should be established prior to validation. Frequency may increase in case of adverse trend (43). Historical data, previous experience on a similar application, and risk analysis can help to define frequency. It could be suggested to start monitoring all parameters at every batch where possible and reduce the frequency in case of repeatable, consistent, and acceptable trend.

The product retention and contaminant passage must not change over time; this is a guarantee of product purity consistency. Any change in these critical parameters indicates a change in the separation and therefore the process.

System integrity should be monitored before and after each use, while clean water flux and process flux should be recorded and trended over time (see Figure 12.18). Consistent water and process fluxes coupled with consistent yields are strong statistics in a membrane reuse validation study. The product retention and contaminant passage must not change over time; this is a guarantee of product purity consistency. Any change in these critical parameters indicates a change in the separation and therefore the process. It is important to have an SOP in place allowing rework in the event of a failure. However, reworking requires validation.

Table 12.7 shows the recommended frequency for several parameters along the reuse validation study.

When making changes to the system, the amount of revalidation required will be dependent upon the impact of the change upon the process dynamics. If extra modules are added to a plate and frame system, this may not require membrane reuse validation or process revalidation, because no impact would be expected on the lifetime of the membranes if there has been no change in the process or cleaning regimen. The process will not be affected if all the operating parameters are equivalent. However, system cleaning will most likely require revalidation due to the change in system design, which may affect the efficiency of the cleaning step. The additional membrane area may require increased cleaning and flush volumes.

12.6 Conclusion

The validation effort is ultimately rewarded by a detailed understanding of the process and the product. Through the validation process, control parameters are put in place at each unit operation. These parameters ensure that the process is reproducible and that the process will generate an equivalent product with every batch. As well as establishing limits of normal operation, the outer limits of the process operation are determined, thereby generating a great deal of data describing how the process behaves under varied conditions. A rugged process is much less likely to fail, and in-depth knowledge of the process and how it reacts to minor perturbations allows for monitoring of the process through multiple runs and understanding of why changes are occurring. This allows the opportunity to compensate for and/or correct for these changes before a batch is lost.

Validation is becoming more of a team approach within drug manufacturing facilities and between drug manufacturers and filter vendors. It is also being incorporated earlier into the drug development phase. The validation department works with R&D, process development, QA and QC, manufacturing, and engineering to ensure that quality is built into the process. Advances in each of these areas are contributing to faster drug development timelines and more quality being built into drug manufacturing processes. For example, the recent integration of process engineering is resulting in improved system designs that minimize system impact on product quality and yield. Advances in assay development are resulting in faster assay turnaround times, which enable faster process times. Vendors are continually providing higher-quality products for use in drug manufacturing processes, such as more robust filter membranes with increased flux to provide faster processing. Disposable products that reduce the burden of cleaning validation, such as disposable capsule filters and bags, are increasing in popularity. Vendors supply validation guides with filter products, which speeds the drug manufacturer's process validation.

The stringency of regulatory requirements is increasing, with emphasis being placed on process characterization and validation. This requires selection of appropriate filtration devices and establishment of acceptable windows of operation for control parameters such as flow rates, pressures, and temperature. These control parameters must meet criteria for yield, quality, purity, and process time. Process development personnel must anticipate the requirements of the full-scale process to ensure that the optimized process is robust, scalable, and validatable. Faster process development while investigating a range of control parameters is enabled through factorial design, contributing to speed to market. As technology improves, assays are developed that are more sensitive, allowing for improved product and contaminant characterization and quantification. This advancement in technology may become an obstacle to drug manufacturers, in that as the ability to characterize and quantitate contaminants improves, regulatory agencies may look to drug manufacturers to design processes resulting in higher product purity levels.

More rigorous process characterization, coupled with faster regulatory approval times and improved technology, are resulting in shorter timelines from drug discovery to market. Yet, even with these shorter timelines, more quality is being built into current drug manufacturing processes, due to the validation process. This trend will ensure that drug manufacturers will continue to provide safe and effective drug products.

Trademarks

Merck, Durapore, Millex, Millipore Express, and OptiScale are registered trademarks of Merck KGaA, Darmstadt, Germany.

The Vibrant M and Vmax are trademarks of Merck KGaA, Darmstadt, Germany.

ATCC is a registered trademark of American Type Culture Collection

ASTM is a registered trademark of American Society for Testing and Materials

References

1. FDA Guidance for industry—Process Validation: General Principles and Practices—January 2011, Current Good Manufacturing Practices (CGMP)—Revision 1
2. ICH Q8(R2)—Pharmaceutical Development—August 2009
3. ICH Q9—Quality Risk Management—November 2005
4. ICH Q10—Pharmaceutical Quality System—June 2008
5. T. H. Meltzer. *Filtration in the Pharmaceutical Industry.*—1st ed. New York: Marcel Dekker, 1987, pp. 532–539.
6. G. M. Morris, J. Rozembersky, L. Schwartz. Validation of filtration. In *Biopharmaceutical Process Validation.* G. Sofer, D.W. Zabriskie, eds. New York: Marcel Dekker, 2000, pp. 213–233.
7. FDA Guidance for Industry—Sterile Drug Products Produced by Aseptic Processing—Current Good Manufacturing Practice. Sept 2004.
8. European Guidelines to Good Manufacturing Practice—Volume 4 Medicinal Products for Human and Veterinary Use—ANNEX 1: Manufacture of Sterile Medicinal Products—Revised 2022.
9. Chinese Guidance of Good Manufacturing Practices for drug—Annex 1: Sterile Medicinal Products—Chapter 12, Sterilization method, Article 75, Revised 2010.
10. Japanese Guidance on the Manufacture of Sterile Pharmaceutical Products by Aseptic Processing, April 2011.
11. Aseptic processing of health care products—Part 2: Sterilizing filtration, ISO 13408–2: 2018.
12. PDA Journal of Pharmaceutical Science & Technology—Technical Report N° 26: Sterilizing Filtration of Liquids.—2008 Supplement, 62(S-5).
13. Millipore Application Note AN1505EN00—Establishing Product Specific Bubble Point Specifications for Sterilizing-Grade (0.22 µm) Durapore® Filters.
14. M. Huang et al. Impact of extractables/leachables from filters on stability of protein formulations. *J. Pharmaceut. Sci.*, 100(11), November 2011.
15. FDA, Code of Federal Regulations, Part 211, "Current Good Manufacturing Practice for Finished Pharmaceuticals", Part 211.65, "Equipment Construction", 2019
16. European Commission, EUDRALEX Volume 4, "Good Manufacturing Practices, Medicinal Products for Human and Veterinary Use", Chapter 3, "Premise and Equipment", 2015
17. Biophorum: Best practices guide for Extractable testing of polymeric single-use components used in the biopharmaceutical manufacturing, 2020.
18. S. K. Mohavenlu, J. Shea. The role of BPOG extractables data in the effective adoption of single-use systems. White paper. Lit. No. MK_WP3014EN 11/2018 Merck
19. Biophorum operations group: Best practices guide for evaluating leachables risk from polymeric single-use systems used in the biopharmaceutical manufacturing, 2018.
20. T. H. Broschard et al. Assessing safety of extractables from materials and leachables in pharmaceuticals and biologics: Current challenges and approaches. *Regulatory Toxicology and Pharmacology*, 81: 201–211, 2016.
21. Guideline on the sterilisation of the medicinal product, active substance, excipient and primary container—EMA/CHMP/CVMP/QWP/850374/2015
22. PDA Journal of Pharmaceutical Science & Technology, Technical Report N° 45 Filtration of liquids using cellulosed-based depth filters. 2008 Supplement, Volume 62 Number S-2.
23. FDA; ICH. Guidance for Industry. Q5A, Quality of Biotechnological Products: Viral Safety Evaluation of Biotechnology Products Derived From Cell Lines of Human or Animal Origin. ICH, 5 March 1997 (Adopted as a guidance document and published in the Federal Register September 1998)
24. The European Agency for the Evaluation of Medicinal Products: Human Medicines Evaluation Unit, Committee for Proprietary Medicinal Products (CPMP).—Guideline on Virus Safety Evaluation of Biotechnological Investigational Medicinal Products—2009. CPMP/BWP/398498/2005.

25. Note for Guidance on Virus Validation Studies: The Design, Contribution and Interpretation of Studies Validating the Inactivation and Removal of Viruses, EMEA CPMP BWP, 268/95, 1996.

26. A. Kundu, K. Reindel. Evaluation of viral clearance in purification processes. In *Process Scale Bioseparations for the Biopharmaceutical Industry*, A. A. Shukla, M. R. Etzel, S. Gadam, eds. Boca Raton, FL: CRC Press, 2006, pp. 419–448.

27. J. Carter, H. Lutz. An overview of viral filtration—in biopharmaceutical manufacturing. *Eur. J. Parenteral Sci.* 7(3): 72–78, 2002.

28. G. Miesegaes, S. Lute, H. Aranha, K. Brorson, M. C. Flickinger. Virus retentive filters. In *Encyclopedia of Industrial Biotechnology*, M. C. Flickinger, ed. New York: John Wiley & Sons, Inc., 2009.

29. R. L. Garnick. Experience with viral contamination in cell culture. *Dev. Biol. Stand.* 88: 49–56, 1996.

30. E. Gefroh, H. Dehghani, M. McClure, L. Connell-Crowley, G. Vedantham. Use of MMV as a single worst-case model virus in viral filter validation studies. *PDA J. Pharm. Sci. Technol.* 68(3): 297–311, 2014. doi: 10.5731/pdajpst.2014.00978. PMID: 25188350.

31. R. V. Levy, M. W. Phillips, H. Lutz. Filtration and the removal of viruses from biopharmaceuticals. In *Filtration in the Biopharmaceutical Industry*, T. H. Meltzer, M. W. Jornitz, eds. New York: Marcel Dekker, 1998, pp. 619–646.

32. Journal of Pharmaceutical Science and Technology, Technical Report N° 41, Virus Filtration, 2005, Supplement Volume 59

33. FDA. *Points to Consider in the Manufacture and Testing of Monoclonal Antibody Products for Human Use.* Rockville, MD: U.S. Department of Health and Human Services, FDA, CBER, 1997.

34. M. Siwak, A. Hong, J.R. Cormier, D. Kinzlmaier.—U.S. Pat. 7, 2008: 465, 397.

35. M. Siwak, A. Hong, J. R. Cormier, D. Kinzlmaier. U.S. Pat. 7, 2006: 118, 675.

36. H. Lutz, W. Chang, T. Blandl et al. Qualification of a novel inline spiking method for virus filter validation. 27 September 2010 in Wiley Online Library

37. G. Bolton, M. Cabatingan, M. Rubino, et al. Normal-flow virus filtration: Detection and assessment of the endpoint in bio-processing. *Biotechnol Appl Biochem.* 42: 133–142, 2005.

38. S. Lute, M. Bailey, J. Combs, M. Sukumar, K. Brorson. Phage passage after extended processing in small-virus-retentive filters. *Biotechnol. Appl. Biochem.* 47: 141–151, 2007.

39. K. Shudipto Ditary, A. Venkiteshwaran, A.L. Zydney.—Probing effects of pressure release on virus capture during virus filtration using confocal microscopy.—*Biotechnol. Bioeng.* 112(10), October 2015.

40. D. Lacasse, P. Genest, K. Pizzelli, P. Greenhalgh, L. Mullin, A. Slocum. Impact of process interruption on virus retention of small-virus filters. *Bioprocess Int.* 11(10): 34–44, 2013.

41. S. L. Michaels, A. S. Michaels, C. Antoniou, et al.—Tangential flow filtration. In *Separations Technology*. W. P. Olson, ed. Buffalo Grove: Interpharm Press, 1995, pp. 57–194.

42. J. Oliver—A 3-D Risk Assessment Model Journal of Validation Technology [Autumn 2008] pp70–76.

43. Parenteral Drug Association Technical Report No. 15, -revised 2009, Validation of Tangential Flow Filtration in Biopharmaceutical Applications.

44. USP XXIII, pp 1699–1703. Section 88—Biological Reactivity Tests, *in vivo*.

45. Merck Millipore, Lit No. TB5882EN00—Recovery Optimization of Process Scale Ultrafiltration/Diafiltration Systems—Ver. 2.0 06/2016

46. M Dawson—ENDOTOXIN LIMITS For Parenteral Drug Products—April, 2017 BET White Paper vol.1 no.2 ©.

47. M. Cheryan.—Membrane properties. In: *Ultrafiltration Handbook*. Lancaster: Technomic Publishing Co., 1986, pp. 53–72.

48. [Code of Federal Regulations] [Title 21, Volume 4] [Revised as of April 1, 2019] [CITE: 21CFR211.67]

47. S. L. Michaels. Clean-water permeability as a determinant of cleaning efficiency in tangential flow filtration systems. *BioPharm* 7: 38–45, 1994.

50. K. Clark. How to develop and validate a total organic carbon method for cleaning applications. *PDA Journal* 55: 290–293, 2001.

51. EMD Millipore Technical Brief TB1502EN00 Ver. 2.0 11/2015.—Techniques for Demonstrating Cleaning Effectiveness of Ultrafiltration Membranes.
52. FDA. Guidance for Industry: Pyrogen and Endotoxins Testing: Questions and Answers. FDA, June 2012.
53. ICH Q7, Good manufacturing practices guide for active pharmaceutical ingredients, 10 November 2000. https://www.ich.org/page/search-index-ich-guidelines.
54. Edward K. White. Risk-based cleaning validation in biopharmaceutical API manufacturing. *BioPharm Int.*, 01 November 2005.
55. Parenteral Drug Association Technical Report No. 29, (revised 2012)—Points to consider in cleaning validation.
56. Millipore Application Note AN1511EN00 Rev. 3/00. Mammalian Cell Culture Clarification.
57. Millipore Technical Note dTN049 Rev. A—3/97. Endotoxin Removal.

13

Validation of Continuous Bioprocesses

Marc Bisschops and Mark Schofield

13.1 Introduction

Over the last 50 years the biopharmaceutical manufacturing industry has been going through different stages of maturation. McLaughlin (1) describes the evolution of biopharmaceutical manufacturing in four stages.

The 1980s were mostly driven by "Plan for Success." The industry was focused on navigating the early biotech product candidates through clinical stages to obtain regulatory approval. The consequence for biopharmaceutical manufacturing was that a lot of effort was dedicated to ensuring patient safety and efficacy. Process development was primarily focused on identification and elimination of impurities. Contract manufacturing organizations (CMOs) became a natural choice for partnering to ensure availability of material for clinical studies and product launch.

The second wave in the industrialization of biopharmaceutical manufacturing happened in the 1990s and is described as "Titer and Yield." Cell culture sciences progressed quite rapidly allowing significant increase in titers for recombinant proteins and monoclonal antibodies. At the same time, the yields in downstream processing operations were gradually improved. The consequence was that the manufacturing facilities that were designed and built based on the assumptions of the 1980s were oversized for their purpose. Some failures of drugs in (late) clinical phases augmented this excess manufacturing capacity. This led to the realization that biopharmaceutical manufacturing would be served with more flexibility to respond to the dynamic supply demands. The adoption of single-use manufacturing solutions was the logical consequence of these trends.

In the third wave of the industrialization in the 2000s the focus shifted predominantly towards "First in Human." Long process development times were identified as a bottleneck with significant financial impact. The adoption of high-throughput experimentation for screening a wide variety of process conditions accelerated the development timelines significantly. At the same time, significant progress was made with analytical technologies to support process development efforts. The concept of manufacturing platforms became very popular. Such platforms would be designed to manufacture a wide range of products with limited variability in materials and process conditions. These efforts addressed most constraints with respect to clinical manufacturing and targeted faster clinical development.

The fourth and last wave that McLaughlin describes is referred to as "Continuous Vision" or "Integrated and Intensified." In the 2010s, the first concepts for continuous downstream processing and connected manufacturing platforms were initiated. This allowed the biopharmaceutical manufacturing sciences to shift their focus towards manufacturing efficiency and costs. Enabling technologies such as continuous multicolumn chromatography,

DOI: 10.1201/9781003143130-13

inline concentration, and inline diafiltration processes gained momentum, and perfusion cell culture went through a revival with more robust cell retention technologies becoming available. Together these developments enabled manufacturing platforms that focused on process efficiency without compromising on efficacy and safety. Various biopharmaceutical companies started exploring the opportunities of fully integrated (end-to-end) continuous manufacturing. In spite of all efforts, it wasn't until 2019 that the first monoclonal antibody produced in a fully integrated (end-to-end) continuous bioprocessing platform entered into clinical studies (2, 3).

The drive to continuous and intensified bioprocessing is facilitated by the manufacturing platforms that were developed in the third wave of industrialization. Although there are quite a few variations among the platforms, a consensus set of unit operations and their place within the process has become defined. These include clarification, capture chromatography via Protein A, virus inactivation (VI), two chromatographic polishing steps, virus retention filtration (VRF), tangential flow filtration (TFF), and finally sterile filtration. Konstantinov and Cooney described various options to transfer the platform to continuous bioprocesses (4). These are schematically represented in Figure 13.1 to Figure 13.5.

The scheme shown in Figure 13.1 represents a manufacturing platform that utilizes perfusion cell culture with a cell retention system attached to it. Subsequent downstream processing steps are conducted in traditional batch systems. This scheme has been employed at commercial scale for many years by companies such as Genzyme, Bayer, Biomarin, Janssen, Merck-Serono, and Shire for the manufacture of less stable biopharmaceuticals, including enzymes and Factor VIII. The initial incentive to consider perfusion upstream was driven by the product stability. With residence times well below 24 hours in the bioreactors, a perfusion process generally delivers better quality and yield for labile products than a fed-batch process.

However, for manufacturing stable biopharmaceutical products, fed-batch cell culture processes have become the dominant platform. This is primarily driven by the fact that they are less susceptible to fouling, making them more robust and generally easier to operate. The increasing titers in fed-batch processes has been an important driver for considering the use of continuous downstream process steps. A fed-batch cell culture process

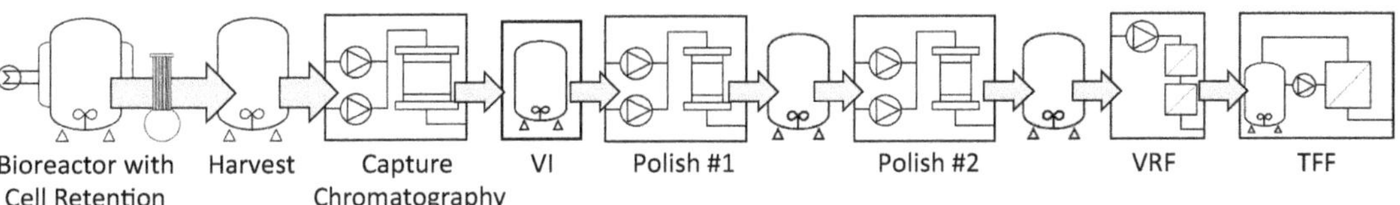

FIGURE 13.1
Schematic representation of a continuous (perfusion) cell culture process with batch downstream processing steps.

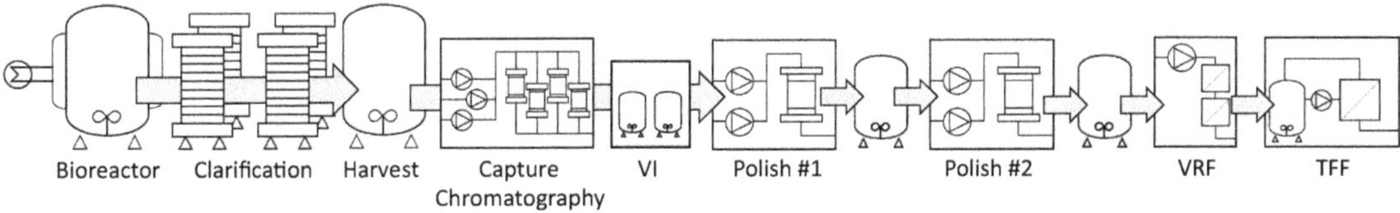

FIGURE 13.2
Schematic representation of a fed-batch cell culture and clarification process with a continuous capture followed by batch polishing steps.

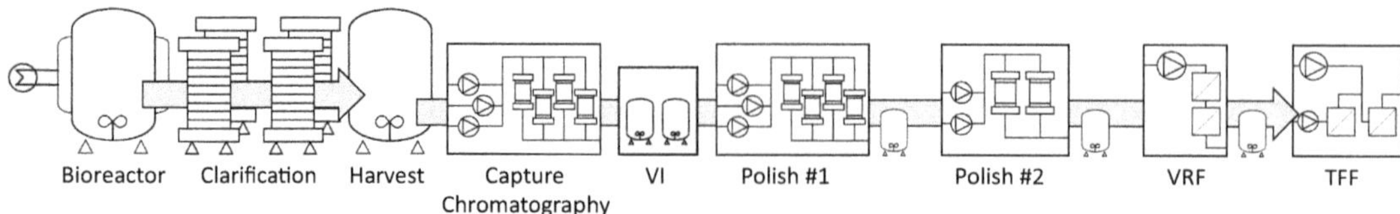

FIGURE 13.3
Schematic representation of a fed-batch cell culture and clarification process with a continuous downstream processing platform.

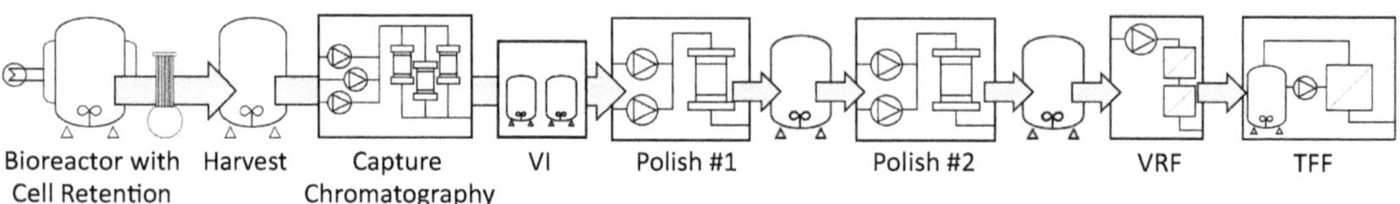

FIGURE 13.4
Schematic representation of a continuous (perfusion) cell culture process with continuous capture and batch polishing steps in the downstream processing.

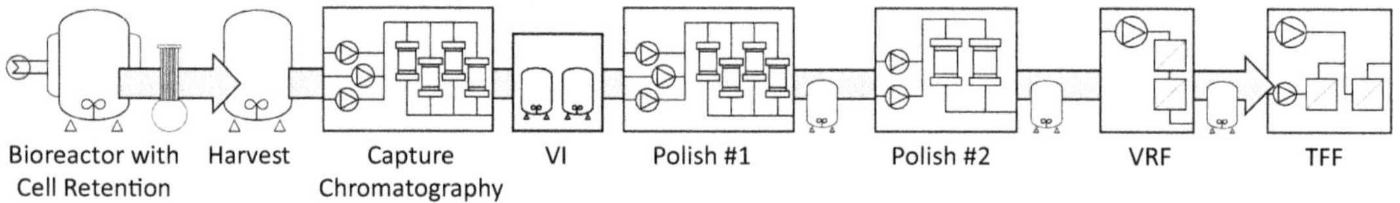

FIGURE 13.5
Schematic representation of a continuous (perfusion) cell culture process with a continuous downstream process platform.

in a 2000-liter bioreactor nowadays can manufacture as much as 10 kg of monoclonal antibody (at 5 g/L expression level). This would require a Protein A capture column of approximately 100 L of traditional affinity adsorbent (3 cycles per batch at 35 g/L binding capacity). The use of continuous multicolumn chromatography has been demonstrated as a very effective approach for significantly reducing the amount of affinity adsorbent that is required by 75% or even more (e.g., 5, 6). Considering the fact that fed-batch cell culture has been the dominant design over the past decades, the schedule as shown in Figure 13.2 has been the first stepping stone for many companies to explore continuous downstream processes.

The logical extension of the continuous capture (and virus inactivation) steps is a continuous polishing step based on multicolumn chromatography as well. This is schematically shown in the scheme presented in Figure 13.3. To establish longer (continuous) manufacturing campaigns, multiple bioreactors could be used to feed into a single continuous downstream processing platform. The scheduling of the cell culture process and the number of bioreactors would then have to be aligned with the feed into the continuous capture process and the intermediate product stability of the clarified cell harvest.

The manufacturing platforms depicted in Figure 13.4 and Figure 13.5 are essentially identical to the preceding ones, albeit that the upstream process is based on a continuous perfusion cell culture bioreactor with a continuous cell retention device. For fully continuous production over longer periods of time (e.g., 30 to 60 days), platforms based on perfusion cell culture are often considered as a preferential approach.

Considering the fact that continuous cell culture processes have been used for licensed products for several decades, the main focus of this chapter is on downstream processing platforms that utilize one or more continuous bioprocessing unit operations.

13.2　Process Validation

13.2.1　Regulatory Requirements

Effective process validation is an inherent part of quality assurance for pharmaceuticals. Process validation is defined as the collection and evaluation of data, from the process design stage through commercial production, which establishes scientific evidence that a process is capable of consistently delivering product at the predefined quality (7). Process validation is a critical requirement since quality, efficacy, and safety of the drug product cannot be adequately assured through in-process and finished-product testing (7). The European Medical Agency (EMA) has issued a similar guideline that follows, to a large extent, the same structure (8). More recently, the FDA has published some conceptual guidance for their expectations with respect to continuous processing in the pharmaceutical industry (9).

There seems to be consensus that these documents provide an adequate guidance for the validation of continuous bioprocessing technologies (10). These guidelines and expectations will be used to evaluate potential strategies for validating manufacturing platforms that include continuous bioprocessing technologies.

In the Validation Guidance (7), the FDA discriminates three stages in the product life cycle that should be considered within the context of process validation: (1) process design, (2) process qualification, and (3) continued process verification. These are discussed next. It should be obvious that these three stages also apply to intensified or continuous processes.

13.2.2　Process Design

During the process design (or process development) stage, the main efforts are focused on building and capturing process knowledge and understanding. Although these steps do not have to follow the current good manufacturing practices (CGMP), it is important to note that the activities are an inherent part of the process characterization and validation work and should therefore be consistent with good scientific practices and good documentation practices. As such, the guidance described in ICH Q10 should also be considered during process design activities.

13.2.2.1　Quality by Design

During process design, it is strongly encouraged to align the efforts with the quality by design philosophy. This includes a sound risk assessment to identify criticality of the product quality attributes (CQAs) and from that identify the critical process parameters (CPPs). The need to establish proven acceptable ranges (PARs) and normal operating ranges (NORs) for the critical process parameters is not any different in a continuous process from in a batch manufacturing platform.

Continuous bioprocessing technologies, however, do not lend themselves easily to high-throughput experimentation in the same way as batch downstream processing systems do. Continuous process steps tend to take some time before they reach steady-state conditions, and as a consequence, a study that follows design of experiments (DOE) may consume more time and material than available during the process design stage.

In order to mitigate this challenge, a model-based quality by design approach has been suggested (11, 12). This strategy relies on the notion that continuous downstream processing steps rely on the same fundamental principles as the traditional batch unit operations. Successful translation of a batch process into a continuous process has often been demonstrated for a variety of steps (e.g., 13, 14). Establishing the design space for a continuous unit operation therefore can be based on combining the knowledge and understanding that is generated in small-scale batch processes.

This approach has been schematically shown for an arbitrary (generic) monoclonal antibody purification platform in Figure 13.6 (15). The example shown in this figure is based on the work reported in the A-Mab case study (16). The approach isolates the identification of the CPPs based on their impact on the CQAs for each individual unit operation, based on risk assessment. The NORs and PARs for these individual process steps can be established in small-scale batch experiments. Based on a risk assessment, suitable process conditions are then selected for verification in a bench-scale continuous system. A recent study that provided a full comparability between batch and continuous manufacturing of a monoclonal antibody supported the vision that batch process unit operations provide a good predictor for the product quality attributes that are delivered in the equivalent continuous platform (17).

13.2.2.2 Process Dynamics

Although the concept presented in Figure 13.6 looks rather straightforward, it is important to note that linking unit operations will result in dynamic process conditions between

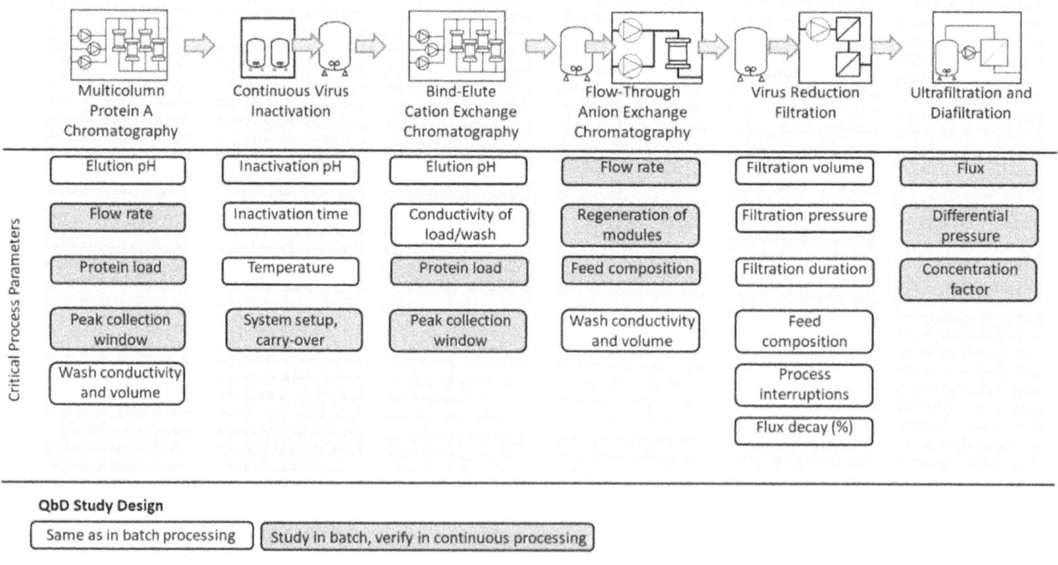

FIGURE 13.6
Conceptual diagram for a quality by design approach in a continuous downstream processing platform (15).

different process steps. As a consequence, studying unit operations in isolation may not be sufficient as such studies often involve homogeneous and stable feed conditions. In a continuous downstream processing platform, the dynamic output of one unit operation becomes the feed solution (typically a critical material attribute, CMA) for the next. Dynamic fluctuations in the feed solution can cause the unit operation to (intermittently) operate outside its proven acceptable ranges even though the average conditions may fall within the PAR and NOR. An example is the conductivity fluctuations coming from a polishing chromatography step going into a next unit operation step. At too high conductivities, the ability of such next unit operation step to, for instance, effectively remove viruses and other impurities may be compromised, even if the (time) averaged conductivity remains within the limits of the design space. This may be resolved by incorporating an adequately sized mixer as a surge container between the steps to dampen the fluctuations in the feed condition. The design and characterization of such a mixer then becomes an integral part of the process design and process validation plan.

Process dynamics during startup and shutdown are also to be considered as an integral part of the process validation. In order to complement experimental studies with process dynamics, one could consider a model-based approach based on adequate numerical simulations, such as presented for multicolumn chromatography processes (18).

13.2.2.3 Virus Safety

Manufacturing of monoclonal antibodies is performed with the use of mammalian cells and hence there is a risk of unintentionally introducing viruses into the process. The regulatory expectation on assuring virus safety includes (a) the selection of materials and cell lines that minimize the risk of introducing adventitious viruses, (b) the need to design the downstream processing of biopharmaceuticals to provide sufficient capacity to clear infectious viruses, and (c) testing appropriate intermediate product for the absence of infectious viruses (19, 20). For in-process virus safety, the regulatory authorities recommend at least two orthogonal steps to be performed. Virus clearance assessments are commonly operated using adequately designed scale-down models and are often performed at dedicated testing laboratories. For continuous bioprocessing, these scale-down validation studies involve various complications.

Model viruses used in virus removal studies are not always stable, which presents a challenge for continuous bioprocessing steps that require some time to reach steady state. In addition to this, the flow rates in continuous bioprocessing are often lower than in the equivalent batch processes. If the auxiliary materials are predominantly designed to process a certain volume rather than handle the flow rate, the lifetime that needs to be validated can be long. An example of this is virus filtration that may be operated for multiple days per filter in continuous mode.

The challenges of validating virus removal in a multicolumn capture chromatography process were recognized early, and initial proof of concept work demonstrated that a single-column could be an adequate mimic for a multicolumn chromatography process to establish virus removal capabilities. Although the initial work was performed with bacteriophages as a model, it provides a promising avenue to relatively straightforward virus safety studies (21). The study demonstrated for various monoclonal antibodies that the log reduction value (LRV) for bacteriophages in a single column was representative for the performance of a continuous multicolumn chromatography system, if operated under the same conditions. The LRV seemed to be independent of the configuration of the continuous chromatography system. A similar study with a two-column chromatography system for Protein A capture confirmed that the virus clearance in a continuous process and batch process are comparable. This study also suggests that a scale-down batch column can be used for assessing virus clearance in a continuous study (22).

Another step that is critical for virus safety is the virus inactivation step. For most monoclonal antibody processes, this is performed at a pH of approximately 3.5 (23). The main critical process parameters for this step are the pH and the incubation time (or exposure time). In a traditional process, this is performed in a (well-designed) mixer and the small-scale validation of this is rather straightforward.

For this reason, a fully automated repetitive batch approach offers an appropriate solution combining efficiency and simplicity. This approach is, for instance, the basis for the Cadence VI system developed by Pall Biotech (Port Washington, NY). The system allows a first single-use mixer to be filled with the eluate of a (continuous) Protein A chromatography step while a second single-use mixer performs the incubation at low pH. Once the incubation and subsequent neutralization have been completed, the second mixer is emptied and becomes available for collecting new eluate. At that point, the product solution in the first mixer will be acidified and the incubation can start. Downstream of the virus inactivation system, a break tank will be required to manage the intermittent release of inactivated process solution. Since this unit operation is essentially based on a repetitive batch process, scale-down validation of the continuous virus inactivation step would rely on the same approaches as the traditional batch virus inactivation process steps.

As an alternative to the dual mixer strategy, approaches based on plug flow have been presented and tested (e.g., 24, 25, 26). These approaches are not trivial, as it can be difficult to assure the virus inactivation pH is achieved for all of the process liquid as it flows past a pH meter due to the hysteresis of pH measurement. This is important because achieving the inactivating pH is essential for virus safety and is a CPP. Moreover, it is not possible to take a representative sample of the inactivation pool. Additionally, there are a variety of

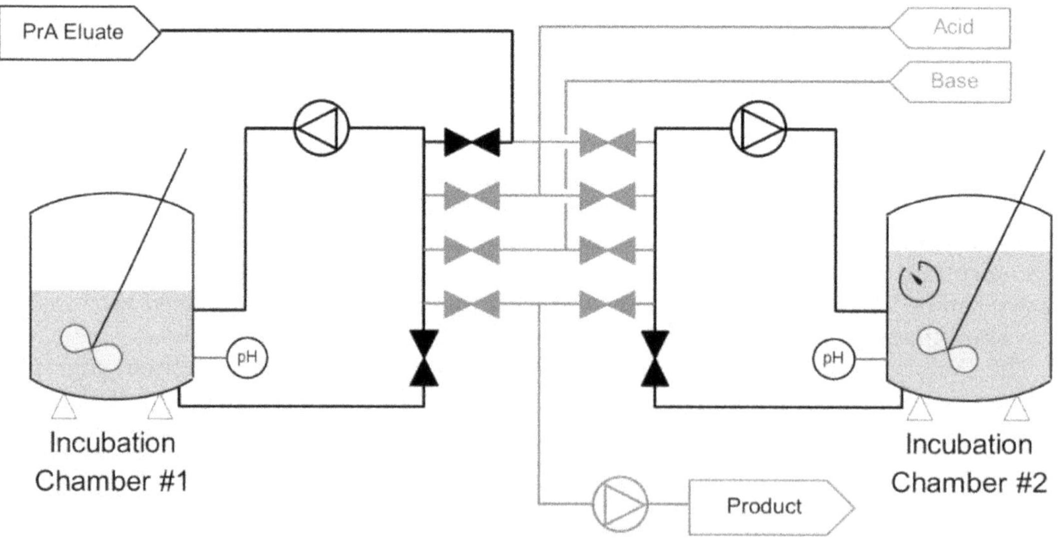

FIGURE 13.7
The repetitive batch approach for virus inactivation relies on two incubation chambers that alternate between collecting the eluate from the Protein A chromatography step (left: Chamber #1) and the incubation/inactivation phase (right: Chamber #2).

uncertainties with respect to scalability, as the residence time distribution depends on the hydrodynamic conditions in the flow contactor. One of the validation strategies for such plug flow devices relies on residence time distribution models (26). If such models can be demonstrated to represent the behavior of the process at commercial manufacturing scale, such model-assisted validation strategies may complement the experimental. An overall validation strategy should include an adequate risk assessment that also covers dynamic behaviors such as flow interruptions and in-process controls.

13.2.2.4 Control Strategy

During the process design stage, the control strategy for the platform should also be established. For the part of the downstream process that involves one or more continuous unit operations, adequate automation and process control is required to ensure uninterrupted operation. This necessity has the advantage that all process parameters, including CPPs, are automatically recorded.

In the continuous downstream processing of intensified biomanufacturing platforms, companies are focusing on a strategy based on a functionally closed fluid path. Very often this involves a single-use flow path. This approach has many advantages with respect to contamination control. If well designed, a flow path based on single-use technologies provides assurance of zero initial bioburden and hence the necessity for cleaning and sanitizing prior to use can be reconsidered. Drawbacks of a functionally closed system are that single-use sensors tend to be less accurate (27) and aseptic sampling is often more challenging. Once integrated in a single-use assembly, recalibrating sensors can be difficult or even impossible, without breaching the functionally closed process environment.

An example of such a challenge was described by BiosanaPharma for their virus inactivation step in the fully integrated continuous bioprocessing platform (28). The CPPs for the virus inactivation process were established offline in small-scale validation studies. With these boundaries, the control strategy was set around the NORs for the incubation pH. The pH was measured, however, with single-use probes, and it was acknowledged that these could be subject to some drift. The mitigation strategy relied on periodically sampling the process while at reduced pH. If the pH of the calibrated offline sensor deviated more than the predefined value, the reading of the internal single-use pH probe was adjusted to match the calibrated pH probe. The control space for the single-use pH was selected such that the operation would fall outside the design space (or PARs), even if significant drift would have been observed.

13.2.3 Process Qualification

The second stage in the product life cycle to be considered for process validation is process qualification. The FDA guidance defines this stage through two elements: (1) design of facility and equipment, and (2) process performance qualification (PPQ).

The first element most likely does not present any challenges that are unique to continuous bioprocessing solutions. With functionally closed process systems, a ballroom facility for manufacturing becomes a viable option. The strict segregation between different parts of the process can be addressed by validating the functionally closed state of the process systems. The process systems themselves need to be qualified to provide evidence that they are fit for use within the intended range. In a connected process, the startup and shutdown may impose a separate challenge, as it may not always be possible to isolate the unit operations from their adjacent systems.

The second element of the process qualification stage is the process performance qualification. It is intended to bring all elements together for a qualified facility, including utilities and equipment, trained operators, controls, procedures, methods, and materials. These elements are brought together to demonstrate that the commercial manufacturing process performs as expected and delivers the product with the predefined quality attributes. This is a critical step in the product life cycle, and commercial distribution of the drug cannot commence without successful completion of the PPQ (7).

Although the PPQ should be done at scale, the cumulative data from all relevant studies, including small-scale experiments, should be used to establish the manufacturing conditions. Where appropriate, the data can be complemented with sound science, data, and know-how, obtained for similar products and processes.

During the PPQ phase, it is expected that a higher frequency of sampling be considered. In systems with dynamic behaviors, such as many continuous downstream processing systems with a cyclic operation, the sampling strategy must be designed to accommodate the process dynamics and to obtain sufficiently statistically relevant data on process performance and product quality.

Many downstream processing unit operations rely on auxiliary materials that are critical to the process performance, such as filters and chromatography adsorbents. These materials typically have a limited performance lifetime, as filters can get clogged and adsorbents are often subject to fouling and/or capacity loss. The usable lifetime of such auxiliary materials can often be established using small-scale (high-throughput) experiments. The lifetime can then be confirmed by an ongoing PPQ protocol during commercial manufacturing.

Processes that rely on PAT for controls may require a different PPQ approach: focus on the effectiveness and dynamics of the controls becomes a critical part of the validation during process design and PPQ.

13.2.4 Continued Process Verification

During the commercial manufacturing stage, the process is subject to continual validation. Trending analysis of the manufacturing process allows for the identification of potential problems early on so that they can be corrected and prevented before the process goes out of specification.

Continuous bioprocessing technologies in general produce far more data than their equivalent batch processes. As an example, one could consider the number of elution peaks that are generated in a typical affinity capture process step. In traditional batch processes, the product produced in one fed-batch bioreactor is typically purified in three to five cycles on a chromatography column. In continuous capture processes, this can easily involve 20 cycles with three to eight columns, resulting in 60 to 160 elution peaks. The use of enhanced statistical methods, such as multivariate data analysis (MVDA), enables the analysis of such large datasets to recognize trends and identify small upsets at a very early stage. This allows manufacturers to implement corrective and preventive actions before the problem causes deviations and/or manufacturing interruptions.

Continued process verification should not be limited to the process information that is generated online but should include any and all information from in-process controls (IPC) and (intermediate) product quality attributes. In addition to this, it should include information on all materials that are being used to the extent that they can be relevant to the product quality and/or data.

Considering the amount of data, the speed at which data is generated, and the diversity in the nature of the data, it is fair to say that continued process verification for continuous bioprocessing technologies will likely end up to be a so-called Big Data problem. Being aware of the associated challenges and opportunities will allow companies to develop adequate strategies to support effective continued process verification (27).

13.3 Validation Examples of Continuous Unit Operations

13.3.1 Multicolumn Capture Chromatography

13.3.1.1 Process Design

Process design strategies for multicolumn chromatography have been presented and published many times. Most CPPs for continuous capture chromatography can be established during small-scale batch experiments (see also Figure 13.6). This is certainly true for all buffer compositions.

The static binding capacity of the adsorbent is used to set the desired operating binding capacity. Together with the protein concentration in the feed solution, this determines the load volume. In continuous multicolumn chromatography, this translates into the load time for the individual columns. The mass transfer kinetics of the separation determine the required contact time to reach the desired operating binding capacity. All these parameters can be evaluated in small-scale batch experiments, followed by a relatively straightforward verification study at small scale in a continuous system. Various approaches for establishing the load volume and contact time in a continuous multicolumn chromatography process have been published (e.g., 13, 29).

The peak collection window in the elution step can best be established during small-scale verification runs. Due to the higher load concentration on the capture columns, the elution volume is sometimes slightly different from in the equivalent batch process. This can be determined through single column overloading experiments or mitigated by designing the elution step with product collection based on the signal of the UV spectrophotometer or any other online monitoring tool that measures protein concentration in the eluate.

Most suppliers of continuous multicolumn chromatography solutions have put in efforts to ensure scalability from bench to commercial manufacturing. This is helped by the fact that chromatography is generally a well-understood process and that large-scale batch chromatography systems have been successfully implemented for many decades. That said, the main challenges with scaling up chromatography have always been related to gradient formation and flow distribution inside columns with large diameters. Multicolumn chromatography for capture processes doesn't typically involve gradients and, due to the high specific productivity, column sizes remain reasonably compact. As a consequence, scale-up risks are quite manageable. Initial evidence to support excellent scalability of multicolumn chromatography processes has been presented or published by various biopharmaceutical companies (e.g., 30, 31).

During process design, the dynamics of the startup and shutdown needs to be considered. Existing data suggests that the purity of the material produced during the startup and shutdown of a multicolumn affinity capture step does not significantly deviate from the product quality produced during steady state, but the product concentration itself is definitely different. Depending on the further downstream processing, the impact of the different product concentration needs to be evaluated.

The lifetime of the adsorbent is not listed as a CPP in the example shown in Figure 13.6. Yet, for late-stage clinical and commercial manufacturing, there is a need to establish the lifetime of the critical materials such as the chromatography adsorbents. An adequate risk assessment needs to be done to determine whether this should involve running the continuous chromatography process over multiple cycles or whether this can be done in a small-scale batch chromatography system. The latter will involve far less product material.

13.3.1.2 System Qualification

The underlying philosophy for the qualification of a multicolumn chromatography system does not deviate significantly from traditional batch systems. One could consider following the traditional approach, where the process system is qualified against the user requirements specification (URS) during the factory acceptance test (FAT) and site acceptance tests (SAT) following the traditional qualification procedures (installation qualification, IQ; operational qualification, OQ; and performance qualification, PQ) as reported by Sanofi-Aventis (32).

13.3.1.3 Process Performance Qualification

A variety of critical steps can be identified that are unique to continuous multicolumn chromatography. Some examples of the risks associated with this unit operation, including some potential mitigation strategies, are summarized in Table 13.1. These mitigation strategies could target either the occurrence (likelihood of the risk to manifest itself), the impact, or the detectability. An adequate risk assessment such as a failure mode and effect analysis (FMEA) should be considered, based on the exact system architecture and the actual process that needs to be operated.

TABLE 13.1

Example of Some Risks, Impacts, and Potential Mitigation Strategies for the Validation of a Continuous Multicolumn Chromatography Process

Risk	Potential Effect	Examples of Potential Mitigation Strategies
Wrong buffer connections	Process doesn't run as intended (wrong buffers are introduced to the columns)	• Operator training • Automated verification before use • Conductivity and pH system inlets
Column connections	Process cannot run because fluid flow is wrong	• Operator training • Automated verification of flow path before use
Process connections (including buffers, columns, outlets, etc.)	Ingress of biocontamination	• Use of single-use flow path with aseptic connections • Sanitization of the entire fluid path before use • Periodic sanitization
Introduction of air bubbles into the columns	Inadequate column performance	• Bubble trap on the inlets • Air alarms • Minimize risks through buffer management
Column-to-column variations	Yield and impurity removal	• Design the process with a safety factor • Establish adequate acceptance criteria for column packing • Continuous monitoring of column performance via frontal analysis

All instruments that relate to the control of critical process parameters and/or that are used for making decisions that directly or indirectly can affect the product quality need to be adequately verified as part of the PPQ. In addition to this, a calibration strategy needs to be defined for these instruments.

Similar to traditional batch chromatography, the column packing is one of the critical material attributes (or critical process parameters) for the process. Columns with a low HETP or high asymmetry are expected to behave differently than adequately packed columns. For high-resolution multicolumn chromatography applications in the purification of small molecules, this had already been acknowledged, and a model-assisted approach for assessing the impact was proposed (33, 34). For capture chromatography based on an affinity ligand, these parameters may not be directly impactful on product quality since the process is inherently robust when it comes to impurity removal. Yet, packing variability can impact the yield for processes that are operated close to the static binding capacity, as these parameters correlate to the packing density and hence to the effective binding capacity (35). For this reason, it is important to establish acceptance criteria for the different columns in the setup. This can be done based on sound science and supported with numerical simulations.

The FDA guidance also recommends characterizing the impact of the process dynamics on the total process performance. All commercially available multicolumn chromatography solutions for biopharmaceutical manufacturing are equipped with sensors to monitor the protein concentration in the eluate. Most commonly this is done with UV spectrophotometers in the eluate outlet line. This is done both in the process-scale systems for commercial operations and for bench-scale systems that are used during process development. The information and data collected during the process design stage can support the PPQ stage, since the time constants and process dynamics are mostly not scale dependent.

Continuous multicolumn chromatography typically results in many more cycles per batch or per campaign than traditional (batch) chromatography processes. As a consequence, the recommendation to establish adequate boundaries for column lifetime is critical for assuring consistent and robust manufacturing process. Since the fundamental mechanisms that limit column fouling are not scale dependent, initial column lifetime can be established in small-scale experiments. During normal process development studies, it is not uncommon to already generate a significant amount of cycles on the columns. If adequately documented, this data can be used to establish initial boundaries for the column lifetime. Some data in the public domain suggests that there is no reason to assume that the column lifetime in continuous multicolumn chromatography would be different from in batch chromatography (5, 6, 36).

13.3.1.4 Continued Process Verification

Continued process verification for intensified bioprocessing platforms and continuous downstream processing technologies follows the same philosophy and approaches as traditional biopharmaceutical manufacturing. Most likely the largest difference between traditional and intensified bioprocessing platform is the amount of data that is generated during manufacturing.

In traditional batch manufacturing, column qualification is typically performed every batch using standard column characterization methods. A next step could be to evaluate column performance based on in-process data, for instance by analyzing the conductivity trace during transitions between subsequent wash steps (37). As stated in section 13.2.4, the amount of data generated during a continuous chromatography step is typically one to two orders of magnitude larger than in traditional batch chromatography. This presents the opportunity to use enhanced statistical evaluation approaches such as multivariate data analysis.

Proof of concept for the use of some of these tools has been presented and discussed for batch chromatography (38) as well as for multicolumn chromatography in biopharmaceutical manufacturing (39, 40). This work mainly focuses on the use of principal components analysis (PCA) to identify small outliers, deviations, and trends in large datasets. This allows comparing datasets from multiple manufacturing runs very accurately to easily identify small discrepancies between different batches. The use of PCA for evaluating large datasets also aligns very well with the concept of review by exception (41).

13.3.2 Example—Virus Inactivation

13.3.2.1 Process Design

In most monoclonal antibody processes, the affinity capture process step is followed by a low pH virus inactivation step. Small-scale virus inactivation studies are commonly conducted to explore the process conditions that provide sufficient virus clearance (typically targeting 5 log reduction values (42)).

A generic quality by design assessment of virus inactivation was presented in the A-Mab case study (16). The critical process parameters that were identified in this work are the pH of the incubation, the duration of the incubation, and temperature. The experimental studies reported in the work suggested that protein concentration (studied up to 35 g/L) was not a critical process parameter. The study suggests that the design space for temperature is around ambient temperatures (15–25 °C). With adequate environmental controls, the process temperature always falls within the proven acceptable ranges. These generic conclusions are confirmed in a more recent review article that covered virus safety data from various different companies (23).

Based on these considerations, it is fair to state that the incubation time and incubation pH are the two main parameters that need to be controlled in the process system.

13.3.2.2 System Qualification

Here we can use the Cadence VI system by Pall Biotech as a case study of a system qualification strategy (43). A thorough risk assessment was performed to identify adequate mitigation strategies related to the system design for a typical virus inactivation step in a monoclonal antibody process. Some of the risks that were identified and the mitigation strategies that were implemented are summarized in Table 13.2.

Further evidence during the system qualification was delivered by an inactivation test with an enveloped bacteriophage (Phi6) to mimic an actual virus inactivation process.

The strategy outlined in the Table 13.2 provides an example of a risk-based system qualification for a continuous virus inactivation step. System qualification very much depends on the exact system architecture as well as on the process that needs to be operated. As the table suggests, small details in the system design can have a significant impact on the risk assessment and hence on the system qualification plan.

13.3.2.3 Process Performance Qualification

In the Cadence VI system, the virus inactivation process essentially relies on a sequential or repetitive batch process. With this, the PPQ follows the traditional approach for

TABLE 13.2

Some Parts of the risk-Based System Qualification for the Cadence Virus Inactivation System Shown in Figure 13.7

Risk	Potential Effect	Examples of Potential Mitigation Strategies
System hold-up volumes (not 100% of material exposed to low pH incubation)	Product can be contaminated with non-inactivated material	• Flow path design for minimum hold-up • Mixer design for zero splashing and minimum foam formation • Hold-up characterized with riboflavin fluorescence test and B diminuta
System hold-up volume (not 100% flushed after cycle)	Material being exposed twice to low pH incubation	•
pH probe stability	Lack of assurance of adequate inactivation conditions	• Demonstrate short wetting time after autoclaving • Establish drift profiles over time • Adjustment to pH reading while in operation
Titration time and transfer time	Potentially impact the time for incubation	• Characterize transfer times • Process design with longer collection windows

virus inactivation process steps. Based on an adequate risk assessment, the critical process parameters need to be verified at scale. This may include mixing studies, foam formation, and the dynamics of pH titration in the actual process solution.

In the PPQ, the process dynamics also need to be considered. For the virus inactivation system, the (continuous) multicolumn chromatography process step will generate a feed flow with dynamic compositions, causing protein concentration and pH to fluctuate over time. In addition to this, the feed flow is often discontinuous as not all processes are operated with fully continuous (uninterrupted) elution flow rates. In a virus inactivation system that is based on repetitive or sequential batch incubation, this challenge is mitigated by collecting the eluate in a mixer and only starting the virus inactivation cycle once the mixer has filled up to the desired level. Homogeneous and consistent starting conditions are obtained by gently mixing the eluate and by designing the process such that each sub-lot in the virus inactivation contains an integer number of elution peaks. A process that is designed and operated in this way inherently mitigates the challenges of fluctuating feed conditions. In the same way, this concept for continuous virus inactivation system also provides homogeneous conditions feeding into the next unit operation. This makes the design and validation of the polishing steps significantly easier.

Another aspect of the system that needs to be considered is the fail-safe state in case a process upset occurs that would demand the (connected) unit operations to be interrupted. The design around two alternating mixers as incubation containers makes it rather easy to implement such a fail-safe procedure. The sublot that is at subject to the low pH incubation needs to complete its incubation cycle and neutralization before the system pauses.

13.3.2.4 Continued Process Verification

Depending on the cadence at which it operates, continuous virus inactivation produces one elution cycle every 1.5 to 2 hours. With that, the continuous virus inactivation generates significantly more process data compared to a single inactivation cycle per batch in a traditional downstream processing platform. With that, continued process verification follows the same philosophy as for multicolumn chromatography (see section 13.3.1.4).

Statistical analysis tools may help in recognizing patterns in the inactivation cycle pH. This can be further enhanced by correlating the acid and base solutions that are added during the titrations. This would provide an extra insight in potential drift of the internal pH probe, addressing the challenge that was reported by BiosanaPharma (28).

13.4 Traceability

The definition of a batch or a lot is a critical element in the definition of the drug manufacturing process, irrespective of whether it involves the drug substance or drug product stage.

One of the most common questions related to continuous bioprocessing technologies is the definition of a lot or a batch. This is driven by the fact that nearly all current biomanufacturing processes rely on batch processing steps, and hence the definition of a lot or a batch is naturally related to the manufacturing schedule. Typically, the material produced in one fed-batch bioreactor is considered as one batch or lot. For perfusion processes, the batch or lot is often defined by the harvest schedule, and hence it is mostly based on time.

The legislation defines a batch as "a specific quantity of a drug or other material that is intended to have uniform character and quality, within specified limits, and is produced according to a single manufacturing order during the same cycle of manufacture," and a lot is defined as

> a batch, or a specific identified portion of a batch, having uniform character and quality within specified limits; or, in the case of a drug product produced by continuous process, it is a specific identified amount produced in a unit of time or quantity in a manner that assures its having uniform character and quality within specified limits.
>
> (44)

The regulatory expectations with respect to a batch or a lot are equally applicable to both batch and continuous manufacturing platforms. A batch therefore can be a defined as a certain (predefined) quantity of product produced, or a quantity of product produced during a (predefined) time interval. The former approach has been adopted in the clinical manufacturing campaign for a monoclonal antibody biosimilar by BiosanaPharma (28). The time-based strategy is being considered for the manufacturing platform developed by Alvotech (45). Both of these examples use a continuous cell culture process based on perfusion. For companies that use fed-batch cell culture, the batch definition is more likely to follow the traditional scheme (46).

The lot definition in a continuous manufacturing platform can also be tuned to correlate with the use of auxiliary materials such as filters and/or single-use assemblies. One could consider relating the time for a lot or a batch to the changeover times for critical single-use materials. Another aspect to consider is the cyclic nature of many continuous unit operations in downstream processing. It makes sense to relate the lot or batch to a multiple of cycles of one or more continuous downstream processing steps. This may be a particularly powerful approach in hybrid platforms. With the continuous virus inactivation step based on the repetitive batch approach as described in section 13.3.2, it makes a lot of sense to correlate the cyclic nature of the affinity capture step and the virus inactivation step. This would then create a natural starting point for repetitive sublots feeding into a repetitive batch downstream process following a scheme as shown in Figure 13.2 or Figure 13.4.

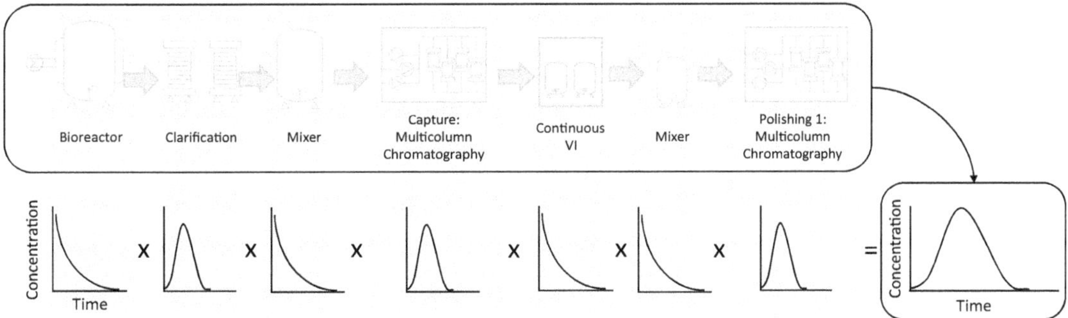

FIGURE 13.8

Schematic representation of building a residence time distribution (RTD) model for an integrated continuous biomanufacturing platform (15).

In fully integrated continuous downstream processing platforms, subsequent lots cannot always be completely segregated. As a consequence, the lot-to-lot carryover of material should be characterized. The most suitable approach for this would be the use of a residence time distribution model. Building such a model can be based on combining the residence time distribution of the individual unit operations, as schematically shown in Figure 13.8 (15).

Experimental proof of concept for the approach was presented by Merck (47). The work included experiments to monitor the propagation of a tracer molecule with exactly the same behavior as the target molecule through the integrated continuous downstream processing platform. These approaches provide meaningful insights into how material that is impacted by a potential deviation in one part of the process propagated into the lots of drug substance.

13.5 Conclusions

During recent conferences on integrated continuous bioprocessing, the topic of validation was discussed among a broader audience (46, 48, 49). Some scientists and engineers in the audience expressed concerns that qualification and validation of continuous bioprocessing technologies would take more time due to the increased complexity of the processing systems involved. Another part of the audience actually believed that continuous bioprocessing would streamline validation efforts because of the larger amounts of data available at an early stage in combination with a reduction in scale-up results. These opinions were expressed before any of the continuous downstream processing solutions were brought into CGMP manufacturing. With the knowledge and experiences today, there seems to be reason for both groups of experts to refresh their opinions.

Validation of intensified bioprocessing platforms that use continuous downstream processing technologies will benefit a lot from data generated during process development. Considering the fact that also at bench scale, intensified bioprocessing solutions will generate more data, one may expect a more solid and statistically relevant data package to be available even before clinical manufacturing. This supports the opinion of the experts that expressed the expectation that validation could potentially be streamlined, albeit that there is no evidence for such claim yet. The success of such strategy likely depends on the digital architecture that supports the intensified manufacturing platform. Having all process information adequately collected in a central data repository certainly makes the "Big Data" challenge more manageable (27).

Systems and solutions for continuous bioprocessing have been implemented in CGMP manufacturing, and the first clinical manufacturing campaigns utilizing continuous downstream processing technologies have been successfully completed (2, 3, 32). This suggests that there are no inherent roadblocks related to the qualification of continuous downstream processing platforms.

References

1. J. McLaughlin: "Integrated continuous and batch operations for efficient initial clinical manufacturing of biopharmaceuticals", presented at *Bioproduction Summit* (12–13 DEC 2016)
2. GEN: "First mab produced via fully continuous biomanufacturing", *Genetic Engineering News* (24 SEP 2019)
3. Biosana Pharma: "BiosanaPharma announces successful outcome of comparative phase I study of BP001, a biosimilar candidate to Xolair® (omalizumab)", *Press Release* (30 MAR 2020)
4. K. B. Konstantinov and C. L. Cooney: "White paper on continuous bioprocessing", *J.Pharm.Sci.*, **104**, 3 (2015) pp. 813–820
5. M. Brower and D. Pollard: "Monoclonal antibody continuous processing enabled by single-use", in: G. Subramanian (ed.): *Continuous Processing in Pharmaceutical Manufacturing*, Wiley VCH (2015)
6. V. Warikoo *et al.*: "Integrated continuous production of recombinant therapeutic proteins", *Biotechnol.Bioengng.*, **109** 12 (2012) pp. 3018–3029
7. US FDA: "Guidance for the industry—process validation: General principles and practices", Rev., **1** (JAN 2011)
8. EMA—CHMP: "Guideline on process validation for the manufacture of biotechnology-derived active substances and data to be provided in the regulatory submission", EMA/CHMP/BWP/187338/2014 (2016)
9. S.L. Lee *et al.*: "Modernizing pharmaceutical manufacturing: From batch to continuous production", *J.Pharm.Innov.*, **10** (2015) pp. 191–199
10. G. Forno and E. Orti: "Quality control and regulatory aspects for continuous biomanufacturing", in: G. Subramanian (ed.): *Continuous Biomanufacturing: Innovative Technologies and Methods*, Wiley VCH (2018) pp. 513–532
11. M. Bisschops: "Quality by design for continuous downstream processing", presented at *Driving Value Through Intensified Bioprocessing*, Oxford UK (27 JUN 2019)
12. M. Bisschops: "Quality by design in a continuous environment—what's required?", presented at *Bioprocess International Conference and Exhibition*, Boston MA (11 SEP 2019)
13. Gjoka X. *et al.*: "A straightforward methodology for designing continuous monoclonal antibody capture multi-column chromatography processes", *J.Chromatogr.A*, **1416** (2016) pp. 38–46
14. A. Utturkar *et al.*: "A direct approach for process development using single column experiments results in predictable streamlined multi-column chromatography bioprocesses", *Biotechnol.J.*, 14 (4) (2019)
15. B. Manser and M. Glenz: "Regulatory and quality considerations of continuous bioprocessing", in G. Subramanian (ed.): *Process Control, Intensification, and Digitalisation in Continuous Biomanufacturing*, Wiley VCH (2021)
16. CMC Biotech Working Group: "A-mab: A case study in bioprocess development", *Version* **2.1** (30 OCT 2009)

17. L. David *et al.*: "Side-by-side comparability of batch and continuous downstream for the production of monoclonal antibodies", *Biotechnol.Bioengngn.*, **117** 4 (2020) pp. 1024–1036
18. M. Bisschops and M. Brower: "Dynamic simulations as a predictive model for a multicolumn chromatography separation", in A. Staby *et al.* (eds.): *Preparative Chromatography for Separation of Proteins*, Wiley & Sons (2017) pp. 457–477
19. International Conference on Harmonization Q5A: "Quality of biotechnological products: Viral safety evaluation of biotechnology products derived from cell lines of human or animal origin", ICH Q5A (R1) (4 STEP 1999)
20. European Commission (Enterprise Directorate General): "EMEA guideline on virus safety evaluation of biotechnological investigational medicinal products", *London* (28 JUN 2006)
21. M-J. Chang *et al.*: "Validation and optimization of viral clearance in a downstream continuous chromatography setting", *Biotechnol.Bioengng.*, **116** 9 (2019) pp 2292–2302
22. J. Angelo *et al.*: "Virus clearance validation across continuous capture chromatography", *Biotechnol.Bioengng.*, **116** 9 (2019) pp 2275–2284
23. J. Mattila *et al.*: "Retrospective evaluation of low-pH viral inactivation and viral filtration data from a multiple company collaboration", *PDA J.Pharm.Sci.Technol.*, **70** 3 (2016) pp. 293–299
24. L. David *et al.*: "Virus study for continuous low pH viral inactivation inside a coiled flow inverter", *Biotechnol.Bioengng.*, **116** 4 (2019) pp 857–869
25. V. Natarajan, J. Pieracci and S. Ghose: "Evaluation of continuous downstream processing: Industrial perspective", in: G. Subramanian (ed.): *Continuous Biomanufacturing: Innovative Technologies and Methods*, Wiley VCH (2018) pp. 571–586
26. D. Martins *et al.*: "Truly continuous low pH viral inactivation for biopharmaceutical process integration", *Biotechnol.Bioengng.*, **117** 5 (2020) pp 1406–1417
27. M. Bisschops and L. Cameron: "Process Intensification and industry 4.0: Two mutually enabling trends", in G. Subramanian (ed.): *Process Control, Intensification, and Digitalisation in Continuous Biomanufacturing*, Wiley VCH (2021)
28. M. Pennings: "Viral clearance validation for a fully continuous manufacturing process for phase 1 studies", presented at *Integrated Continuous Biomanufacturing V*, Brewster MA (7 OCT 2019)
29. M. Bisschops and M. Brower: "The impact of continuous multicolumn chromatography on biomanufacturing efficiency", *Pharm.Bioprocess.*, **1** 4 (2013) pp. 361–372
30. M. Brower: "Scale-up of continuous chromatography using cadence BioSMB process system", presented at *Bioprocess International Conference*, Boston MA (04 OCT 2016)
31. O. Ötes *et al.*: "Scale-up of continuous multicolumn chromatography for the Protein A capture step: From bench to clinical manufacturing", *Journal of Biotechnology*, **281** (2018) pp. 168–174
32. O. Ötes *et al.*: "Moving to CoPACaPAnA: Implementation of a continuous Protein A capture process for antibody applications within an end-to-end single-use GMP manufacturing downstream process", *Biotechnol.Rep.*, **26** (JUN 2020) p. e00465
33. K. Milhbachler: "Effect of the homogeneity of the column set on the performance of a simulated moving bed unit, I. Theory", *J.Chromatogr.A*, **908** (2001) pp. 49–70
34. K. Michlbachler *et al.*: "Effect of the homogeneity of the column set on the performance of a simulated moving bed unit, II. Experimental study", *J.Chromatogr.A*, **944** (2002) pp. 3–22
35. M. Bisschops and M. Brower: "The impact of variations in column packing on the performance of BioSMB multicolumn chromatography", presented at *PREP Conference*, Boston MA (16 JUL 2012)
36. M. Brower: "Stepping into the future, continuously working towards an integrated antibody purification process", presented at *Biopharmaceutical Manufacturing and Development Conference*, San Diego CA (SEP 2011)
37. T. M. Larson, J. Davis, H. Lam and J. Cacia: "Use of process data to assess chromatographic performance in production-scale protein purification columns", *Biotechnol. Prog.*, **19** 2 (2003) pp. 485–492
38. Y. Hou, C. Jiang, A. A. Shukla and S. M. Cramer: "Improved process analytical technology for Protein A chromatography using predictive principal component analysis tools", *Biotechnology and Bioengineering*, **108** 1 (2011) pp. 59–68.
39. M. Bisschops, M. Strawn, B. To and J. Coffman: "Multivariate data analysis: Managing large amounts of data coming from continuous biomanufacturing processes", presented at *Recovery of Biological Products XVI*, Rostock, Germany (28 JUL 2014)

40. E. Ayturk and M. Bisschops: "Capturing the value of continuous bioprocessing through MVDA", presented at *Integrated Continuous Bioprocessing II*, Berkeley CA (04 NOV 2015)

41. J. Zebib: "Review by exception: Connecting the dots for faster batch release", *Biopharm Int.* (12 NOV 2019)

42. ASTM E2888–12: "Standard practice for process for inactivation of rodent retrovirus by pH", *ASTM International*, West Conshohocken, PA (2012)

43. M. Schofield and K. Jones: "Assessing viral inactivation for continuous processing", *Biopharm Int.* (28 SEP 2018)

44. Code of Federal Regulations: 21 CFR 210.3 (Definitions), Title 21, volume 4 (revised April 1, 2018)

45. A. Falconbridge: "Implementation and development of continuous processing, challenges and lessons learned the story so far", presented at *Continuous Biomanufacturing: Driving value through intensified bioprocessing*, Oxford UK (26 JUN 2019)

46. A. Sinclair and K. Sitney: "Workshop: Regulatory gaps in continuous processing", Report from *Integrated Continuous Biomanufacturing IV*, Brewster MA (06 OCT 2019)

47. N. Pinto: "Automated material traceability in end-to-end continuous biomanufacturing for batch disposition", presented at *Bioprocess International Conference and Exhibition*, Boston MA (23 SEP 2020)

48. S. Farid, B. Thompson and A. Davidson: "Continuous bioprocessing: The real thing this time?" *mAbs*, **6** 6 (2014) pp. 1357–136

49. ECI: "Integrated continuous biomanufacturing IV conference", Brewster MA (06–10 OCT 2019)

14

Role of Multivariate Analysis in Process Validation

Anurag S. Rathore and Vishwanath Hebbi

14.1 Multivariate Data Analysis (MVDA)

Manufacturing biopharmaceuticals is a complex process. With the advent of quality by design (QbD), implementation of data analysis-based process characterization and data-driven quality control in the manufacturing environment are gaining popularity (Sahoo, Choudhury, and Manchikanti 2009). Operators can take swift decisions with real-time monitoring and control of critical unit operations, which in turn results in higher consistency in process performance and product quality (FDA 2004). The role of statistics and MVDA is becoming increasingly indispensable for analysis of the otherwise complex manufacturing data (Steinwandter, Borchert, and Herwig 2019; Glassey et al. 2011; CMC-Biotechgroup 2009; A. Rathore and Singh 2015).

Process validation is a major part of the product life cycle. It consists of three major steps: process design, process qualification, and continued process verification (Sofer 2005). Across these steps various data analytical activities are performed to facilitate process control, which ultimately impact product quality. Some of the applications of MVDA include monitoring and control of granulation in pharmaceutical industries (Liu et al. 2017; Hansuld and Briens 2014), reaction monitoring of active pharmaceutic ingredients (Steinwandter, Borchert, and Herwig 2019), process comparability in biopharmaceutical drug manufacturing (Godavarti et al. 2005; A. S. Rathore and Singh 2017), and in-process monitoring in refolding unit operation (Hebbi, Thakur, and Rathore 2019). Advancements in analysis and measurements have enabled a continuous flow of data from a variety of analytical instruments (such as spectrophotometers, online sampling units, sensors, and data loggers). More recently, the biotech industry is looking to convert batch to continuous operation for manufacturing of biotherapeutics (Ayturk 2015). A typical bioprocess has tens of thousands of variables, data for which are collected efficiently and logged into the data servers. MVDA recognize patterns relating the variables through projection techniques such as principal component analysis (PCA) or correlations established through partial least squares (PLS). However, data in general consists of unwanted information that needs to be cleaned, normalized before performing analysis (Eriksson et al. 2013).

In this chapter, we explore the role that MVDA can play in process validation. Types of manufacturing data, pre-processing of data, and the various MVDA modeling approaches are discussed briefly.

DOI: 10.1201/9781003143130-14

14.1.1 Data Types

During routine product development and manufacturing, various measurements are taken, which can be grouped into input and output variables. These measurements together are called data. Each datum consists of two components associated with itself, one is actual value and the other is noise that might have resulted from the nonlinearity of the system.

Data are typically categorized into two types as illustrated in Figure 14.1, factors and responses. In every unit operation, these two data are measured, collected in an organized manner, and logged in data storage using advanced communication systems in the manufacturing environment. To get meaningful outcome from MVDA, the data must be amenable to analysis.

The complexity of data is one of the important factors to consider during analysis. As long as the number of variables is less than or equal to five, data can be analyzed using commonly used statistical attributes (mean, variance, and covariance). If the number of variables is more than five, traditional statistical analysis is not possible. A key feature that MVDA offers is to reduce the dimensionality of the data, thus paving the way for further analysis (Eriksson et al. 2013).

Data can also be categorized as continuous, discrete, and categorical. Table 14.2 lists examples of different data types. It is important to note that fully logged raw data are needed for determining conclusive results using MVDA. Analyzing only a subset of the data or one variable at a time might result in partial or incomplete analysis and erroneous conclusions. When more variables are analyzed, it is very possible that many of them are related, which thus directly or indirectly affects the output of the process.

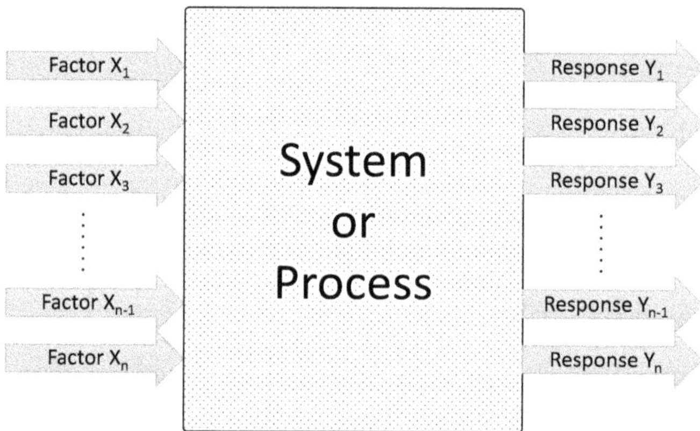

FIGURE 14.1

Factors (1 to n) act as input and responses (1 to n) are output variables obtained by performing a process. Factors (X) can be related to (Y) using MVDA techniques.

TABLE 14.1

Classification of Data Based on Number of Variables. Adapted from Eriksson et al. 2013.

Complexity of Data	Number of variables, K
Univariate	$K = 1$
Bivariate	$K = 2$
Low variate	$K \leq 5$
Multivariate	$K \geq 6$

TABLE 14.2

Different Data Types and Their Examples in Biopharmaceutical Manufacturing Processes (Liland 2011)

Data Types	Factors
Categorical	Batch ID, manufacturing line, pre-seed vessel, appearance, identity, viability, sterility
Discrete	Time of operation, pH, conductivity, spectroscopic, volume, temperature, protein content (%)
Continuous	pH, conductivity, pressure, UV 280 data during chromatography, pressure and flux values during ultrafiltration/diafiltration operation, NIR data in process monitoring

14.1.2 Pre-Processing of Manufacturing Data

Data are collected in increasingly complex ways during biopharmaceutical manufacturing. Various communications (MODBUS, RS232, RS485, USB) are utilized, each having different capabilities to transfer the data to and from the data source (Clapp, Castan, and Lindskog 2018). Errors during data collection can result in undesired values being collected in the data historian. Examples for such cases include "pH value:−10" (pH range is 1–14), unwanted combinations like "Expression system: Mammalian type: gram −ve" (where bacterial expression only can be classified into Gram negative and positive), and missing values. Such erroneous data make it difficult to perform MVDA (Eriksson et al. 2013). Hence, representation of raw data into meaningful data is a prerequisite for data analysis. Common data pre-processing methods include cleaning, scaling, mean centering, and transformation.

14.1.2.1 Scaling and Transformations

Scaling involves taking each value in the data and multiplying it by a scaling vector to make them of similar magnitude across variables. Often, there may be large variation in numerical values, and hence their variance (pH and RPM). Since multivariate methods are maximum variance projection methods, if variables are not scaled optimally, the model can give false predictions (A. Rathore and Singh 2017).

Unit variance (UV) scaling is a widely used method for scaling the data. Each data value is multiplied by a factor of $\dfrac{1}{S_k}$ where S_k represents standard deviation of variable "k." This type of scaling is employed to obtain the scaled variable with similar variance amongst large and small numerical values (Eriksson et al. 2013).

Pareto scaling is intermediate between no scaling and UV scaling. The data are scaled by scaling weight of $\dfrac{1}{\sqrt{S_k}}$. Pareto scaling results in the variable equal to the standard deviation. This comes as default pre-processing method in software packages such as SIMCA (Liland 2011).

Mean centering is one of the widely used methods to improve model interpretability. It is preferred in most of the cases that mean centering is combined with unit variance scaling. However, it is important to know the skewness of the variables that may adversely affect the model quality.

In an ideal scenario, data should be normally distributed. However, in reality data may be skewed due to the presence of nonlinear terms. Skewness of the data might adversely affect the model validity and interpretability (A. Rathore and Singh 2015; Eriksson et al.

2013). Distribution plots for variables and tests such as the Anderson–Darling test are often used to verify normality. If the data are not normal, various transformations such as logarithmic, square root, fourth root inverse, and power transformations can be performed to convert them into normal distribution (Zink 2018).

14.1.3 MVDA Modeling Approaches

Typical biopharmaceutical manufacturing involves running the process at optimal set points of the input parameters. This means that the process data that is collected in the manufacturing facility has only minor deviations in the input parameters with a lot of critical process parameters (CPP) held constant. Statistical analysis of this data is unlikely to unravel the correlations between the critical quality attributes (CQA) and the CPP (A. S. Rathore and Singh 2017). For example, if pH of a process chromatography column is held constant during manufacturing of the various batches, it is not possible to find the impact it has on the column performance and product quality. An additional challenge is the large number of input parameters that typically exist for complex unit operations such as bioreactor and process chromatography. The resulting manufacturing data will have multiple parameters varying in every batch, making it difficult to identify the CPP and also to correlate the CQA to the CPP.

Principal component analysis (PCA):

> PCA can be defined as a process of finding the lines and planes which fits closely to the points which are located in the space. Statistically it finds the components which are aligned to have maximum variance of projection of coordinates (Eriksson et al. 2013; A. S. Rathore et al. 2014). PCA is performed to reduce the dimension of data without losing any crucial information as follows:
>
> - Arranging the data into matrix array which is suitable for the software used for analyzing the data.
> - Pre-processing of the data (generally autoscaling followed by mean centering).
> - Calculation of covariance matrix, which gives an overall idea of the spread of the data.
> - Determination of Eigenvalues and Eigenvectors called principal components.
> - Generating a model with the number of principal components as many as three to four, which describes the variation of the data and predictability.

The PCA model can used to analyze the data with many tools such as score plots, loading plots, and Hotelling T^2 charts (Bhushan, Hadpe, and Rathore 2012).

Partial least square regression (PLS): PLS is used generally to relate two matrices of data, one is input variable called as X matrix and other is output variables called as Y matrix. In general Y is dependent on X. The equations for determining models for PCA and PLS are described elsewhere (A. S. Rathore et al. 2014; A. S. Rathore and Singh 2017; Eriksson et al. 2013).

14.1.3.1 Case Study I: Application of MVDA for Qualification of Scale-Down Model

Process analytical technology (PAT) is gaining interest in biopharmaceutical manufacturing. In 2004, the US FDA introduced the guidance on PAT implementation in pharmaceutical manufacturing as a means to achieve greater consistency in product quality (CMC-Biotechgroup 2009; FDA 2004). The underlying basis of PAT is process understanding, and this primarily comes from what is called process characterization studies. As large-scale experimentation is expensive, most of the knowledge generation is performed using laboratory-scale setups (F. Li et al. 2006). Researchers have reported that successful scale-up from bench scale to manufacturing scale requires that the manufacturer uses an accurate scale-down model for performing process characterization studies (Chen et al. 2009). However, developing a scale-down model that accurately represents manufacturing-scale unit operation is a challenging task (Rathore et al. 2005).

Cell culture is the primary step for manufacturing monoclonal antibodies and a complex unit operation (Tsang et al. 2014; Kirdar et al. 2007). As such it is non-trivial to identify all interactions between the numerous parameters and the correlations between the inputs and output variables. On the other hand, MVDA is quite capable of analyzing these complex datasets and hence is increasingly being used for analyzing scale-down data during process characterization studies.

This case study showcases application of MVDA towards comparison of a 2 L cell culture set with a 2000 L scale bioreactor. Numerous parameters, such as pCO_2, pO_2, glucose, pH, lactate, and ammonium ions, and output attributes, such as purity, viable cell density, viability, and osmolality, were modeled and evaluated in detail. The flow chart of the steps involved in the MVDA modeling and analysis in this case study is as shown in Figure 14.2. First, unwanted data are removed and the data are imported into SIMCA. A suitable pre-processing method is then applied to normalize the data and then modeled using the modeling wizard of SIMCA P +11 software.

As seen in Figure 14.3, score plots give vital information about the nature of batches between the two groups, 2 L and 2000 L. The score plots have been plotted between score vectors t[1] and t[2], corresponding to the first and second principal components (A. S. Rathore, Bhushan, and Hadpe 2011). It has been advised to remove the outliers outside the Hotelling T^2 95% confidence during pre-processing to obtain a more meaningful

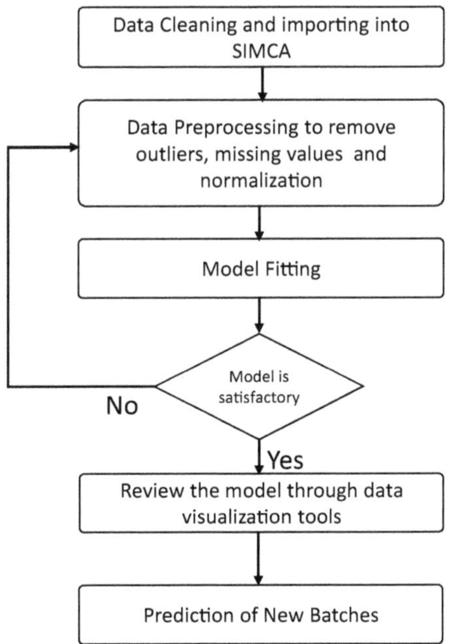

FIGURE 14.2
Flow chart for multivariate data analysis.

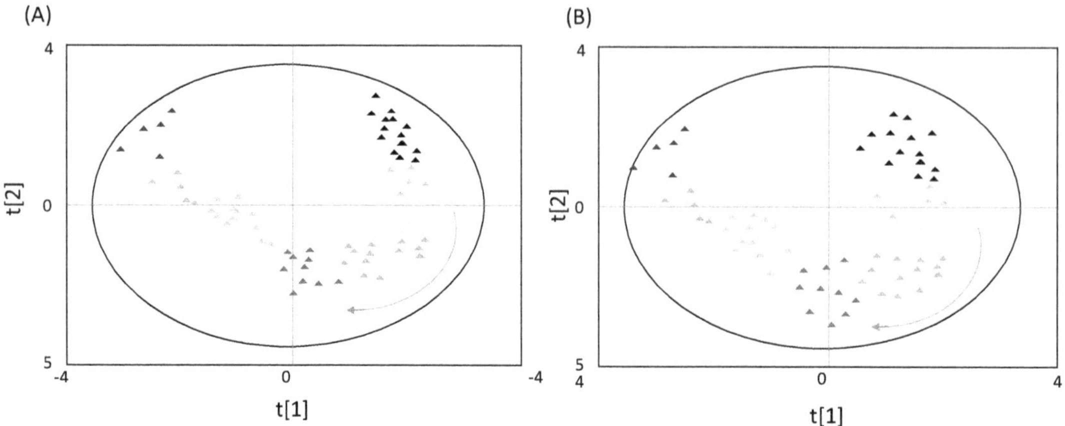

FIGURE 14.3
Scores scatter plot (t[1] vs. t[2]) for 2 L (A) and for 2000 L (B) batches along with the 95% tolerance ellipse. Trajectory of the batch is shown by an arrow. Figure adapted from Kirdar et al. 2007.

model (Eriksson et al. 2013). It is observed that all the batches are within the ellipse, which indicates that the model fit is satisfactory. However, another possibility of outliers is heteroscedastic data that might need transformations (log or power transformations) (Liland 2011). It advised that model building is an iterative process that involves the creation of an observation-level model to observe the outliers, agreement with the technical interpretation of the model and, if not satisfactory, a refitting of the model with the desired modifications in the data, as shown in Figure 14.3. Color coding also helps to identify the trajectory of process across the scales (B. Li, Morris, and Martin 2002). This kind of representation can be clearly an advantage in the monitoring and control of process. Different kinds of variables can be correlated and compared across the scales to validate the scale-down models.

Further, PLS loadings plots are shown in Figure 14.4. Loading vectors were calculated by correlating the input variables and significant principal components, in this case p[1] and p[2]. Therefore, these plots reveal how the variables are correlated to t[1] and t[2]. Variables in the same quadrant show positive correlation. On the other hand, those in opposite quadrants show negative correlations. It can be seen from the loading plots in Figure 14.4A and B that the viable cell density (VCD) and titer are positively correlated. At the initial stages of culture, VCD is low and so is the titer. With the increase in the VCD, the number of viable cells increases, and as cells reach production stage, the product concentration also gradually increases. Similarly, VCD and viability are negatively correlated. At the early stages of cell culture, cells are more viable but VCD is less. With time, as VCD increases, viability of cells decreases.

Gas transfer kinetics is a major concern during scale-up of cell culture process. Even though the score plots in Figure 14.3 suggest satisfactory comparison between data from 2 L and 2000 L reactors, a closer look in the loading plots reveals that there are some discrepancies with regard to $NH4^+$ and osmolality. The reason behind this can be understood by the mechanism of pH control in bioreactor. As scale-up limits the gas transfer rate, CO_2 increasingly accumulates, and this lowers the pH (Kirdar et al. 2007). The controller has to pump more base to maintain the pH required at the set point and as a result osmolality increases gradually. The comparison presented in this case study not only helps in gaining a deeper understanding of the process across the scales but can also offer guidance for potential process improvements.

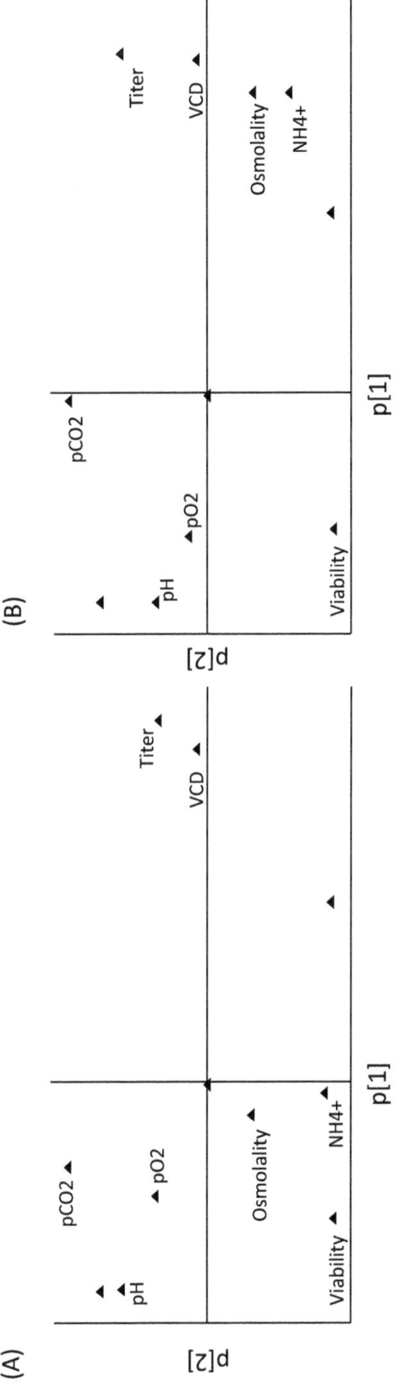

FIGURE 14.4
PLS loadings plot for 2 L (A) and 2000 L (B) batches. Figure adapted from Kirdar et al. 2007.

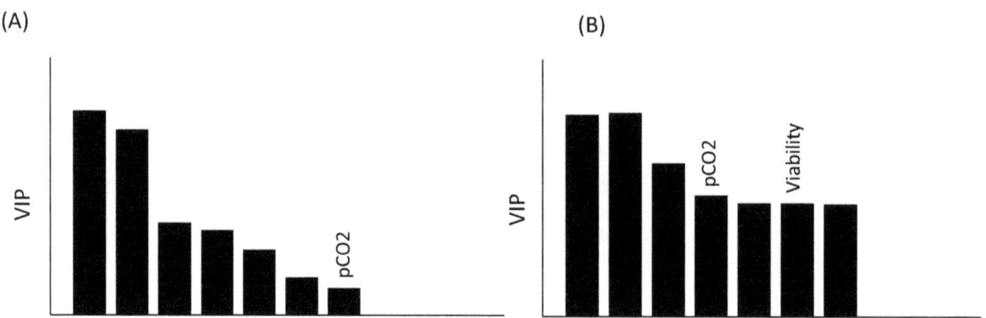

FIGURE 14.5
Variable importance for the projection plot for 2 L (A) and 2000 L (B) batches. Figure adapted from Kirdar et al. 2007.

Variable of importance to projection (VIP) plot summarizes the variables that are important to be considered in the model. The majority of variables represented in VIP plots in Figure 14.5A (2 L) and B (2000 L) are in good agreement with the loading and score plots. Also, it can be noted that pO_2 is important at larger scale, as its effect is negligible in the 2 L scale.

14.1.3.2 Case Study II: Application of MVDA for Analysis of DOE Data

Implementation of PAT requires establishment of process models, mathematical relationships between the CQA and the CPP, and the feed material attributes. While mechanistic models are desirable, most biotech unit operations are too complex to create mechanistic models that are simple enough for use in the manufacturing environment. Use of empirical models or hybrid models is more of a norm in the industry (Glassey et al. 2011). It is commonplace to use statistical approaches like multivariate data analysis, multivariate statistical process control (MSPC), and batch statistical process control (BSPC). It is well accepted that multivariate-based approaches offer better monitoring of manufacturing processes than do univariate approaches (Eriksson et al. 2013).

Design of experiments (DOE)-based approaches are now commonplace in process development and process characterization studies. The generated data are analyzed to define design space as well as to propose a suitable monitoring and control strategy. Finally, the designed process and the controls are demonstrated at commercial scale (process validation) followed by monitoring and control during routine manufacturing. SPC not only helps in ensuring consistency in product quality but also can identify opportunities for continuous process improvement (Undey et al. 2011).

A multivariate partial least squares (MPLS) multivariate approach was used to develop statistical process control charts using the online data from the oxidation-reduction potential (ORP), pH, and temperature probes (Nomikos and MacGregor 1995). MPLS is an extension of PLS used to handle data in three-dimensional arrays, as a historical dataset of batch trajectory data consists of i number of runs with each having online measurements of j variables over k time intervals throughout the batch (Nomikos and MacGregor 1995). Since there is a significant amount of data, the data must be unfolded into a two-dimensional structure before regression techniques can be applied. For batch evolution modeling, the unfolding technique involves putting the complete time profiles of the first batch in the first k rows and continue adding rows vertically underneath for each subsequent run.

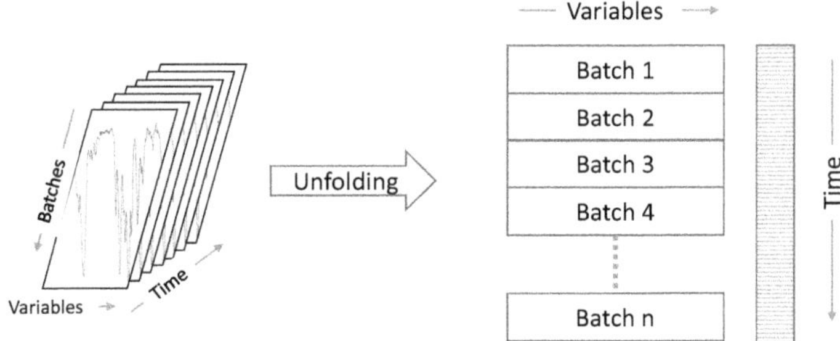

FIGURE 14.6
Unfolding method and MPLS matrix for 3D batch evolution data. Figure adapted from Hebbi, Thakur, and Rathore 2019.

Thus, the X-matrix has dimension of $J \times IK$. The corresponding Y-matrix is a single-column vector with $k \times i$ total rows, with time increasing from zero to the final time in each set of k rows, corresponding to the batch information in the X-matrix. The Y-matrix serves as a dummy time variable for regression in order to chart the progress of the batch (Eriksson et al. 2013). Figure 14.6 illustrates the approach that was applied.

After unfolding the matrix, each column of X was mean centered and scaled to unit variance. Thus, MPLS in this framework explains the variation of a process variable about its average trajectory at each point of time. The nonlinear iterative partial least squares (NIPALS) algorithm was applied to decompose the unfolded, scaled matrix into a summation of "R" score vectors "t" and loading vectors "p" and "q," and residual matrices "E" and "F," as per the decomposition equations $X = \sum_{r=1}^{R} t_r p_r' + E$ and $Y = \sum_{r=1}^{R} t_r q_r' + F$. Umetrics Simca software (Sartorius Stedim Data Analytics AB) was used for the MVDA calculations.

This study leverages the characteristic ORP, pH, and temperature profiles produced by stepwise addition of buffer and protein components in a pre-characterized refolding process to develop a PAT monitoring tool for protein refolding unit operation. These time profiles are treated as batch evolution variables and combined using MVDA. The refolding unit operation is treated as a semi-batch evolution process that should be maintained within operating ranges at each point in time. SPC charts are used to detect deviations from normal operation and to identify the process step causing the deviation, thereby allowing unrecoverable batches to be disposed of early and record a fingerprint for each run for future reference when CQAs are not reached. The PAT tool is developed using the MPLS technique and utilizes online operational information in real time to recognize deviations from previously characterized average trajectories of in-control batches. The study demonstrates that the online variables pH, ORP, and temperature can be used to obtain signature profiles of successful addition of each component during refolding unit operation. The combination of these variables along with MVDA can be used to develop an SPC-based monitoring system.

Figure 14.7 shows the characteristic ORP, pH, and temperature profiles and the changes caused by the addition of each subsequent component in the solubilization and refolding buffers. The periodic addition of buffer components resulted in consistent changes in the ORP, temperature, and pH profiles in a range of +280 mV (pure water) to −510 mV (after addition of DTT).

Multivariate tools can be used to create a control scheme that allows us to monitor CPPs as well as other relevant process parameters that affect process performance and

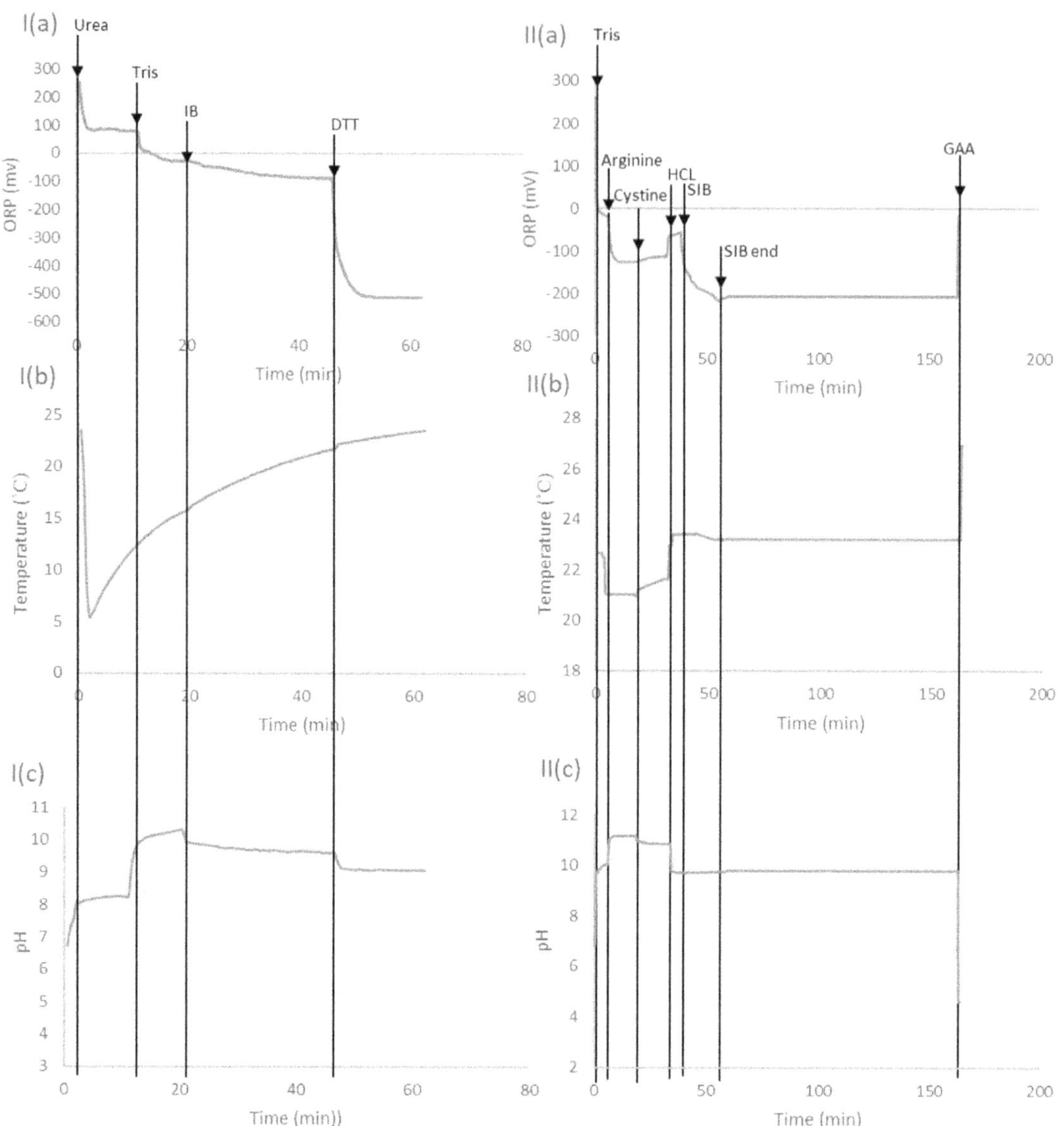

FIGURE 14.7

(a) ORP, (b) pH, and (c) temperature characteristic profiles for (i) solubilization and (ii) refolding showing time-wise component addition. Figure adapted from Hebbi, Thakur, and Rathore 2019.

product quality (Pieracci et al. 2018; Rüdt, Briskot, and Hubbuch 2017). Another benefit that such an approach offers is the reduction of idle time at the end of the process step by eliminating the need for offline analysis, as well as time between component additions due to lack of knowledge of the time required for solubilization and/or reaction. Statistical process control (SPC) charts combine time series analysis methods with graphical presentation of data in the form of control charts. In our case, the process monitoring limits were established on SPC charts using a batch evolution modeling (BEM) approach (Nomikos and MacGregor 1995). The solubilization process was considered to be a semi-batch reaction system with periodic addition of buffer components, inclusion bodies,

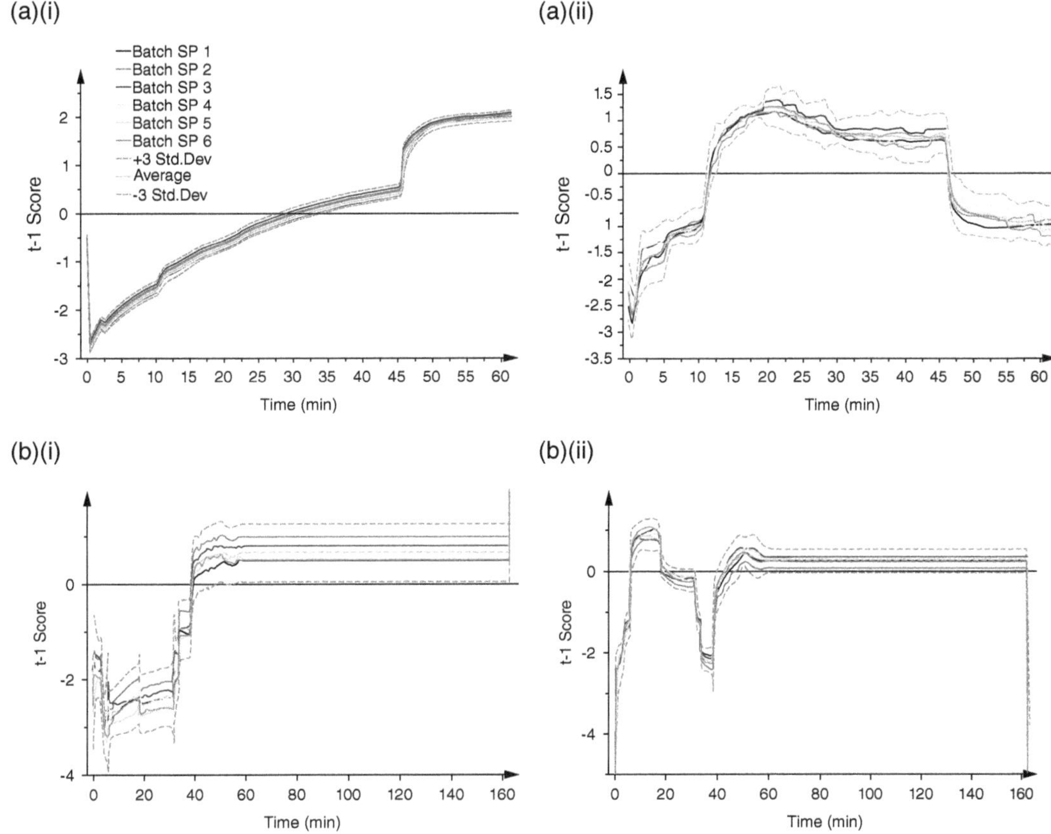

FIGURE 14.8
SPC control charts of MPLS (i) t-1 scores and (ii) t-2 scores from operating space control batches for (a) solubilization and (b) refolding. Figure adapted from Hebbi, Thakur, and Rathore 2019 with copyright permission.

and the chaotropic agent. Similarly, the refolding process was considered as a semi-batch system with periodic addition of buffer components, the solubilized inclusion bodies (SIB), and the components for pH adjustment and refold quenching. The multivariate BEM approach used MPLS modeling to correlate continuous ORP, temperature, and pH data with time, yielding two-component batch evolution profiles that are used for monitoring of refolding in real time.

Cumulative R^2 values for the x-variables in both solubilization and refolding are approximately 0.95, signifying that the time-regression profiles for the set-point runs are very consistent and well captured by the weighted combination of variables used in the model (Hebbi, Thakur, and Rathore 2019). The SPC charts ensure that CPPs lie within the design space, and deviation in these variables will be impacting the signature line in SPC chart.

The SPC charts were validated by performing five batches in which CPPs were deliberately deviated. Two batches were conducted outside the set point but still within the design space (SP Dev 1–2) and three batches were conducted outside the design space (DS Dev 1–3). Table 14.3 lists the deviations applied to the CPPs in each case and the resultant CQAs. It is evident that the set-point deviations maintain CQAs, unlike the design space deviations. However, the system should still raise an alarm, as the manufacturing process

TABLE 14.3

CPPs and Resultant CQAs of Batches Deviated from Operating Space and Design Space. Adapted from Hebbi, Thakur, and Rathore 2019.

	CPP			CQA			
	Urea (M)	DTT (mM)	Refold pH	Yield (%)	Conc. (mg/mL)	Oxidized impurity (%)	Reduced impurity (%)
OS Range	6.2–6.4	13.1–14.1	9.6–9.8	> 70	> 0.18	< 15	< 10
SP Dev 1 (Urea+DTT+pH)	6.5	9.5	9.5	77.21	0.22	3.92	4.72
SP Dev 2 (Urea+DTT+pH)	6.6	10.0	9.6	74.47	0.20	5.25	2.88
DS Range	6.0–6.6	9.2–18	9.5–10	> 70	> 0.18	< 15	< 10
DS Dev 1 (Urea only)	3.0	18.0	10.0	40.11	0.11	4.70	2.85
DS Dev 2 (DTT only)	6.0	3.0	10.0	78.37	0.23	16.36	0.00
DS Dev 3 (pH only)	6.6	18.0	7.0	55.97	0.20	5.81	3.11

is not operating within prescribed limits. When the CPPs are outside the design space itself, the batch output should be kept in quarantine for further investigation or discarded.

An online application was created by combining the SPC charts and the MPLS model using a LabVIEW (National Instruments Laboratory Virtual Instrument Engineering Workbench) and Python-based MVDA and graphing script. Online data from the ORP, temperature, and pH probes were received in real time and processed with the MPLS model from the control runs. Plots of the t-score components (t-1 and t-2) were obtained in real time. Figure 14.9 shows the points of deviation as they appeared in the real-time Python monitoring application.

For validation purposes, the deviation batches were followed through to the end in order to compare the output process parameters and CQAs of the refolded protein. Figure 14.9 shows the t-score SPC charts and the profiles of the complete deviation batches. The moment of deviation is clearly reflected in the SPC charts. Low concentrations of urea led to both higher ORP and higher temperature than for control batches. Low concentrations of DTT led to significantly higher ORP at the end of the solubilization buffer, which is expected as DTT is a strong reducing agent. Lowering the pH of the refolding buffer to 7 leads to a significant change in the pH profile and small changes in the ORP profile. The combined effect of these variables is captured in the MPLS t-scores calculated based on the training batch data and plotted on the SPC charts in real time.

Figure 14.10 shows the Hotelling's T^2 plots of the deviation batches plotted using the training model. The deviated batches can be seen clearly to lie outside the statistical thresholds of normal operation. Since time is the y-variable in the regression, the deviations on the SPC charts and Hotelling's T^2 plots begin to take place at the exact time when the deviated component is added to the batch. Thus, for the solubilization buffer, the urea deviation appears at zero minutes and the DTT deviation appears at 46 minutes. In the refolding buffer, the pH deviation appears at 37 minutes, and the deviations caused by the addition of improperly solubilized inclusion bodies (IBs)—due to low urea or DTT—appear when the SIB is added at 40 minutes. Once a deviation is present in the system, the errors in the SPC charts persist for future time intervals, as the changes in the online measured variables of ORP, pH, and temperature persist in future stages of the batch.

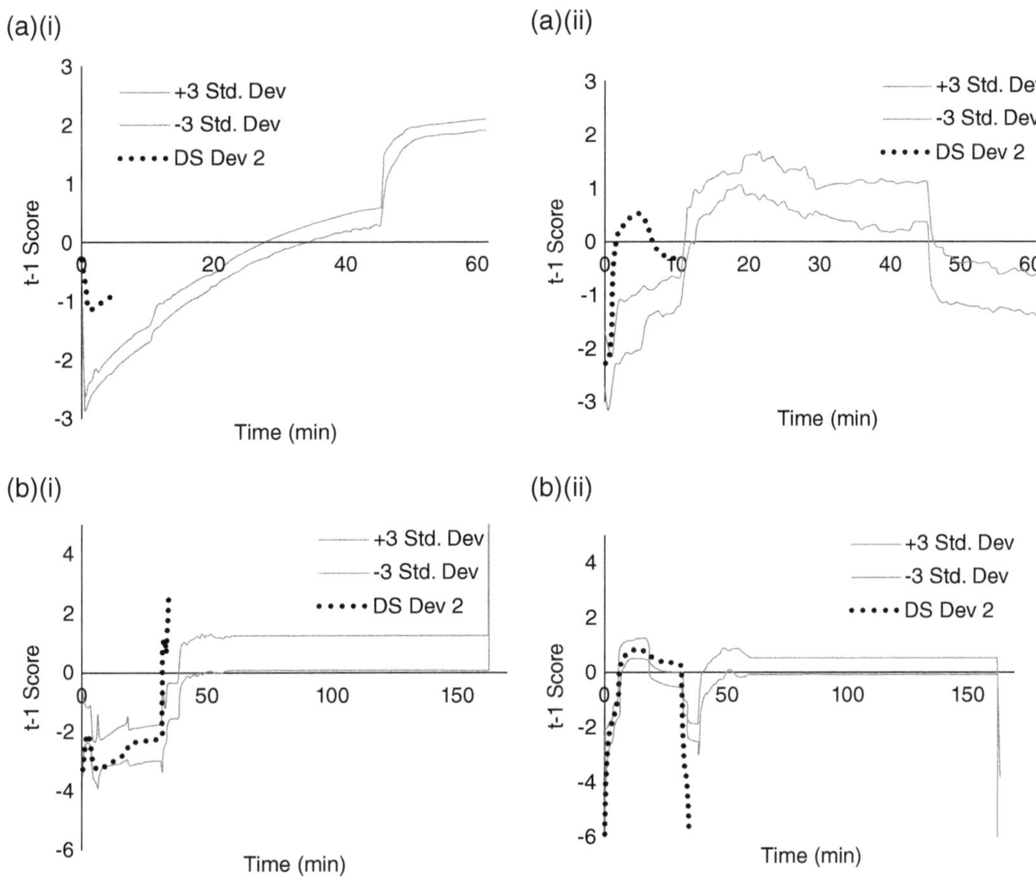

FIGURE 14.9

Two examples of real-time deviation detection by online monitoring application using SPC charts of (i) t-1 and (ii) t-2 scores for (a) low urea in solubilization and (b) low pH in refolding. The black dots are plotted every 30 seconds using online data. Figure adapted from Hebbi, Thakur, and Rathore 2019.

14.3.3.3 Case Study III: Application of MVDA for Assessing Product and Process Comparability

Biopharmaceutical manufacturing is a very complex process involving 15–20 unit operations or more in series or parallel. Hundreds of variables are involved in manufacturing, which affects its quality and safety (Hernández-Cuevas 2007). Process comparability studies during the development of drug therapeutics are mandatory as per regulatory guidelines. However, the large number of variables makes this exercise difficult (CMC-Biotechgroup 2009). Therefore, many times manufacturers use empirical comparisons of operational variables. The method is not efficient enough for a multivariate dataset and also does not reveal interaction effects, different correlations, and critical variables that are affected by scales. MVDA can handle more than one variable at a time. Projection methods are powerful in handling a multidimensional dataset, missing values, and variability from process data (Bhushan, Hadpe, and Rathore 2012).

In this case study, two different sets of batches are compared using MVDA: 28 batches from the first dataset and 93 batches from the second dataset. The process was divided into a total of seven individual steps, and in each step were many variables as listed in

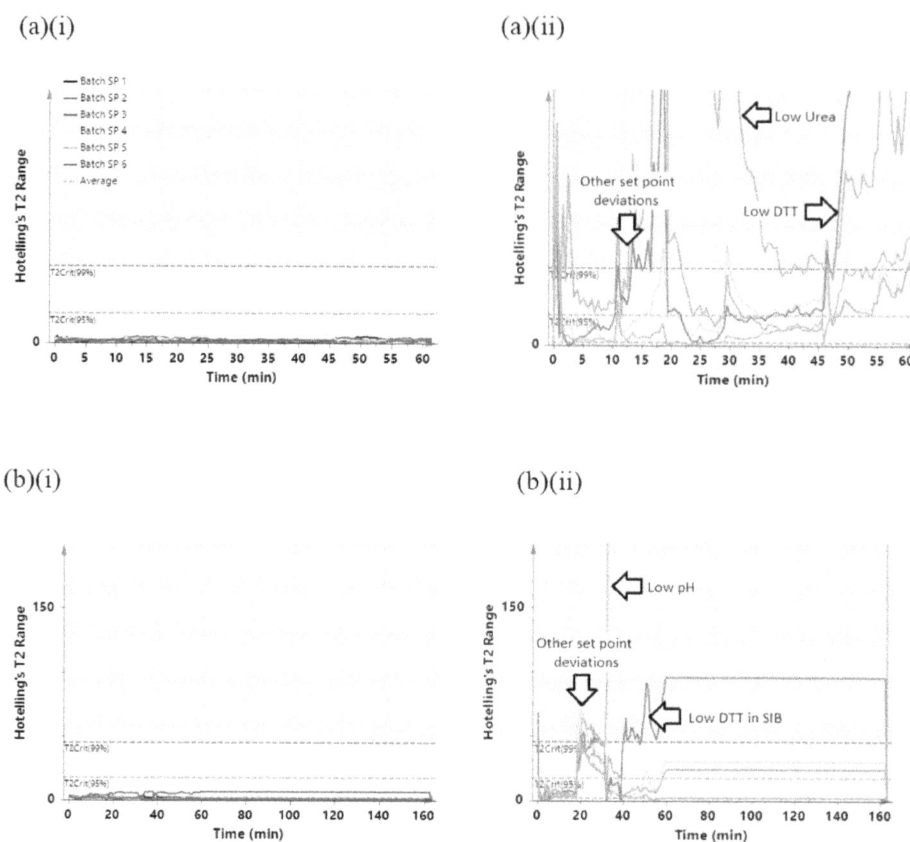

FIGURE 14.10

Hotelling's T^2 charts for (a) solubilization and (b) refolding for (i) control batches and (ii) deviation batches. Figure adapted from Hebbi, Thakur, and Rathore 2019 with copyright permission.

Table 14.3. Two main differences between the sets were different scale of operation and different lot size of a drug substance (Bhushan, Hadpe, and Rathore 2012).

Initial examination of data was performed by two sample t-tests using JMP software. Since two datasets share the same number of sets, t-test was performed assuming equal variance. Variables in the table were categorized into two groups: one with p value > 0.05 (statistically insignificant are shown in white background) and another with p < 0.05 (statistically significant are shown in gray background). However, t-tests compare two different datasets, considering only one variable per test. This is ineffective and difficult to quantitively analyze the data. MVDA can handle this type of multidimensional data and can produce a meaningful model.

Three approaches have been used in this study. The first approach involved the PCA and PLS-DA of the integrated dataset of two scales. Apparently, the PCA revealed the clustering for Steps 1, 3, and 6. It was expected that Step 6 was grouped separately because pooling decisions were different between the scales. A closer look was gained by performing the PLS-DA model, which again revealed that there was a strong Q^2 value for Steps 3, 6, and 7 as listed in Table 14.3. As expected, group differences were justifiable due to pooling differences at different scales; however, differences for Step 3 were unexpected, as that step performed identically at different scales. Upon close observation of data, it was hypothesized that the effect was due to some of the variables that are linked to the amount/volume of the product.

The second approach involved the normalization of the variables by dividing each variable for all the batches by the batch that had the highest value. This approach was specially assessed to normalize the lot size that arose due to different pooling decisions. After this treatment, data was in the range of 0 to 1. However, this type of pre-treatment led to clustering of Step 1 along with Steps 6 and 7 as listed in Table 14.4. Closer examination of pretreated data revealed that a significantly large number of outliers were present in Step 1. Hence, the normalization technique was not able to provide a meaningful model of the data.

The third approach consisted of dividing each variable with its mean value, which is also called mean scaling. Clustering was observed only in Steps 6 and 7 as listed in Table 14.4. None of the other steps were clustered, suggesting that mean scaling is better for analysis of process comparability compared to other scaling of normalization.

The overall approach for assessing process comparability is shown in Figure 14.11. Users should first clean the data and pre-process it using mean scaling. Two datasets at different scales after scaling are integrated into one file for analysis. It was advised to first perform PCA-X and then PLS-DA to get score plots and model over plots or values. The distinct pattern of clustering in both the PCA and PLS-DA for certain steps are identified. Score plots and Q^2 values are used for analysis data from PCA and PLS-DA respectively. These facts should align with the process knowledge. MVDA as such is a technique that gives spurious results based on the quality of data and pre-processing. It is up to the user to interpret the model results based on extensive process knowledge.

TABLE 14.4

Table Variables for Each Process Step Are Categorized as per Their p-Values. Adapted from Bhushan, Hadpe, and Rathore 2012.

Step	Variables							
1	V1	V2	V3					
2	V3	V4	V5	V6	V7			
3	V7	V8	V9	V10				
4	V10	V11	V12	V13				
5	V12	V13	V14	V15	V16	V17	V18	
6	V17	V18	V19	V20	V21	V22	V23	
7	V22	V23	V24	V25	V26	V27	V28	V29

TABLE 14.5

PLS-DA Q2 Values for the Three Approaches Used in the Study. Adapted from Bhushan, Hadpe, and Rathore 2012.

PLS-DA Q2	Approach 1: Integrated	Approach 2: Normalization	Approach 3: Mean Scaling
All variables	**0.81**	**0.87**	**0.83**
Step 1	0.05	0.74	0.02
Step 2	0.21	0.22	0.22
Step 3	0.67	0.24	0.30
Step 4	0.42	0.38	0.19
Step 5	0.04	0.48	0.01
Step 6	0.86	0.84	0.86
Step 7	0.71	0.82	0.55

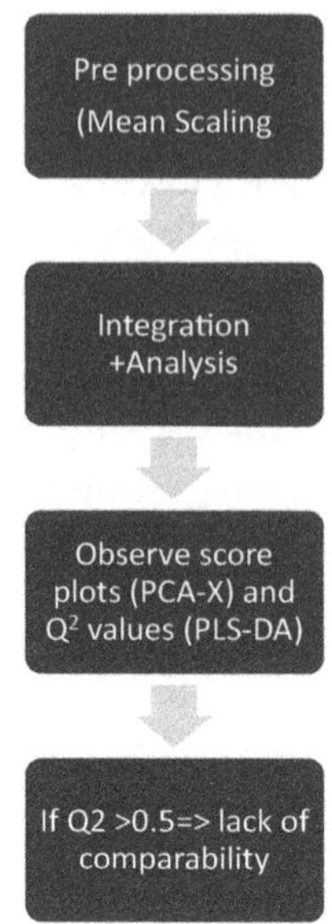

FIGURE 14.11
Flow chart showing the methodology for process/product comparability. Adapted from Bhushan, Hadpe, and Rathore 2012.

14.2 Summary

Multivariate data analysis is a versatile tool and can be used to assess different kinds of operations such as monitoring, control charts, analyzing manufacturing data, scale-up and scale-down qualification, process comparability across the scales, process characterization, and fault diagnosis. Hotelling T^2 charts can be used to ensure the batches run within the limits. An alarm-based system can be easily developed that can be implemented at manufacturing sites for process control. It is evident from the case studies presented that as advancements occur in process automation, sensors, and process analytics, MVDA has a prominent role in all aspects of process validation, be it process characterization, process validation, or process monitoring.

References

Ayturk, Engin. 2015. "Leveraging Large Data Sets in Continuous Chromatography Applications: Monitoring Critical Process Parameters Using MVDA." In Christopher Hwang, Genzyme-Sanofi Karol Lacki, Novo Nordisk (eds), *Integrated Continuous Biomanufacturing II", Chetan Goudar, Amgen Inc. Suzanne Farid, University College London. ECI Symposium Series.* https://dc.engconfintl.org/biomanufact_ii/84.

Bhushan, Nitish, Sandip Hadpe, and Anurag S. Rathore. 2012. "Chemometrics Applications in Biotech Processes: Assessing Process Comparability." *Biotechnology Progress* 28 (1): 121–28.

Chen, Aaron, Rajesh Chitta, David Chang, and Ashraf Amanullah. 2009. "Twenty-Four Well Plate Miniature Bioreactor System as a Scale-Down Model for Cell Culture Process Development." *Biotechnology and Bioengineering* 102 (1): 148–60.

Clapp, Kenneth P., Andreas Castan, and Eva K. Lindskog. 2018. "Upstream Processing Equipment." In *Biopharmaceutical Processing*, 457–76. Elsevier.

CMC-Biotechgroup. 2009. *A-Mab Case Study.* https://ispe.org/publications/guidance-documents/a-mab-case-study-in-bioprocess-development.

Eriksson, Lennart, Tamara Byrne, E. Johansson, Johan Trygg, and C. Vikström. 2013. *Multi-and Megavariate Data Analysis Basic Principles and Applications.* Vol. 1. Umetrics Academy.

FDA. 2004. *Guidance for Industry: PAT—a Framework for Innovative Pharmaceutical Development, Manufacturing, and Quality Assurance.* FDA.

Glassey, Jarka, Krist V Gernaey, Christoph Clemens, Torsten W Schulz, Rui Oliveira, Gerald Striedner, and Carl-Fredrik Mandenius. 2011. "Process Analytical Technology (PAT) for Biopharmaceuticals." *Biotechnology Journal* 6 (4): 369.

Godavarti, Ranga, J. Petrone, Jeff Robinson, Richard Wright, and Brian D. Kelley. 2005. "Scale-Down Models for Purification Processes: Approaches and Applications." *Biotechnology and Bioprocessing Series* 29: 69.

Hansuld, E. M., and L. Briens. 2014. "A Review of Monitoring Methods for Pharmaceutical Wet Granulation." *International Journal of Pharmaceutics* 472 (1–2): 192–201.

Hebbi, Vishwanath, Garima Thakur, and Anurag S. Rathore. 2019. "Process Analytical Technology Implementation for Protein Refolding: GCSF as a Case Study." *Biotechnology and Bioengineering* 116 (5): 1039–52.

Hernández-Cuevas, Cristián. 2007. "Collaborative Innovation: The Future of Biopharma." *Journal of Technology Management \& Innovation* 2 (3): 1–3.

Kirdar, Alime Ozlem, Jeremy S. Conner, Jeffrey Baclaski, and Anurag S. Rathore. 2007. "Application of Multivariate Analysis toward Biotech Processes: Case Study of a Cell-Culture Unit Operation." *Biotechnology Progress* 23 (1): 61–67.

Li, Baibing, Julian Morris, and Elaine B. Martin. 2002. "Model Selection for Partial Least Squares Regression." *Chemometrics and Intelligent Laboratory Systems* 64 (1): 79–89.

Li, Feng, Yasunori Hashimura, Robert Pendleton, Jean Harms, Erin Collins, and Brian Lee. 2006. "A Systematic Approach for Scale-Down Model Development and Characterization of Commercial Cell Culture Processes." *Biotechnology Progress* 22 (3): 696–703.

Liland, Kristian Hovde. 2011. "Multivariate Methods in Metabolomics—from Pre-Processing to Dimension Reduction and Statistical Analysis." *TrAC Trends in Analytical Chemistry* 30 (6): 827–41.

Liu, Ronghua, Lian Li, Wenping Yin, Dongbo Xu, and Hengchang Zang. 2017. "Near-Infrared Spectroscopy Monitoring and Control of the Fluidized Bed Granulation and Coating Processes—A Review." *International Journal of Pharmaceutics* 530 (1–2): 308–15.

Nomikos, Paul, and John F. MacGregor. 1995. "Multi-Way Partial Least Squares in Monitoring Batch Processes." *Chemometrics and Intelligent Laboratory Systems* 30 (1): 97–108.

Pieracci, John P., John W. Armando, Matthew Westoby, and Jorg Thommes. 2018. "Industry Review of Cell Separation and Product Harvesting Methods." In *Biopharmaceutical Processing*, 165–206. Elsevier.

Rathore, Anurag S., Nitish Bhushan, and Sandip Hadpe. 2011. "Chemometrics Applications in Biotech Processes: A Review." *Biotechnology Progress* 27 (2): 307–15.

Rathore, Anurag S., Raj Krishnan, Stephanie Tozer, Dave Smiley, Steve Rausch, and Jim Seely. 2005. "Scaling down of Biopharmaceutical Unit Operations: Part I: Fermentation." *Biopharm International* 18 (3): 60–68.

Rathore, Anurag S., Shachi Mittal, Mili Pathak, and Arushi Arora. 2014. "Guidance for Performing Multivariate Data Analysis of Bioprocessing Data: Pitfalls and Recommendations." *Biotechnology Progress* 30 (4): 967–73.

Rathore, A. S., and S. K. Singh. 2015. "Use of Multivariate Data Analysis in Bioprocessing." *BioPharm International* 28 (6): 26–31.

Rathore, Anurag S., and Sumit K. Singh. 2017. "Chemometrics Applications in Process Chromatography." *Preparative Chromatography for Separation of Proteins*: 479–500.

Rüdt, Matthias, Till Briskot, and Jürgen Hubbuch. 2017. "Advances in Downstream Processing of Biologics—Spectroscopy: An Emerging Process Analytical Technology." *Journal of Chromatography A* 1490: 2–9.

Sahoo, Niharika, Koel Choudhury, and Padmavati Manchikanti. 2009. "Manufacturing of Biodrugs." *BioDrugs* 23 (4): 217–29.

Sofer, Gail. 2005. *Process Validation in Manufacturing of Biopharmaceuticals: Guidelines, Current Practices, and Industrial Case Studies*. CRC Press.

Steinwandter, Valentin, Daniel Borchert, and Christoph Herwig. 2019. "Data Science Tools and Applications on the Way to Pharma 4.0." *Drug Discovery Today* 24 (9): 1795–805.

Tsang, Valerie Liu, Angela X. Wang, Helena Yusuf-Makagiansar, and Thomas Ryll. 2014. "Development of a Scale down Cell Culture Model Using Multivariate Analysis as a Qualification Tool." *Biotechnology Progress* 30 (1): 152–60.

Undey, Cenk, Duncan Low, Jose C. Menezes, and Mel Koch. 2011. *Pat Applied in Biopharmaceutical Process Development and Manufacturing: An Enabling Tool for Quality-by-Design*. CRC Press.

Zink, Richard C. 2018. "Uncovering Fraud, Misconduct, and Other Data Quality Issues in Clinical Trials." In K. Peace, D. G. Chen, S. Menon (eds), *Biopharmaceutical Applied Statistics Symposium*, 261–76. ICSA Book Series in Statistics. Springer. https://doi.org/10.1007/978-981-10-7820-0_13.

15

Process Development for Plasmid DNA Production

Frank Agbogbo, Renija George, Ganesh Krishnamoorthy, Carole Davis,
and Allison Farrand

15.1 Introduction

The term "plasmid" was originally coined by Joshua Lederberg in 1952 to describe any bacterial genetic element that exists in extrachromosomal state for at least part of its replication cycle (Lederberg, 1952). The description includes bacterial viruses that are genetic elements in extrachromosomal state for part of their replication, so the term "plasmid" needed to be refined. In 2003, Fibar Hayes (Hayes, 2003) subsequently refined the definition to describe a plasmid as "exclusively or predominantly extrachromosomal genetic elements that replicate autonomously." This definition clearly excludes bacterial viruses that cannot replicate autonomously inside the cell. Naturally occurring plasmids are known to be present in most species of Eubacteria that have been examined (Summers, 1996) but are non-essential for the survival of the host bacterium.

Plasmids have served as invaluable model systems for the study of processes such as DNA replication, segregation, conjugation, and evolution (Thomas, 2000). Since plasmids were first utilized in the early 1970s, they have played a vital role in modern recombinant DNA technology as gene-cloning and gene-expression vehicles (Hayes, 2003). Plasmids carrying recombinant DNA have been applied to develop therapeutics and vaccines for various ailments such as cancer, diabetes, diphtheria, rubella, poliomyelitis, pneumonia, and meningitis (Khan et al., 2016). Plasmid DNA (pDNA) technology offers an opportunity to develop vaccines and therapeutics faster than conventional processes that rely on chicken eggs or cell culture. Other advantages of pDNA include lower production cost in comparison to antibodies or microorganism-based vaccines, longer and higher immune response, low infection risk, and stability of the product for storage and shipping. On the other hand, pDNA vaccines suffer from low immunogenicity in humans and large animals, poor DNA cellular uptake and potential to alter host genes (Abdulrahman & Ghanem, 2018).

Recent advances in gene therapy and nucleic acid vaccinations have resulted in an increased demand for pDNA. Some of these products are made using viral vectors or non-viral vectors such as naked pDNA. The global demand for pDNA and viral vectors is expected to grow from US$ 1.5 billion in 2019 to US$ 5.2 billion in 2027 (Figure 15.1, Maximize Market Research, 2020). The increase in pDNA demand is driven by technological advancement in viral vector production, an increase in the number of products in clinical trials, and vaccine development for COVID-19, some of which use viral vectors and mRNA produced from pDNA (for example, Astrazeneca, Johnson & Johnson, Moderna, Pfizer). The Alliance for Regenerative Medicine (ARM) report shows that there were 1220

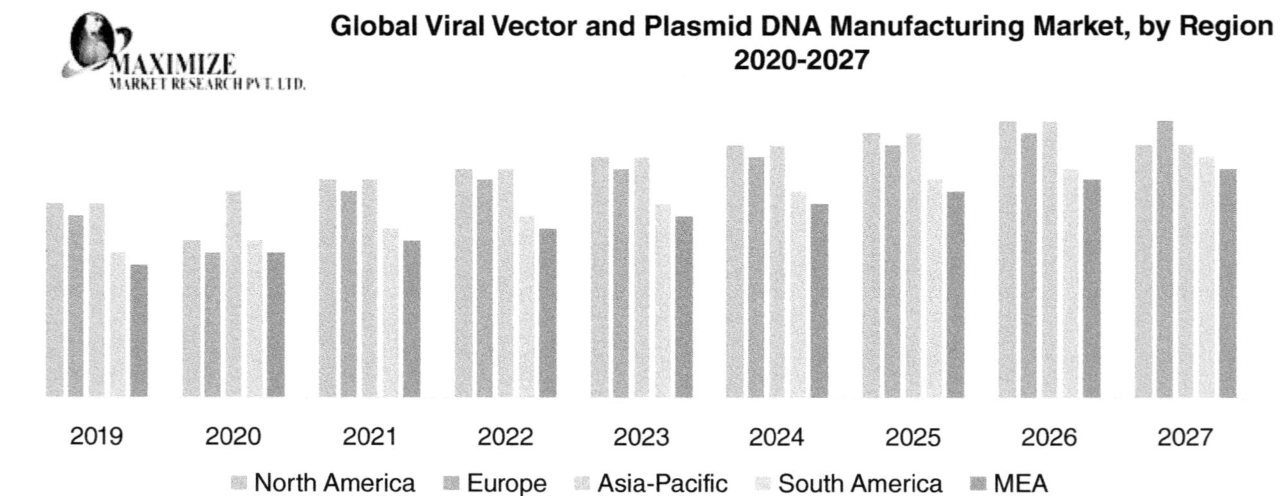

FIGURE 15.1
Maximize Market Research PVT LT, global viral vector and plasmid DNA manufacturing market—industry analysis and forecast (2019–2027)—by type, application, disease, end-user and region.

clinical trials ongoing at the end of 2020 (Alliance for Regenerative Medicine Report, 2020). The demand for pDNA is expected to surge in the coming years and it requires an efficient process that can be used to produce and purify pDNA to meet the regulatory requirements for purity, safety, and efficacy. The upstream process development, primary recovery, purification, and analytical development for pDNA production are discussed in this chapter.

15.2 Upstream Process

The upstream process development comprises of strain selection and plasmid optimization, media composition, and optimization of fermentation conditions that are used during the production of pDNA. Biomass yield, plasmid yield and plasmid quality can be improved by optimizing the growth environment of the plasmid producing organism. Volumetric yield (mg/L) and specific yield (mg/g) of supercoiled pDNA should be the primary focus for designing the fermentation process (Carnes & Williams, 2007). Supercoiled pDNA is the product of choice, and the FDA recognizes the open end, linear and nicked forms as less effective therapeutically. Developing the necessary analytics to measure the percent supercoiled is very important in the production of plasmids. The fermentation process should be developed and designed to produce a high percentage of supercoiled pDNA (> 80%). This section covers recent developments in pDNA upstream process.

15.2.1 Strain Selection and Plasmid Optimization

E. coli strains are genetically engineered to produce the plasmid of interest at high concentrations. Strain engineering through overexpression of rationally selected genes and or knock-out of some genes leads to higher pDNA titers (F. Silva et al., 2012). *E coli* strain K-12 is usually preferred due to its non-pathogenic characteristics but other strains such as *E coli* B derivatives have also been used. Mutations in *endA* and *recA* genes are present in the host strains selected for pDNA production. The *endA* mutation prevents plasmid degradation after cell lysis, and the *recA* mutation ensures insert stability and decreases mutation frequency, preventing plasmid structural instability (F. Silva et al., 2012). Another recommended mutation is in the *relA* gene which causes DNA amplification under amino acid limitation or starvation conditions, thus increasing the specific pDNA yields (F. Silva et al., 2012).

Hosts such as DH5 derivatives and XL1 Blue are the popular choices when it comes to pDNA production (Ciccolini et al., 1999; F. Silva et al., 2012). The BL21 strain has been widely used for the overexpression of recombinant proteins but has not been a good candidate for pDNA production because of the intact *endA* and *recA* genes. Comparing K-12 and B strain metabolically, B strains produce less acetate when grown in a medium rich in glucose because of a more active glyoxylate shunt, which means more active acetate uptake pathways (Bower & Prather, 2009). Phue et al. studied BL21 for pDNA production by deleting both *recA* and *endA* and observed that higher pDNA yields were obtained with BL21 compared to DH5α (K-12 derivative), and the plasmid properties were identical to the plasmids currently produced by DH5α. Consistent with the observation that BL21 has a more active glyoxylate shunt and fared better than K-12 DH5α strain, disruption of *fruR*, the positive regulator of the glyoxylate shunt, in DH5α led to increased pDNA production by up to 21% in specific pDNA yield from a fed-batch fermentation (Bower & Prather, 2009).

Other ways to improve plasmid quality and quantity are through plasmid optimization by incorporating the desired origin of replication, sequences, antibiotic resistance markers or other selection markers in combination with optimized media composition and fermentation process parameters. The origin of replication determines the plasmid copy number. Most therapeutic plasmids contain origins of replication derived from ColE1 or pMB1 (Carnes et al., 2006; Twigg & Sherratt, 1980). Mutations, such as *rop* gene (repressor of primer gene) deletion and G-to-A point mutations can be incorporated to induce selective plasmid amplification by growth at high temperature (42 °C) with pUC and pMM1 replication origins (Wong et al., 1982; Carnes et al., 2006). Temperature sensitive origins such as pUC, pMM1, and pMM7 are useful in fermentation as they allow for 30-to-40-fold increase in plasmid copy number when the temperature is shifted from 30 °C to 42 °C (Wong et al., 1982; Carnes et al., 2006).

The sequences in the plasmid backbone must be optimized so that nicking (associated with AT rich regions that are susceptible to endogenous single stranded nucleases), stability issues caused by palindrome sequences, and direct or inverted repeats are avoided. During plasmid backbone optimization, dimer formation in the pUC plasmid caused by the increase in oligo-pyrimidine or oligo-purine sequences can result in unusual DNA structures such as triple helix (Cato, 1993; Prazeres et al., 1999; Carnes& Williams, 2007). Antibiotic resistance genes are commonly used as plasmid selection markers. The gene encoding kanamycin resistance is the most widely used antibiotic resistance marker (Williams et al., 2009b). Resistance markers for β-lactam antibiotics such as ampicillin are not acceptable due to hypersensitivity in some patients.

The tetracycline resistance marker is also avoided since it is toxic to *E. coli* at high copy number (Chiang & Bremer, 1988; Carnes & Williams, 2007). To avoid dissemination of antibiotic resistance genes to a patient's enteric bacteria as well as to reduce the size of the therapeutic plasmid, alternative selection strategies have been designed (Williams et al., 2009b). Alternative selection systems include:

1. Auxotrophy complementation where an essential gene is maintained on the plasmid with a corresponding chromosomal deletion or suppressible mutation (Williams et al., 2009b).

2. Repressor titration where an operator sequence is placed on a multicopy plasmid that depresses the chromosomal gene (Williams et al., 2009b).

3. RNA based selection marker where the regulated gene can be an antibiotic resistance marker, transcriptional repressor, or a lethal gene (Williams et al., 2009b). In 2009, Luke et al. developed a novel antibiotic free selection system where vectors incorporate and express a 150 bp RNA-OUT antisense RNA. Sac B (counter selectable marker) expression is repressed by RNA-OUT which allows plasmid selection on sucrose. High yields of > 1 g/L pDNA have been observed when previously kanamycin resistant DNA vaccine plasmids were converted to antibiotic free sucrose selection (Carnes et al., 2010).

15.2.2 Fermentation Media

The quality of the plasmid and yield are affected by the composition of the medium. A balanced medium is required for energy, biomass, and cell maintenance during high cell density fermentation. pDNA production medium should support high nucleotide pools in cells and supply energy for replication while minimizing other cellular activities (Carnes & Williams, 2007). The impact of the fermentation medium components on plasmid yield

TABLE 15.1

Bacterial Elemental Composition (Stanbury et al., 1995)

Element	Dry Weight (%)
Carbon	50–53
Hydrogen	7
Nitrogen	12–15
Phosphorous	2–3
Sulfur	0.2–1.0
Potassium	1.0–4.5
Sodium	0.5–1.0
Calcium	0.01–1.10
Magnesium	0.1–0.5
Chloride	0.5
Iron	0.02–0.20

and quality, biomass yield, and their potential interference with purification and regulatory concerns should be kept in mind while formulating media for plasmid production (Carnes & Williams, 2007). Balanced media is designed based on energy requirements and elemental composition (Table 15.1) of bacteria (Stanbury et al., 1995).

The nutritional requirements are met using either complex or chemically defined media. *E. coli* can grow on minimal media with simple carbon, nitrogen, and various salts. The fermentation process in minimal media is highly reproducible and high plasmid copy numbers are obtained. Complex media contain yeast extracts and peptones, which supply growth factors, amino acids, purines, and nucleic acids and support higher cell densities, but reproducibility is a concern due to lot-to-lot variability of the yeast extract or peptone (O'Kennedy, 2000).

The carbon source that provides energy and builds biomass is usually the limiting nutrient in a fed-batch fermentation process. Glucose is the common choice as it is inexpensive and is metabolized very efficiently. The only problem with glucose is the acetate production due to metabolic overflow. Glycerol is often used as a carbon source in batch cultures, and due to its slightly lower cellular yield than glucose, glycerol does not cause as much acetate production nor is it inhibitory at higher concentrations. The nitrogen source in minimal media is ammonia and ammonium salts, while yeast extracts and peptones supply nitrogen in complex media (Carnes & Williams, 2007). Salts and minerals necessary for bacterial growth, metabolism and enzymatic reactions are magnesium, phosphorus, potassium, sulfur, calcium, copper, cobalt, iron, manganese, molybdenum, and zinc. The salts and minerals are provided by adding trace metal solution (Lugo et al., 2016).

15.2.3 Fermentation Process

Plasmid DNA can be produced in shake flasks, or in fermenters as a batch, or fed-batch fermentation process (O'Kennedy et al., 2003). In the production of pDNA, fed-batch fermentation is desired for increasing yield and productivity while decreasing manufacturing costs. The fermentation process has an impact on the quality of pDNA at harvest as well as the quality of the plasmid after lysis and purification (Williams et al., 2009a; Long et al., 2018). Fed-batch fermentation produces higher quality plasmids than the batch fermentation (O'Kennedy et al., 2003). Carbon/Nitrogen ratio in the growth medium affects cell harvest and lysis (O'Kennedy et al., 2000). Process optimization (modification of process

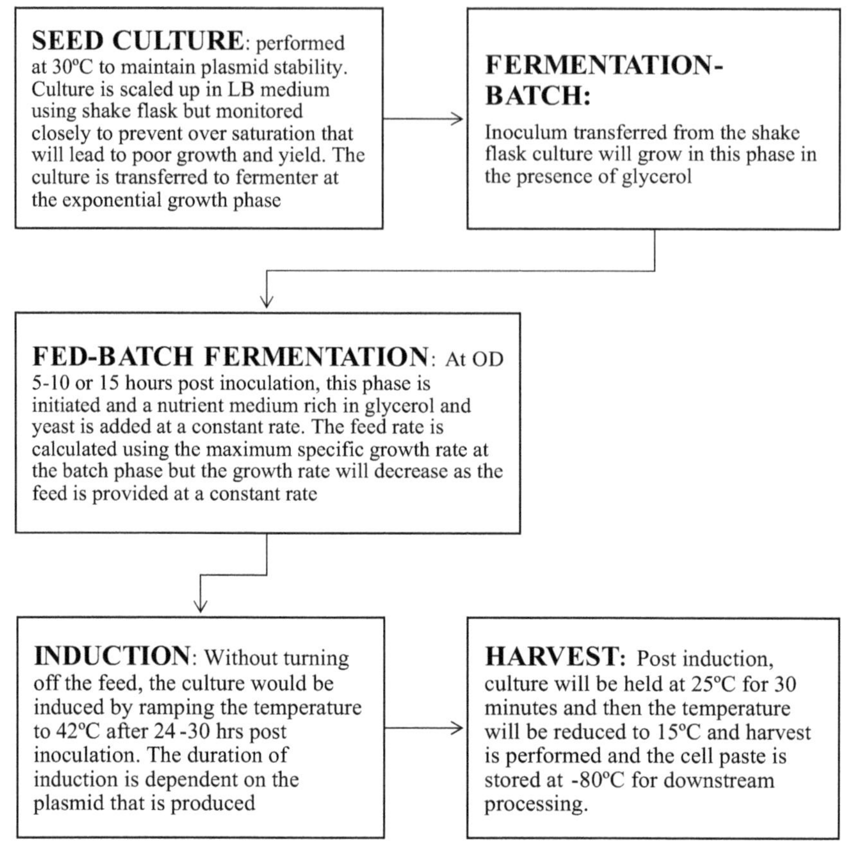

FIGURE 15.2
pDNA upstream process flow diagram (Carnes & Williams, 2007).

conditions such as temperature or feed rate) is needed for better plasmid yield, a greater supercoiled percentage, and the compatibility of the resulting cells for harvest and primary recovery steps (Williams et al., 2009a).

Different temperatures and feed rates have impact on plasmid yield and quality. Lower temperatures (~30°C) for example results in slower cell growth and could lead to better plasmid replication and reduced mutation rates. Higher temperatures (~37°C) generally lead to a faster cell growth and higher plasmid yields. However, increased growth rate can lead to higher mutation rates and lower quality of plasmids. Higher feed rates involve supplying nutrients to support faster cell growth, which can lead to accumulation of by-products and low quality of plasmids. The optimal temperature and feed rate will depend on the specific strain, plasmid, and the expression system.

A typical pDNA fermentation process flow is shown in Figure 15.2. The Carnes group (Williams et al., 2009a), developed a pUC plasmid fermentation process that reduces the metabolic burden by utilizing conditions that both restrict growth rate and maintain reduced plasmid copy number during the biomass accumulation phase prior to a plasmid accumulation phase before harvest.. Figure 15.3 shows the pictorial representation of the pUC plasmid inducible fed-batch fermentation.

In this process, plasmid containing E. coli cells are grown at a reduced temperature of 30 °C during batch phase, and part of the fed-batch phase during which the plasmid is

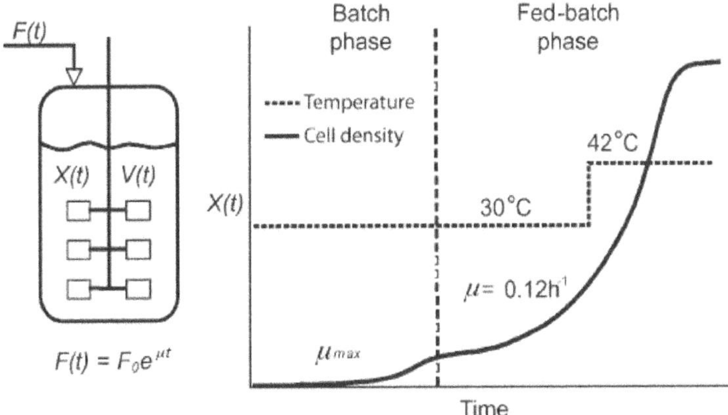

FIGURE 15.3
Inducible fed-batch plasmid fermentation process (Williams et al., 2009a).

maintained stably while the biomass accumulates (Williams et al., 2009a; O'Kennedy et al., 2000). An exponential feeding strategy is maintained during the fed-batch phase maintaining the specific growth rate at 0.12 per hour (Williams et al., 2009a). Induction of plasmid accumulation is performed by increasing the temperature to 42 °C (Lahijani et al., 1996; Williams et al., 2009a). To maintain a high percentage of supercoiled plasmid, a temperature shift is performed at restricted growth rate. Plasmid yield improvement, and a reduction of nicked and dimer plasmid percentages, was seen after holding the culture at 25 °C for 0.5 to 3 h prior to dropping the temperature to 15 °C for harvest (Williams et al., 2009a). One of the advantages of the inducible fed-batch process is the amplification of plasmid copy number after biomass accumulation and this preserves the quality while maximizing the overall yield (Carnes & Williams, 2007).

In the production of plasmids, a high yield and high percentage of supercoiled pDNA are desired. The growth rate of the cells during fermentation has an impact on whether these desired characteristics can be achieved. With high growth rates in batch and fed-batch fermentations, a lower percentage of supercoiled pDNA was observed (O'Kennedy et al., 2000; Carnes & Williams, 2007). Other factors that are known to have an impact on the percent supercoiled pDNA are oxygen levels and temperature (Dorman et al., 1988; Goldstein & Drlica, 1984). Temperature, pH, dissolved oxygen, nutrient concentration, and growth rate can cause formation of nicked plasmids and multimers (Durland & Eastman, 1998). Therefore, these should be considered for inclusion in the critical performance parameter (CPP) list. Temperature should be kept low during growth to keep the cells at a lower specific growth rate (Carnes & Williams, 2007).

Compared to batch fermentation, fed-batch fermentation is more useful for plasmid production. To control the growth rate below the maximum growth rate of the cells, controlled addition of limiting nutrient such as glucose or glycerol should be implemented (Durland & Eastman, 1998). In a fed-batch fermentation, the feed rate needs to match the rate of substrate consumption so that overfeeding such as the accumulation of glucose or glycerol does not occur. The conversion of substrate to biomass is very efficient when there is no accumulation of feed and the residual substrate is zero, preventing metabolic overflow, thus avoiding the formation of inhibitory products such as acetate (Carnes & Williams, 2007).

At the end of fermentation, the cells are usually harvested by a variety of centrifugation methods to separate the cell pellets from the supernatant (Theodossiou et al, 1997). The simplest and most translatable from small-scale to large-scale is bucket centrifugation. However, bucket centrifugation does have its limitations such as processing logistics, large volume bucket rotors, and removal of the paste from the centrifugation containers. At larger scales, it is common to utilize continuous centrifuges like tubular bowl or disc stack. These methodologies allow for operational efficiencies at large scale, but at the sacrifice of recovery as there is always a percentage of the cell paste lost to waste during discharges. These methods also don't produce a paste but a cell slurry instead, in which case the residual medium have to be considered for its impact to the process. If the residual medium is unsuitable, then other means can be used to exchange the residual medium components to another buffer like tangential flow filtration (TFF) or bucket centrifugation of the smaller concentrated cell slurry volume to achieve a cell paste. The paste is usually stored at -20°C or -80°C or processed immediately through downstream processing. Storage of the paste is preferred since it serves as a good hold point prior to downstream processing.

The process development for plasmids could be performed by using Quality by Design principles (QbD), which provides a deep scientific understanding of the process. QbD involves the use of a systematic experimental approach in evaluating the process parameters to ensure the quality of the plasmids. The key aspects of this approach includes defining Critical Quality Attributes (CQA's) such as purity and safety, Critical Process Parameters (CPP) such as temperature and feed rate, Design of Experiments (DoE) to evaluate the impact of different process parameters on plasmid quality. The remaining aspects of QbD include risk assessment to identify and access potential risks, Process Analytical Technologies (PAT) to monitor key process parameters and control strategy on controlling critical process parameters, implementing appropriate process controls and conducting regular quality checks.

15.3 Downstream Process

The focus of the downstream process in pDNA production is to maximize the yield of supercoiled pDNA with high purity (US Food and Drug Administration, 2007). The amount of pDNA required can vary from milligrams to several grams and the quality, from research grade to cGMP grade (Prather et al., 2003). Typical plasmid sizes range from under 1000 base pairs (minicircles) to several thousand base pairs depending on the application. To meet the strict FDA standards (FDA, 2016), it is critical to design a process that is (1) robust in removal of impurities and process additives, (2) scalable without compromising on the yield and purity, and (3) cost effective and less complicated with a minimum number of steps. Though small-scale purifications of pDNA are well established, pharmaceutical grade pDNA production at industrial scale poses several challenges that can impact the final yield and quality.

pDNA being a minor component of *E. coli* cell (< 1% of dry cell mass) (Diogo et al., 2005) must be isolated from other cellular contaminants such as genomic DNA and RNA with similar chemical properties, cellular proteins, membrane lipids (including endotoxin), sugars and other molecules (Abdulrahman & Ghanem, 2018). To produce pharmaceutical grade pDNA for gene therapy, the industry-wide focus is to improve the Critical Quality Attributes such as yield, homogeneity, purity, identity, maintain potency and efficacy, and exclude the usage of animal derived enzymes (including Proteinase K, RNAse, and Lysozyme). Moreover, the use of organic solvents such as phenol, chloroform, ethanol, and isopropyl alcohol is avoided in large-scale manufacturing due to their toxicity, flammability, and corrosive properties. Contaminants from column chromatography due to leakage of ligands (such as Quaternary ammonium) must also be avoided (Abdulrahman & Ghanem, 2018). Though very useful, enzymes or organic solvents add extra manufacturing costs and increase the burden on the purification steps. Their usage also necessitates analytical tests to show the clearance of these extraneous materials to safe levels.

Isolation of plasmids from *E. coli* cells, small or large scale, involves a primary recovery step, which removes most of the genomic DNA, endotoxin, host-cell proteins, and RNA and the residual contaminants. The primary recovery step is designed to facilitate efficient purification of pDNA through concentration and buffer exchange step followed by a two-step chromatographic purification and final formulation (Figure 15.4). This portion of the chapter will focus mainly on the different avenues for scalable methods for purification, risks, limitations, and challenges of large-scale manufacturing, and how the choice of process steps can affect the quality of pDNA meant for gene therapy.

Plasmid DNA manufacturing challenges are not commonly seen in the production of therapeutic proteins due to the unique physio-chemical properties of pDNA (Urthaler et al., 2005). These properties influence capital costs, the choice of downstream process steps, cost of consumables, scalability, and timeline. In addition, there are several bottlenecks in the purification of pDNA (Table 15.2). To overcome these challenges, it is necessary to understand the properties of pDNA. Thus, it is imperative that the pDNA isolation process is specifically designed by incorporating the QbD with deep scientific understanding of the process, technology, and final product qualities. This involves the development of a systematic approach, the application of controls over the process design, the use of the latest technologies and analytical tests for in-process analysis.

TABLE 15.2

Challenges During Large-Scale Purification of Plasmid DNA (Prazeres et al., 1999)

Primary Recovery	Challenges
a) High cell density.	a) Difficult to lyse resulting in Low-yield.
b) High Viscosity and Shear Thickening of Lysate.	b) Insufficient mixing leading to 1. Increased power requirements. 2. Incomplete lysis, which results in low recovery. 3. Local high concentration of NaOH leads to irreversible denaturation of pDNA.
c) Handling of very large volumes.	c) Affects reproducibility and recovery due to large scale.
d) Shear sensitivity and fragmentation of gDNA.	d) Due to high shear linear gDNA pieces in the product are difficult to separate from pDNA by chromatography.
e) Irreversible denaturation of pDNA.	e) Low yield of Sc pDNA due to loss during lysis.
f) Contamination with nucleases.	f) Degradation of Sc pDNA into Oc form.
Purification	
a) Poor selectivity, Low binding capacity and Poor recovery.	a) Copurification of contaminants due to similarities in properties to pDNA affecting final product yield and quality. Poor binding capacity leads to multiple cycles of purification resulting in cost escalation.
b) Poor resolution of isoforms.	b) Low homogeneity due to presence of high amounts of undesirable isoforms.
c) Reduced flow rate due to high viscosity and long process times.	c) Slower turn around times can hold-up production capacity.
d) Large column volumes, cleaning and regeneration.	d) Consume large quantities of buffer and requires cumbersome cleaning-validation.

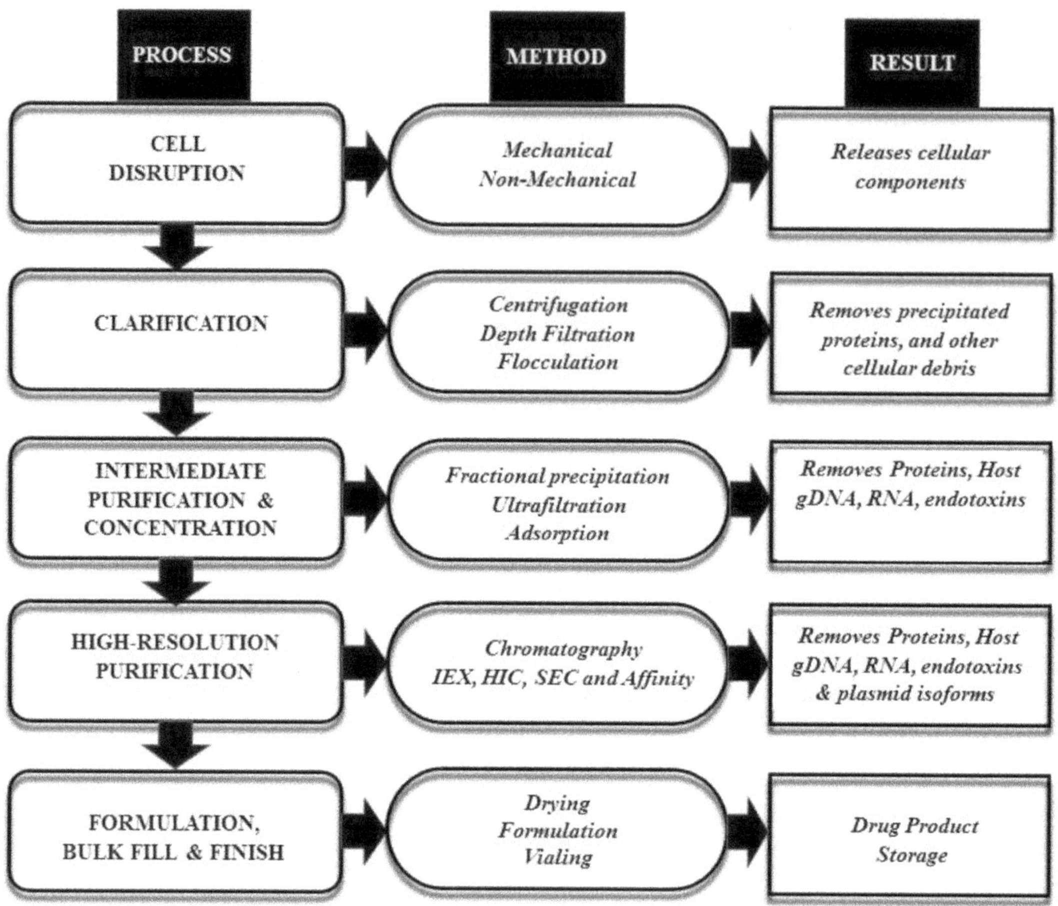

FIGURE 15.4

Process Flow Chart for Large scale production of Plasmid DNA (Ferreira et al., 2000b, Levy at al., 2000).

15.3.1 Cell Disintegration

The first critical step in pDNA purification is cell disintegration. The method of lysis is selected based on the properties of the biomolecule under consideration, purification steps and the quality of the final product desired. Cell Lysis methods include mechanical or non-mechanical methods (Shehadul et al., 2017) (Figure 15.4). Common mechanical cell lysis methods such as homogenization, micro-fluidization or bead milling are highly efficient at lysing cells (Carlson et al., 1995; Middelberg, 1995). However, the size of the gDNA makes it fragile and susceptible to breakage and these methods cannot be used for pDNA isolation, which necessitates the use of non-mechanical methods (Lengsfeld & Anchordoquy 2002; Levy at al., 1999).

Non-mechanical methods such as osmotic shock, enzymatic lysis, thermal lysis, (Zhu et al., 2005; Zhu et al., 2006) or sonication, have their own limitations due to inefficient cell lysis. In addition, inadequate denaturation of gDNA, decreased yield of supercoiled pDNA, safety concerns, and cost factors make them unfavorable for scale-up. Other restrictions include the use of enzymes of animal origin and toxic reagents that do not meet the stringent regulatory requirements (Carnes et al., 2009; Eon-Duval & Burke, 2004; Meacle et al., 2004).

Chemical method such as alkaline lysis, which was first described by Birnboim and Doly (1979) and Sambrook et al. (1989), is commonly used due to its simplicity and ability to selectively denature the proteins and genomic DNA, leaving the plasmid intact (Li et al., 2007). However, in other methods of lysis such as thermal or enzymatic lysis or a combination of the two, large amounts of impurities are not removed from the pDNA and pose problems during chromatography (Zhu et al., 2005; Zhu et al., 2006; Murphy et al., 2006; Holmes & Quigley, 1981). Although alkaline lysis is very common, it has its own limitations with regards to (1) stringent duration of lysis, (2) lack of efficient mixing technique, (3) high viscosity of the lysate, and (4) corrosive nature of the lysis solution.

In the alkaline lysis method, cell suspension (0.1 g: 1 mL) is mixed with lysis buffer containing the base, sodium hydroxide (200 mM) and Sodium Dodecyl Sulfate (SDS) (1% w/v SDS at volume ratios of 1:1, typically at pH 12–12.5) (Meacle et al., 2004; Birnboim & Doly, 1979; Sambrook et al., 1989; Sambrook & Russell, 2006; Urthaler et al., 2007; Ballantyne, 2006). Under such conditions, the cell envelope is solubilized, releasing the cellular components. During this step, proteins are denatured at the same time due to the action of SDS. At this point the solution becomes very viscous due to the release of gDNA. Further, addition of cold and concentrated acidic neutralization buffer (3M potassium acetate, pH 5.5) at this stage causes precipitation of SDS due to formation of Potassium Dodecyl Sulfate (KDS), which has poor solubility (Sasagawa, 2019). As a result, denatured proteins, gDNA, High Molecular Weight RNA, endotoxins, and other cellular debris co-precipitate along with KDS out of the solution (Urthaler et al., 2007).

During lysis, the hydrogen bonds between the DNA strands are disrupted and cause gDNA to denature irreversibly at a narrow pH range of 12-12.5 and at a final concentration of not more than 0.1 N NaOH. Unlike gDNA, the pDNA stays intact with only partial denaturation under these conditions. However, above pH 12.5, pDNA too is irreversibly denatured and results in the decrease of supercoiled plasmid yield and gets worse with an increase in incubation time. Thus, homogeneity of the pDNA depends primarily on the pH of the lysis reaction rather than the timing of the lysis, which can extend up to 30 minutes (Meacle et al., 2004; Urthaler et al., 2007; Cloninger et al., 2008). Owing to its smaller size than gDNA, pDNA gets easily renatured during neutralization with Potassium Acetate solution at pH 5.5. However, gDNA cannot renature by correcting base pairing and thus stays in the flocculate, which can be removed by "low shear" clarification. This selective denaturation allows the removal of impurities from pDNA.

Despite the advantages of alkaline lysis, one of the major risks in the process is the presence of local high concentrations of alkali that can irreversibly damage the pDNA (Prazeres et al., 1999). Uneven high local concentrations are a result of poor mixing during lysis. At laboratory scale, lysis is done in small tubes or bottles, which do not require special equipment, thus mixing is not a critical factor. However, in large-scale operations, due to a lack of proper mixing technology, it is common to use manual mixing in Carboys or polypropylene tanks with Rushton impellers (Clemson & Kelly, 2003), mixtainers or even Teflon paddles, albeit with poor results (Theodossiou et al., 1997; Wright et al., 2001; Thatcher et al., 1999; Marquet et al., 1995). These manual methods suffer from poor mixing, lack of reproducibility, and scalability problems and in addition generate hydrodynamic shear stress producing smaller pieces of gDNA that result in poor yield and quality of pDNA.

Moreover, these problems are amplified due to rheological changes during and after cell lysis. A study on rheological properties of the lysis reaction showed a rapid increase in viscosity due to the fast action of the SDS, which revealed the need for "micro- and macromixing" (Ciccolini et al., 1999). This study indicated that the lysis reactor must mix at a rate faster than the time taken for the cell lysis reaction, which occurs in a matter of seconds. In addition, it also showed that after neutralization the mixing process must be thorough yet gentle enough to prevent fragmentation of the gDNA in the gel-like flocculated material due to the drastic changes in viscoelastic behavior of the flocculate.

It is a common misinterpretation that plasmids are not as shear sensitive as the gDNA. However, it has been shown that under certain circumstances, shear stress can induce damage to the tertiary structure of supercoiled plasmid (Lengsfeld & Anchordoquy 2002; Levy at al., 1999). The damaging effect of shear stress has been shown to increase with the size of the plasmid and decrease with increasing salt concentration but remain independent of plasmid concentration over a wide range. In recent years, a few manufacturers have come up with ingenious designs to overcome this mixing issue to efficiently lyse cells in large scale set up (Urthaler et al., 2007; Levy et al., 2000; Ciccolini et al., 1999) while at the same time being gentle on the gDNA.

15.3.2 RNA and Endotoxin Removal

Another important consideration in pDNA manufacturing is the removal of RNA, host cell proteins (HCP) and endotoxins, the major contaminants in pDNA production. Neutralized lysate contains a significant amount of High Molecular Weight (HMW) and Low Molecular Weight (LMW) RNA that does not separate into the precipitated gel-like flocculate due to its small size. Instead, it partitions into the clear lysate along with the pDNA. FDA regulation stipulates that RNA must be below detectable levels on agarose gel. Moreover, the presence of RNA can slow down the flow rates in chromatography and reduce column capacity. If RNA is not removed at the early stages of purification, it competes with pDNA for binding to chromatographic supports and affects the yield as well the purity of pDNA product.

The most common and traditional way to remove RNA is through the addition of RNAse enzyme during the lysis step as it is very stable even at pH 12.5 and high temperature. While RNase easily and efficiently removes RNA, it can linger in the pDNA drug product and is very difficult to remove. RNAse is often sourced from bovine pancreas, but enzyme free processes are preferred to avoid material of animal origin and reduce production costs. Therefore, there have been considerable efforts in developing non-enzymatic procedures to separate pDNA from RNA at an early stage of the process.

Non-enzymatic techniques to remove RNA range from the use of spermidine (Murphy et al., 1999), selective precipitation of DNA by CTAB (Lander et al., 2002), precipitation by calcium chloride or lithium chloride (Sauer et al., 2008; Lev, 1987), size-exclusion chromatography (Norgard, 1981), and (TFF) tangential flow filtration (Eon-Duval et al., 2003). Due to very narrow resolution between pDNA and undegraded RNA during ion exchange or hydrophobic interaction chromatography (HIC), it is advantageous to remove high molecular weight RNA by precipitation with salts like calcium chloride and subsequently remove low molecular weight RNA using TFF along with gDNA and endotoxins. Studies have shown that calcium chloride can prevent formation of the gel layer during TFF. In addition, with careful selection of membrane type, pore size (Molecular Weight Cut Off, MWCO), and proper adjustment of the TFF conditions, it is possible to get good clearance of RNA (Eon-Duval et al., 2003). This also shows that TFF becomes necessary at this stage to remove calcium chloride as it will affect the downstream chromatographic steps.

The envelope of *E. coli* contains phospholipids and endotoxin (LPS, Lipopolysaccharide) that is present as a contaminant in biologicals. LPS is an amphipathic molecule with a hydrophilic polysaccharide chain and a hydrophobic fatty acid (Steimle et al., 2016), which comprise a monomer ~10–30 kDa in size. In the presence of divalent cations such as Ca^{2+} and Mg^{2+}, LPS exist as micelles, vesicles and aggregates and can interact with basic proteins through ionic and hydrophobic interactions (Jang et al., 2009). LPS aggregates of ~200 kDa in size are excluded in the void volume of size exclusion chromatography column (Jang et al., 2009). However, when stripped of the metal ion or diluted, LPS is destabilized and predominantly exists as a monomeric free form. During alkaline lysis, the presence of metal chelators like Na-EDTA (Sodium salt of Ethylene Diamine Tetra Acetic acid) in combination with Tris buffer can cause destabilization of LPS structure (Vaara, 1992). Under diluted conditions, the equilibrium can shift towards monomeric form and can be easily eliminated through ultrafiltration.

Endotoxin is particularly toxic in the free form and can cause serious illness if administered to humans and animals. High rejection rates of biopharmaceuticals occur when unacceptable levels of pyrogens like endotoxin or other cellular components are present in the final product. Moreover, the presence of LPS in pDNA preparations can inhibit transfection of eukaryotic cells (Friedrich & Pfeiffer, 2009). Thus, it is necessary to keep the endotoxin levels below the stipulated FDA limits of < 40 EU/mg (FDA, 2007). However, the main challenge in its removal comes from its negative charge, which results in its co-purification with DNA. Though considerable amount of lipids and endotoxins are removed along with the flocculate material during the neutralization step by dropping the pH (Aida & Pabst, 1990), there is still a substantial amount left over that can decrease the purity of the pDNA product.

There are several techniques that have been shown to remove endotoxins, which include (1) chromatographic separation (Ongkudon et al., 2012), (2) ultra-filtration (Jang et al., 2009; Nunes et al., 2012), (3) detergent treatment (Rozkov et al., 2008, Adam et al., 1995), (4) two-phase partitioning (Lopes et al., 2010), and (5) selective salt precipitation (Ongkudon & Danquah, 2011). Detergent treatment with triton X-114 is an attractive way to remove endotoxin, but traces of the detergent must also be removed from the final product, making it not a desirable method for large-scale pDNA production (Aida & Pabst, 1990) due to environmental concerns from the disposal of large amounts of triton waste created in this step. Two-phase partitioning systems are employed for small-scale systems; however, it is not suitable for scale-up due to the presence of organic solvents or alcohols (Manzano et al., 2015). Thus, ultra-filtration and chromatography are favored and are commonly used in large-scale manufacturing of pDNA.

15.3.3 Clarification of Lysate

After neutralization of the alkaline lysate, the gel-like flocculant must be removed before proceeding to purification. This step is necessary as the gel-like material can foul the chromatographic column and cause high back pressure. Several methods are available to clarify the neutralized lysate. These include: (1) centrifugation, (2) hollow-fiber microfiltration, (3) ammonium bicarbonate flocculation, (4) depthfiltration, and (5) expanded bed chromatography. A comparison of these methods is shown in Table 15.3. The main considerations for clarification are the high viscosity of the lysate, gentle handling of fragile flocculate with gDNA, and handling very large volumes. Centrifugation with a fixed-angle rotor is not a method of choice in large-scale operations due to low throughput and increased process times. Continuous disk stack or tubular bowl centrifuges are avoided due to the generation of shear forces (Levy et al., 2000).

Depth filtration is usually preferred instead of centrifugation; however, depending on the type of filter material, pDNA can either be adsorbed by depth filters and cause loss during clarification (Khanal et al., 2019) or efficiently remove particulates from lysate. Studies have shown that passing the lysate through diatomaceous earth and cellulose pre-coat materials resulted in loss of pDNA (Theodossiou et al., 1999). Use of filter aids or

TABLE 15.3

Comparison of Stationary Phase Properties of Different Commercial Anion-Exchange Matrices. (Table Adapted from Diogo et al, 2005 and Urthaler et al, 2005)

Matrix	Bead Size (µm)	Pore Size (µm)	Linear Velocity (cm/h)	Pressure Drop (MPa)	Static and Dynamic[*] Binding Capacity mg/ml		References
					pDNA	Proteins	
Beads or			——300—	——0.30—			
Particles	90	0.19—	500——	0.37——	1.3/0.72[*]	120	Ferreira et al., 2000
Q-Sepharose FF	34	0.30—			2.5	70—>110	Ferreira et al., 2000
Q-Sepharose	80	<0.80			1.2/3.3[*], >5.3[*]	60	Theodossiou et al., 2000
Q HyperD	200				3		Ferreira et al., 2000
Streamline QXL	50				~10.0/2.12[*]		Eon-Duval & Burke, 2004
Poros 50HQ							
Monoliths	-	0.01–4[a]	500–1000	1.20–1.30	10	-	Branovic et al., 2004
DEAE-CIM							
Tentacles	40–90	0.80	100–500	0.61–0.71	2.45[*], 5.4[*]	100	Diogo et al., 2005, Urthaler et al., 2005, Branovic et al., 2004
Fractogel							
DMAE							
Membranes	-	0.8	-	-	10, 15[*]	56	Diogo et al., 2005
Mustang Q							

[a] Channel radius, [*] Dynamic binding capacity

removal of bulk of the flocculent by floatation does not alleviate the problem of plasmid loss. Rather, the study showed that the leakage of solid material increased through the filter with overall improvement in flux.

In another case, flocculent removal is done using solid chunks of ammonium bicarbonate, which releases carbon-dioxide gas and ammonia that lifts the gel-like flocculent to the top and enables the drainage of the clear lysate from the bottom (Blom et al., 2010). The reduction of solids by this method can improve the throughput of the filters. However, this method has been tested only at pilot scale, and its use in large-scale GMP manufacturing can be challenging. To circumvent this clarification problem, some manufacturers have built proprietary clarification devices filled with glass beads and porous plastic plates that allow the retention of solids at the bottom while permitting the clear lysate to be drained from the bottom (Urthaler et al., 2007). Thus, a technique that combines the ability to reduce solids levels and facilitate high throughput of filtration would be ideal for large-scale manufacturing.

15.3.4 Tangential-Flow Filtration (TFF)

TFF has gained importance in biopharmaceutical manufacturing as a method of choice during primary recovery. It has several advantages over other methods such as two-phase liquid separation, centrifugal filtration, or even chromatography (Freitas et al., 2009). Due to the very large volumes handled in manufacturing, a method that would reduce process time, cost of operations, and be easily scaled up is preferred. TFF can be employed to remove not only contaminants from the pDNA but also get rid of unwanted buffer components such as salts and base. It is simple to install, can be modular and effectively reduce large volumes to manageable levels within a short period of time. The volumes usually range from several hundreds to thousands of liters, which is a cumbersome, time consuming and prohibitively expensive to process by just chromatography alone.

Tangential flow ultrafiltration has been used to either recover pDNA directly from fermentation broth in microfiltration mode (Nunes et al., 2012) or from lysate post-neutralization using ultrafiltration membranes (Manzano et al., 2015; Kahn et al., 2000). After lysis and neutralization, the conductivity and pH of the lysate must be adjusted to facilitate binding of pDNA to the chromatographic column. For this, ultrafiltration (UF) is a very convenient method to exchange into a buffer system with desired conductivity and pH for chromatography. Typically, TFF membranes with a MWCO of 100–500 kDa are used, which is large enough to remove proteins, endotoxins, and LMW RNA, while retaining pDNA (Manzano et al., 2015; Figueroa-Rosette et al., 2013). Though TFF is a straightforward method, it is critical to select conditions that will support the retention of pDNA while removing contaminants.

Studies show that plasmid size, buffer conditions and flux rate influence the passage of pDNA through the pores of the semi-permeable TFF membrane (Arkhangelsky et al., 2008). Despite being negatively charged and displaying significant electrostatic repulsion with the membrane, pDNA can slither through the pores owing to its elongated and flexible tertiary structure due to shear forces generated by the flow near the pore entrances. This occurs even though the pDNA is at least 25 times larger in size than the membrane pores. Such behavior is attributed to the ionic strength, pH of the buffer, filtrate flux, and the type of salt used (NaCl vs $MgCl_2$). An increase from 0.005 M to 0.1 M salt concentration caused a reduction of about 25% in the radius of gyration of the plasmid (Latulippe & Zydney, 2008). Studies also show that there appears to be a critical point in pH (~11) and ionic strength (0.1 M) for plasmid penetration. Apart from employing TFF for removal of contaminants, it is also used for isolating different isoforms of pDNA based on the fundamental differences in their elongational flexibility in the highly converging flow field during filtration. This separation is different from that of size exclusion chromatography and agarose gel electrophoresis (Arkhangelsky et al., 2008; Borochov et al., 1981; Latulippe & Zydney, 2008; Latulippe et al., 2007; Latulippe & Zydney, 2011; Li et al., 2016) and it is more efficient.

15.3.5 Chromatographic Purification

Chromatography has been central to the purification of biomolecules including pDNA. Although several other techniques such as TFF and two-phase extractions are efficient, none of these techniques are used as a stand-alone method to purify pDNA, especially to separate the supercoiled form from non-supercoiled isoforms of pDNA. During pDNA production, it is common to incorporate a primary recovery step, an intermediate purification step, and a high- resolution purification step. This high-resolution purification is achieved by chromatography, which is efficient, cost-effective, scalable and has become a method of choice. Chromatography has been used to recover biologicals either from fermentation broth directly or post-primary recovery. In the case of pDNA, recovery after neutralization is typically done but only after reducing the burden of impurities and volume, thus facilitating the removal of residual contamination by chromatography.

Though it may appear to be straightforward, there are several limitations in the purification of pDNA by chromatography due to the (1) similarities in physio-chemical properties between pDNA and the contaminants such as endotoxins, gDNA and RNA, all of which are negatively charged, (2) large size of pDNA (~0.2 μm) (Fishman & Patterson, 1996) that influences the diffusion coefficients and mass transfer, (3) steric effects that decrease binding capacity of the resin, affects resolution and increases process times due to low flow rates caused by high viscosity (Diogo et al., 2005; Sousa et al., 2010; Prazeres et al., 1998), and (4) stationary phase characteristics. Moreover, poor selectivity is a major problem in pDNA purification because contaminating molecules often compete for binding sites in the stationary phase and reduce binding capacity of pDNA.

Using technologies such as membrane chromatography, beads that are super porous and have tentacles, monolith matrices, using small bead size etc., come as relief from some of the limitations of pDNA purification (Giovannini et al, 1998, Pereira et al, 2010). Plasmids are several orders of magnitude larger than proteins and are negatively charged. The Stokes radius is about 100 - 300 nm, 20-50 times larger than proteins (Urthaler et al., 2005). Due to the large size of pDNA, the viscosity of the medium is comparatively very high and becomes extremely difficult to handle at large scale (Ciccolini et al., 1999). These properties also affect the capacity and smooth operation of the chromatographic columns.

Further, removal of impurities such as endotoxin, proteins, RNA, gDNA and undesired isoforms of pDNA depends on the chemistry or mode of chromatographic separation employed. Usually, two or more types of chromatographic steps are employed to achieve this. Most common methods are anion-exchange (IEX) and HIC, which utilize the negative charge and hydrophobic properties of the DNA, respectively. Other types include size-exclusion, affinity, and hydroxyapatite chromatography (Abdulrahman and Ghanem, 2018). Table 15.3 shows a comparison of some of the properties of commercially available anion-exchange resins that are commonly used for pDNA purification.

Although packed-bed bead resins are very common in protein purification, they are limited by low dynamic binding capacity when compared to monoliths and membranes, especially for large molecules like DNA, because of their lower diffusivity constants and small pore size (Diogo et al., 2005). As the bead size decreases, it becomes difficult to pack and often requires special column hardware with axial compression. During purification, these matrices exhibit increased back pressure at higher flow rates due to the viscous nature of DNA, which in turn compels reduction of flow rates, thus increasing process times. In some cases, expanded bed supports are also used to recover pDNA. Although expanded bed supports offer advantages such as saving time, it requires special column hardware and resins (Theodossiou et al., 2001).

The first step in pDNA chromatography is a capture and concentration step. Anion-exchange chromatography using positively charged diethyl amino ethyl (DEAE) or quaternary ammonium stationary phase is well suited for this purpose since it separates molecules based on negative charge. A single capture step with AEC can result in ~40-fold increase in purity of pDNA (Ferreira et al., 2000a). Due to the difference in net charge, pDNA and contaminants such as RNA and proteins bind to the matrix with different affinities and can be washed with buffers under increasing salt gradient (Abdulrahman & Ghanem, 2018). RNA is first removed at a lower salt concentration and DNA at a higher salt concentration. The eluted DNA fraction contains gDNA and all isoforms of pDNA, including supercoiled, linear, and relaxed-circular forms.

A polishing step is often employed to remove gDNA and undesired pDNA isoforms. For this purpose, HIC or SEC is commonly employed. HIC takes advantage of the difference in the hydrophobicity of the gDNA, supercoiled pDNA and other pDNA isoforms (Abdulrahman & Ghanem, 2018; Bo et al., 2013). During downstream processing, pDNA can become linearized or form a relaxed circular conformation (Open circular, OC). The HIC step is used to separate supercoiled pDNA from these unwanted isoforms with decreasing salt concentration (Abdulrahman & Ghanem, 2018). For a HIC step, pDNA from the capture step is adjusted with ammonium sulfate or other salts to facilitate binding of the plasmid to the resin.

Upon binding, the open circular form of pDNA and supercoiled pDNA are separated using a decreasing salt gradient, with the supercoiled form eluting later. gDNA, being more hydrophobic than the pDNA, is washed off the column in buffer with no salt. In some cases, size exclusion chromatography is also employed as a polishing step by taking advantage of the differences in the sizes of the different DNA species and RNA and can be used for exchange into formulation buffer. However, SEC has its own limitations with a need for increased column length for effective resolution, leading to long chromatographic run times especially for resins with smaller particle size, dilution of the product and difficulty in scaling up (Diogo et al., 2005). Other forms of chromatography such as affinity chromatography, thiophilic adsorption chromatography, multi-modal chromatography, and protein-DNA affinity chromatography have been explored and developed for pDNA purification. However, these methods have not gained popularity due to either poor yield or difficultly in scaling up.

15.3.6 Formulation and Storage

After chromatographic purification, pDNA is essentially devoid of contaminating molecules. In small-scale processes, precipitation of pDNA by organic alcohols or polyethylene glycol is employed, which will require resuspension into a buffer. However, at large scale manufacturing intended for therapeutic purposes, use of organic solvents is avoided. Moreover, depending on the chromatographic step, the buffer systems may contain high concentrations of salts like NaCl or ammonium sulfate, EDTA, and minor impurities that must be removed prior to formulation. In this case, tangential-flow filtration is a preferred method for the final formulation step. It serves as an important tool to concentrate pDNA, remove salts and exchange into a desired formulation buffer suitable for further use.

Typically, the final formulation of pDNA depends on the duration of storage as well as its intended downstream application, which can be either for transfection in mammalian cells to produce viral vectors or direct injection of naked DNA for therapeutic purposes. Formulation into a suitable buffer such as Tris-EDTA (TE buffer), phosphate buffered saline (PBS) or nuclease free pure water along with excipients is preferred depending on the stability and intended storage condition. pDNA is often mixed with lipids and polypeptides to prevent shear damage and aid in transfection (Levy et al., 2000). After formulation of pDNA, sterile filtration is done through a 0.2 μm filter before final bulk fill. Though at lower concentrations, the pDNA solution readily passes through the 0.2 μM filter, at higher concentrations ($\geq$ 6–7 mg/mL), the viscosity of the solution increases dramatically and retards the rate of filtration due to high back pressure. Sterile filtered pDNA with certain excipients can be stored as a liquid up to 12 months at 4 °C and 3 years at –20 °C, and pDNA intended for long term storage can be freeze dried but needs to be reconstituted before use (Przybylowski et al., 2007; Walther et al., 2003). Though supercoiled topology of pDNA can be maintained at –80 °C, it is typically shipped at –20 °C due to lack of infrastructure in certain regions of the world. Although freeze dried pDNA has a long shelf life, it has the drawback of getting damaged during freezing and drying steps due to removal of water of hydration surrounding the DNA. Therefore, sugars such as sucrose, mannitol, trehalose, or other oligosaccharides that do not freeze but maintain hydration around DNA are used as excipients (Quaak et al., 2010). In some cases, pDNA is formulated in cationic lipids, polymers and microparticles, which are shown to improve immune responses after vaccination (D'Souza et al., 2002; Greenland & Letvin, 2007).

15.4 Analytical Development

As with protein and chemical or synthetic pharmaceuticals, manufactured pDNA must undergo rigorous testing to demonstrate the product's quantity, purity, identity, potency, quality, and safety (US Food and Drug Administration, 2007; World Health Organization,

2005, 2020). Several analytical methods have been developed to assess each of these attributes, including classical molecular biology techniques. Analytical testing is important to inform process development and is critical to monitoring pDNA yield and quality throughout the manufacturing process and for product release after final formulation. Most release testing specifications are based on guidelines for DNA vaccines established by the FDA, while some may be product-specific (US Food and Drug Administration, 2007). Testing requirements for a pDNA product depend on the grade of material being produced: R&D, critical reagent, or cGMP grade. Critical reagent and cGMP grade pDNA must pass strict safety and purity specifications (product and process related) as well as sequence confirmation in addition to the full list of testing required for R&D grade pDNA.

15.4.1 Quantity/Concentration

In-process and release testing of pDNA products require determination of the quantity, or concentration, of the sample. pDNA concentration is determined by measuring the absorbance at 260 nm using a spectrophotometer and is calculated using a modified version of the Beer-Lambert equation:

$$\text{Concentration} = (\text{A260 absorbance*ds-DNA factor})/\text{pathlength}$$

where A260 = absorbance at 260 nm; ds-DNA (double stranded DNA) factor = an industry standard of 50 ng-cm/μL (Barbas III, 2007), based as the amount of nucleic acid contained in 1 mL that produces an OD of 1; and pathlength is specific to the spectrophotometer used. A baseline correction at 320 nm is used as a normalization factor to offset any particulates that may be in the sample.

15.4.2 Purity

Sample purity must be assessed to ensure effective removal of process- and product-related impurities. Measurement of pDNA concentration by spectrophotometer can provide an indication of sample purity by determining the A260/A280 ratio. This ratio is impacted by the presence of DNA, RNA, and protein in the sample. An A260/A280 ds-DNA purity ratio of 1.80-2.00 is the widely accepted range throughout the industry for purity determination. If the sample contains significant proteinaceous impurities, the ratio will be below 1.80, whereas if there is significant RNA contamination present, the ratio will typically be above 2.00.

pDNA purity can also be assessed by analyzing the physical forms, or topoisomers, of the plasmid. pDNA can exist in multiple topoisomers, with the supercoiled configuration being most desirable configuration for gene therapy due to its increased stability compared to the other conformations; however, different criteria may be used if data indicate "that the selected criteria are predictive of immunogenicity or other intended biological activities" (US Food and Drug Administration, 2007). pDNA can also be observed in an open-circular, or nicked form, as well as linear DNA. The open-circular/nicked configuration is as the name implies; the pDNA is in a circular configuration that has undergone partial cleavage that causes the supercoiling to relax. The linear isoform occurs due to the complete cleavage of both DNA strands that causes the pDNA to uncoil and become linear. Analysis of pDNA topoisomer homogeneity can be achieved by several methods, with the most common being agarose gel electrophoresis (AGE), high pressure liquid chromatography (HPLC), and capillary electrophoresis utilizing a laser induced fluorescent detector (CE-LIF) (Figures 15.5–15.7). pDNA isoforms are separated based on the different rates at which they migrate through the gel or column matrix, which is related to their comparative shape and level of compactness. Supercoiled pDNA will migrate through agarose gel faster since it is more compact, while the less compact open-circular and linear

FIGURE 15.5

(a) pDNA homogeneity by agarose gel electrophoresis (AGE). pDNA samples were analyzed by AGE to separate plasmid isoforms. Supercoiled pDNA (~6 kB band in pDNA uncut) migrates through agarose gel faster since it is more compact in size while the open-circular and linear migrate at slower rates. (b) The figure illustrates supercoiled (pDNA Uncut), nicked (pDNA Open-Circular) and linearlized DNA (pDNA linear). We could change the titles of the lanes from pDNA Uncut to Supercoiled, pDNA Open-Circular to Nicked and pDNA linear to Linearized DNA.

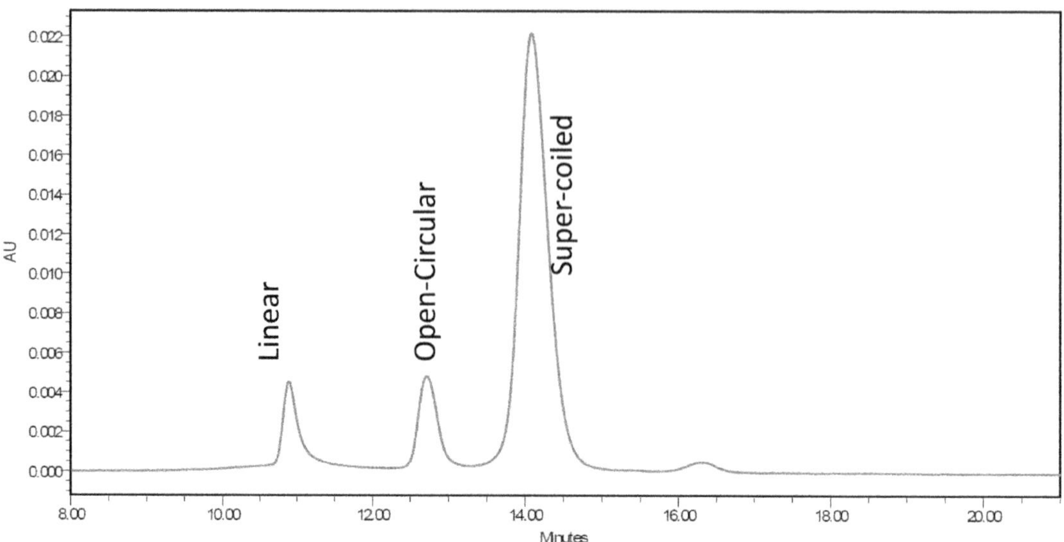

FIGURE 15.6

pDNA homogeneity by HPLC. Topoisomers of pDNA were separated by anion exchange HPLC. Supercoiled pDNA is retained longer than open-circular or linear DNA due to its higher negative charge density interacting more strongly with the positive-charged stationary phase.

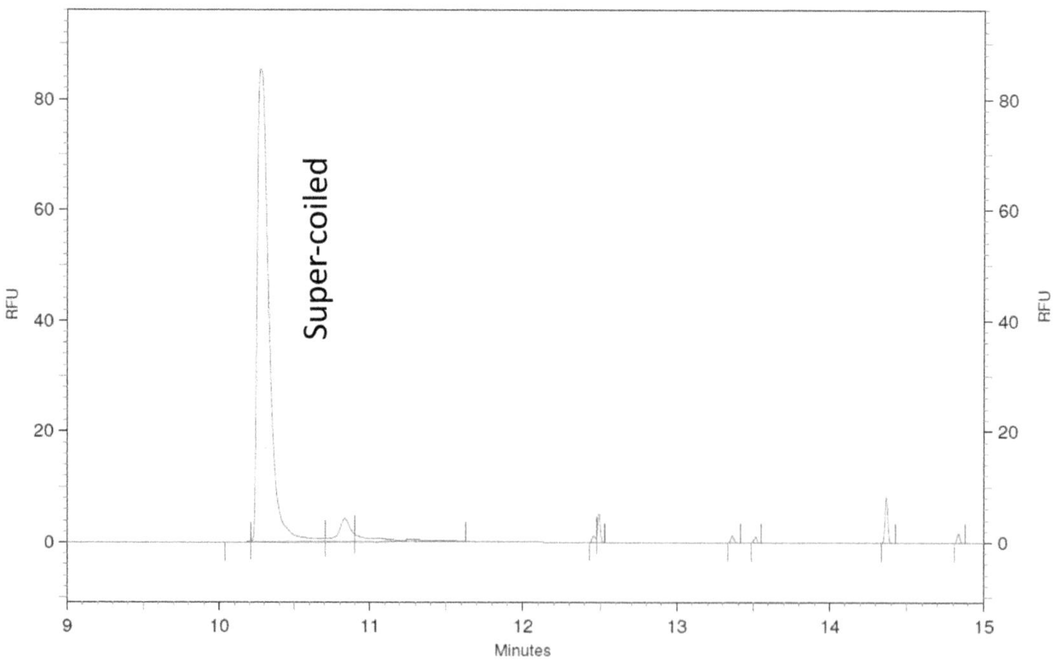

FIGURE 15.7

pDNA homogeneity by CE-LIF. Labeled isoforms of pDNA were separated by capillary electrophoresis using laser-induced fluorescence detection. Supercoiled pDNA migrates through the capillaries first.

pDNA forms migrate at slower rates. The same pattern will be seen on the CE-LIF method with the supercoiled DNA being detected first, then the open-circular and the linear afterward. Linear and open-circular pDNA are often detected before the supercoiled isoform by HPLC analysis based upon the column chemistry. Migration of all pDNA isomers in a sample can assist in determining the purity and identifying the plasmid when compared to a plasmid reference sample. The FDA recommends that the purity of manufactured pDNA is > 80% supercoiled (US Food and Drug Administration, 2007).

15.4.3 Identity

pDNA identity determination is critical to supporting process and analytical development as well as the release of manufactured products. Plasmid sequencing performed early in development will aid in the identification of sequence mutations or modifications that may occur during the upstream or downstream process and is used for identity confirmation of the final product. Plasmid sequence also informs the selection of appropriate restriction enzymes that recognize specific sequences in the DNA strand. Digestion of the pDNA sample with selected enzymes will produce DNA fragments of specific size that can be separated by agarose gel electrophoresis (Figure 15.8). This method, known as restriction enzyme mapping,

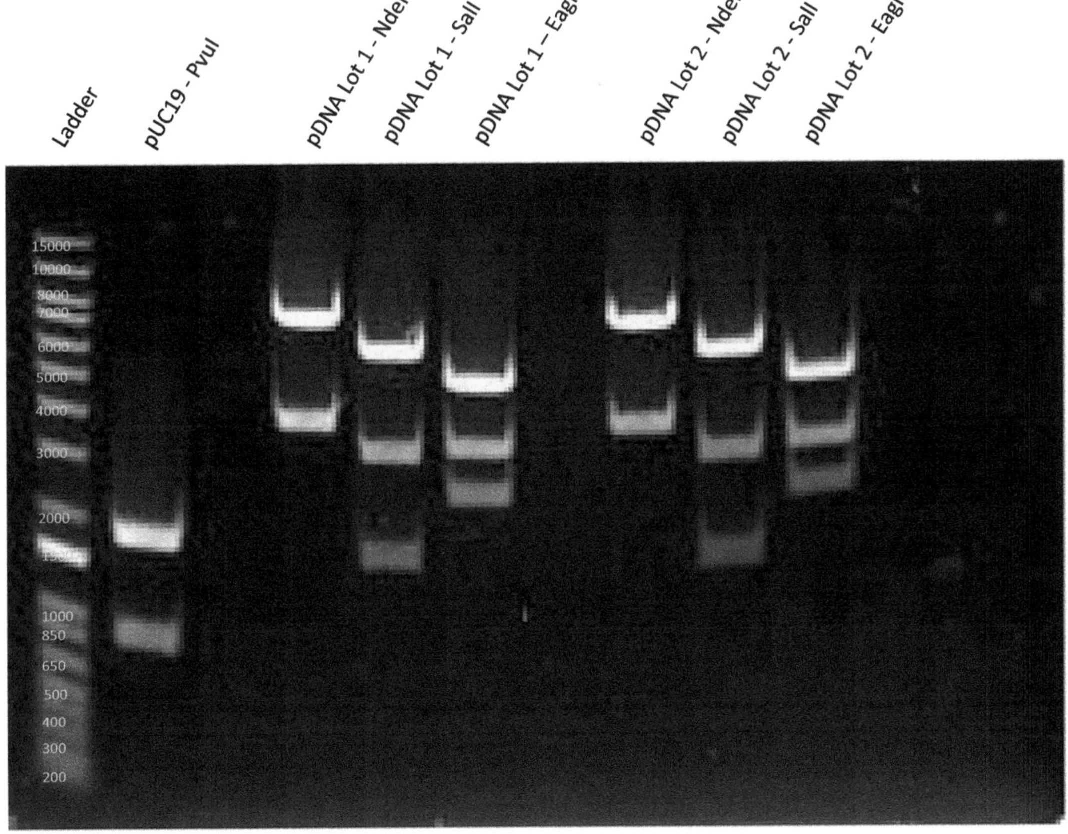

FIGURE 15.8
pDNA identity by restriction enzyme digest. Two lots of pDNA product were digested with NdeI, SalI, or EagI and separated by agarose gel electrophoresis.

determines pDNA identity by comparing the sample digestion pattern to that of the plasmid reference sample. This identity confirmation can be determined regardless of mutations occurring during the manufacturing process; however, mutations in the cleavage sites can be problematic, causing failure of the identity test. In addition, the specific migration pattern of undigested pDNA isomers in a sample can also assist in determining the identity of a plasmid when compared to the isomer migration pattern of the plasmid reference sample.

15.4.4 Potency

Potency is defined as the "qualitative measure of the specific ability or capacity of a product to achieve a defined biological effect" (FDA ICH Harmonised Tripartite Q6B, 1999). Potency of pDNA products should be determined to demonstrate the function of the product. Potency of early phase pDNA products can often be tested by *in vitro* assays measuring cellular uptake of pDNA, expression of the gene of interest, and the resulting cellular output, such as production of the protein product encoded in the plasmid. In later phases, it is recommended to assess pDNA potency using a quantitative method and demonstrate a direct relationship between potency and *in vivo* activity using appropriate animal model systems. pDNA product potency is assessed by comparison to the potency of the plasmid reference standard. The appropriate cell type and functional readout are dependent on the desired use of the pDNA product, and acceptance criteria are product specific.

15.4.5 Quality

Like other biopharmaceuticals, the quality of a pDNA product is assessed by measuring the sample appearance, pH, and osmolality. Appearance testing is performed by visual inspection of the product to ensure the appropriate color, turbidity, and presence or absence of visible particles. pH and osmolality measurements are typically reserved for cGMP grade pDNA and are performed according to USP <791> and <785>, respectively. Only final cGMP pDNA product formulated in Tris EDTA (TE) or PBS require pH and osmolality testing; these methods are not needed for pDNA formulations in water for injection (WFI).

Microbial contamination of the pDNA product is assessed by determining sterility and bioburden of the sample. This is achieved by testing for microbial growth as well as bacteriostasis and/or fungistasis according to USP <71>. Identification of *Mycoplasma*, a common bacterial contaminant in biopharmaceuticals, is also critical for the safety of gene therapies produced in mammalian cell culture, but not for pDNA produced from *E. coli* specifically (Armstrong et al., 2010). Measurement of bacterial endotoxin is performed to ensure the absence of the highly inflammatory bacterial cell wall component in the sample. Endotoxin levels are commonly quantified using the limulus amebocyte lysate (LAL) assay, a kinetic assay that is based upon the reaction of horseshoe crab amoebocytes with Gram-negative bacterial endotoxin, or lipopolysaccharide (LPS), within a sample. In the presence of endotoxin, a chromogenic substrate is cleaved, causing a color change that can be measured at 405 nm. The endotoxin level will be plasmid specific and the use/grade of the plasmid will determine what the level of acceptable endotoxin should be. This is a critical measurement in any biopharmaceutical produced in bacterial cells, with acceptable limits for cGMP grade less than 40 EU/mg of plasmid (US Food and Drug Administration, 2007). These safety tests must be performed for release of critical reagent or cGMP grade pDNA.

15.4.6 Residual Testing

The final pDNA product should be tested for specific product- or process-related residual molecules in the sample, including host cell DNA, RNA, protein, and antibiotics. Host cell protein is commonly measured using a microBCA (bicinchoninic acid) assay. Protein in a sample causes a color change by inducing the BCA-copper biuret reaction, while no color change is seen in the presence of DNA alone. In addition to protein, host-derived nucleic acid (gDNA) may be present in the pDNA product. Quantitative polymerase chain reaction (qPCR) methods are employed to measure residual *E. coli* genomic DNA using primers specific for *E. coli* genes, which will amplify the contaminant if residual host DNA is present. Residual RNA is determined by agarose gel electrophoresis using the SYBR green stain (Figure 15.9). The FDA recommends the limit for these host-derived residuals to be < 1% (US Food and Drug Administration, 2007). Process-related residuals, such as antibiotics used as selection agents during fermentation, must be demonstrated to be absent from the final pDNA product. HPLC or liquid chromatography/mass spectrometry (LC/MS) methods are frequently used to measure most residual antibiotics, with each method developed based on specific molecule attributes. Acceptable limits for process-related residuals are often set based upon the detection limit of the assay.

Each method used for pDNA release testing should be qualified or validated according to guidelines established by the appropriate regulatory agencies to demonstrate appropriate method parameters, including linearity, accuracy, precision, and specificity such as ICH guidelines for Validation of Analytical Procedures: Text and Methodology Q2 (R1) (ICH, 1995a). In addition, care should be taken to establish appropriate release specifications for each test to ensure that the final product is functional and safe for gene therapy applications.

15.4.7 Stability Testing

It is important to establish stability of a pDNA product to determine the optimal storage conditions and shelf-life of the product. The FDA and WHO recommend monitoring stability early in development and prior to clinical testing (FDA, 2007; WHO, 2005). Several factors can impact pDNA stability including the buffer composition and storage temperature/conditions. Stability-indicating methods described in this chapter should be employed to monitor changes in the pDNA product during storage in specified conditions. Plasmid homogeneity is an important stability-indicating method used to monitor changes in pDNA isoforms. A reduction in percent supercoiled pDNA is indicative of decreased stability, and has been shown to occur upon exposure to higher storage temperatures ($\geq$ 4°C) or plasmid damage due to lyopholization in the absence of stabilizing excipients such as certain sugars (Walther, et al., 2003; Quaak, et al., 2010). Additionally, purity analyses and restriction enzyme mapping can identify pDNA degradation or cleavage of the product. Potency analyses should be utilized in stability testing where available to track changes in activity, which could be indicative of a loss of product integrity. Stability testing should be performed according to appropriate regulations, such as ICH Q5C (ICH, 1995b), to establish an analytical package that has been demonstrated to monitor stability-related product changes.

15.5 Summary

Plasmid DNA technologies offer an opportunity to develop vaccines and therapeutics faster compared to conventional processes. This work analyzed the strain background, fermentation medium and conditions that can lead to high plasmid yield. The downstream process evaluated the cell disintegration where approaches that lead to low shear and retention of high supercoiling percentage is preferred. The purification looks at various

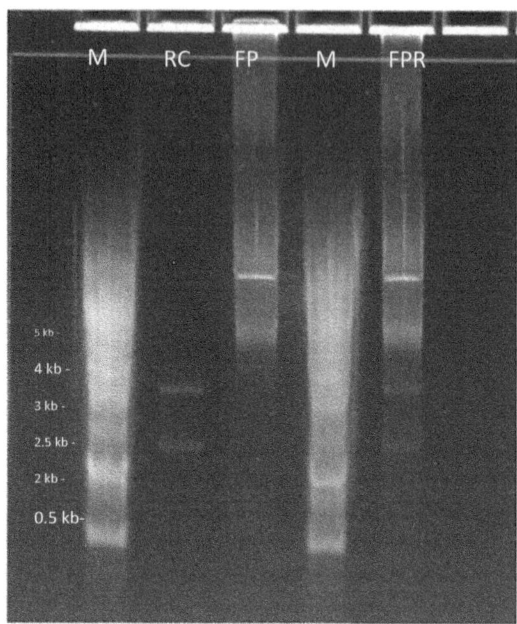
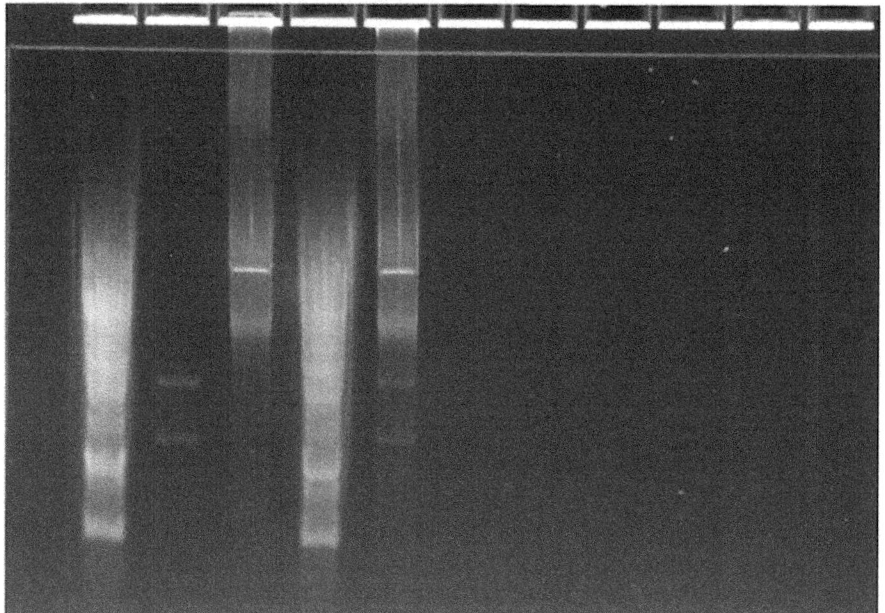

FIGURE 15.9

Residual RNA determination by AGE. pDNA final product was separated by agarose gel electrophoresis alone (FP) or spiked with RNA (FPR) to demonstrate RNA contamination. RNA control (RC) and RNA marker (M) were separated for comparison.

approaches for purifying the plasmid including RNA and endotoxin removal and the formulation of the final product. The analytics required for the product depends on the grade (R&D Grade, Reagent grade or cGMP) of pDNA that is desired and the various methods such as A260, A260/A280, micro-BCA and qPCR, that can be used to measure the safety, quality, and potency of the final pDNA product.

References

Abdulrahman, A., and Ghanem, A. (2018) Recent advances in chromatographic purification of plasmid DNA for gene therapy and DNA vaccines: A review, *Anal Chim Acta 1025*, 41–57.

Adam, O., Vercellone, A., Paul, F., Monsan, P. F., and Puzo, G. (1995) A nondegradative route for the removal of endotoxin from exopolysaccharides, *Anal Biochem 225*, 321–327.

Aida, Y., and Pabst, M. J. (1990) Removal of endotoxin from protein solutions by phase separation using Triton X-114, *J Immunol Methods 132*, 191–195.

Alliance for Regenerative Medicine Annual Report Highlights. (2020) The Alliance for Regenerative Medicine. Available from: https://www.globenewswire.com/en/news-release/2021/03/16/2193839/0/en/Alliance-for-Regenerative-Medicine-Annual-Report-Highlights-Record-Sector-Growth-and-Resilience-in-2020.html.

Arkhangelsky, E., Steubing, B., Ben-Dov, E., Kushmaro, A., and Gitis, V. (2008) Influence of pH and ionic strength on transmission of plasmid DNA through ultrafiltration membranes, *Desalination 227*, 111–119.

Armstrong, S.E., J. A. Mariano, and Lundin, D. J. (2010) The scope of mycoplasma contamination within the biopharmaceutical industry. *Biologicals*, 38(2):211–213.

Ballantyne, J. (2006) Practical methods for supercoiled pNDA production, *Methods Mol Med 127*, 311–337.

Barbas III, C. D. (2007) Quantitation of DNA and RNA. *Cold Spring Harb Protoc.* 2007 Nov 1;2007:pdb. ip47. doi: 10.1101/pdb.ip47. PMID: 21356961.

Birnboim, H. C., and Doly, J. (1979) A rapid alkaline extraction procedure for screening recombinant plasmid DNA, *Nucleic Acids Res 7*, 1513–1523.

Blom, H., Bennemo, M., Berg, M., and Lemmens, R. (2010) Flocculate removal after alkaline lysis in plasmid DNA production, *Vaccine 29*, 6–10.

Bo, H., Wang, J., Chen, Q., Shen, H., Wu, F., Shao, H., and Huang, S. (2013) Using a single hydrophobic-interaction chromatography to purify pharmaceutical-grade supercoiled plasmid DNA from other isoforms, *Pharm Biol 51*, 42–48.

Borochov, N., Eisenberg, H., and Kam, Z. (1981) Dependence of DNA conformation on the concentration of salt, *Biopolymers 20*, 231–235.

Bower, D. M., and Prather, K. L. (2009) Engineering of bacterial strains and vectors for the production of plasmid DNA, *Appl Microbiol Biotechnol 82*(5), 805–813. doi:10.1007/s00253-009-1889-8.

Branovic, K., Forcic, D., Ivancic, J., Strancar, A., Barut, M., Kosutic Gulija, T., Zgorelec, R., and Mazuran, R. (2004) Application of short monolithic columns for fast purification of plasmid DNA, *J Chromatogr B Analyt Technol Biomed Life Sci 801*, 331–337.

Carlson, A., Signs, M., Liermann, L., Boor, R., and Jem, K. J. (1995) Mechanical disruption of *Escherichia coli* for plasmid recovery, *Biotechnol Bioeng 48*, 303–315.

Carnes, A. E., Hodgson, C. P., Luke, J. M., Vincent, J. M., and Williams, J. A. (2009) Plasmid DNA production combining antibiotic-free selection, inducible high yield fermentation, and novel autolytic purification, *Biotechnol Bioeng 104*, 505–515.

Carnes, A. E., Hodgson, C. P., and Williams, J. (2006) Inducible Escherichia coli fermentation for increased plasmid DNA production. *Biotechnol Appl Biochem 45*(3), 155. doi:10.1042/ba20050223.

Carnes, A. E., Luke, J. M., Vincent, J. M., Anderson, S., Schukar, A., Hodgson, C. P., and Williams, J. A. (2010) Critical design criteria for minimal antibiotic-free plasmid vectors necessary to combine robust RNA Pol II and Pol III-mediated eukaryotic expression with high bacterial production yields. *J Gene Med 12*(10), 818–831. doi:10.1002/jgm.1499.

Carnes, A., and Williams, J. (2007) Plasmid DNA manufacturing technology, *Recent Patents Biotechnol 1*(2), 151–166. doi:10.2174/187220807780809436.

Cato, M. (1993) Polypyrimidine/polyurine sequence in plasmid DNA enhances formation of dimer molecules in *Escherichis coli, Mol Biol Rep 18*, 183–187.

Chiang, C. S., and Bremer, H. (1988) Stability of pBR322-derived plasmids, *Plasmid 20*, 207–220.

Ciccolini, L. A. S., Ayazi Shamlou, P., Titchener-Hooker, N. J., Ward, J. M., and Dunnill, P. (1999) Rheological properties of chromosomal and plasmid DNA during alkaline lysis reaction, *Bioprocess Eng 21*, 231–237.

Clemson, M., and Kelly, W. J. (2003) Optimizing alkaline lysis for DNA plasmid recovery, *Biotechnol Appl Biochem 37*, 235–244.

Cloninger, C., Felton, M., Paul, B., Hirakawa, Y., and Metzenberg, S. (2008) Control of pH during plasmid preparation by alkaline lysis of *Escherichia coli, Anal Biochem 378*, 224–225.

Diogo, M. M., Queiroz, J. A., and Prazeres, D. M. (2005) Chromatography of plasmid DNA, *J Chromatogr A 1069*, 3–22.

Dorman, C. J., Barr, G. C., Bhriain, N. N., and Higgins, C. F. (1988) DNA supercoiling and the anaerobic and growth phase regulation of ton B gene expression, *J Bacteriol 179*, 2816–2826.

D'Souza, S., Rosseels, V., Denis, O., Tanghe, A., De Smet, N., Jurion, F., Palfliet, K., Castiglioni, N., Vanonckelen, A., Wheeler, C., and Huygen, K. (2002) Improved tuberculosis DNA vaccines by formulation in cationic lipids, *Infect Immun 70*, 3681–3688.

Durland, R. H., and Eastman, E. M. (1998) Manufacturing and quality control of plasmid-based gene expression systems, *Adv Drug Deliver Rev 30*, 33–48.

Eon-Duval, A., and Burke, G. (2004) Purification of pharmaceutical-grade plasmid DNA by anion-exchange chromatography in an RNase-free process, *J Chromatogr B Analyt Technol Biomed Life Sci 804*, 327–335.

Eon-Duval, A., MacDuff, R. H., Fisher, C. A., Harris, M. J., and Brook, C. (2003) Removal of RNA impurities by tangential flow filtration in an RNase-free plasmid DNA purification process, *Anal Biochem 316*, 66–73.

FDA. (2016) Recommendations for Microbial Vectors used for Gene Therapy Guidance for Industry, U.S. Department of Health and Human Services Food and Drug Administration Center for Biologics Evaluation and Research.

Ferreira, G. N., Cabral, J. M., and Prazeres, D. M. (2000a) Studies on the batch adsorption of plasmid DNA onto anion-exchange chromatographic supports, *Biotechnol Prog 16*, 416–424.

Ferreira, G. N., Monteiro, G. A., Prazeres, D. M., and Cabral, J. M. (2000b) Downstream processing of plasmid DNA for gene therapy and DNA vaccine applications, *Trends Biotechnol 18*, 380–388.

Figueroa-Rosette, J. A., Guerrero-Germán, P., Montesinos-Cisneros, R. M., Mártin-García, A. R., and Tejeda-Mansir, A. (2013) A straightforward method to scale-up plasmid DNA intermediate recovery by tangential flow ultrafiltration, *Biotechnol Biotechnol Equip 27*, 4014–4017.

Fishman, D. M., and Patterson, G. D. (1996) Light scattering studies of supercoiled and nicked DNA, *Biopolymers 38*, 535–552.

Freitas, S., Canario, S., Santos, J. A., and Prazeres, D. M. (2009) Alternatives for the intermediate recovery of plasmid DNA: Performance, economic viability and environmental impact, *Biotechnol J 4*, 265–278.

Friedrich, R., and Pfeiffer, J. (2009) Relevance of endotoxin contamination in plasmid DNA. https://api.semanticscholar.org/CorpusID:30413932.

Giovannini, R., Freitag, R., and Tennikova, T. B. (1998) High-performance membrane chromatography of supercoiled plasmid DNA, *Anal Chem 70*, 3348–3354.

Goldstein, E., and Drlica, K. (1984) Regulation of bacterial DNA supercoiling: plasmid linking numbers vary with growth temperature. *Proc Natl Acad Sci USA 81*, 4046–4050.

Greenland, J. R., and Letvin, N. L. (2007) Chemical adjuvants for plasmid DNA vaccines, *Vaccine 25*, 3731–3741.

Hayes, F. (2003) Chapter 1 – The Function and Organization of Plasmids. *In Nicola Casali, Andrew Presto (eds) E. Coli Plasmid Vectors: Methods and Applications. Methods in Molecular Biology, 235. Humana Press. pp. 1–5.*

Holmes, D. S., and Quigley, M. (1981) A rapid boiling method for the preparation of bacterial plasmids, *Anal Biochem 114*, 193–197.

International Conference on Harmonization of Technical Requirements for Registration of Pharmaceuticals for Human Use. (1995a) Q2 (R1) Validation of Analytical Procedures: Text and Methodology. Available from: https://www.fda.gov/regulatory-information/search-fda-guidance-documents/q2-r1-validation-analytical-procedures-text-and-methodology.

International Conference on Harmonization of Technical Requirements for Registration of Pharmaceuticals for Human Use. (1995b) Q5C Quality of Biotechnological Products: Stability Testing of Biotechnological/Biological Products. Available from: https://www.fda.gov/regulatory-information/search-fda-guidance-documents/q5c-quality-biotechnological-products-stability-testing-biotechnologicalbiological-products.

International Conference on Harmonization of Technical Requirements for Registration of Pharmaceuticals for Human Use. (1999) Q6B. Specifications: Test Procedures and Acceptance Criteria for Biotechnological/Biological Products. Available from: https://www.fda.gov/regulatory-information/search-fda-guidance-documents/q6b-specifications-test-procedures-and-acceptance-criteria-biotechnologicalbiological-products.

Jang, H., Kim, H. S., Moon, S. C., Lee, Y. R., Yu, K. Y., Lee, B. K., Youn, H. Z., Jeong, Y. J., Kim, B. S., Lee, S. H., and Kim, J. S. (2009) Effects of protein concentration and detergent on endotoxin reduction by ultrafiltration, *BMB Rep 42*, 462–466.

Kahn, D. W., Butler, M. D., Cohen, D. L., Gordon, M., Kahn, J. W., and Winkler, M. E. (2000) Purification of plasmid DNA by tangential flow filtration, *Biotechnol Bioeng 69*, 101–106.

Khan, S., Ullah, M. W., Siddique, R., Nabi, G., Manan, S., Yousaf, M., and Hou, H. (2016) Role of recombinant DNA technology to improve life, Int J Genom 1–14, Article ID 2405954.

Khanal, O., Xu, X., Singh, N., Traylor, S. J., Huang, C., Ghose, S., Li, Z. J., and Lenhoff, A. M. (2019) DNA retention on depth filters, *J Membr Sci 570–571*, 464–471.

Lahijani, R., Hulley, G., Soriano, G., Horn, N. A., and Marquet, M. (1996) High yield production of pBR-322-derived plasmids intended for human gene therapy by employing a temperature-controlled point mutation, *Hum Gene Ther, 7*, 1971–1980.

Lander, R. J., Winters, M. A., Meacle, F. J., Buckland, B. C., and Lee, A. L. (2002) Fractional precipitation of plasmid DNA from lysate by CTAB, *Biotechnol Bioeng 79*, 776–784.

Latulippe, D. R., Ager, K., and Zydney, A. L. (2007) Flux-dependent transmission of supercoiled plasmid DNA through ultrafiltration membranes, *J Membr Sci 294*, 169–177.

Latulippe, D. R., and Zydney, A. L. (2008) Salt-induced changes in plasmid DNA transmission through ultrafiltration membranes, *Biotechnol Bioeng 99*, 390–398.

Latulippe, D., and Zydney, A. (2011) Separation of plasmid DNA isoforms by highly converging flow through small membrane pores, *J Colloid Interface Sci 357*, 548–553.

Lederberg, J. (1952) Cell genetics and hereditary symbiosis, **Physiol Rev** 32, 403–430.

Lengsfeld, C. S., and Anchordoquy, T. J. (2002) Shear-induced degradation of plasmid DNA, *J Pharm Sci 91*, 1581–1589.

Lev, Z. (1987) A procedure for large-scale isolation of RNA-free plasmid and phage DNA without the use of RNase, *Anal Biochem 160*, 332–336.

Levy, M. S., Collins, I. J., Yim, S. S., Ward, J. M., Titchener-Hooker, N., Ayazi Shamlou, P., and Dunnill, P. (1999) Effect of shear on plasmid DNA in solution, *Bioprocess Eng 20*, 7–13.

Levy, M. S., O'Kennedy, R. D., Ayazi-Shamlou, P., and Dunnill, P. (2000) Biochemical engineering approaches to the challenges of producing pure plasmid DNA, *Trends Biotechnol 18*, 296–305.

Li, Y., Butler, N., and Zydney, A. L. (2016) Size-based separation of supercoiled plasmid DNA using ultrafiltration, *J Colloid Interface Sci 472*, 195–201.

Li, X., Jin, H., Wu, Z., Rayner, S., Yin, J., Yu, Y., Zhang, W., He, Z., Wang, C., and Wang, B. (2007) An automated process to extract plasmid DNA by alkaline lysis, *Appl Microbiol Biotechnol 75*, 1217–1223.

Long, J., Zhao, X., Liang, F., Liu, N., Sun, Y., and Xi, Y (2018) Optimization of fermentation conditions for an *Escherichia coli* strain engineered using response surface method to produce a novel therapeutic DNA vaccine for rheumatoid arthritis, *J Biol Eng 12*(22), 1–10.

Lopes, A. M., Magalhães, P. O., Mazzola, P. G., Rangel-Yagui, C. O., de Carvalho, J. C., Penna, T. C., and Pessoa, A., Jr. (2010) LPS removal from an *E. coli* fermentation broth using aqueous two-phase micellar system, *Biotechnol Prog 26*, 1644–1653.

Lugo, F. I., Vega-Estrada, J., Alvis, C. A., Ortega-Lopez, J., and Montes-Horcasitas, M. C. (2016) Developing strategies to increase pDNA production in *Eschericha coli* DH5α using batch culture, *J Biotechnol 233*, 66–73.

Luke, J., Carnes, A. E., Hodgson, C. P., and Williams, J. A. (2009) Improved antibiotic-free DNA vaccine vectors utilizing a novel RNA based plasmid selection system, *Vaccine 27*(46), 6454–6459. doi:10.1016/j.vaccine.2009.06.017.

Manzano, I., Guerrero-German, P., Montesinos-Cisneros, R. M., and Tejeda-Mansir, A. (2015) Plasmid DNA pre-purification by tangential flow filtration, *Biotechnol Biotechnol Equip 29*, 586–591.

Marquet, M., Horn, N. A., and Meek, J. A. (1995) Process development for the manufacture of plasmid DNA vectors for use in gene therapy, *Biopharm Technol Bus Biopharm 8*, 26.

Maximize Market Research PVT LT. Global Viral Vector and Plasmid DNA Manufacturing Market – Industry Analysis and Forecast (2019–2027) – By Type, Application, Disease, End-User and Region.

Meacle, F. J., Lander, R., Ayazi Shamlou, P., and Titchener-Hooker, N. J. (2004) Impact of engineering flow conditions on plasmid DNA yield and purity in chemical cell lysis operations, *Biotechnol Bioeng 87*, 293–302.

Middelberg, A. P. (1995) Process-scale disruption of microorganisms, *Biotechnol Adv 13*, 491–551.

Murphy, J. C., Wibbenmeyer, J. A., Fox, G. E., and Willson, R. C. (1999) Purification of plasmid DNA using selective precipitation by compaction agents, *Nat Biotechnol 17*, 822–823.

Murphy, J. C., Winters, M. A., and Sagar, S. L. (2006) Large-scale, nonchromatographic purification of plasmid DNA, *Methods Mol Med 127*, 351–362.

Norgard, M. V. (1981) Rapid and simple removal of contaminating RNA from plasmid DNA without the use of RNase, *Anal Biochem 113*, 34–42.

Nunes, J. C., Morão, A. M., Nunes, C., Pessoa de Amorim, M. T., Escobar, I. C., and Queiroz, J. A. (2012) Plasmid DNA recovery from fermentation broths by a combined process of micro- and ultrafiltration: Modeling and application, *J Membr Sci 415–416*, 24–35.

O'Kennedy, R. (2000) Effects of growth medium selection on plasmid DNA production and initial processing steps, *J Biotechnol 76*(2–3), 175–183. doi:10.1016/s0168-1656(99)00187-x.

O'Kennedy R.D., Baldwin, C., and Keshavarz-Moore, E. (2000) Effects of growth medium selection of plasmid DNA production and initial processing steps. *J Biotechnol 76*, 175–183.

O'Kennedy, R. D., Ward, J. M., and Keshavarz-Moore, E. (2003) Effects of fermentation strategy on the characteristics of plasmid DNA production. *Biotechnol Appl Biochem 37*(1), 83. doi:10.1042/ba20020099.

Ongkudon, C. M., Chew, J. H., Liu, B., and Danquah, M. K. (2012) Chromatographic removal of endotoxins: A bioprocess engineer's perspective, *ISRN Chromatogr 2012*, 649746.

Ongkudon, C. M., and Danquah, M. K. (2011) Analysis of Selective metal-salt-induced endotoxin precipitation in plasmid DNA purification using improved limulus amoebocyte lysate assay and central composite design, *Anal Chem 83*, 391–397.

Pereira, L. R., Prazeres, D. M., and Mateus, M. (2010) Hydrophobic interaction membrane chromatography for plasmid DNA purification: Design and optimization, *J Sep Sci 33*, 1175–1184.

Phue, J., Lee, S. J., Trinh, L., and Shiloach, J. (2008) Modified *Escherichia coli* B (BL21), a superior producer of plasmid DNA compared with *Escherichia coli* K (DH5α), *Biotechnol Bioeng 101*(4), 831–836. doi:10.1002/bit.21973.

Prather, K. J., Sagar, S., Murphy, J., and Chartrain, M. (2003) Industrial scale production of plasmid DNA for vaccine and gene therapy: plasmid design, production, and purification, *Enzyme Microb Technol 33*, 865–883.

Prazeres, D. M. F., Ferreira, G. N. M., Monteiro, G. A., Cooney, C. L., and Cabral, J. M. S. (1999) Large-scale production of pharmaceutical-grade plasmid DNA for gene therapy: problems and bottlenecks, *Trends Biotechnol 17*, 169–174.

Prazeres, D. M. F., Schluep, T., and Cooney, C. (1998) Preparative purification of supercoiled plasmid DNA using anion-exchange chromatography, *J Chromatogr A 806*, 31–45.

Przybylowski, M., Bartido, S., Borquez-Ojeda, O., Sadelain, M., and Rivière, I. (2007) Production of clinical-grade plasmid DNA for human Phase I clinical trials and large animal clinical studies, *Vaccine 25*, 5013–5024.

Quaak, S. G. L., Haanen, J. B. A. G., Beijnen, J. H., and Nuijen, B. (2010) Naked plasmid DNA formulation: Effect of different disaccharides on stability after lyophilisation, *AAPS PharmSciTech 11*, 344–350.

Rozkov, A., Larsson, B., Gillström, S., Björnestedt, R., and Schmidt, S. R. (2008) Large-scale production of endotoxin-free plasmids for transient expression in mammalian cell culture, *Biotechnol Bioeng 99*, 557–566.

Sambrook, J., Fritsch, E. F., and Maniatis, T. (1989) *Molecular Cloning: A Laboratory Manual*, Cold Spring Harbor Laboratory Press.

Sambrook, J., and Russell, D. W. (2006) Purification of plasmid DNA by chromatography, *CSH Protoc 2006*; doi:10.1101/pdb.prot3926.

Sasagawa, N. (2019) Plasmid Purification. In *Plasmid*. IntechOpen. doi:10.5772/intechopen.76226.

Sauer, M. L., Kollars, B., Geraets, R., and Sutton, F. (2008) Sequential CaCl2, polyethylene glycol precipitation for RNase-free plasmid DNA isolation, *Anal Biochem 380*, 310–314.

Shehadul Islam, M., Aryasomayajula, A., and Selvaganapathy, P. R. (2017) A review on macroscale and microscale cell lysis methods, *Micromachines (Basel) 8*, 83.

Silva, F., Queiroz, J. A., and Domingues, F. C. (2012) Plasmid DNA fermentation strategies: Influence on plasmid stability and cell physiology, *Appl Microbiol Biotechnol 93*(6), 2571–2580. doi:10.1007/s00253-011-3668-6.

Sousa, F., Passarinha, L., and Queiroz, J. A. (2010) Biomedical application of plasmid DNA in gene therapy: A new challenge for chromatography, *Biotechnol Genet Eng Rev 26*, 83–116.

Stanbury, P. F., Whitaker, A., Hall, S. J. (1995) *Principles of Fermentation Technology* (Second Edition), Butterworth-Heinemann.

Steimle, A., Autenrieth, I. B., and Frick, J.-S. (2016) Structure and function: Lipid A modifications in commensals and pathogens, *Int J Med Microbiol 306*, 290–301.

Summers, D. K. (1996) *The Biology of Plasmids, Blackwell Science.*

Thatcher David, R., Hitchcock, A., Hanak Julian, A. J., and Varley Diane, L. (1999) *Method of Plasmid DNA Production and Purification,* Cobra Therapeutics Ltd.

Theodossiou, I., Collins, I. J., Ward, J., Thomas, O. R. T., and Dunnill, P. (1997) The processing of a plasmid-based gene from *E. Coli.* Primary recovery by filtration, *Bioprocess Eng 16,* 175–183.

Theodossiou, I., Søndergaard, M., and Thomas, O. R. (2001) Design of expanded bed supports for the recovery of plasmid DNA by anion exchange adsorption, *Bioseparation 10,* 31–44.

Theodossiou, I., Thomas, O. R. T., and Dunnill, P. (1999) Methods of enhancing the recovery of plasmid genes from neutralised cell lysate, *Bioprocess Eng 20,* 147–156.

Thomas, C. M. (ed) (2000) *The Horizontal Gene Pool, Harwood Academic.*

Twigg, A. J., Sherratt, D. (1980) Trans-complementable copy number mutants of plasmid ColE1, *Nature 283,* 216–218.

Urthaler, J., Ascher, C., Wohrer, H., and Necina, R. (2007) Automated alkaline lysis for industrial scale cGMP production of pharmaceutical grade plasmid-DNA, *J Biotechnol 128,* 132–149.

Urthaler, J., Buchinger, W., and Necina, R. (2005) Improved downstream process for the production of plasmid DNA for gene therapy, *Acta Biochim Pol 52,* 703–711.

US Food and Drug Administration. (2007, November) Guidance for Industry: Considerations for Plasmid DNA Vaccines for Infectious Disease Indications. Available from: www.ema.europa. eu/docs/en_GB/document_library/Scientific_guideline/2009/10/WC500003987.pdf.

Vaara, M. (1992) Agents that increase the permeability of the outer membrane, *Microbiol Rev 56,* 395–411.

Walther, W., Stein, U., Voss, C., Schmidt, T., Schleef, M., and Schlag, P. M. (2003) Stability analysis for long-term storage of naked DNA: Impact on nonviral in vivo gene transfer, *Anal Biochem 318,* 230–235.

Williams, J. A., Carnes, A. E., and Hodgson, C. P. (2009b) Plasmid DNA vaccine vector design: Impact on efficacy, safety and upstream production, *Biotechnol Adv 27*(4), 353–370. doi:10.1016/j. biotechadv.2009.02.003.

Williams, J. A., Luke, J., Langtry, S., Anderson, S., Hodgson, C. P., and Carnes, A. E. (2009a) Generic plasmid DNA production platform incorporating low metabolic burden seed-stock and fed-batch fermentation processes. *Biotechnol Bioeng 103*(6), 1129–1143. doi:10.1002/bit.22347.

Wong, E. M., Muesing, M. A., and Polisky, B. (1982) Temperature-sensitive copy number mutants of ColE1 are located in an untranslated region of the plasmid genome. *PNAS 79*(11), 3570–3574.

World Health Organization. (2005) Guidelines for Assuring the Quality and Nonclinical Safety Evaluation of DNA Vaccines. Available from: ECBS 2005 Annex 1 DNA.pdf.

World Health Organization. (2020) ECBS Guidelines for Assuring the Quality, Safety, and Efficacy of Plasmid DNA Vaccines. Available from: DNA_Vaccines_R_WHO.BS.2020.2390_12_May_2020.pdf.

Wright, J. L., Jordan, M., and Wurm, F. M. (2001) Extraction of plasmid DNA using reactor scale alkaline lysis and selective precipitation for scalable transient transfection, *Cytotechnology 35,* 165–173.

Zhu, K., Jin, H., He, Z., Zhu, Q., and Wang, B. (2006) A continuous method for the large-scale extraction of plasmid DNA by modified boiling lysis, *Nat Protoc 1,* 3088–3093.

Zhu, K., Jin, H., Ma, Y., Ren, Z., Xiao, C., He, Z., Zhang, F., Zhu, Q., and Wang, B. (2005) A continuous thermal lysis procedure for the large-scale preparation of plasmid DNA, *J Biotechnol 118,* 257–264.

E

E. coli, 53, *54*, 63, 78, 81, 84, 146, 154, 176, 355–358,
    360, 364, 374
electrophoretic assay, **183**
ELISA, 161, 175, 176, 182, 185, 308
endogenous virus, 284
endotoxin, 158, 299, 308
    identical to DNA clearance, 54
    limits/level of, 292, 299, 374
    measurement of bacterial, 374
    microorganisms, 147
    removal, 299, 363–365
enzymatic assay, **183**
equipment user requirements and
        specifications, 222–223
European Commission, 1
European Medical Agency (EMA), 319
European Union
    Committee for Proprietary Medicinal
        Products (CPMP), 99
    concern for BPyV in, 200
    PIC/s Guide to Good Manufacturing
        Practice, 213, 224
extractables, 296–297
    defined, 279
    and leachables patient safety evaluation,
        278–280
    and leachables product impact
        evaluation, 278

F

facility design, 225–234
    ancillary areas, 230–231
    API and, 231
    biotechnology manufacturing areas and, 231
    design considerations, 226–229
    laboratory/quality control areas, 230
    non-sterile drug product production areas,
        231–232
    packaging areas, 230
    sterile manufacturing areas, 232–234
    warehousing/shipping areas, 229
factory acceptance test (FAT), 224, 326
failure chain, 133, *133–134*, *137*
failure modes and effects analysis (FMEA),
        36–44, 131, 133
    applications of, 37
    custom-made risk category rating, **47**
    evaluation, 39–41, **40**
    line item for a routine bioprocess operation,
        136

process transfer, 41–43, **42**
removal of detergent from protein using
        ion-exchange chromatography, 48, **49**
report, 43–44
risk analysis, 36–37
worksheet, **38**, 38–39
fed-batch cell culture, *317–318*
Federal Food, Drug, and Cosmetic Act, 4
FEL, *see* front-end loading
fermentation media, 356–357, **357**
fermentation process, 357–359, *358*
fiber shedding, 283
filter capacity, 283
filter integrity test (FIT), 270, *273*, 273–275, *276*
filter interaction within process fluid, 276–280
    adsorption/transmission study, 277–278
    chemical compatibility, 277
    extractables and leachables patient safety
        evaluation, 278–280
    extractables and leachables product impact
        evaluation, 278
filtration media, experimental approaches for
        determining and validating, 158–161
    blank runs, 161
    clearance of impurities, 160
    concurrent validation at pilot and/or full
        scale, 161
    filter analysis, 160–161
    filter integrity measurements, 160
    normalized water permeability (NWP), 159
    parameters to measure, **159**, 159–161
    product yield and purity, 160
    scale-down modeling and qualification, **161**
    small-scale models, **158**, 158–159
    transmembrane pressure (TMP) *vs.* flux
        curves, 160
filtration system flushing, 280–281, *281*
filtration system sterilization, 282
filtration validation, 266–312
    clarification/prefiltration filters, 282–284
    overview, 266
    phases of, 266–267
    scales of, 267, *267*
    sterilizing grade, 268–282, *see also* sterilizing
        grade filtration validation
    tangential flow filters (TFF), 292–311
    viral clearance filter, 284–292
final design review, 224
fit for duty, 282
flasks and culture bags, inoculum expansion
        and maintenance in, 79
flush water residuals, 306–308
FMEA, *see* failure modes and effects analysis

For Product Safety Concerns and Information please contact our EU
representative GPSR@taylorandfrancis.com
Taylor & Francis Verlag GmbH, Kaufingerstraße 24, 80331 München, Germany

www.ingramcontent.com/pod-product-compliance
Ingram Content Group UK Ltd.
Pitfield, Milton Keynes, MK11 3LW, UK
UKHW050153110726
473146UK00013B/917